Methods in Molecular Biology

Series Editor
John M. Walker
School of Life and Medical Sciences
University of Hertfordshire
Hatfield, Hertfordshire, UK

For further volumes:
http://www.springer.com/series/7651

For over 35 years, biological scientists have come to rely on the research protocols and methodologies in the critically acclaimed *Methods in Molecular Biology* series. The series was the first to introduce the step-by-step protocols approach that has become the standard in all biomedical protocol publishing. Each protocol is provided in readily-reproducible step-by-step fashion, opening with an introductory overview, a list of the materials and reagents needed to complete the experiment, and followed by a detailed procedure that is supported with a helpful notes section offering tips and tricks of the trade as well as troubleshooting advice. These hallmark features were introduced by series editor Dr. John Walker and constitute the key ingredient in each and every volume of the *Methods in Molecular Biology* series. Tested and trusted, comprehensive and reliable, all protocols from the series are indexed in PubMed.

The Integrated Stress Response

Methods and Protocols

Edited by

Daniel Matějů and Jeffrey A. Chao

Friedrich Miescher Institute for Biomedical Research, Basel, Switzerland

Humana Press

Editors
Daniel Matějů
Friedrich Miescher Institute for Biomedical Research
Basel, Switzerland

Jeffrey A. Chao
Friedrich Miescher Institute for Biomedical Research
Basel, Switzerland

ISSN 1064-3745 ISSN 1940-6029 (electronic)
Methods in Molecular Biology
ISBN 978-1-0716-1974-2 ISBN 978-1-0716-1975-9 (eBook)
https://doi.org/10.1007/978-1-0716-1975-9

Cover Illustration Caption: Cells with stress granules and P-bodies.

This Humana imprint is published by the registered company Springer Science+Business Media, LLC, part of Springer Nature.
The registered company address is: 1 New York Plaza, New York, NY 10004, U.S.A.

Preface

Cells often have to deal with changes in environmental conditions such as oxidative stress, osmotic stress, temperature fluctuations, hypoxia, or viral infections. The integrated stress response (ISR) is an evolutionarily conserved mechanism that enables eukaryotic cells to adapt to these stresses and alter their gene expression programs. The four eIF2α kinases (GCN2, PERK, PKR, and HRI) are activated by different stimuli and initiate a signaling cascade that results in a reduction in ternary complex (eIF2+GTP+tRNAiMET). This in turn leads to global translation inhibition, the induction of stress response genes, and the formation of cytoplasmic RNA-protein condensates termed stress granules and processing bodies. This signaling pathway is activated in diverse biological contexts ranging from cancer to learning and memory.

A variety of experimental techniques have been applied to understand the ISR. Biochemical approaches and structural biology have been instrumental in describing the formation and recycling of ternary complex that is at the core of this pathway. The consequent reprogramming of translation can be investigated by a wide range of methods including polysome profiling, ribosome profiling, or proteomics. Various imaging-based techniques have been applied to study stress granules and their interactions with mRNAs in single cells, while transcriptomic and proteomic techniques can be used to analyze the reorganization of RNA–protein interaction networks in stressed cells. Furthermore, screening approaches have been developed to identify novel factors and regulators of the ISR pathway.

This volume aims to provide an up-to-date collection of protocols describing some of the key methods to investigate the ISR. This collection is a diverse mixture of experimental approaches, including detailed descriptions of classic techniques as well as recently developed methods. We hope this volume will help accelerate research into the complex and fascinating biology of the ISR. Since many of these protocols can be applied to other biological questions, we believe it will also benefit the wider scientific community. We thank all the authors for making this possible.

Basel, Switzerland

Daniel Matějů
Jeffrey A. Chao

Contents

PART II ANALYSIS OF INTERACTION NETWORKS AND RNP GRANULES

Contributors

MICHAEL BACKLUND • *Department of Pediatric Oncology, Hematology and Immunology, Heidelberg University, Heidelberg, Germany; Hopp Children's Cancer Center, National Center for Tumor Diseases (NCT), Heidelberg, Germany; Molecular Medicine Partnership Unit (MMPU), Heidelberg University, Heidelberg, Germany; European Molecular Biology Laboratory (EMBL), Heidelberg, Germany*

JÜRGEN BENEKE • *Advanced Biological Screening Facility, BioQuant, Heidelberg University, Heidelberg, Germany; CellNetworks Cluster of Excellence, Heidelberg University, Heidelberg, Germany*

ANNE BERTOLOTTI • *MRC Laboratory of Molecular Biology, Cambridge, UK*

FLORENCE BESSE • *Université Côte d'Azur, CNRS, Inserm, Institut de Biologie Valrose, Nice, France*

ROLAND BRANDT • *Department of Neurobiology, Osnabrück University, Osnabrück, Germany; Center of Cellular Nanoanalytics, Osnabrück University, Osnabrück, Germany; Institute of Cognitive Science, Osnabrück University, Osnabrück, Germany*

DEZERAE COX • *Department of Biochemistry and Pharmacology, Bio21 Molecular Science and Biotechnology Institute, The University of Melbourne, Parkville, VIC, Australia*

FABIENNE DE GRAEVE • *Université Côte d'Azur, CNRS, Inserm, Institut de Biologie Valrose, Nice, France*

ERIC DEBREUVE • *Université Côte d'Azur, CNRS, Inria, Laboratoire I3S, Sophia Antipolis, France*

XAVIER DESCOMBES • *Université Côte d'Azur, Inria, CNRS, Laboratoire I3S, Sophia Antipolis, France*

KHANH DAO DUC • *Department of Mathematics, University of British Columbia, Vancouver, BC, Canada*

NINA EIERMANN • *Division of Biochemistry, Medical Faculty Mannheim, Mannheim Institute for Innate Immunoscience (MI3), Heidelberg University, Mannheim, Germany; Center for Molecular Biology of Heidelberg University (ZMBH), German Cancer Research Center (DKFZ)-ZMBH Alliance, Heidelberg, Germany*

SARA ELMSAOURI • *Department of Cellular and Molecular Medicine, University of California, San Diego, La Jolla, CA, USA*

ALYSSA M. ENGLISH • *Department of Human Genetics, University of Michigan, Ann Arbor, MI, USA; Center for RNA Biomedicine, University of Michigan, Ann Arbor, MI, USA*

HOLGER ERFLE • *Advanced Biological Screening Facility, BioQuant, Heidelberg University, Heidelberg, Germany; CellNetworks Cluster of Excellence, Heidelberg University, Heidelberg, Germany*

NADIA FORMICOLA • *Université Côte d'Azur, CNRS, Inserm, Institut de Biologie Valrose, Nice, France*

YULIA GONSKIKH • *Department of Chemistry, Biochemistry and Pharmaceutical Sciences, University of Bern, Bern, Switzerland*

XIAOYAN GUO • *Institute for Neurodegenerative Diseases, University of California, San Francisco, CA, USA*

Danny M. Hatters • *Department of Biochemistry and Pharmacology, Bio21 Molecular Science and Biotechnology Institute, The University of Melbourne, Parkville, VIC, Australia*

Katharina Hoerth • *Division of Biochemistry, Medical Faculty Mannheim, Mannheim Institute for Innate Immunoscience (MI3), Heidelberg University, Mannheim, Germany; Center for Molecular Biology of Heidelberg University (ZMBH), German Cancer Research Center (DKFZ)-ZMBH Alliance, Heidelberg, Germany*

Michael J. Holmes • *Department of Biochemistry and Molecular Biology, Indiana University School of Medicine, Indianapolis, IN, USA; Department of Pharmacology and Toxicology, Indiana University School of Medicine, Indianapolis, IN, USA*

Nicholas Ingolia • *Department of Molecular and Cell Biology, University of California, Berkeley, Berkeley, CA, USA*

Martin D. Jennings • *Division of Molecular and Cellular Function, School of Biological Sciences, Faculty of Biology, Medicine and Health, Manchester Academic Health Science Centre, The University of Manchester, Manchester, UK*

Martin Kampmann • *Institute for Neurodegenerative Diseases, University of California, San Francisco, CA, USA; Chan Zuckerberg Biohub, San Francisco, CA, USA; Department of Biochemistry and Biophysics, University of California, San Francisco, CA, USA*

Kevin Klann • *Institute of Biochemistry II, Faculty of Medicine, Goethe University, Frankfurt am Main, Germany*

Philipp Klein • *Department of Infectious Diseases, Molecular Virology, Center for Integrative Infectious Disease Research (CIID), Heidelberg University, Heidelberg, Germany*

Agnieszka Krzyzosiak • *MRC Laboratory of Molecular Biology, Cambridge, UK*

Andreas E. Kulozik • *Department of Pediatric Oncology, Hematology and Immunology, Heidelberg University, Heidelberg, Germany; Hopp Children's Cancer Center, National Center for Tumor Diseases (NCT), Heidelberg, Germany; Molecular Medicine Partnership Unit (MMPU), Heidelberg University, Heidelberg, Germany; European Molecular Biology Laboratory (EMBL), Heidelberg, Germany*

Carine Legrand • *Independent Researcher, Mannheim, Germany*

Vera Magg • *Department of Infectious Diseases, Molecular Virology, Center for Integrative Infectious Disease Research (CIID), Heidelberg University, Heidelberg, Germany*

Sebastian Markmiller • *Department of Cellular and Molecular Medicine, University of California, San Diego, La Jolla, CA, USA*

Gemma E. May • *Department of Biological Sciences, Carnegie Mellon University, Pittsburgh, PA, USA*

C. Joel McManus • *Department of Biological Sciences, Carnegie Mellon University, Pittsburgh, PA, USA; Computational Biology Department, Carnegie Mellon University, Pittsburgh, PA, USA*

Britnie Santiago Membreno • *RNA Biology Laboratory, Center for Cancer Research, National Cancer Institute, National Institutes of Health, Frederick, MD, USA*

Jagannath Misra • *Department of Biochemistry and Molecular Biology, Indiana University School of Medicine, Indianapolis, IN, USA*

Stephanie L. Moon • *Department of Human Genetics, University of Michigan, Ann Arbor, MI, USA; Center for RNA Biomedicine, University of Michigan, Ann Arbor, MI, USA*

Tatsuya Morisaki • *Department of Biochemistry and Molecular Biology, Colorado State University, Fort Collins, CO, USA*

Christian Münch • *Institute of Biochemistry II, Faculty of Medicine, Goethe University, Frankfurt am Main, Germany; Frankfurt Cancer Institute, Frankfurt am Main, Germany; Cardio-Pulmonary Institute, Frankfurt am Main, Germany*

Akira Nakamura • *Department of Germline Development, Institute of Molecular Embryology and Genetics, and Graduate School of Pharmaceutical Sciences, Kumamoto University, Kumamoto, Japan*

Alejandro Padrón • *Department of Molecular and Cell Biology, University of California, Berkeley, Berkeley, CA, USA*

Graham D. Pavitt • *Division of Molecular and Cellular Function, School of Biological Sciences, Faculty of Biology, Medicine and Health, Manchester Academic Health Science Centre, The University of Manchester, Manchester, UK*

Valentina Pecoraro • *Department of Chemistry, Biochemistry and Pharmaceutical Sciences, University of Bern, Bern, Switzerland*

Manuela Pérez-Berlanga • *Department of Quantitative Biomedicine, University of Zurich, Zurich, Switzerland*

Aleksandra P. Pitera • *MRC Laboratory of Molecular Biology, Cambridge, UK*

Norbert Polacek • *Department of Chemistry, Biochemistry and Pharmaceutical Sciences, University of Bern, Bern, Switzerland*

Magdalini Polymenidou • *Department of Quantitative Biomedicine, University of Zurich, Zurich, Switzerland*

Kavya Vinayan Pushpalatha • *Université Côte d'Azur, CNRS, Inserm, Institut de Biologie Valrose, Nice, France*

Mona Radwan • *Department of Biochemistry and Pharmacology, Bio21 Molecular Science and Biotechnology Institute, The University of Melbourne, Parkville, VIC, Australia*

David Ron • *Cambridge Institute for Medical Research (CIMR), University of Cambridge, Cambridge, UK*

Alessia Ruggieri • *Department of Infectious Diseases, Molecular Virology, Center for Integrative Infectious Disease Research (CIID), Heidelberg University, Heidelberg, Germany*

Sonu Sahadevan • *Department of Quantitative Biomedicine, University of Zurich, Zurich, Switzerland*

Yusuke Sekine • *Division of Endocrinology and Metabolism, Department of Medicine, Aging Institute, University of Pittsburgh, Pittsburgh, PA, USA*

Mohammad Shafieinouri • *RNA Biology Laboratory, Center for Cancer Research, National Cancer Institute, National Institutes of Health, Frederick, MD, USA*

Hadjara Sidibé • *Department of Neurosciences, Université de Montréal and CHUM Research Center, Montreal, QC, Canada*

Anna-Carina Söhnel • *Department of Neurobiology, Osnabrück University, Osnabrück, Germany*

Timothy J. Stasevich • *Department of Biochemistry and Molecular Biology, Colorado State University, Fort Collins, CO, USA; World Research Hub Initiative, Institute of Innovative Research, Tokyo Institute of Technology, Yokohama, Kanagawa, Japan*

Georg Stoecklin • *Division of Biochemistry, Medical Faculty Mannheim, Mannheim Institute for Innate Immunoscience (MI3), Heidelberg University, Mannheim, Germany; Center for Molecular Biology of Heidelberg University (ZMBH), German Cancer Research Center (DKFZ)-ZMBH Alliance, Heidelberg, Germany; CellNetworks Cluster of Excellence, Heidelberg University, Heidelberg, Germany*

Xiaojing Sui • *Department of Molecular Biosciences, Rice Institute for Biomedical Research, Northwestern University, Evanston, IL, USA*
Shuying Sun • *Department of Pathology, Physiology, Brain Science Institute, Johns Hopkins University School of Medicine, Baltimore, MD, USA*
Nataliya I. Trushina • *Department of Neurobiology, Osnabrück University, Osnabrück, Germany*
Francesca Tuorto • *Division of Biochemistry, Mannheim Institute for Innate Immunoscience (MI3), Medical Faculty Mannheim, Heidelberg University, Mannheim, Germany; Center for Molecular Biology of Heidelberg University (ZMBH), DKFZ-ZMBH Alliance, Mannheim, Germany*
Christine Vande Velde • *Department of Neurosciences, Université de Montréal and CHUM Research Center, Montreal, QC, Canada*
Shaopeng Wang • *Department of Pathology, Physiology, Brain Science Institute, Johns Hopkins University School of Medicine, Baltimore, MD, USA*
Ronald C. Wek • *Department of Biochemistry and Molecular Biology, Indiana University School of Medicine, Indianapolis, IN, USA*
Colin Chih-Chien Wu • *RNA Biology Laboratory, Center for Cancer Research, National Cancer Institute, National Institutes of Health, Frederick, MD, USA*
Gene W. Yeo • *Department of Cellular and Molecular Medicine, University of California, San Diego, La Jolla, CA, USA*
Alisa F. Zyryanova • *Cambridge Institute for Medical Research (CIMR), University of Cambridge, Cambridge, UK*

Part I

Analysis of mRNA Translation Regulation

Chapter 1

An Overview of Methods for Detecting eIF2α Phosphorylation and the Integrated Stress Response

Agnieszka Krzyzosiak, Aleksandra P. Pitera, and Anne Bertolotti

Abstract

Phosphorylation of the translation initiation factor eIF2α is an adaptive signaling event that is essential for cell and organismal survival from yeast to humans. It is central to the integrated stress response (ISR) that maintains cellular homeostasis in the face of threats ranging from viral infection, amino acid, oxygen, and heme deprivation to the accumulation of misfolded proteins in the endoplasmic reticulum. Phosphorylation of eIF2α has broad physiological, pathological, and therapeutic relevance. However, despite more than two decades of research and growing pharmacological interest, phosphorylation of eIF2α remains difficult to detect and quantify, because of its transient nature and because substoichiometric amounts of this modification are sufficient to profoundly reshape cellular physiology. This review aims to provide a roadmap for facilitating a robust evaluation of eIF2α phosphorylation and its downstream consequences in cells and organisms.

Key words Integrated stress response, Signaling, eIF2α phosphorylation, eIF2α dephosphorylation, Translation, Stress signaling, Unfolded protein response, ATF4, CHOP, PPP1R15A/GADD34, PPP1R15B/CReP

1 Introduction

Phosphorylation of the α-subunit of the heterotrimeric translation initiation factor eIF2 is an essential and evolutionarily conserved defense mechanism against many insults, stresses, and changes in the cellular environment [1, 2]. When bound to GTP, eIF2, which is composed of subunits α, β, and γ, delivers initiator methionyl tRNA to the 40S ribosomal subunit to form the 43S pre-initiation complex. This complex scans the 5′ untranslated region (UTR) of mRNAs for AUG initiation codons [3]. After base-pairing of the start codon with methionyl tRNA, eIF2 hydrolyses GTP, leading to the release of eIF2-GDP and initiation of translation with the

Agnieszka Krzyzosiak and Aleksandra P. Pitera should be considered equal first authors.

Daniel Matějů and Jeffrey A. Chao (eds.), *The Integrated Stress Response: Methods and Protocols*, Methods in Molecular Biology, vol. 2428, https://doi.org/10.1007/978-1-0716-1975-9_1,

recruitment of the 60S ribosomal subunit and formation of the 80S complex [3]. To enable translation initiation, eIF2-GDP must be reactivated, a step catalyzed by the rate-limiting GTP-exchange factor eIF2B [4]. Phosphorylated eIF2α acts as a non-competitive eIF2B inhibitor [4, 5]. Because the abundance of eIF2B is limited in cells, small amounts of eIF2α phosphorylation reduce the functional pool of eIF2B considerably. As a consequence, eIF2α phosphorylation results in the attenuation of bulk protein synthesis. Because protein synthesis consumes the vast majority of cellular resources, a small reduction is sufficient to shift resources from mass production of proteins to adaptations to change in the environment, or damage repair.

1.1 Evolutionary Perspective

Phosphorylation of eIF2α, the central event of the ISR, is catalyzed by specific protein kinases (Fig. 1). Yeast has only one eIF2α kinase, GCN2, which is activated by amino acid shortage. Mammals have evolved three additional eIF2α kinases that sense a variety of signals [6]. The higher number of eIF2α kinases from yeast to mammals attests an increased reliance on this pathway with evolution. Importantly, as is generally the case in stress signaling, eIF2α

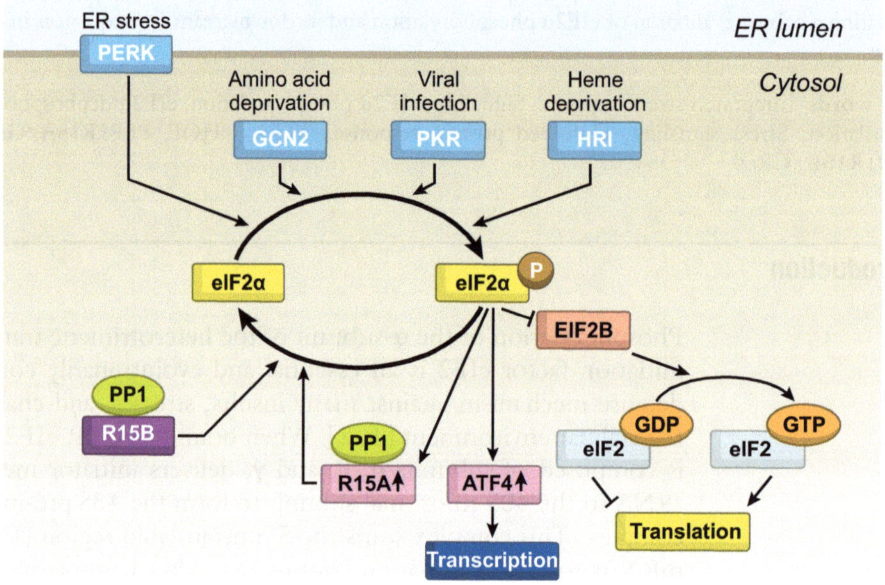

Fig. 1 The integrated stress response. Integrated stress response (ISR) is a protective response to different cellular stresses. Under stress or physiological or pathological changes, eIF2α is phosphorylated by PERK, GCN2, PKR, or HRI sensing respectively, misfolded proteins in the ER, amino acid deprivation, viral infection, or heme deprivation. Phosphorylated eIF2α inhibits eIF2B, which is required to recycle eIF2-GDP to eIF2-GTP. When eIF2α is phosphorylated, the bulk of translation is attenuated up to ~30–40%. In parallel, some mRNAs are preferentially translated such as ATF4, R15A. R15A-PP1, and R15B-PP1 are eIF2α phosphatases. Unlike R15A which is stress-inducible, R15B is expressed in the absence of stress, and its translation is resistant to eIF2α phosphorylation

phosphorylation is transient, avoiding the risk of persistent inhibition of protein synthesis. In mammals, dephosphorylation of eIF2α is catalyzed by two specific protein phosphatases, which are composed of the catalytic subunit PP1 bound to one of two related substrate receptors: PPP1R15A (R15A, also known as GADD34) and PPP1R15B (R15B, also known as CReP). The former is stress-inducible and expressed only when eIF2α is phosphorylated, whereas the latter is expressed constitutively [7, 8]. eIF2α phosphorylation leads to a reduction of protein synthesis by up to 40% and does not affect all cellular transcripts to the same extent, but preferentially decreases translation of highly stable proteins [9]. Thus, this adaptive response to many forms of insults has been evolutionarily optimized to maximize benefit and minimize cost [9].

All four mammalian eIF2α kinases have similar catalytic domains, but different regulatory regions, enabling them to detect distinct stimuli [10] (Fig. 1). GCN2 (encoded by *EIF2AK4*) is activated by uncharged tRNAs and ribosome stalling under conditions such as amino acid starvation [3, 11]. HRI (encoded by *EIF2AK1*) is a heme-regulated kinase involved in balancing the production of globin with the level of heme [1]. PERK (or PEK, encoded by *EIF2AK3*) responds to misfolded proteins within the endoplasmic reticulum and is a component of both the ISR and the unfolded protein response (UPR) [1]. PKR (encoded by *EIF2AK2*) is activated by double-stranded RNA during viral infection [1].

While phosphorylation of eIF2α results in a reduction in bulk protein synthesis, some transcripts are preferentially translated when eIF2α is phosphorylated. The first of such transcripts identified, GCN4, was discovered in yeast [3], and it shares similar function and regulation to mammalian ATF4 (Fig. 1). Transcripts resistant to translational attenuation by eIF2α contain short upstream open reading frames (uORFs) [12]. GCN4 and ATF4 are transcription factors that regulate the expression of a large number of genes, which are predominantly involved in amino acid metabolism and synthesis. This is probably because the yeast ISR evolved as an adaptive mechanism to amino acid starvation, a crucial pathway for the survival of unicellular organisms [13].

1.2 Importance of eIF2α Phosphorylation and the ISR for Cell and Organismal Survival

The essential role for eIF2α phosphorylation was revealed in model organisms. The absence of a phosphorylable allele of eIF2α is lethal in both yeast [14] and mammals [15]. Ablation of PERK is also lethal to cells undergoing endoplasmic reticulum stress [16]. In mice, it results in diabetes and early postnatal death due to pancreatic β-cell loss [17, 18], a phenotype similar to that of mice lacking a phosphorylable allele of eIF2α [15]. In addition, PERK knock-out mice exhibit skeletal defects and postnatal growth retardation [17, 19]. When PERK ablation is induced in adult mice, it also results in severe pancreatic defects [20]. Although not lethal under normal conditions, ablation of the amino acid sensing GCN2 reduces the expected number of pups in mice deprived of leucine during gestation [21]. Deregulation of ISR termination also leads to severe developmental defects, with inactivation of R15A and R15B, the selective substrate receptors of eIF2α phosphatases, resulting in early embryonic lethality [22]. This demonstrates that ISR regulation is vital, with both a loss and an excess of eIF2α phosphorylation being incompatible with life.

1.3 Hypophosphorylation of eIF2α in Human Diseases Due to Mutations Impairing the ISR

Consistent with an essential role in maintaining cellular homeostasis, deregulation of the ISR has been associated with a number of human diseases. Nonsense and splice-site mutations in *EIF2AK3* result in a loss of PERK function and give rise to Wolcott—Rallison syndrome (WR), an autosomal recessive disorder characterized by neonatal or early infancy insulin-dependent diabetes, as well as epiphyseal dysplasia, microcephaly, and short stature [23, 24]. A mutation in *PPP1R15B* was also found to cause diabetes, short stature, intellectual disability, and microcephaly [25, 26]. Thus, the loss of an eIF2α kinase or its constitutive phosphatase results in similar developmental defects. Genome-wide association studies have also identified PERK genetic variants as risk factors for progressive supranuclear palsy (PSP)—a neurodegenerative disease associated with abnormal tau deposition [27]. PSP risk variants are different from WR mutations and result in hypomorphic forms of PERK, with an impaired ability to induce the ISR [28]. Signs of neurodegeneration and tau hyperphosphorylation have also been described in a child with WR syndrome [29]. Mutations in the *EIF2AK4* gene encoding GCN2 have been associated with a rare, severe form of pulmonary arterial hypertension (PAH), pulmonary veno-occlusive disease (PVOD) [30, 31]. The PVOD-causing mutations lead to decreased expression of GCN2, and biallelic *EIF2AK4* mutations are now used to diagnose PVOD [32]. Interestingly, a decrease in the abundance of GCN2 has also been associated with PAH in patients lacking a mutation in *EIF2AK4*, suggesting that the loss of GCN2 function may be of broader relevance for PAH [33]. Recently, genome-wide association studies have shown that increased expression of *EIF2AK1* encoding HRI delays the onset of Huntington's disease [34], consistent with the notion that the ISR is a protective pathway. Another study reported an association of a SNP in *EIF2AK2* encoding PKR with the onset of Alzheimer's disease [35], but the effects of this SNP on PKR function are unknown.

1.4 Hyperphosphorylation of eIF2α as a Cause of Human Diseases

There is a large body of literature reporting increased ISR signaling in human diseases, including Alzheimer's disease and other neurodegenerative diseases [10], giving rise to the view that excessive ISR signaling is maladaptive and pathological. An alternative explanation to ISR induction in diseases is that the ISR may be a defense mechanism that is mounted as an attempt but unable to rescue cellular homeostasis and organismal health. A correlation between eIF2α phosphorylation and memory formation was initially reported in *GCN2* knock-out mice [36]. The relationship between memory formation and eIF2α phosphorylation is complex, with both an increase and a decrease affecting memory [37–39]. While a loss of *PERK* function leads to diabetes in mice and humans, alteration in memory such as that described in *GCN2* knock-out mice [36] has not been reported in humans lacking GCN2.

With eIF2α phosphorylation playing a central role in the maintenance of cell and organismal fitness, this pathway has generated significant interest. Compounds that prolong (Guanabenz, Sephin1, Raphin1) or inhibit ISR signaling (ISRIB) have been identified and are being evaluated in preclinical disease models, and some are developed in human. These compounds have been extensively reviewed elsewhere [10, 13]. Here, we present the challenges in accurately measuring ISR activation by monitoring its components (Fig. 1) and provide an overview of the strengths and weaknesses of a number of assays.

2 Measuring ISR Activation

2.1 Measuring Kinase Activity

Phosphorylation of eIF2α is catalyzed by four kinases. Thus, measuring the activities of PERK, GCN2, PKR, and HRI can provide a readout for pathway activation. Activation of each ISR kinase leads to its autophosphorylation, which can be assessed by virtue of the reduced electrophoretic mobility of the protein [40–42]. This method detects stoichiometric changes but may be insensitive to minor, physiological changes in phosphorylation. Phostag-based western blotting can also be used, with the same limitations [43, 44].

Sensitive detection of phosphorylation of proteins can be obtained by labeling cells with ^{32}P-orthophosphate and immunoprecipitation of the kinase of interest, followed by radiodetection of labeled proteins. Alternatively, antibodies have been produced to detect the phospho-variants of eIF2α kinases. When using such antibodies, it is critical to assess their specificity. In addition to monitoring the activation status of ISR kinases, it is essential to monitor the activation of downstream components in parallel.

2.2 Monitoring eIF2α Phosphorylation

Phosphorylation of eIF2α on serine 51 is the central event of the ISR; it is often used as a first-line readout to assess ISR activation. However, eIF2α phosphorylation is the most difficult readout to measure, because small changes can trigger the pathway. Despite being tricky, immunoblotting is widely used for detecting phosphorylated eIF2α. Positive controls are essential, such as tunicamycin, which blocks N-linked glycosylation and causes the misfolding

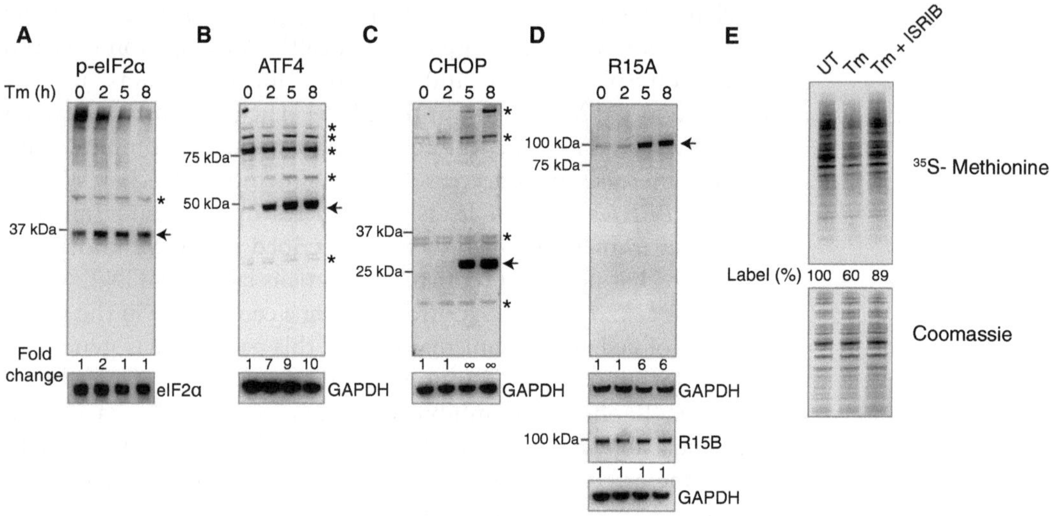

Fig. 2 Typical example of ISR activation by tunicamycin. (**a–d**) Immunoblots of the indicated proteins (p-eIF2α, ATF4, CHOP, R15A) in HeLa cells after treatment with 2.5 μg/ml tunicamycin (Tm) for 0, 2, 5, 8 h. Arrow indicates the band corresponding to the protein of interest, whereas asterisks indicate unspecific bands. The expression of protein was normalized to the level of total eIF2α (for p-eIF2α) or GAPDH (for ATF4, CHOP, and R15A) and the fold change increase of the protein over time is indicated. (**d**) Also shows immunoblot of constitutively expressed R15B which is not induced upon ISR activation. (**e**) Autoradiogram of ^{35}S-methionine-labeled proteins in HeLa cells treated for 2.5 h with 2.5 μg/ml tunicamycin (Tm), 2.5 μg/ml tunicamycin with 2.5 μM ISRIB (Tm + ISRIB) or untreated (UT). The Coomassie stained gel is shown to reveal equal loading of the samples

of proteins in the endoplasmic reticulum, or L-histidinol, a GCN2 inducer. These stressors induce phosphorylation of eIF2α and therefore guide the identification of the specific p-eIF2α signal on immunoblots; they also establish the dynamic range for the experimental setup. As shown in a typical experiment, the detected increase in eIF2α phosphorylation was ~2-fold after treatment with tunicamycin (Fig. 2a). As is the case for phosphorylation studies in general, phosphatase inhibitors are required in lysis buffers, unless denaturing buffers are used. The level of eIF2α phospho-signal should be compared to that of total eIF2α.

It is important to keep in mind that phosphorylation of eIF2α is not only substoichiometric (in the range of a maximum 5–10% in our hands), but also transient. In the example provided, increased eIF2α phosphorylation peaked after 2 h of tunicamycin treatment, and decreased after 5 or 8 h (Fig. 2a), when R15A was expressed, which catalyzes the rapid dephosphorylation of eIF2α (Fig. 2d). Time-course studies are essential, and where possible, dose—response curves are also recommended.

Alternative methods to immunoblotting can be used to detect p-eIF2α. Phospho-specific eIF2α antibodies are often used in immunohistochemistry experiments. It is essential to control for the specificity of the signal and to keep in mind that the levels of phosphorylation of eIF2α are low in cells or tissues. Therefore, we do not recommend monitoring phosphorylation of eIF2α as a stand-alone method, but to add additional readouts. Flow cytometry provides a fast, quantitative detection method of phospho-signal in cells [45, 46]. However, the specificity caveats associated with use of antibodies also apply. Alpha (Amplified Luminescent Proximity Homogeneous Assay) Lisa immunoassays are attractive high-throughput methods that can be used for screening purposes [47, 48].

Monitoring eIF2α phosphorylation using phospho-specific antibodies relies on straightforward and commonly used methods, but is not robust because of the low levels of phosphorylation and limited antibody specificity. Therefore, we recommend testing other downstream components of the pathway, because they provide more robust readouts, since the signal is amplified.

2.3 In Vitro eIF2α Phosphorylation and Dephosphorylation Assays

Phosphorylation and dephosphorylation assays can use recombinant proteins. The amino-terminal fragment of eIF2α can be expressed in large quantities [49] and phosphorylated with the effector domain of an eIF2α kinase, such as PERK [50]. The eIF2α kinases are relatively specific when used at catalytic concentrations, which are substoichiometric to the substrate. Optimal concentrations of kinase to substrate can be established in titration experiments, followed by mass spectrometry (MS) to monitor phosphorylation sites and to avoid non-selective phosphorylation at spurious sites when high concentrations of kinase relative to substrate are used. Phosphorylated eIF2α can be dephosphorylated with stoichiometric amounts of purified PP1 [50]. At high concentrations, PP1 is non-selective and can dephosphorylate many substrates [51]. Selectivity for recognition of eIF2α is encoded by the amino-terminal regions of R15A and R15B and dephosphorylation can be achieved with PP1 in complex with R15A or R15B [50]. Others have performed dephosphorylation of eIF2α using PP1 and a small fragment of R15 containing the PP1-binding part but lacking the substrate-recruitment region in the presence of actin, proposed to be required for selectivity [52]. However, the relevance of actin for eIF2α dephosphorylation in the presence of full-length R15 containing the substrate-binding region remains unknown.

2.4 Tracking Downstream Effectors of the ISR

ISR induction results in a decrease of bulk translation activity, as well as in an increase in the translation of some proteins, such as ATF4 and R15A. Induction of these two proteins is robust, providing ideal readouts for ISR activation.

ATF4 is one of the most studied effectors of the ISR, and it can be used as a readout for activation of all four ISR kinases [1]. Assessing ATF4 induction is straightforward with specific antibodies; it is one of the most robust readouts, because unstressed cells express only low levels of ATF4, and its induction is readily detectable. Because ATF4 induction strictly depends on eIF2α phosphorylation, it is a selective ISR indicator. Thus, ATF4 is our first-in-line marker for ISR activation. A typical experiment is shown in Fig. 2b. Although it was initially reported that ATF4 is regulated predominantly at the translational level [53], ATF4 mRNA also accumulates upon ISR induction [9]. Thus, ATF4 induction can be robustly monitored at both mRNA and protein levels. Reporter assays have also been developed using the 5′UTR of ATF4 fused to the luciferase coding sequence [54]. This system has been implemented in vivo in mice by expressing the construct in which ATF4 coding region was replaced with luciferase [55]. ATF4 is a transcription factor, and its activity can be assessed indirectly by measuring the expression of specific target genes, such as *Asns* and *Fgf21* [56]. An ATF4-target gene reporter has been generated using ATF4-specific response element found in specific target genes to drive the expression of luciferase [57].

R15A and CHOP are also prominently induced by the ISR (Fig. 2c, d). CHOP is expressed at low levels and is highly induced upon stress and thus represents a powerful readout [58, 59]. However, it has been shown that ATF6-associated signaling can drive the induction of CHOP upon endoplasmic reticulum stress [56]. Therefore, CHOP levels are not a pure ISR readout, unlike ATF4.

2.5 UPR or ISR?

The UPR and the ISR partly overlap but are different. The UPR comprises three branches that are controlled by IRE1, PERK, and ATF6 [58, 60]. Distinguishing the activation of the ISR from that of the UPR can be achieved by probing these different branches. One of the most robust and quantitative readouts of UPR induction is monitoring of the splicing of XBP1 by IRE1 using qPCR or end-point PCR followed by gel electrophoresis [58, 60]. ATF6 gets cleaved when activated, and this can serve as a UPR readout, although it is less robust than others [61]. Induction of the endoplasmic reticulum-resident chaperones BiP and GRP94 can also be used to measure UPR induction. If immunoblotting or immunohistochemistry is used, it is important to ensure that experiments are in the linear range. Because chaperones are abundant, it is easy to miss their induction. Quantitative PCR can circumvent this problem because it has a larger dynamic range. XBP1, ATF6, BiP, and GRP94 are UPR-specific markers.

2.6 Measuring Translation

Phosphorylation of eIF2α by ISR kinases leads to a decrease in bulk translation. Measuring translation is thus another robust way to investigate ISR activation, keeping in mind that there are many ways by which translation can be inhibited.

One of the best-established quantitative methods for measuring the rates of protein synthesis is metabolic labeling consisting of a pulse treatment with radiolabeled amino acids (such as ^{35}S-methionine) [62]. Quantifying radioactivity enables measuring the incorporation of the radioisotope into nascent proteins. Quantification can be done either on gels or upon protein precipitation. A typical example of translation attenuation is provided in Fig. 2e.

To avoid radioactive amino acids, methods using stable isotopes have been developed. One approach relies on tagging nascent proteins with an artificial amino acid containing azide moiety (azidohomoalanine, AHA), a method known as BONCAT (bio orthogonal non-canonical amino-acid tagging). After incorporation of the modified amino acid, the AHA-labeled groups are coupled via Click chemistry to fluorescent compounds (FUNCAT) or affinity tags. Nascent proteins can be visualized on gels or affinity-purified and analyzed using MS [63, 64]. While more convenient than the radioactive method, it is not clear whether modified amino acids perturb protein translation or folding. This must be assessed in each experimental setup. SILAC (pulsed stable isotope labeling of amino acids in culture) relies on labeling newly synthesized proteins with stable isotopologs of arginine or lysine for further analysis using MS and can also be used to measure translation [63]. Quantitative non-canonical amino acid tagging (QuaNCAT), a combination of SILAC and BONCAT, relies on the simultaneous pulsing with SILAC and BONCAT amino acids and subsequent purification of nascent proteins by Click chemistry. SILAC labeling is then used to identify proteins by MS [63, 65].

Experiments aimed at measuring translation with labeled amino acids are usually preceded by depletion of the relevant amino acids from cell culture media. Because amino acid depletion is a robust and rapid inducer of the ISR, it is essential to avoid this step when measuring stress signaling. This can be achieved by adding ^{35}S-methionine to media without prior methionine depletion [66–68]. The same approach can be considered with the other modified amino acids. Titration experiments are required for identifying an optimal ratio of modified/unmodified amino acids to achieve high labeling efficiency without inducing stress pathways.

Ribopuromycylation is another method that can be used to measure translation. Puromycin mimics charged tRNAs and can thus get incorporated into the ribosome. It results in the premature termination of translation which can be quantified using anti-puromycin antibodies and immunoblotting, fluorescence-activated cell sorting (FACS) or fluorescence microscopy [69]. In a modification called PUNCH-P (puromycin-associated nascent chain pro teomics), puromycin is linked to biotin, allowing immunoprecipitation of newly synthesized proteins using immobilized streptavidin for subsequent MS analysis [70]. Puromycin has been commonly used for studying protein translation, including the studies investigating ISR activation [71, 72]. However, it has been suggested that this method is not suitable for measuring changes of translation in energy-deprived cells [73]. Importantly, it is less sensitive than ^{35}S-methionine labeling of proteins, and in our hands, not sensitive enough to monitor translation recovery. Because puromycin interferes with translation, it may also have direct or indirect effects on the ISR.

The aforementioned approaches rely on the analysis of nascent proteins. Measurement of mRNA bound to ribosomes can also be used to analyze translation. These methods are more labor-intensive, but provide high-resolution and comprehensive analysis of translation. Ribosome profiling is a method based on the extraction of RNA fragments that are associated with translating ribosomes. These RNA sequences can be then identified using high-throughput sequencing to identify ribosome-protected RNA fragments that can be mapped to the genome [74]. However, ribosome profiling identifies RNA fragments that are protected, not only by translating polysomes, but also by not efficiently translating light polysomes and monosomes. To obtain information about transcriptome fragments that are highly translated in polysomes, polysome profiling can be used. This method is based on the separation of mRNAs bound to different numbers of ribosomes by sedimentation in a sucrose gradient. Fractions containing efficiently translated mRNAs can then be sequenced or subjected to DNA microarrays [75]. Polysome profiling and RNA sequencing methods have been successfully applied to measure translational changes, also those related to the ISR [9, 76–78].

Our preferred method for measuring translation changes remains metabolic labeling because it is so far the most sensitive method and does not perturb cellular homeostasis as long as the label is added to the media without exposing cells to amino acid deprivation, which induces the ISR. If a genome-wide analysis is needed, polysome- or ribosome-profiling can be used for an in-depth analysis of translational changes.

2.7 Optimal Conditions to Measure ISR

One important aspect to take into consideration when investigating the ISR in cells in culture is to ensure that the cells are kept under optimal conditions to avoid spurious ISR induction. Because the ISR is activated by various stressors, it is not trivial to culture cells with minimal ISR induction. We recommend using cultures at subconfluency (70–80% maximum) to avoid depletion of nutrients in the media. It is also important to remember that confluent cells use up glucose and that glucose deprivation was among the first known inducers of the UPR [6].

As mentioned above, when selecting a method to study ISR activation, it is also important to ensure that the method itself is not inducing or interfering with the pathway. This can be assessed by measuring ATF4, R15A, and CHOP induction. We also recommend addition of a positive control, such as tunicamycin, to provide a benchmark for maximal levels of induction of ISR markers.

2.8 Detection of the ISR In Vivo

Because of the small changes of some ISR markers, the transient nature of their induction and the partial selectivity of most antibodies, it is often difficult to detect induction of the ISR in tissues. Immunohistochemistry is a popular method, but the selectivity of the signal is difficult to control, and this method is prone to false-positives. While immunostaining can be a valuable addition, we recommend validating the results by additional methods, such as qPCR. It will be a major challenge in the years to come to develop robust and quantitative readouts to study the ISR in health and disease.

2.9 Roadmap to ISR Detection

Here we provide a roadmap for detecting ISR induction. Commercially available antibodies are commonly used to assess the levels of ISR targets. We present raw images of typical immunoblots revealed by chemiluminescence (Fig. 2). Increased eIF2α phosphorylation upon tunicamycin treatment was transient and visible after 2 h (Fig. 2a). Because of the small increase in eIF2 phosphorylation with maximal ISR induction with tunicamycin and its transient nature, we recommend testing other markers, such as ATF4, CHOP, or R15A (Fig. 2b–d). Tunicamycin caused a prominent increase in the abundance of these ISR targets with fold changes ranging from 5 for R15A and infinite for CHOP (Table 1). It is important to validate ISR induction using multiple readouts. Thus, we recommend combining immunodetection methods with qPCR experiments using several targets. As summarized in Table 1, tunicamycin increased ATF4 mRNA by ~3.5-fold, CHOP by ~50-fold, and R15A by ~4-fold. It is also important to note that different targets are maximally induced at different times. Thus, eIF2α phosphorylation precedes downstream target induction (Fig. 2 and Table 1). In our hands, the kinetics and fold changes are largely consistent in various cell types (data not shown).

Translation assays, such as ^{35}S-methionine labeling, are powerful readouts of ISR activation. Phosphorylation of eIF2α and ISR signaling does not lead to the complete shutdown of protein synthesis, but to a decrease of 30–40%. ISRIB, an ISR inhibitor [10, 13], can be used to assess if translational attenuation is mediated by eIF2α phosphorylation (Fig. 2e). ISRIB can also be used to block induction of the ISR markers described above, such as ATF4 [48], as a validation that the induction is mediated by the ISR.

Table 1
Typical fold change in selected ISR markers. The fold change in ISR markers were obtained in experiments such as the one presented in Fig. 2. Note that the pathway is dynamic and some changes transient

	Expected fold change upon tunicamycin treatment	
Marker	**Protein**	**mRNA**
p-eIF2α 2 h	~2	–
p-eIF2α 8 h	0	–
ATF4 5 h	~9	~3.5
CHOP 5 h	∞	~50
R15A 5 h	~6	~4
Translation 2.5 h	0.6	–

3 Conclusion

eIF2α signaling is dynamic, and small changes can lead to dramatic effects in cellular physiology and pathology. Since the signal is amplified between the sensor and its effectors, monitoring downstream components provides robust readouts.

It is best to use a combination of readouts when assessing the ISR, since single assay can be misleading. For instance, an inability to detect eIF2α phosphorylation is not evidence of a lack of induction, because the changes can be small and therefore difficult to detect. Many antibodies are commercially available and provide useful tools to assess ISR activation, so long as they are used with suitable controls and the dynamic range of the experimental setup is taken into consideration.

Assessment of ISR markers at various levels (mRNA and protein) prevents inaccurate conclusions. However, the in vivo assessment of ISR and UPR remains difficult. With a growing therapeutic interest in this pathway, a major challenge for the future is to develop readouts that are also useful for translational research.

Acknowledgments

We are grateful to the Bertolotti lab members for discussions and comments, D. Alessi and M. Goedert for comments on the manuscript, and J. Westmoreland for the illustration. A.K. was supported by the European Molecular Biology Organization (EMBO, ALTF 1171-2013) and Human Frontier Science Program (LT000888/2014-L). A.P.P. and A.B. are supported by the Medical Research Council (MC_U105185860).

References

1. Wek RC (2018) Role of eIF2alpha kinases in translational control and adaptation to cellular stress. Cold Spring Harb Perspect Biol 10(7): a032870. https://doi.org/10.1101/cshperspect.a032870
2. Cao SS, Kaufman RJ (2012) Unfolded protein response. Curr Biol 22(16):R622–R626. https://doi.org/10.1016/j.cub.2012.07.004
3. Sonenberg N, Hinnebusch AG (2009) Regulation of translation initiation in eukaryotes: mechanisms and biological targets. Cell 136(4):731–745. https://doi.org/10.1016/j.cell.2009.01.042
4. Merrick WC, Pavitt GD (2018) Protein synthesis initiation in eukaryotic cells. Cold Spring Harb Perspect Biol 10(12):a033092. https://doi.org/10.1101/cshperspect.a033092
5. Marintchev A, Ito T (2020) eIF2B and the integrated stress response: a structural and mechanistic view. Biochemistry 59(13): 1299–1308. https://doi.org/10.1021/acs.biochem.0c00132
6. Pilla E, Schneider K, Bertolotti A (2017) Coping with protein quality control failure. Annu Rev Cell Dev Biol 33:439–465. https://doi.org/10.1146/annurev-cellbio-111315-125334
7. Jousse C, Oyadomari S, Novoa I et al (2003) Inhibition of a constitutive translation initiation factor 2alpha phosphatase, CReP, promotes survival of stressed cells. J Cell Biol 163(4):767–775. https://doi.org/10.1083/jcb.200308075
8. Novoa I, Zeng H, Harding HP et al (2001) Feedback inhibition of the unfolded protein response by GADD34-mediated dephosphorylation of eIF2alpha. J Cell Biol 153(5): 1011–1022. https://doi.org/10.1083/jcb.153.5.1011
9. Schneider K, Nelson GM, Watson JL et al (2020) Protein stability buffers the cost of translation attenuation following eIF2alpha phosphorylation. Cell Rep 32(11):108154. https://doi.org/10.1016/j.celrep.2020.108154
10. Costa-Mattioli M, Walter P (2020) The integrated stress response: from mechanism to disease. Science 368(6489):eaat5314. https://doi.org/10.1126/science.aat5314
11. Ishimura R, Nagy G, Dotu I et al (2016) Activation of GCN2 kinase by ribosome stalling links translation elongation with translation initiation. Elife 5:e14295. https://doi.org/10.7554/eLife.14295
12. Young SK, Wek RC (2016) Upstream open Reading frames differentially regulate gene-specific translation in the integrated stress response. J Biol Chem 291(33): 16927–16935. https://doi.org/10.1074/jbc.R116.733899
13. Luh LM, Bertolotti A (2020) Potential benefit of manipulating protein quality control systems in neurodegenerative diseases. Curr Opin Neurobiol 61:125–132. https://doi.org/10.1016/j.conb.2020.02.009
14. Dever TE, Feng L, Wek RC et al (1992) Phosphorylation of initiation factor 2 alpha by protein kinase GCN2 mediates gene-specific translational control of GCN4 in yeast. Cell 68(3):585–596. https://doi.org/10.1016/0092-8674(92)90193-g
15. Scheuner D, Song B, McEwen E et al (2001) Translational control is required for the unfolded protein response and in vivo glucose homeostasis. Mol Cell 7(6):1165–1176. https://doi.org/10.1016/s1097-2765(01)00265-9
16. Harding HP, Zhang Y, Bertolotti A et al (2000) Perk is essential for translational regulation and cell survival during the unfolded protein response. Mol Cell 5(5):897–904. https://doi.org/10.1016/s1097-2765(00)80330-5
17. Harding HP, Zeng H, Zhang Y et al (2001) Diabetes mellitus and exocrine pancreatic dysfunction in perk−/− mice reveals a role for translational control in secretory cell survival. Mol Cell 7(6):1153–1163. https://doi.org/10.1016/s1097-2765(01)00264-7
18. Zhang W, Feng D, Li Y et al (2006) PERK EIF2AK3 control of pancreatic beta cell differentiation and proliferation is required for postnatal glucose homeostasis. Cell Metab 4(6): 491–497. https://doi.org/10.1016/j.cmet.2006.11.002
19. Zhang P, McGrath B, Li S et al (2002) The PERK eukaryotic initiation factor 2 alpha kinase is required for the development of the skeletal system, postnatal growth, and the function and viability of the pancreas. Mol Cell Biol 22(11):3864–3874. https://doi.org/10.1128/mcb.22.11.3864-3874.2002
20. Gao Y, Sartori DJ, Li C et al (2012) PERK is required in the adult pancreas and is essential for maintenance of glucose homeostasis. Mol Cell Biol 32(24):5129–5139. https://doi.org/10.1128/MCB.01009-12
21. Zhang P, McGrath BC, Reinert J et al (2002) The GCN2 eIF2alpha kinase is required for

adaptation to amino acid deprivation in mice. Mol Cell Biol 22(19):6681–6688. https://doi.org/10.1128/mcb.22.19.6681-6688.2002

22. Harding HP, Zhang Y, Scheuner D et al (2009) Ppp1r15 gene knockout reveals an essential role for translation initiation factor 2 alpha (eIF2alpha) dephosphorylation in mammalian development. Proc Natl Acad Sci U S A 106(6):1832–1837. https://doi.org/10.1073/pnas.0809632106
23. Delepine M, Nicolino M, Barrett T et al (2000) EIF2AK3, encoding translation initiation factor 2-alpha kinase 3, is mutated in patients with Wolcott-Rallison syndrome. Nat Genet 25(4): 406–409. https://doi.org/10.1038/78085
24. Julier C, Nicolino M (2010) Wolcott-Rallison syndrome. Orphanet J Rare Dis 5:29. https://doi.org/10.1186/1750-1172-5-29
25. Abdulkarim B, Nicolino M, Igoillo-Esteve M et al (2015) A missense mutation in PPP1R15B causes a syndrome including diabetes, short stature, and microcephaly. Diabetes 64(11):3951–3962. https://doi.org/10.2337/db15-0477
26. Kernohan KD, Tetreault M, Liwak-Muir U et al (2015) Homozygous mutation in the eukaryotic translation initiation factor 2alpha phosphatase gene, PPP1R15B, is associated with severe microcephaly, short stature and intellectual disability. Hum Mol Genet 24(22):6293–6300. https://doi.org/10.1093/hmg/ddv337
27. Hoglinger GU, Melhem NM, Dickson DW et al (2011) Identification of common variants influencing risk of the tauopathy progressive supranuclear palsy. Nat Genet 43(7):699–705. https://doi.org/10.1038/ng.859
28. Yuan SH, Hiramatsu N, Liu Q et al (2018) Tauopathy-associated PERK alleles are functional hypomorphs that increase neuronal vulnerability to ER stress. Hum Mol Genet 27(22):3951–3963. https://doi.org/10.1093/hmg/ddy297
29. Bruch J, Kurz C, Vasiljevic A et al (2015) Early neurodegeneration in the brain of a child without functional PKR-like endoplasmic reticulum kinase. J Neuropathol Exp Neurol 74(8): 850–857. https://doi.org/10.1097/NEN.0000000000000224
30. Eyries M, Montani D, Girerd B et al (2014) EIF2AK4 mutations cause pulmonary veno-occlusive disease, a recessive form of pulmonary hypertension. Nat Genet 46(1):65–69. https://doi.org/10.1038/ng.2844
31. Hadinnapola C, Bleda M, Haimel M et al (2017) Phenotypic characterization of EIF2AK4 mutation carriers in a large cohort of patients diagnosed clinically with pulmonary arterial hypertension. Circulation 136(21): 2022–2033. https://doi.org/10.1161/CIRCULATIONAHA.117.028351
32. Galie N, Humbert M, Vachiery JL et al (2016) 2015 ESC/ERS Guidelines for the diagnosis and treatment of pulmonary hypertension: The Joint Task Force for the Diagnosis and Treatment of Pulmonary Hypertension of the European Society of Cardiology (ESC) and the European Respiratory Society (ERS): Endorsed by: Association for European Paediatric and Congenital Cardiology (AEPC), International Society for Heart and Lung Transplantation (ISHLT). Eur Heart J 37(1): 67–119. https://doi.org/10.1093/eurheartj/ehv317
33. Nossent EJ, Antigny F, Montani D et al (2018) Pulmonary vascular remodeling patterns and expression of general control nonderepressible 2 (GCN2) in pulmonary veno-occlusive disease. J Heart Lung Transplant 37(5): 647–655. https://doi.org/10.1016/j.healun.2017.09.022
34. Wright GEB, Caron NS, Ng B et al (2020) Gene expression profiles complement the analysis of genomic modifiers of the clinical onset of Huntington disease. Hum Mol Genet 29(16):2788–2802. https://doi.org/10.1093/hmg/ddaa184
35. Bullido MJ, Martinez-Garcia A, Tenorio R et al (2008) Double stranded RNA activated EIF2 alpha kinase (EIF2AK2; PKR) is associated with Alzheimer's disease. Neurobiol Aging 29(8):1160–1166. https://doi.org/10.1016/j.neurobiolaging.2007.02.023
36. Costa-Mattioli M, Gobert D, Harding H et al (2005) Translational control of hippocampal synaptic plasticity and memory by the eIF2alpha kinase GCN2. Nature 436(7054): 1166–1173. https://doi.org/10.1038/nature03897
37. Costa-Mattioli M, Gobert D, Stern E et al (2007) eIF2alpha phosphorylation bidirectionally regulates the switch from short- to long-term synaptic plasticity and memory. Cell 129(1):195–206. https://doi.org/10.1016/j.cell.2007.01.050
38. Trinh MA, Kaphzan H, Wek RC et al (2012) Brain-specific disruption of the eIF2alpha kinase PERK decreases ATF4 expression and impairs behavioral flexibility. Cell Rep 1(6): 676–688. https://doi.org/10.1016/j.celrep.2012.04.010
39. Zhu PJ, Huang W, Kalikulov D et al (2011) Suppression of PKR promotes network excitability and enhanced cognition by interferon-gamma-mediated disinhibition. Cell 147(6):

1384–1396. https://doi.org/10.1016/j.cell.2011.11.029

40. Harding HP, Zhang Y, Ron D (1999) Protein translation and folding are coupled by an endoplasmic-reticulum-resident kinase. Nature 397(6716):271–274. https://doi.org/10.1038/16729
41. Lu L, Han AP, Chen JJ (2001) Translation initiation control by heme-regulated eukaryotic initiation factor 2alpha kinase in erythroid cells under cytoplasmic stresses. Mol Cell Biol 21(23):7971–7980. https://doi.org/10.1128/MCB.21.23.7971-7980.2001
42. Zhang F, Romano PR, Nagamura-Inoue T et al (2001) Binding of double-stranded RNA to protein kinase PKR is required for dimerization and promotes critical autophosphorylation events in the activation loop. J Biol Chem 276(27):24946–24958. https://doi.org/10.1074/jbc.M102108200
43. Kinoshita E, Kinoshita-Kikuta E, Takiyama K et al (2006) Phosphate-binding tag, a new tool to visualize phosphorylated proteins. Mol Cell Proteomics 5(4):749–757. https://doi.org/10.1074/mcp.T500024-MCP200
44. Yang L, Xue Z, He Y et al (2010) A Phos-tag-based approach reveals the extent of physiological endoplasmic reticulum stress. PLoS One 5(7):e11621. https://doi.org/10.1371/journal.pone.0011621
45. Krutzik PO, Irish JM, Nolan GP et al (2004) Analysis of protein phosphorylation and cellular signaling events by flow cytometry: techniques and clinical applications. Clin Immunol 110(3):206–221. https://doi.org/10.1016/j.clim.2003.11.009
46. Rabouw HH, Langereis MA, Anand AA et al (2019) Small molecule ISRIB suppresses the integrated stress response within a defined window of activation. Proc Natl Acad Sci U S A 116(6):2097–2102. https://doi.org/10.1073/pnas.1815767116
47. Eglen RM, Reisine T, Roby P et al (2008) The use of AlphaScreen technology in HTS: current status. Curr Chem Genomics 1:2–10. https://doi.org/10.2174/1875397300801010002
48. Sidrauski C, Acosta-Alvear D, Khoutorsky A et al (2013) Pharmacological brake-release of mRNA translation enhances cognitive memory. Elife 2:e00498. https://doi.org/10.7554/eLife.00498
49. Ito T, Marintchev A, Wagner G (2004) Solution structure of human initiation factor eIF2alpha reveals homology to the elongation factor eEF1B. Structure 12(9):1693–1704. https://doi.org/10.1016/j.str.2004.07.010
50. Carrara M, Sigurdardottir A, Bertolotti A (2017) Decoding the selectivity of eIF2alpha holophosphatases and PPP1R15A inhibitors. Nat Struct Mol Biol 24(9):708–716. https://doi.org/10.1038/nsmb.3443
51. Bertolotti A (2018) The split protein phosphatase system. Biochem J 475(23):3707–3723. https://doi.org/10.1042/BCJ20170726
52. Chen R, Rato C, Yan Y et al (2015) G-actin provides substrate-specificity to eukaryotic initiation factor 2alpha holophosphatases. Elife 4:e04871. https://doi.org/10.7554/eLife.04871
53. Harding HP, Novoa I, Zhang Y et al (2000) Regulated translation initiation controls stress-induced gene expression in mammalian cells. Mol Cell 6(5):1099–1108. https://doi.org/10.1016/s1097-2765(00)00108-8
54. Vattem KM, Wek RC (2004) Reinitiation involving upstream ORFs regulates ATF4 mRNA translation in mammalian cells. Proc Natl Acad Sci U S A 101(31):11269–11274. https://doi.org/10.1073/pnas.0400541101
55. Iwawaki T, Akai R, Toyoshima T et al (2017) Transgenic mouse model for imaging of ATF4 translational activation-related cellular stress responses in vivo. Sci Rep 7:46230. https://doi.org/10.1038/srep46230
56. Fusakio ME, Willy JA, Wang Y et al (2016) Transcription factor ATF4 directs basal and stress-induced gene expression in the unfolded protein response and cholesterol metabolism in the liver. Mol Biol Cell 27(9):1536–1551. https://doi.org/10.1091/mbc.E16-01-0039
57. Chaveroux C, Carraro V, Canaple L et al (2015) In vivo imaging of the spatiotemporal activity of the eIF2alpha-ATF4 signaling pathway: insights into stress and related disorders. Sci Signal 8(374):rs5. https://doi.org/10.1126/scisignal.aaa0549
58. Ron D, Walter P (2007) Signal integration in the endoplasmic reticulum unfolded protein response. Nat Rev Mol Cell Biol 8(7):519–529. https://doi.org/10.1038/nrm2199
59. Oyadomari S, Mori M (2004) Roles of CHOP/GADD153 in endoplasmic reticulum stress. Cell Death Differ 11(4):381–389. https://doi.org/10.1038/sj.cdd.4401373
60. Mori K (2000) Tripartite management of unfolded proteins in the endoplasmic reticulum. Cell 101(5):451–454. https://doi.org/10.1016/s0092-8674(00)80855-7
61. Ye J, Rawson RB, Komuro R et al (2000) ER stress induces cleavage of membrane-bound ATF6 by the same proteases that process

SREBPs. Mol Cell 6(6):1355–1364. https://doi.org/10.1016/s1097-2765(00)00133-7

62. Wong WL, Brostrom MA, Kuznetsov G et al (1993) Inhibition of protein synthesis and early protein processing by thapsigargin in cultured cells. Biochem J 289(Pt 1):71–79. https://doi.org/10.1042/bj2890071
63. Dermit M, Dodel M, Mardakheh FK (2017) Methods for monitoring and measurement of protein translation in time and space. Mol BioSyst 13(12):2477–2488. https://doi.org/10.1039/c7mb00476a
64. Dieterich DC, Lee JJ, Link AJ et al (2007) Labeling, detection and identification of newly synthesized proteomes with bioorthogonal non-canonical amino-acid tagging. Nat Protoc 2(3):532–540. https://doi.org/10.1038/nprot.2007.52
65. Howden AJ, Geoghegan V, Katsch K et al (2013) QuaNCAT: quantitating proteome dynamics in primary cells. Nat Methods 10(4):343–346. https://doi.org/10.1038/nmeth.2401
66. Das I, Krzyzosiak A, Schneider K et al (2015) Preventing proteostasis diseases by selective inhibition of a phosphatase regulatory subunit. Science 348(6231):239–242. https://doi.org/10.1126/science.aaa4484
67. Krzyzosiak A, Sigurdardottir A, Luh L et al (2018) Target-based discovery of an inhibitor of the regulatory phosphatase PPP1R15B. Cell 174(5):1216–1228. e1219. https://doi.org/10.1016/j.cell.2018.06.030
68. Tsaytler P, Harding HP, Ron D et al (2011) Selective inhibition of a regulatory subunit of protein phosphatase 1 restores proteostasis. Science 332(6025):91–94. https://doi.org/10.1126/science.1201396
69. Schmidt EK, Clavarino G, Ceppi M et al (2009) SUnSET, a nonradioactive method to monitor protein synthesis. Nat Methods 6(4):275–277. https://doi.org/10.1038/nmeth.1314
70. Aviner R, Geiger T, Elroy-Stein O (2013) Novel proteomic approach (PUNCH-P) reveals cell cycle-specific fluctuations in mRNA translation. Genes Dev 27(16):1834–1844. https://doi.org/10.1101/gad.219105.113
71. Follo C, Vidoni C, Morani F et al (2019) Amino acid response by Halofuginone in cancer cells triggers autophagy through proteasome degradation of mTOR. Cell Commun Signal 17(1):39. https://doi.org/10.1186/s12964-019-0354-2
72. Perry BD, Rahnert JA, Xie Y et al (2018) Palmitate-induced ER stress and inhibition of protein synthesis in cultured myotubes does not require toll-like receptor 4. PLoS One 13(1):e0191313. https://doi.org/10.1371/journal.pone.0191313
73. Marciano R, Leprivier G, Rotblat B (2018) Puromycin labeling does not allow protein synthesis to be measured in energy-starved cells. Cell Death Dis 9(2):39. https://doi.org/10.1038/s41419-017-0056-x
74. Ingolia NT (2014) Ribosome profiling: new views of translation, from single codons to genome scale. Nat Rev Genet 15(3):205–213. https://doi.org/10.1038/nrg3645
75. Piccirillo CA, Bjur E, Topisirovic I et al (2014) Translational control of immune responses: from transcripts to translatomes. Nat Immunol 15(6):503–511. https://doi.org/10.1038/ni.2891
76. Gonen N, Sabath N, Burge CB et al (2019) Widespread PERK-dependent repression of ER targets in response to ER stress. Sci Rep 9(1):4330. https://doi.org/10.1038/s41598-019-38705-5
77. Greenman IC, Gomez E, Moore CE et al (2007) Distinct glucose-dependent stress responses revealed by translational profiling in pancreatic beta-cells. J Endocrinol 192(1):179–187. https://doi.org/10.1677/joe.1.06898
78. Guan BJ, van Hoef V, Jobava R et al (2017) A unique ISR program determines cellular responses to chronic stress. Mol Cell 68(5):885–900. e886. https://doi.org/10.1016/j.molcel.2017.11.007

Chapter 2

CRISPR-Based Screening for Stress Response Factors in Mammalian Cells

Xiaoyan Guo and Martin Kampmann

Abstract

In the presence of different physiological and environmental stresses, cells rapidly initiate stress responses to re-establish cellular homeostasis. Stress responses usually orchestrate both transcriptional and translational programs via distinct mechanisms. With the advance of transcriptomics and proteomics technologies, transcriptional and translational outputs to a particular stress condition have become easier to measure; however, these technologies lack the ability to reveal the upstream regulatory pathways. Unbiased genetic screens based on a transcriptional or translational reporter are powerful approaches to identify regulatory factors of a specific stress response. CRISPR/Cas-based technologies, together with next-generation sequencing, enable genome-scale pooled screens to systematically elucidate gene function in mammalian cells, with a significant reduction in the rate of off-target effects compared to the previously used RNAi technology. Here, we describe our fluorescence-activated cell sorting (FACS)-based CRISPR interference (CRISPRi) screening platform using a translational reporter to identify novel genetic factors of the mitochondrial stress response in mammalian cells. This protocol provides a general framework for scientists who wish to establish a reporter-based CRISPRi screening platform to address questions in their area of research.

Key words CRISPRi, Genetic screens, Mitochondrial stress response, Transcriptional reporter, Translational reporter, FACS, Mammalian cells

1 Introduction

Cells deploy unique and complex response pathways to survive many different stress conditions. Stress sensors initiate these response pathways, which result in specific outputs to cope with the encountered stress (Fig. 1a). Depending on the types of stress, these outputs usually are alterations of the levels of particular proteins via transcriptional and/or translational regulation. Oftentimes, there are multiple factors transducing a signal from the stress sensor to the final stress response output (Fig. 1a) [1]. Uncovering these factors not only provides a better mechanistic understanding of the specific stress response pathway, but can also advance the

Daniel Matějů and Jeffrey A. Chao (eds.), *The Integrated Stress Response: Methods and Protocols*,
Methods in Molecular Biology, vol. 2428, https://doi.org/10.1007/978-1-0716-1975-9_2,

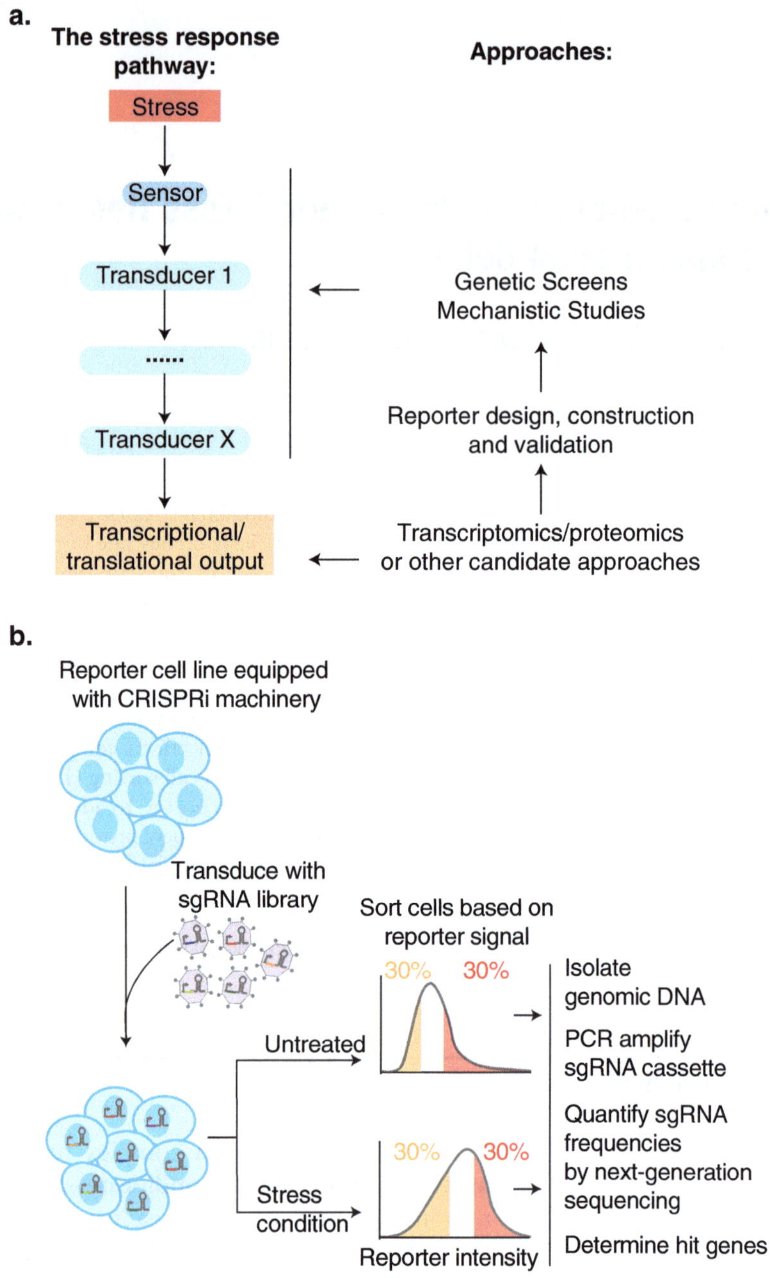

Fig. 1 Overall strategies to study the stress response. (**a**) A typical stress response pathway and approaches to characterize it. (**b**) The workflow of a FACS-based CRISPRi screen

development of therapeutics for diseases associated with a misregulated stress response [2, 3].

Unbiased genetic screening approaches have revealed some of the most important stress–response regulators. An ethyl methanesulfonate (EMS) mutagenesis screen in yeast using a lacZ reporter under the control of the promoter of *KAR2* (the homolog of mammalian endoplasmic reticulum (ER) chaperone gene *BiP*), which is upregulated upon ER stress, identified IRE1 as a critical upstream regulator of the ER unfolded protein response (UPR^{ER}) [4, 5]. Mammalian IRE1 plays the same role [6, 7]. The molecular mechanisms of how IRE1 senses and transduces the ER stress were characterized by several subsequent studies (reviewed in [8, 9]). Two additional branches of the UPR^{ER} in multicellular eukaryotes are transduced through PERK and ATF6. PERK was identified and cloned through a search for the nucleotide sequence that shares similarity to the kinase domain of PKR (one of the eIF2α kinases) and the transmembrane domain of IRE1 [10]. ATF6 was discovered in an effort to identify human cDNAs that bind to the *BiP* promoter using yeast one-hybrid screening [11]. Furthermore, RNAi screens defined the mitochondrial unfolded protein response (UPR^{mito}) pathway in *C. elegans*: using a transcriptional reporter of mitochondrial chaperone *hsp-60* in RNAi screens, Cole Haynes and coworkers successfully identified that several genetic factors including *clpp-1*, *dve-1*, *ubl-5*, and *atfs-1* participate in the signal transduction in response to mitochondrial stress [12, 13]. Subsequent studies uncovered the mechanisms of how these factors upregulate the expression of mitochondrial specific homeostatic factors to re-establish mitochondrial homeostasis [13–16].

In contrast to the well-characterized UPR^{ER}, which is mostly conserved from yeast to mammals, and UPR^{mito} in *C. elegans*, the mitochondrial stress response in mammalian cells was less defined until we [17] and Lucas Jae's group [18] independently uncovered the molecular mechanism of how mammalian cells relay the mitochondrial stress to the cytosol. Unlike the *C. elegans* UPR^{mito}, we and others showed that the main response in mammalian cells triggered by mitochondrial stress seems to be mediated by ATF4 [17–20], a transcriptional factor, which is translationally upregulated by the integrated stress response (ISR). The ISR is triggered by many different stress conditions [21]. Briefly, these stress conditions can activate four different eIF2α kinases (GCN2, PKR, PERK, and HRI), which eventually converge to phosphorylate eIF2α, a major stress response signaling hub that regulates protein translation [22]. GCN2 is activated by uncharged tRNAs due to amino acid deprivation [23, 24]. Double-stranded RNAs from viral infections activate PKR [25]. Accumulation of misfolded/unfolded proteins in the ER activates PERK [10, 26]. HRI is canonically activated by heme deprivation in the erythrocytes [27]; however, it can also be activated by other stress conditions including oxidative stress, osmotic stress, and proteasome inhibition in a heme-independent manner [28]. Phosphorylated eIF2α inhibits the guanine nucleotide exchange factor (GEF) activity of eIF2B, which reduces the formation of translation initiation ternary complexes and thus attenuates global translation [29]. However, a group of mRNAs, including ATF4, that contain 5′ upstream open reading frames (uORFs), can be selectively translated under these stress conditions [30]. Therefore, ATF4 protein levels, monitored either by immunostaining or reporters, can be used as a readout for the integrated stress response.

To uncover how mammalian cells relay mitochondrial stress to the integrated stress response, we established an ATF4 translational reporter based on its translational regulation by uORFs. Using this reporter and immunoblotting of the endogenous ATF4, we first determined that HRI is the eIF2α kinase that mediates the integrated stress response triggered by the mitochondrial dysfunction. Further, we performed FACS-based CRISPR (clustered regularly interspaced short palindromic repeats) interference (CRISPRi) screens to reveal genes knockdown of which abolish mitochondrial stress-induced ATF4 translation. From such an ATF4 translational reporter-based screen, we have successfully identified a novel pathway involving the proteins OMA1, DELE1, and HRI that relays mitochondrial stress to the integrated stress response [17].

The CRISPR/CRISPR-associated (Cas) system, originally discovered as a bacterial adaptive immune system protecting against bacteriophage infection [31–33], provides a versatile tool kit enabling genome editing [34, 35]. Heterologous expression of the key components of the CRISPR/Cas system, including the nuclease Cas9 and a single-guide RNA (sgRNA) composed of a programmable crispr RNA (crRNA) and a constant transactivating crRNA (tracrRNA), enables precise genome editing in a wide range of cells and organisms.

In addition to the use of Cas9 for genome editing, nuclease-dead Cas9 (dCas9) can be fused to different functional domains, such as a transcriptional repressors or activators, to regulate gene expression. For example, dCas9-KRAB or dCas9-VP64 is used to suppress (CRISRPi) or activate (CRISPRa) gene expression, respectively, when recruited to the transcriptional start site by specific sgRNAs [36]. Based on these tools, CRISPR knock out (KO) and CRISPRi or CRISPRa pooled genetic screening platforms have been developed and serve as powerful and broadly accessible tools for systematically elucidation of gene functions at genome scale in mammalian cells [37–42]. Compared to CRISPR KO, CRISPRi significantly decreases the non-specific cellular toxicity associated with DNA double-strand breaks [43]. In addition, CRISPRi has dramatically reduced off-target effects [36], enabling much more robust screens compared to previously conducted genome-wide screens in human cells using large si/shRNA libraries [44]. Here, we describe a detailed protocol for FACS-based CRISPRi screens using an ATF4 translational reporter to study the mitochondrial stress response (Fig. 1b). This protocol uses the human CRISPRi-V2 sgRNA libraries targeting each human gene with five independent sgRNAs, established and validated by Horlbeck et al. [45]. This protocol provides general guidance for scientists who wish to establish a FACS-based CRISPRi screen platform to address questions in their areas of research.

2 Materials

2.1 Cell Culture

Cell culture procedures are performed using standard sterile techniques that prevent contamination from bacteria, fungi, mycoplasma, and cross contamination with other cell lines. Cells are cultured in 37 °C incubators with 5% CO_2. Cell lines are tested for mycoplasma contamination once every 3 months using the Universal Mycoplasma Detection Kit (ATCC, 30-1012KTM).

1. HEK293T (ATCC, CRL-3216).
2. HEK293T reporter cell line (or other cell type of interest).
3. DPBS without Mg^{2+}/Ca^{2+}.
4. Cell culture medium: DMEM, 10% FBS, L-Glutamine, penicillin–streptomycin.
5. 0.05% Trypsin–EDTA.
6. Puromycin (Sigma-Aldrich).
7. Oligomycin (Sigma-Aldrich) (replace oligomycin with other stress conditions based on the specific research interest).
8. Tissue culture microscope.

2.2 Plasmids for Generating the Reporter Cell Line

1. CRISPRi plasmids (pMH0006, Addgene, #127968 for lentiviral integration or pC13N-dCas9-BFP-KRAB, Addgene, #127968 for TALEN method).
2. CLYBL TALEN expression vectors (Addgene, #62196 and #62197).
3. Reporter plasmids (pXG237, and pXG260, Addgene, #141281 and #141282, or other reporters).

2.3 Plasmids for Generation of Lentivirus

Third generation of Lentiviral packaging plasmids (Addgene, #12251, #12253, and #12259).

2.4 CRISPRi Libraries

hCRISPRi-v2 sub-library h1-h7 (Addgene, #83971-83977).

2.5 Validation of the Activity of CRISPRi

1. sgRNA plasmid targeting *N-Cadherin* (or other cell surface marker) (*see* **Note 1**).
2. Non-targeting control (NTC) sgRNA plasmid (*see* **Note 1**).
3. Non-enzymatic cell dissociation buffer (Gibco).

4. Blocking IgG (BD Bioscience).
5. PE-Cy7-CDH2 (Biolegend).
6. Flow cytometer.

2.6 Reagents for Generation of Lentivirus

1. TransIT-Lenti Transfection Reagent (Mirus).
2. Opti-MEM (Gibco).
3. Lentiviral precipitation solution (Alstembio).
4. Syringes (5 mL, 50 mL).
5. 0.45 μm PVDF filters (Millipore).
6. 10% bleach.

2.7 FACS

1. FACS media: D-PBS + 0.5% FBS.
2. FACS tubes.
3. FACS sorter (BD FACSaria Fusion or any other available sorters with compatible setups for the reporter used).

2.8 Sample Preparation

2.8.1 Genomic DNA Extraction

1. Nucleospin Blood kits in mini (Blood), midi (Blood L), and maxi (Blood XL) forms (Macherey Nagel).
2. Molecular-grade ethanol.

2.8.2 PCR Enrichment of sgRNAs

1. Primers (*see* **Note 2**).
2. NEBNext® Ultra™ II Q5® Master Mix or NEB-Q5.
3. PCR tubes.
4. Standard PCR thermal cycler.

2.8.3 Home-Made SPRI Beads

Alternatively, use commercially available SPRIselect (Beckman Coulter).

1. Sera-Mag SpeedBeads (Fisher).
2. PEG-8000 (Amresco).
3. 0.5 M EDTA, pH 8.0.
4. 1 M Tris–HCl, pH 8.0.
5. Tween 20 (Amresco).
6. 5 M NaCl.
7. GeneRuler 50 bp DNA ladder (Thermo fisher).

2.8.4 SPRI Beads Purification of sgRNA

1. Magnetic stand (Dyna mag 2, Life Technologies).
2. Low retention/nonstick tubes.
3. 80% ethanol (freshly made).
4. DNA Elution buffer: 10 mM Tris–HCl, 1 mM EDTA, pH 8.0.
5. Novex TBE 4–20% Gel (Invitrogen).

6. TBE buffer (Bio-Rad).
7. Qubit Fluorometer.
8. Qubit™ dsDNA HS Assay Kit (ThermoFisher).
9. Qubit tubes (ThermoFisher).

2.9 Next-Generation Sequencing

1. Bioanalyzer or Tape Station.
2. Illumina sequencer.
3. Sequencing primers (*see* **Note 2**).

2.10 Individual sgRNA Cloning

1. 1× Annealing Buffer: 100 mM Potassium acetate, 30 mM HEPES-KOH (pH 7.4), 2 mM Mg acetate.
2. sgRNA oligos ordered from IDT.
3. pCRISPRia-v2 (Addgene, #84832).
4. Restriction enzymes: Bpu1102I, BstXI.

3 Methods

3.1 Construction of the Reporter Cell Line

3.1.1 Establishment of a CRISPRi Cell Line

CRISPRi machinery (dCas9-BFP-KRAB) is integrated to the cells for loss-of-function phenotypic screens, either lentivirally (pMH0006) or via TALEN technology. Here we describe the protocol for integration of dCas9-BFP-KRAB into the CLYBL safe harbor locus in HEK293T cells using TALEN.

Integration of the CRISPRi into the Cells

1. Day 0: Seed 0.3 million HEK293T cells in one well of a 12-well plate.
2. Day 1: Transfect HEK293T cells according to the manufacture protocol of the Mirus transfection reagent:
   (a) Warm TransIT-Lenti Reagent and Opti-MEM to room temperature and vortex gently.
   (b) Pipette 100 μL of Opti-MEM reduced serum medium into a 1.5-mL Eppendorf tube.
   (c) Pipette 400 ng of pC13N-dCas9-BFP-KRAB, 400 ng of CLYBL-L, 400 ng of CLYBL-R to the tube containing Opti-MEM. Mix gently by pipetting.
   (d) Add 2.5 μL of TransIT-Lenti Reagent. Mix gently by pipetting.
   (e) Incubate the mixture at room temperature for 10 min to allow transfection complexes to form.
   (f) During the incubation time, gently replace the growth medium with fresh medium for HEK293T cells.
   (g) Add the mixture from **step 4** dropwise to different areas of the well. Gently rock the plate to ensure the even distribution of complexes.

3. Day 3: Expand all the cells to a T75 flask.
4. Day 5 or Day 6: FACS sort for the BFP+ cells (*see* **Note 3**).

Validation of the Activity of CRISPRi

Once the stable CRISPRi cell line is established, functional validation of the CRISPRi activity is achieved by testing the knockdown efficiency of a selected gene after introducing a sgRNA targeting this gene. This can be done through different approaches, including RT-PCR, Western blot, or immunostaining. We recommend the immunostaining method using a cell surface maker. Flow cytometry analysis allows to compare the expression level of the gene at the single-cell level in both sgRNA- (BFP-, serves as negative control) and sgRNA+ (BFP+) populations simultaneously.

Here, we describe a protocol using an sgRNA targeting *N-Cadherin* (*CDH2*) to evaluate the CRISPRi activity. The detailed sgRNA cloning protocol and the transfection/transduction protocol can be found in Subheading 3.7. After lentiviral transduction of the CRISPRi cell line established in Subheading "Integration of the CRISPRi into the Cells" with the *N-Cadherin* sgRNA or a non-targeting control sgRNA, immunostaining of N-Cadherin is performed at day 3, day 5, and day 8 post transduction according to the following steps:

1. Harvest cells: prepare a single cell solution and aliquot 1×10^6 cells/100 μL into FACS tubes. Do not use trypsin for harvesting. Use non-enzymatic cell dissociation medium instead.
2. Wash cells with 1× FACS buffer (PBS + 0.05% FBS).
3. Incubate cells with blocking IgG (5 μL (2.5 μg) of 0.25 mg/mL for one million cells) for 15 min at room temperature. Do not wash excess blocking IgG from this reaction.
4. Add conjugated antibody (suggested amount from the antibody data sheet or a previously titrated amount—5 μL of PE-Cy7-CDH2 antibody to 100 μL of one million cells) and mix by flicking the tube or pipetting up and down.
5. Incubate cells on ice for 30 min in the dark.
6. Remove any unbound antibody by washing the cells with 1× FACS buffer.
7. Centrifuge the suspended cells at 400 × *g* for 5 min and discard the buffer.
8. Resuspend the cells by adding 500 μL of 1× FACS buffer.
9. For a negative control, use an unstained control (Fc Block—no antibody).
10. Flow cytometry analysis to compare the N-Cadherin level between sgRNA- (BFP-) and sgRNA+ (BFP+) populations.

3.1.2 Construction of the Reporter Plasmid(s)

To study how mammalian cells relay mitochondrial stress to the integrated stress response, an ATF4 translational reporter containing two upstream open reading frames followed by a fluorescent gene mApple (pXG237) was constructed into a lentiviral vector following standard molecular cloning protocols. The ATF4 translational reporter is under the control of CMV promoter. To control for any transcriptional regulation of the CMV promoter, a secondary reporter with only EGFP driven by CMV (pXG260) was also constructed. The ratio between mApple and EGFP is used as an indicator for ATF4 translation.

3.1.3 Production Lentivirus of the Reporter Plasmids

1. Day 0: Seed 0.8 million HEK293T cells per well on a six-well plate (one well per reporter plasmid).
2. Day 1: Transfect HEK293T cells following the transfection steps described in Subheading 3.1.1 except adding the following reagents to a 1.5-mL Eppendorf tube in the order listed: 200 μL OptiMEM, 750 ng Lentiviral packaging mix (1:1:1 mixture of three lentiviral packaging plasmids), 750 ng Reporter plasmid, 5 μL of TransIT-Lenti Reagent.
3. Day 2: Observe the plate to check the growth condition of cells. If the color of the medium turns to yellow, collect the supernatant and store it at 4 °C, and supplement with fresh medium in each well.
4. Day 3: Combine the supernatants from Days 2 and 3, filter virus-containing supernatant through a 0.45-μm PVDF filter to remove any cells. Virus can be stored at 4 °C for a week before long-term storage at −80 °C.

3.1.4 Establishment of the Reporter Cell Line

1. Day 0: Transduce HEK293T cells with the lentivirus of both pXG237 (500 μL) and pXG260 (500 μL), by mixing the two viruses with resuspended HEK293T equipped with CRISPRi (500 μL, around 0.05 million cells) simultaneously. Seed the mixture in one well of a 24-well plate.
2. Day 1: Observe the transduced cells. If the color of the medium is turning yellow, do partial replacement of the medium.
3. Day 3 or 4: Disassociate the cells and perform flow cytometry analysis to measure the percentage of double-positive cells (pXG237 expresses a red fluorescent protein, while pXG260 expresses a green one). Expand all the remaining cells to a T-25 flask.
4. Day 6 or when cells reach confluency: Perform monoclonal FACS sorting to collect around 180 individual cells with both reporters on two 96-well plates. Make sure to sort the polyclonal population as well. The polyclonal cells can be expanded and frozen down just in case it is necessary to repeat the monoclonal selection. Alternatively, a limited dilution method can be used with these polyclonal cells to make monoclonal reporter cells.

5. 2 weeks post monoclonal sorting: Observe under the microscope to identify wells with one round colony (the medium in these wells usually also starts turning orange). Disassociate cells from wells with a single colony and transfer them to wells of a 96-well plate for future high-throughput characterizations and selections (*see* **Note 4**).

3.2 Introduction of the sgRNA Library to the Reporter Cell Line

3.2.1 Production of Virus with sgRNA Libraries

1. Day 0: Seed 15 million HEK293T cells with a total volume of 25 mL per 15 cm dish (one dish is sufficient for one sub-library).
2. Day 1: Transfect HEK293T cells following the transfection step described in Subheading 3.1.3 except adding the following reagents to a 15-mL falcon tube in the order listed: 3 mL Opti-MEM, 15 μg Lentiviral packaging mix, 15 μg sub-library plasmids, 40 μL of TransIT-Lenti Reagent.
3. Day 2: Observe the dish under a microscope to check the growth condition and whether the cells start expressing the BFP, as the sgRNA construct expresses a BFP marker.
4. Day 3: Harvest the virus: collect supernatant through a 0.45-μm PVDF filter. To precipitate the virus, add ¼ volume of the lentiviral precipitation solution. Keep the mixture in 4 °C overnight.
5. Day 4: Centrifuge the virus suspension at 1500 × g for 30 min at 4 °C. Discard the supernatant and resuspend the white pellet in about 2 mL of cold DMEM (or other desired medium depending on cell types). Aliquot 0.5 mL per tube and store the virus in the −80 °C freezer.

3.2.2 Transduction of the sgRNA Library Virus to the Reporter Cell Line

It is recommended to use a multiplicity of infection (MOI) of 0.3 (70% of the cells remain untransduced) to maximize the number of cells with a single sgRNA integration. Prior to a large-scale transduction, a small-scale experiment can be performed to evaluate the transduction efficiency of the virus (*see* **Note 5**).

Small-Scale Transduction of the sgRNA Library Virus

1. Thaw out one vial of the frozen virus generated in Subheading 3.2.1 at room temperature.
2. Resuspend the reporter cells and determine the cell concentration.
3. Seed about 0.375 million cells per well of a 12-well plate, 12 wells in total.
4. Pipette 0 μL, 0.5 μL, 1 μL, 1.5 μL, 2 μL, 2.5 μL, 3 μL, 3.5 μL, 4 μL, 4.5 μL, 5 μL, and 5.5 μL of virus to the individual well of the 12-well plate, respectively.
5. Supplement the fresh medium to a total volume of 1 mL.

6. 48 h post transduction, perform the flow cytometry to determine the transduction efficiency by analyzing the percentage of BFP+ population as the sgRNA vector expresses a BFP marker.

Large-Scale Transduction of sgRNA Library Virus

1. Day 0: Perform the transduction based on the following calculation. The large-scale transduction will be done using the 15-cm dish, which has a surface area around 30-fold of a well on a 12-well plate.
   (a) To ensure about 1000× coverage (1000 cells infected per sgRNA), the total cell number required at the time of transduction is the number of sgRNAs included in the sub-library × 1000/30% transduction rate (for example, the h1-sublibrary contains 16,000 sgRNA, and the total number of cells needed at the time of transduction is 16,000 × 1000/30% = 48 million cells).
   (b) Based on the result from the viral titering (Subheading "Small-Scale Transduction of the sgRNA Library Virus"), 30× scale-up is recommended for a 15-cm dish: 12 million cells per dish (for h1, four 15 cm dishes will be needed), with a total of 30 mL medium, plus v × 30 μL amount of virus for 30% infection rate (v = the amount of virus, which gives about 30% infection rate in the small-scale experiment. For example, if in the small-scale experiment, 2 μL virus gave rise to 30%, here 60 μL per 15-cm dish should be used).
2. Day 2: Initiate the puromycin selection:
   (a) Trypsinize all the cells and combine them together.
   (b) Use a flow cytometer to measure the percentage of BFP+ cells. Based on this percentage, determine how many cells to proceed to the puromycin selection step with 1000× coverage (the number of cells = the number of sgRNAs in the library × 1000/percentage of BFP+ cells).
   (c) Count the cell density, seed the total number of cells calculated from the step above (about 12 million cells per 15 cm dish), with puromycin at a final concentration of 2 μg/mL.
3. Day 4 or Day 5: Stop the puromycin selection once the percentage of the BFP+ cells are over 80%. Continue culturing the cells at 1000× coverage and allow cells to recover for 2–3 days after puromycin selection (*see* **Note 6**).

3.3 FACS-Based CRISPRi Screens

Ideally, FACS sorting should be performed between 8 and 10 days post transduction, considering the time required for the expression of sgRNAs, the turn-over rate of the proteins and the dropout of cells expressing sgRNAs targeting essential genes (*see* **Note** 7).

1. Day 1: Seed 2.5 million cells on a 10-cm dish with a total volume of 7.5 mL medium. Eight dishes for each condition.

2. Day 2: Add 7.5 mL medium to untreated condition, and 7.5 mL medium with 2.5 ng/mL oligomycin for a final concentration of 1.25 ng/mL.
3. Day 3: Perform FACS sorting:
   (a) After about 16 h of treatment, disassociate all the cells and combine the cells with the same treatment.
   (b) Pellet the cells at 200 × *g* for 5 min, discard the supernatant and resuspend the cells in the FACS buffer.
   (c) Store the cells on ice.
   (d) Gate the cells with top 30% or bottom 30% of the reporter level (here we use the ratio between mApple and EGFP).
   (e) Follow the procedure of FACS sorting and sort about ten million cells for each gate under both conditions. Ten million cells give rise to an approximately 500× coverage for each CRISPRi-V2 sub-library. For HEK293T cells, sorting for one condition using the BD FACSaria Fusion machine takes about 2 h at a speed of 10,000 cells/s.
   (f) Centrifuge the sorted cells, 200 × *g* for 5 min, and freeze down the pellet in −80 °C until the sample preparation step.

3.4 Sample Preparation

It is very important to avoid contamination of genomic DNA with enriched sgRNA PCR product and sgRNA plasmids in the lab before the step of enrichment PCR. Therefore, it is critical to separate the work area for steps prior and post enrichment PCR. If possible, use a dedicated hood only for steps prior to the PCR. All steps before the samples are put into the PCR machines are carried out with equipment only dedicated for pre-PCR steps. Always rinse possibly contaminated surfaces with water before performing protocol. The overall workflow of sample preparation is outlined in Fig. 2a.

3.4.1 Genomic DNA Extraction

Nucleospin blood kits from Macherey Nagel are recommended for genomic DNA extraction. Determine which version of the kit to use based on cell number (cell numbers should not exceed: mini: 5×10^6, midi: 2×10^7; maxi: 1×10^8). Strictly follow the manufacture protocol. 10–12 μg of genomic DNA per million of HEK293T cells is expected.

3.4.2 PCR Enrichment of sgRNA

Test PCR

The test PCR is performed to verify that PCR conditions work before proceeding with the whole sample. Compatible index primers should be chosen to facilitate the de-multiplex after next-generation sequencing (Fig. 2b, *see* **Note 2**). Never use the same index for two libraries with any overlap of sgRNAs. Two NEB DNA polymerases (NEB-Q5 or NEB-ULTRA Q5) can be used, and the difference between the two is the amount of DNA template.

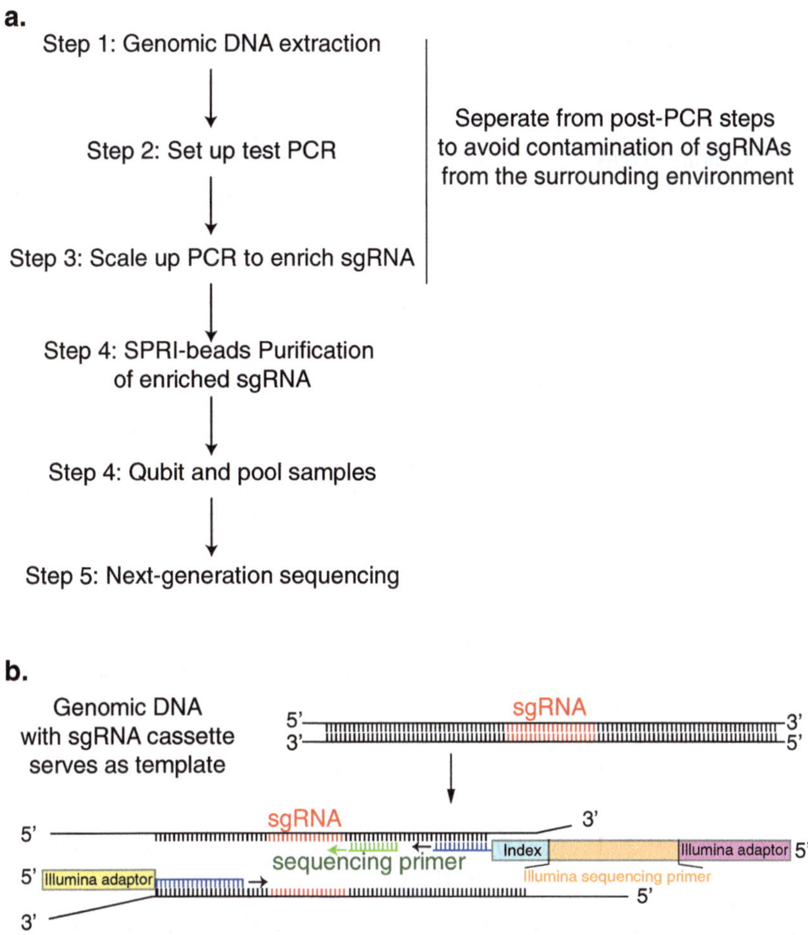

Fig. 2 The workflow of sample preparation after the cell enrichment step for a CRISPRi screen. (**a**) Step-by-step procedure for sgRNA enrichment. (**b**) A diagram of enrichment PCR for sgRNA cassette using primers, from which the next-generation sequencer will output the sequence of sgRNA in the reverse complementary orientation

Table 1 provides a general guideline for using these two enzymes. After the test PCR, check the PCR product using electrophoresis. The expected PCR product is 275 bp. Run samples (3 μL of test PCR) on a 4–20% TBE acrylamide gel in 1× TBE buffer at 120 V for 45 min. The gel is then stained in 0.3 μg/mL ethidium bromide solution for 5 min before imaging. Save the remaining test PCRs in −20 °C and combine them with the final PCRs.

Scale-up PCR

Once the test PCR is confirmed to be successful, proceed with all the remaining genomic DNA with a total volume of PCR reaction of 50 μL (scale up 2.5× of 20 μL reaction) or 100 μL (scale up 5× of 20 μL reaction) depending on the types of the PCR thermal cycler. Use the same primers and PCR conditions as the test PCRs.

Table 1
PCR conditions used for sgRNA enrichment

		NEB-Q5	NEB-ULTRA Q5
20 μL test PCRs	Genomic DNA	400–800 ng	1–2 μg
	Primer 1	0.5 μL (10 μM stock)	0.5 μL (100 μM stock)
	Primer 2	0.5 μL (10 μM stock)	0.5 μL (100 μM stock)
	2× DNA polymerase master mix	10 μL	10 μL
	Water	To a final volume of 20 μL	To a final volume of 20 μL
PCR conditions		Step 1: 98 °C, 30 s Step 2: 98 °C, 30 s Step 3: 60 °C, 15 s Step 4: 72 °C, 15 s Step 5: Go to step 2, 22× Step 6: 72 °C, 10 min Step 7: 12 °C hold	Step 1: 98 °C, 30 s Step 2: 98 °C, 10 s Step 3: 65 °C, 75 s Step 4: Go to Step 2, 22× Step 5: 65 °C, 5 min Step 6: 12 °C hold

Combine all the PCRs including the remaining reactions from the test PCR for each sample in either 2-mL or 15-mL tubes and mix well.

3.4.3 SPRI Beads Purification of the Enriched PCR Product

Home-Made SPRI Beads Solution (Skip this Step if Using Commercially Available SPRI Beads)

1. Prepare 1 M Tris by dissolving 7.88 g of solute and bring to 50 mL with dH_2O. Adjust pH to 8.0.
2. Prepare 0.5 M EDTA by dissolving 7.3 g of solute and bring to 50 mL with dH_2O. Adjust pH to 8.0.
3. Prepare TE buffer by mixing the following: 500 μL 1 M Tris, 100 μL 0.5 M EDTA, 49.4 mL dH_2O.
4. Prepare PEG solution in a 50-mL conical tube: 9 g of PEG-8000, 10 mL 5 M NaCl (or 2.92 g NaCl), 500 μL Tris–HCl, and 100 μL 0.5 M EDTA. Fill the conical tube to about 49 mL with sterile dH_2O. Mix the conical tube for 3–5 min until PEG is dissolved. The PEG solution should be clear after settling. Add 27.5 μL of Tween 20 to the conical tube and mix gently.
5. Mix Sera-Mag beads thoroughly and transfer 1 mL of the beads to a 1.5-mL microtube.
6. Place beads on a magnetic stand until the supernatant is clear.
7. Carefully remove the supernatant.
8. Add 1 mL of TE to the beads and remove the tube from the magnet.
9. Thoroughly resuspend the beads and transfer the beads to the PEG solution.

10. Fill the conical tube with dH_2O to 50 mL and gently mix until it is homogenously brown.
11. Validate the SPRI beads by purification of the DNA between 200 and 300 bp from a GeneRuler 50 bp DNA ladder (proceed with 2 μL of GeneRuler with 18 μL dH_2O and follow the protocol described in Subheading "SPRI Bead Purification").
12. Aliquot the SPRI beads to 15-mL conical tubes and store the beads in dark at 4 °C.

SPRI Bead Purification

A double SPRI beads purification is used to select for the 275 bp PCR product.

1. Equilibrate SPRI beads to room temperature before use.
2. Transfer 100 μL of PCR pool (from Subheading "Scale-up PCR") to a low-retention tube for SPRI beads purification.
3. Mix SPRI beads by vortexing.
4. Add 0.65 × (65 μL) SPRI beads to 100 μL reaction. Mix well and spin down briefly. At this ratio, fragments >300 will bind to the beads.
5. Incubate 10 min at room temperature.
6. Place tubes on a magnetic stand for 5 min or until clear.
7. Transfer supernatant to a new low-retention tube (*keep the supernatant*).
8. Add 1.1× (100 (1.1–0.65×) = 45 μL) SPRI beads. Mix well (vortex or pipetting).
9. Incubate for 5–10 min at room temperature.
10. Place tubes on a magnetic stand for 5 min or until clear.
11. Remove supernatant (*keep the beads*).
12. Wash beads with 500 μL FRESH 80% ethanol. Incubate for 30 s.
13. Remove the ethanol. Repeat for a total of two washes.
14. Air dry for 5–15 min. The beads will look glossy after your remove the ethanol. The texture will turn from glossy (wet) to matte (dry, looking like a leaf). *Tip:* To dry the beads faster, spin tubes down. Put tubes on a magnet stand and wait until liquid is clear. Remove residual ethanol with a 20-μL pipette tip.
15. Elute with 30 μL DNA elution buffer at room temperature for 2 min.
16. Save 25 μL supernatant. Do not carry over any beads.
17. Quantify the concentration of each sample by Qubit fluorometer.
18. Run 1–2 μL of sample on a 4–20% gradient TBE gel as before to assess the quality of the purified product.

3.5 Submission for Next-Generation Sequencing (NGS)

Based on the Qubit concentration and the proportion of the sequencing clusters for each sample, prepare a pooled sample with the correct ratio between each sample for submission. For pooling the samples, here is a general guideline to follow:

1. To reduce the technical error, make sure all the samples which will be pooled together are measured by Qubit fluorometer at the same time.
2. Make sure the indices are compatible.
3. Use low-retention tubes.
4. Adjust the molarity according to the instructions of the sequencing facility. Typical values are:
   (a) 2 nM in 20 μL for the HiSeq 2500.
   (b) 5–10 nM in 20 μL for the HiSeq 4000.
5. Consider the complexity of the samples. For pooled sample with low diversity (for example, all the samples are sgRNA PCR products), a spike-in of 5–10% PhiX should be requested when submitting the sample to the sequencing facility.
6. Depending on the requirements of the sequencing facility, results from either Bioanalyzer or Tape-station might be required for sample submission.

3.6 Data Analysis

This protocol does not cover the bioinformatic analysis, and we highly recommend the MAGeCK pipeline (https://sourceforge.net/p/mageck/wiki/Home/) developed and actively maintained by Dr. Xiaole Shirley Liu's laboratory for the analysis of the CRISPR screens [46].

3.7 Cloning Individual sgRNA for Post-screen Validation

Once a list of hit genes is obtained, Enrichr (https://maayanlab.cloud/Enrichr/) and string-db (https://string-db.org) are useful online tools to discover the potential signaling pathways involved. For validation of top hits from screens, sgRNAs for individual hit gene can be cloned using the following protocol:

1. Order pairs of oligonucleotides for each sgRNA (*see* **Note 8**). Order standard non-phosphorylated oligos; gel-purification or HPLC-purification is not necessary.
2. Anneal the oligo pairs.
   (a) Mix the following components:
   Top oligo (100 μM): 1 μL.
   Bottom oligo (100 μM): 1 μL.
   1× Annealing buffer: 48 μL.
   (b) Incubate the reaction for 5 min at 95 °C and then gradually decrease temperature (0.5 °C/s) to room temperature to allow oligos to anneal.

3. Prepare the vector backbone.
   (a) Set up the digestion reaction for the vector backbone:
       10× Thermo FastDigest Buffer: 10 μL.
       Vector: 5 μg.
       BstXI: 2.5 μL.
       Bpu1102I: 2.5 μL.
       dH_2O: up to 100 μL.
   (b) Digest for 4 h at 37 °C.
   (c) Run the digest reaction on a 1% agarose gel with a wide comb. Excise the linearized vector and purify using agarose gel purification kit.
   (d) Quantify the concentration of the linearized vector by Nanodrop, dilute to 20 ng/μL, aliquot and store at −20 °C.
4. Ligate the annealed oligos with the linearized vector.
   (a) Dilute the annealed oligos 1:20 with dH_2O.
   (b) In a PCR tube, mix the following components:
       Linearized Vector (20 ng/μL): 1 μL.
       1:20 diluted annealed oligos: 0.5 μL.
       10× T4 Ligase Buffer: 1 μL.
       T4 Ligase: 0.5 μL.
       dH_2O: up to 5 μL.
   (c) Include a negative control with no insert, mix the following components in a PCR tube:
       Linearized Vector (20 ng/μL): 1 μL.
       10× T4 Ligase Buffer: 1 μL.
       T4 Ligase: 0.5 μL.
       dH_2O: up to 5 μL.
   (d) Incubate both ligation reaction and control reaction at room temperature for 30 min.
5. Following the standard protocol for transformation, inoculation, mini-prep, and Sanger sequencing to select for correct plasmid.
6. Follow the transfection and transduction protocol described in Subheading 3.1.3 to make a stable cell line with a specific gene knocked down. Since the sgRNA vector expresses a puromycin-resistant gene (puromycin-*N*-acetyltransferase), 2 μg/mL of puromycin can be added to select for sgRNA-positive cells.

4 Notes

1. *N-Cadherin* or NTC sgRNA plasmids can be requested from the Kampmann lab or cloned following the Subheading 3.7 using the sgRNA sequence listed below:
   (a) sgRNA for *N-Cadherin*: GGGCCGAGCGAA GAGCCGG.
   (b) sgRNA for non-targeting control: GTCCACCCTTATC TAGGCTA.
2. The following set of PCR primers are used in our lab to enrich the sgRNA cassettes (Fig. 2b):
   (a) Index Primer (5′ to 3′): **aatgatacggcgaccaccga**GATCTA CAC **GATCGGAAGAGCACACGTCTGAACTC-CAGTCACNNNNNN** cgactcggtgccactttttc (**black** = Illumina adapter; **BLACK**=Illumina sequencing primer for index; **NNNNNN**=barcode placeholder for the index listed in Table 2; black=sequence complementing sgRNA cassette for enrichment).
   (b) Constant Primer (5′ to 3′): **caagcagaagacggcatacga**-GAT**GCACAAAAGGAAACTCACCCT** (**black** = Illumina adapter; **BLACK**=sequence complementing sgRNA cassette for enrichment).
   (c) NGS-sequencing-primer (5′ to 3′): Ccactttttcaagttgataacggactagccttatttaaacttgctatgctgt.

      Next-generation sequencing results in the reverse complementary sequence of the sgRNA amplified with this set of PCR primers.
3. To obtain a pure cell population with dCas9-BFP-KRAB stably integrated, more than one round of cell sorting might be necessary, especially if the integration rate is low.
4. A thorough validation of the reporter cell line before proceeding to screens is strongly recommended. For the ATF4 translational reporter cell line, we tested and selected the monoclonal cell line that gave us the best induction of the reporter in response to many different mitochondrial toxins.
5. Even though a low MOI is recommended, we have successfully performed screens with MOI as high as 70%, potentially because the chance of separate cells being transduced with the same combinations of sgRNAs is very low.
6. *This* is a good time point to freeze down a large number of cells just in case of a repetition or other types of screens will be performed in the future.

Table 2
Options for designing the index primers described in Note 2

Trueseq single indices#	Barcode sequence	Trueseq single indices#	Barcode sequence
Index1	ATCACG	Index13	AGTCAA
Index2	CGATGT	Index14	AGTTCC
Index3	TTAGGC	Index15	ATGTCA
Index4	TGACCA	Index16	CCGTCC
Index5	ACAGTG	Index18	GTCCGC
Index6	GCCAAT	Index19	GTGAAA
Index7	CAGATC	Index20	GTGGCC
Index8	ACTTGA	Index21	GTTTCG
Index9	GATCAG	Index22	CGTACG
Index10	TAGCTT	Index23	GAGTGG
Index11	GGCTAC	Index25	ACTGAT
Index12	CTTGTA	Index27	ATTCCT

7. Considering the timeline and large amount of time needed for sorting, it is important to secure the availability of a FACS sorter well ahead of the expected date of the sort.
8. Oligos for sgRNA cloning:
   (a) Top Oligo: 5′-TGG + sgRNA sequence (starting with G) + GTTTAAGAGC- 3′.
   (b) Bottom Oligo: 5′-TTAGCTCTTAAAC + reverse complement of sgRNA sequence + CAACAAG-3′.

Acknowledgments

We thank many of the current (Lydia Lee, Sydney Sattler, Stephanie See, Nina Drager, Avi Samelson, Emmy Li, Kun Leng, Jaime Leong, Merissa Chen, Giovanni Aviles, Athony Abarientons, Brandan Rooney, Greg Mohl) and previous members (Diane Nathaniel, Connor Ludwig, Jason Hong, Poornima Ramkumar, John Chen, Ruilin Tian) of the Kampmann lab for optimizing and organizing many protocols included here. We thank Ruilin Tian and Avi Samelson for comments on the manuscript. This work was supported by the Larry L. Hillblom Foundation to X.G. and by the National Institutes of Health grants GM119139 to M.K.. M.K. is a Chan Zuckerberg Biohub Investigator.

References

1. Simmons SO, Fan CY, Ramabhadran R (2009) Cellular stress response pathway system as a sentinel ensemble in toxicological screening. Toxicol Sci 111:202–225. https://doi.org/10.1093/toxsci/kfp140
2. Kourtis N, Tavernarakis N (2011) Cellular stress response pathways and ageing: intricate molecular relationships. EMBO J 30:2520–2531. https://doi.org/10.1038/emboj.2011.162
3. Fulda S, Gorman AM, Hori O, Samali A (2010) Cellular stress responses: cell survival and cell death. Int J Cell Biol 2010:214074. https://doi.org/10.1155/2010/214074
4. Cox JS, Shamu CE, Walter P (1993) Transcriptional induction of genes encoding endoplasmic reticulum resident proteins requires a transmembrane protein kinase. Cell 73:1197–1206. https://doi.org/10.1016/0092-8674(93)90648-A
5. Morl K, Ma W, Gething MJ, Sambrook J (1993) A transmembrane protein with a cdc2+ CDC28-related kinase activity is required for signaling from the ER to the nucleus. Cell 74:743–756. https://doi.org/10.1016/0092-8674(93)90521-Q
6. Wang XZ, Harding HP, Zhang Y, Jolicoeur EM, Kuroda M, Ron D (1998) Cloning of mammalian Ire1 reveals diversity in the ER stress responses. EMBO J 17:5708–5717. https://doi.org/10.1093/emboj/17.19.5708
7. Tirasophon W, Welihinda AA, Kaufman RJ (1998) A stress response pathway from the endoplasmic reticulum to the nucleus requires a novel bifunctional protein kinase/endoribonuclease (Ire1p) in mammalian cells. Genes Dev 12:1812–1824. https://doi.org/10.1101/gad.12.12.1812
8. Korennykh A, Walter P (2012) Structural basis of the unfolded protein response. Annu Rev Cell Dev Biol 28:251–277. https://doi.org/10.1146/annurev-cellbio-101011-155826
9. Adams CJ, Kopp MC, Larburu N, Nowak PR, Ali MMU (2019) Structure and molecular mechanism of ER stress signaling by the unfolded protein response signal activator IRE1. Front Mol Biosci 6:1–12. https://doi.org/10.3389/fmolb.2019.00011
10. Harding HP, Zhang Y, Ron D (1999) Protein translation and folding are coupled by an endoplasmic- reticulum-resident kinase. Nature 397:271–274. https://doi.org/10.1038/16729
11. Yoshida H, Haze K, Yanagi H, Yura T, Mori K (1998) Identification of the cis-acting endoplasmic reticulum stress response element responsible for transcriptional induction of mammalian glucose- regulated proteins: involvement of basic leucine zipper transcription factors. J Biol Chem 273:33741–33749. https://doi.org/10.1074/jbc.273.50.33741
12. Haynes CM, Petrova K, Benedetti C, Yang Y, Ron D (2007) ClpP mediates activation of a mitochondrial unfolded protein response in C. elegans. Dev Cell 13:467–480. https://doi.org/10.1016/j.devcel.2007.07.016
13. Haynes CM, Yang Y, Blais SP, Neubert TA, Ron D (2010) The matrix peptide exporter HAF-1 signals a mitochondrial UPR by activating the transcription factor ZC376.7 in C. elegans. Mol Cell 37:529–540. https://doi.org/10.1016/j.molcel.2010.01.015
14. Nargund AM, Pellegrino MW, FioreseLister C, Baker MJ, Haynes CM (2012) Mitochondrial import efficiency of ATFS-1 regulates mitochondrial UPR activation. Science 337:587–591
15. Zhu D, Wu X, Zhou J, Li X, Huang X, Li J, Wu J, Bian Q, Wang Y, Tian Y (2020) NuRD mediates mitochondrial stress–induced longevity via chromatin remodeling in response to acetyl-CoA level. Sci Adv 6:1–13. https://doi.org/10.1126/sciadv.abb2529
16. Shao LW, Peng Q, Dong M, Gao K, Li Y, Li Y, Li CY, Liu Y (2020) Histone deacetylase HDA-1 modulates mitochondrial stress response and longevity. Nat Commun 11:1–12. https://doi.org/10.1038/s41467-020-18501-w
17. Guo X, Aviles G, Liu Y, Tian R, Unger B, Lin Y-H, Wiita A, Xu K, Correia A, Kampmann M (2020) Mitochondrial dysfunction is signaled to the integrated stress response by OMA1, DELE1 and HRI. Nature. https://doi.org/10.1101/715896
18. Fessler E, Eckl EM, Schmitt S, Mancilla IA, Meyer-Bender MF, Hanf M, Philippou-Massier J, Krebs S, Zischka H, Jae LT (2020) A pathway coordinated by DELE1 relays mitochondrial stress to the cytosol. Nature 579:433–437. https://doi.org/10.1038/s41586-020-2076-4
19. Bao XR, Ong SE, Goldberger O, Peng J, Sharma R, Thompson DA, Vafai SB, Cox AG, Marutani E, Ichinose F, Goessling W, Regev A, Carr SA, Clish CB, Mootha VK (2016) Mitochondrial dysfunction remodels one-carbon metabolism in human cells. Elife 5:1–24. https://doi.org/10.7554/eLife.10575

20. Quirós PM, Prado MA, Zamboni N, D'Amico D, Williams RW, Finley D, Gygi SP, Auwerx J (2017) Multi-omics analysis identifies ATF4 as a key regulator of the mitochondrial stress response in mammals. J Cell Biol 216:2027–2045. https://doi.org/10.1083/jcb.201702058
21. Costa-Mattioli M, Walter P (2020) The integrated stress response: From mechanism to disease. Science 368(6489):eaat5314. https://doi.org/10.1126/science.aat5314
22. Wek RC (2018) Role of eIF2α kinases in translational control and adaptation to cellular stress. Cold Spring Harb Perspect Biol 10: a032870. https://doi.org/10.1101/cshperspect.a032870
23. Wek RC, Jackson BM, Hinnebusch AG (1989) Juxtaposition of domains homologous to protein kinase and histidyl-tRNA synthetases in GCN2 protein suggests a mechanism for coupling GCN4 expression to amino acid availability. Proc Natl Acad Sci U S A 86:4579–4583. https://doi.org/10.1073/pnas.86.12.4579
24. Wek SA, Zhu S, Wek RC (1995) The histidyl-tRNA synthetase-related sequence in the eIF-2 alpha protein kinase GCN2 interacts with tRNA and is required for activation in response to starvation for different amino acids. Mol Cell Biol 15:4497–4506. https://doi.org/10.1128/mcb.15.8.4497
25. Meurs E, Chong K, Galabru J, Thomas NSB, Kerr IM, Williams BRG, Hovanessian AG (1990) Molecular cloning and characterization of the human double-stranded RNA-activated protein kinase induced by interferon. Cell 62: 379–390. https://doi.org/10.1016/0092-8674(90)90374-N
26. Harding HP, Zhang Y, Bertolotti A, Zeng H, Ron D (2000) Perk is essential for translational regulation and cell survival during the unfolded protein response. Mol Cell 5:897–904. https://doi.org/10.1016/S1097-2765(00)80330-5
27. Chen JJ (2007) Regulation of protein synthesis by the heme-regulated eIF2α kinase: relevance to anemias. Blood 109:2693–2699. https://doi.org/10.1182/blood-2006-08-041830
28. Suragani RNVS, Zachariah RS, Velazquez JG, Liu S, Sun CW, Townes TM, Chen JJ (2012) Heme-regulated eIF2α kinase activated Atf4 signaling pathway in oxidative stress and erythropoiesis. Blood 119:5276–5284. https://doi.org/10.1182/blood-2011-10-388132
29. Hinnebusch AG, Lorsch JR (2012) The mechanism of eukaryotic translation initiation: new insights and challenges. Cold Spring Harb Perspect Biol 4:1–25. https://doi.org/10.1101/cshperspect.a011544
30. Vattem KM, Wek RC (2004) Reinitiation involving upstream ORFs regulates ATF4 mRNA translation in mammalian cells. Proc Natl Acad Sci U S A 101:11269–11274. https://doi.org/10.1073/pnas.0400541101
31. Bolotin A, Quinquis B, Sorokin A, Dusko Ehrlich S (2005) Clustered regularly interspaced short palindrome repeats (CRISPRs) have spacers of extrachromosomal origin. Microbiology 151:2551–2561. https://doi.org/10.1099/mic.0.28048-0
32. Pourcel C, Salvignol G, Vergnaud G (2005) CRISPR elements in Yersinia pestis acquire new repeats by preferential uptake of bacteriophage DNA, and provide additional tools for evolutionary studies. Microbiology 151: 653–663. https://doi.org/10.1099/mic.0.27437-0
33. Mojica FJM, Díez-Villaseñor C, García-Martínez J, Soria E (2005) Intervening sequences of regularly spaced prokaryotic repeats derive from foreign genetic elements. J Mol Evol 60:174–182. https://doi.org/10.1007/s00239-004-0046-3
34. Jinek M, Chylinski K, Fonfara I, Hauer M, Doudna JA, Charpentier E (2012) A programmable dual-RNA—guided DNA endonuclease in adaptive bacterial immunity. Science 337: 816–822. https://doi.org/10.1126/science.1225829
35. Cong L, Ran F, Cox D, Ling S, Barretto R, Habib N, Hsu P, Wu X, Jiang W, Marraffini L, Zhang F (2013) Multiplex genome engineering using CRISPR/Cas systems. Science 339: 819–824
36. Gilbert LA, Larson MH, Morsut L, Liu Z, Brar GA, Torres SE, Stern-Ginossar N, Brandman O, Whitehead EH, Doudna JA, Lim WA, Weissman JS, Qi LS (2013) CRISPR-mediated modular RNA-guided regulation of transcription in eukaryotes. Cell 154: 442. https://doi.org/10.1016/j.cell.2013.06.044
37. Zhou Y, Zhu S, Cai C, Yuan P, Li C, Huang Y, Wei W (2014) High-throughput screening of a CRISPR/Cas9 library for functional genomics in human cells. Nature 509:487–491. https://doi.org/10.1038/nature13166
38. Koike-Yusa H, Li Y, Tan EP, Velasco-Herrera MDC, Yusa K (2014) Genome-wide recessive genetic screening in mammalian cells with a lentiviral CRISPR-guide RNA library. Nat Biotechnol 32:267–273. https://doi.org/10.1038/nbt.2800
39. Shalem O, Sanjana NE, Hartenian E, Shi X, Scott DA, Mikkelsen TS, Heckl D, Ebert BL, Root DE, Doench JG, Zhang F (2014) Genome-scale CRISPR-Cas9 knockout

screening in human cells. Science 343:84–87. https://doi.org/10.1126/science.1247005

40. Wang T, Wei JJ, Sabatini DM, Lander ES (2014) Genetic screens in human cells using the CRISPR-Cas9 system. Science 343: 80–84. https://doi.org/10.1126/science.1246981

41. Gilbert LA, Horlbeck MA, Adamson B, Jacqueline E, Chen Y, Whitehead EH, Guimaraes C, Ploegh HL, Bassik MC, Qi LS, Kampmann M, Weissman JS (2015) Genome-scale CRISPR-mediated control of gene repression and activation. Cell 159:647–661. https://doi.org/10.1016/j.cell.2014.09.029. Genome-Scale

42. Konermann S, Brigham MD, Trevino AE, Joung J, Abudayyeh OO, Barcena C, Hsu PD, Habib N, Gootenberg JS, Nishimasu H, Nureki O, Zhang F (2015) Genome-scale transcriptional activation by an engineered CRISPR-Cas9 complex. Nature 517: 583–588. https://doi.org/10.1038/nature14136

43. Aguirre AJ, Meyers RM, Weir BA, Vazquez F, Zhang CZ, Ben-David U, Cook A, Ha G, Harrington WF, Doshi MB, Kost-Alimova M, Gill S, Xu H, Ali LD, Jiang G, Pantel S, Lee Y, Goodale A, Cherniack AD, Oh C, Kryukov G, Cowley GS, Garraway LA, Stegmaier K, Roberts CW, Golub TR, Meyerson M, Root DE, Tsherniak A, Hahn WC (2016) Genomic copy number dictates a gene-independent cell response to CRISPR/Cas9 targeting. Cancer Discov 6:914–929. https://doi.org/10.1158/2159-8290.CD-16-0154

44. Smith I, Greenside PG, Natoli T, Lahr DL, Wadden D, Tirosh I, Narayan R, Root DE, Golub TR, Subramanian A, Doench JG (2017) Evaluation of RNAi and CRISPR technologies by large-scale gene expression profiling in the connectivity map. PLoS Biol 15:1–23. https://doi.org/10.1371/journal.pbio.2003213

45. Horlbeck MA, Gilbert LA, Villalta JE, Adamson B, Pak RA, Chen Y, Fields AP, Park CY, Corn JE, Kampmann M, Weissman JS (2016) Compact and highly active next-generation libraries for CRISPR-mediated gene repression and activation. Elife 5:1–20. https://doi.org/10.7554/eLife.19760

46. Li W, Xu H, Xiao T, Cong L, Love MI, Zhang F, Irizarry RA, Liu JS, Brown M, Liu X (2014) MAGeCK enables robust identification of essential genes from genome-scale CRISPR/Cas9 knockout screens. Genome Biol 15:554. https://doi.org/10.1186/preaccept-1316450832143458

Chapter 3

Multiplexed Analysis of Human uORF Regulatory Functions During the ISR Using PoLib-Seq

Gemma E. May and C. Joel McManus

Abstract

Protein synthesis is a highly regulated essential process. As such, it is subjected to substantial regulation in response to stress. One hallmark of the Integrated Stress Response (ISR) is the immediate shutdown of most translation through phosphorylation of the alpha subunit of translation initiation factor eIF2 and activation of eIF4E binding proteins. While these posttranslational modifications largely inhibit cap-dependent translation, many mRNA resist this inhibition by alternative translation mechanisms involving *cis*-regulatory sequences and structures in 5′ transcript leaders, including upstream Open Reading Frames (uORFs), Internal Ribosome Entry Sites (IRESes), and Cap-Independent Translation Elements (CITEs). Studies of uORF and IRES activity are often performed on a gene-by-gene basis; however, high-throughput methods have recently emerged. Here, we describe a protocol for *Po*lysome *Lib*rary *Seq*uencing (PoLib-Seq; Fig. 1), a multiplexed assay of reporter gene translation that can be used during the ISR. A designer library of reporter RNAs are transfected into tissue-culture cells, and their translation is assayed via sucrose gradient fractionation followed by high-throughput sequencing. As an example, we include PoLib-seq results simultaneously assaying translation of wildtype and uORF mutant human ATF4 reporter RNAs, recapitulating the known function of uORF1 in resisting translational inhibition during the ISR.

Key words mRNA translation, Polysome gradient fractionation, uORFs, Massively parallel reporter assay

1 Introduction

Inhibition of translation is a hallmark of the ISR. Multiple kinases, responding to a variety of stresses, phosphorylate the alpha subunit of eIF2, reducing the availability of ternary complex for translation initiation [1, 2]. While this globally represses mRNA translation, some stress-responsive transcripts continue to be translated through alternative initiation mechanisms controlled by *cis*-acting sequences and RNA structures located in 5′ transcript leaders [3]. For example, the transcript leader of human *ATF4* encodes three uORFs (uORFs 0, 1, and 2; Fig. 1). Previous studies have shown that uORFs 1 and 2 in the homologous mouse *ATF4* gene

Daniel Matějů and Jeffrey A. Chao (eds.), *The Integrated Stress Response: Methods and Protocols*, Methods in Molecular Biology, vol. 2428, https://doi.org/10.1007/978-1-0716-1975-9_3,

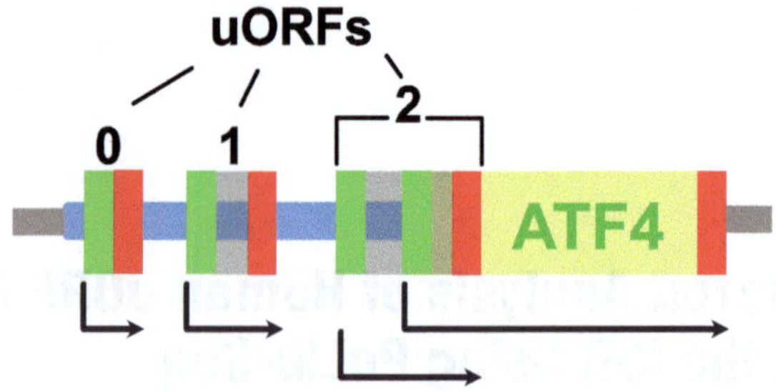

Fig. 1 Human ATF4 transcript. The start (green boxes) and stop codons (red boxes) are indicated for the three ATF4 uORFs (transparent gray boxes) and the ATF4 CDS (yellow). The arrows indicate the location and frame of translation for the uORFs, and for ATF4, which is overlapped by the translation of uORF 2

function to activate translation during the ISR [4]. After translation of the first uORF, which is only three amino acids in length, the small ribosomal subunit resumes scanning and quickly reacquires a new ternary complex. This allows subsequent reinitiation at uORF2, which overlaps the ATF4 CDS in an alternate reading frame and precludes translation of the ATF4 CDS. During the ISR, reduced levels of ternary complex delay translation reinitiation, allowing ribosomes to bypass uORF 2 and translate *ATF4.* Many other genes involved in the ISR are regulated via single uORFs, including CHOP, and GADD34 [1]. Thus, uORFs play a prominent role in the translational response to stress.

Ribosome profiling studies have shown that uORFs are extremely common, being found at ~15% of yeast and −50% of human genes, respectively [5, 6]. Although the number of predicted uORFs has increased dramatically, relatively few have been experimentally tested for functions in regulating translation, especially during the ISR. Historically, uORF activity has been tested on a gene-by-gene basis primarily through the use of luciferase reporter assays. Usually, matched wild-type and uORF mutant transcript leaders are cloned upstream of luciferase and separately transfected to compare the amount of functional protein from each reporter, relative to reporter mRNA levels. Although effective, the relatively low throughput of these reporter assays is the chief reason why most uORFs have not been tested.

Over the past decade, several labs have developed Massively Parallel Reporter Assays (MPRAs) for high-throughput measurements of transcription [7, 8], splicing [9, 10], and translation [11–15]. Some translation MPRAs involve FACS-binning cells based on fluorescent reporter expression, while others separate reporter mRNAs based on their association with ribosomes through polysome sucrose-gradient fractionation. Separation by polysome fractionation has the advantage that it can be used to investigate acute translational responses to stress, as are seen in the ISR. However, this has not been reported previously. Here, we describe our protocol for Polysome Library Sequencing (PoLib-seq) in human tissue culture cells. We include PoLib-seq results using three ATF4 wild-type and uORF-mutant reporters as a proof-of-principle, recapitulating the established roles of uORFs 1 and 2 in the ATF4 translational response to stress (Fig. 2). By extension, large designer libraries cloned from pools of synthetic oligos [16] could be used to apply PoLib-seq to thousands of designer reporter RNAs simultaneously.

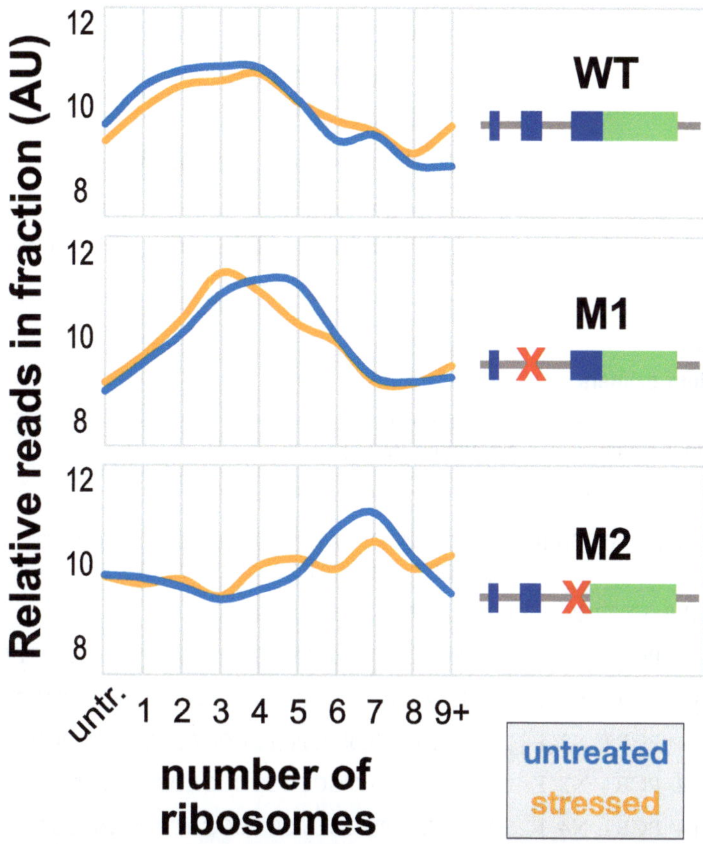

Fig. 2 The effect on translation of the 5′ leader sequence of ATF4 during stress. HEK 293T cells were simultaneously transfected with wild-type (WT) and start codon mutants of the second (M2) and third (M3) uORFs (red Xs, uORFs 1 and 2) upstream of MGFP (green box). The cells were either subjected to sodium arsenite stress (orange 40 μM; 30 min) or left untreated (blue). The number of reads for each variant were counted in each fraction and normalized by the number of total reads in each fraction. Data shown are the averages of two biological replicates for three transcript variants assayed. As expected, the wild-type ATF4 transcript leader shifts into the polysome in response to stress (top, compare blue to gold). This translational induction does not occur in transcript leaders mutated for uORF1 (middle). Mutation of the inhibitory uORF2 (bottom) increases translation under both unstressed and stressed conditions (compare to WT, top)

2 Materials

1. Pipettes with filter tips (P1000, P200, P20, P2).
2. Monster Green® Fluorescent Protein phMGFP vector (Promega).
3. Custom primers (Table 1).
4. 1.5-ml and 2.0-ml microcentrifuge tubes.
5. 0.2-ml tubes for PCR.
6. Q5 Polymerase (New England Biolabs).
7. 10 mM dNTP mix.
8. Nuclease-free water.
9. Thermocycler, any model.
10. Agarose.
11. TAE buffer (Tris–acetate EDTA buffer for gel electrophoresis).

Table 1
Custom primers

Primer name	Sequence (5′–3′)
phMGFP-T7-F	TAATACGACTCACTATAGG
phMGFP-GFP3p-R	TTTTTTTTTTTTTTTTTTTTTTTTTTTTTTTTTTTTTTTTTTT TTTTTTTTTTTTTTTTTTTTTTTTTTTTGCTCGAAGCATTAACCCTC
MGFP-52-72-R	ACGAATTTGTGGCCGTTCAC
ATF4_PCR1_N0_F	TTCAGACGTGTGCTCTTCCGATCTTTTCTACTTTGCCCGCCCAC
ATF4_PCR1_N1_F	TTCAGACGTGTGCTCTTCCGATCTNTTTCTACTTTGCCCGCCCAC
ATF4_PCR1_N2_F	TTCAGACGTGTGCTCTTCCGATCTNNTTTCTACTTTGCCCGCCCAC
ATF4_PCR1_N3_F	TTCAGACGTGTGCTCTTCCGATCTNNNTTTCTACTTTGCCCGCCCAC
ATF4_PCR1_N4_F	TTCAGACGTGTGCTCTTCCGATCTNNNNTTTCTACTTTGCCCGCCCAC
ATF4_PCR1_N5_F	TTCAGACGTGTGCTCTTCCGATCTNNNNNTTTCTACTTTGCCCGCCCAC
ATF4_PCR1_N6_F	TTCAGACGTGTGCTCTTCCGATCTNNNNNNTTTCTACTTTGCCCGCCCAC
ATF4_PCR1_N7_F	TTCAGACGTGTGCTCTTCCGATCTNNNNNNNTTTCTACTTTGCCCGCCCAC
ATF4_PCR1_N0_R	GCGACCACCGAGATCTACACTCTTTCCCTACACGACGCTCTTCCGATC TAGCAACGCTGCTGCTGAATG
ATF4_PCR1_N2_R	GCGACCACCGAGATCTACACTCTTTCCCTACACGACGCTCTTCCGATC TNNAGCAACGCTGCTGCTGAATG
ATF4_PCR1_N4_R	GCGACCACCGAGATCTACACTCTTTCCCTACACGACGCTCTTCCGATC TNNNNAGCAACGCTGCTGCTGAATG
ATF4_PCR1_N6_R	GCGACCACCGAGATCTACACTCTTTCCCTACACGACGCTCTTCCGATC TNNNNNNAGCAACGCTGCTGCTGAATG

12. SYBR safe DNA gel stain (or equivalent).
13. DpnI enzyme and CutSmart® buffer (New England Biolabs).
14. DNA Clean and Concentrator kit (Zymo Research).
15. NanoDrop™ 2000 Spectrophotometer or similar.
16. HiScribe T7 Quick High Yield RNA Synthesis kit (New England Biolabs).
17. 3 M sodium acetate pH 5.2.
18. Acid-Phenol:Chloroform pH 4.5 (with IAA, 125:24:1).
19. Chloroform:Isoamyl alcohol 24:1.
20. Ethanol, absolute 200 proof (EtOH, molecular biology grade).
21. 80% Ethanol: 80% absolute ethanol, 20% nuclease-free water.
22. Refrigerated centrifuge with rotor for 1.5- to 2.0-ml microcentrifuge tubes.
23. Thermomixer with 1.5-ml tube capacity (or similar).
24. Vaccinia Capping System (New England Biolabs).
25. 2-propanol (isopropanol).
26. HEK 293T cell line.
27. Fetal bovine serum (FBS).
28. Dulbecco's Modified Eagle Media (DMEM).
29. Disposable pipettes (25 ml, 10 ml, 5 ml).
30. 10-cm tissue culture–treated culture dish.
31. Incubator suitable for tissue culture, supplied with 5% CO_2.
32. 50-ml conical tube.
33. Steriflip-GP sterile 50-ml centrifuge tube top filter unit (MilliporeSigma™).
34. Polysome Lysis Buffer (PLB): 20 mM Tris–HCl pH 7.5, 250 mM NaCl, 15 mM $MgCl_2$, 0.5% Triton™ X-100. Prepare 50 ml in nuclease-free water. Filter sterilize using Steriflip-GP sterile 50-ml centrifuge tube top filter unit according to the manufacturer's instructions. Store at 4 °C.
35. Opti-MEM Reduced Serum Media.
36. Lipofectamine® MessengerMAX™ transfection reagent (Life Technologies).
37. 50 mM sodium arsenite: Mix 0.065 g sodium arsenite in 10 ml of nuclease-free water. Store at room temperature.
38. 50 mg/ml cycloheximide: Mix 50 mg cycloheximide in 1 ml ethanol. Store at −20 °C.
39. 1 M DL-dithiothreitol (DTT).
40. SUPERase-In™ RNase inhibitor, 20 U/μl (Invitrogen™).

41. TURBO™ DNase, 2 U/ml (Invitrogen™).
42. Phosphate-buffered saline (PBS).
43. Cell scraper.
44. 10-ml syringe with Luer Tip.
45. 20 gauge needle.
46. TRIzol™ Reagent.
47. RNA Clean and Concentrator kit (Zymo Research).
48. RNA reagents ScreenTape (Agilent Technologies).
49. TapeStation Instrument (Agilent Technologies, model 4400 or similar).
50. Sucrose.
51. 10% Polysome Sucrose-Gradient Solution (10% PLB): 5 g of sucrose, 1 ml of 1 M Tris–HCl pH 7.5, 3.125 ml of 4 M NaCl, 750 μl of 1 M $MgCl_2$, 250 μl of Triton™ X-100, and nuclease-free water to 50 ml. Agitate until the sucrose is fully dissolved into solution. Filter sterilize using Steriflip-GP sterile 50-ml centrifuge tube top filter unit according to the manufacturer's instructions. Store at 4 °C.
52. 50% Polysome Sucrose-Gradient Solution (50% PLB): 25 g of sucrose, 1 ml of 1 M Tris–HCl pH 7.5, 3.125 ml of 4 M NaCl, 750 μl of 1 M $MgCl_2$, 250 μl of Triton™ X-100, and nuclease-free water to 50 ml. Agitate until the sucrose is fully dissolved into solution. Filter sterilize using Steriflip-GP sterile 50-ml centrifuge tube top filter unit according to the manufacturer's instructions. Store at 4 °C.
53. Polyclear™ open top ultracentrifuge tubes 9/16″ × 3 ¾″ (Seton Scientific 7031).
54. Silicone grease.
55. Stainless steel 304 syringe needle.
56. Lint-free wipes.
57. Ultracentrifuge with Beckman Coulter SW 40 Ti swinging bucket rotor (Beckman Coulter).
58. Gradient station™ with gradient forming attachments (Biocomp Instruments).
59. Gilson Fraction collection system with Triax flow cell and fractionating tube mount (Biocomp Instruments).
60. Sodium dodecyl sulfate (SDS).
61. 70% ethanol (70% EtOH): 70% ethanol, 30% ddH_2O.
62. SuperScript™ IV Reverse Transcriptase (Thermo Fisher).
63. Magnetic separation rack for 0.2-ml tubes.

64. AMPure® XP magnetic PCR purification beads (Beckman Coulter).
65. 100 μM forward N primer mix: 10 μl of each of the following: 100 μM ATF4_PCR1_N0-F, 100 μM ATF4_PCR1_N1-F, 100 μM ATF4_PCR1_N2-F, 100 μM ATF4_PCR1_N3-F, 100 μM ATF4_PCR1_N4-F, 100 μM ATF4_PCR1_N5-F, 100 μM ATF4_PCR1_N6-F, and 100 μM ATF4_PCR1_N7-F (Table 1).
66. 100 μM reverse N primer mix: 10 μl of each of the following: 100 μM ATF4_PCR1_N0_R, 100 μM ATF4_PCR1_N2_R, 100 μM ATF4_PCR1_N4_R, and 100 μM ATF4_PCR1_N6_R (Table 1).
67. NEBNext® Multiplex Oligos for Illumina (unique dual index primer pairs, New England Biolabs).
68. D1000 ScreenTape and sample buffer (Agilent).
69. Qubit™ dsDNA HS Assay Kit (Invitrogen).
70. 0.5-ml optically clear individual PCR tube.

3 Methods

3.1 In Vitro RNA Transcription Template Preparation

In this section, a template for in vitro transcription is produced by PCR amplification of the phMGFP vector (Fig. 3). The forward primer, phMGFP-T7-F (Table 1), anneals to the T7 polymerase site upstream of the 5′ LS, and the reverse primer, phMGFP-GFP3p-R (Table 1), anneals downstream of the GFP (green fluorescent protein) sequence. The reverse primer also incorporates a 70-nucleotide-long poly A tail sequence into the PCR product. For this step, the vector template can contain either a single 5′ leader sequence or a library of up to thousands of leader sequences.

1. Dilute the plasmid vector containing the 5′ leader sequence (s) of interest to 1 ng/μl in nuclease-free water.
2. Set up the following reaction in a 0.2-ml PCR tube on ice.

Reagent	Volume (μl)
phMGFP-5′LS plasmid (1 ng/μl)	2
Nuclease-free water	64
5× Q5 reaction buffer	20
10 mM dNTPs	4
10 μM phMGFP-T7-F	4
10 μM phMGFP-GFP3p-R	4
Q5 polymerase	3

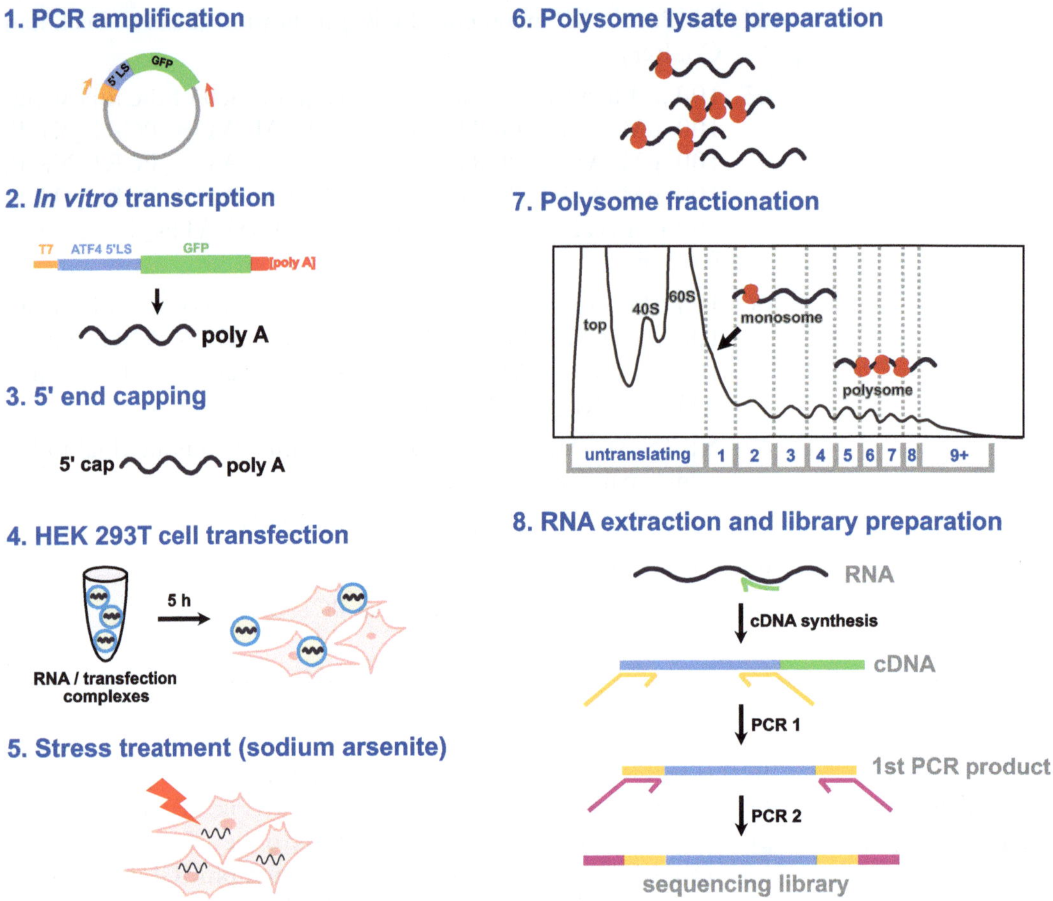

Fig. 3 Polib-seq protocol overview. (Step 1) The cloned library is amplified to generate a template for in vitro transcription. A T7 polymerase site and a poly A sequence are incorporated during this step. (Step 2) Uncapped in vitro transcribed RNA is produced using the PCR product as the template. (Step 3) A 5′ cap is enzymatically added to the RNA. (Step 4) HEK 293T cells are transfected for 5 h with the RNA. (Step 5) The cells are subjected to sodium arsenite for 30 min to induce the integrated stress response. (Step 6) Cell polysome lysates are prepared and fractionated over a sucrose gradient into 10 fractions (step 7). (Step 8) RNA is extracted from the fractions and sequencing libraries are prepared by targeted RT-PCR

3. Mix the reaction by setting a pipettor to 50 μl and pipetting up and down several times.
4. Put the reaction in a thermocycler and run the program:
   (a) 98 °C for 30 s.
   (b) 35 cycles of 98 °C for 30 s, 55 °C for 30 s, and 72 °C for 45 s (or 40 s per kb of product).
5. Run 10 μl of the PCR product on a 1% agarose gel in 1×TAE to check for successful amplification.
6. Purify the PCR product using the Zymo DNA clean & concentrator kit (or similar) according to the manufacturer's

instructions. Elute the PCR product in 45 μl of nuclease-free water.

7. Set up the following reaction on ice.

Reagent	Volume (μl)
PCR product (**step 6**)	44
10× CutSmart buffer	5
DpnI (20 U/μl)	1

8. Mix the reaction by setting pipettor to 25 μl and pipetting up and down several times.
9. Incubate the reaction for 1 h at 37 °C in a thermocycler.
10. Purify the PCR product using the Zymo DNA clean & concentrator kit according to the manufacturer's instructions. Elute the PCR product in 10 μl of nuclease-free water.
11. Measure the concentration of the PCR product using a NanoDrop™ spectrophotometer or something similar.

3.2 In Vitro RNA Transcription

1. In a 0.2-ml PCR tube, dilute the PCR product from Subheading 3.1, **step 12** to 1 μg in 8 μl of nuclease-free water.
2. Add 10 μl of NTP Buffer mix followed by 2 μl of T7 RNA polymerase mix from the HiScribe™ T7 Quick High Yield RNA Synthesis Kit.
3. Mix by setting a pipettor to 10 μl and pipetting up and down several times. The total volume of the reaction should be 20 μl.
4. Incubate the reaction for 2 h at 37 °C in a thermocycler.
5. Add 30 μl of water to the reaction, followed by 2 μl DNase I (supplied in the kit) to the reaction. Incubate for 15 min at 37 °C in a thermocycler.
6. Transfer the reaction to a 1.5-ml tube. Add 150 μl of nuclease-free water, and 20 μl of 3 M sodium acetate, pH 5.2. Mix well using a pipettor, or briefly vortex.
7. Add 250 μl of acid-phenol:chloroform pH 4.5 (with IAA, 125:24:1). Vortex on high for 15 s.
8. Centrifuge at 14,000 × *g* for 2 min at room temperature. Remove the supernatant to a new 1.5-ml tube.
9. Add 300 μl of chloroform:isoamyl alcohol 24:1 to the supernatant. Vortex on high for 15 s. Repeat **step 8**.
10. Add 600 μl of ethanol to the supernatant. Briefly vortex to mix and precipitate at −20 °C overnight, or for at least several hours.
11. Centrifuge the RNA at 20,000 × *g* for 30 min at 4 °C.

12. Carefully remove the supernatant as to not disturb the pellet.
13. Wash the pellet by adding 1 ml of 80% ethanol. Briefly vortex and centrifuge the RNA at 20,000 × *g* for 10 min at 4 °C.
14. Carefully remove the supernatant as to not disturb the pellet. Leave the lid open and let the remaining ethanol evaporate (roughly 2–5 min). A clear or slightly white pellet should be visible at the bottom of the tube.
15. Resuspend the pellet in 50 μl of nuclease-free water.
16. Measure the concentration of the RNA using a nanodrop. This transcription reaction produces up to about 180 μg of RNA.
17. Save the RNA at −20 °C overnight or continue on to Subheading 3.3.

3.3 5′ Capping Reaction

1. Combine 100 μg of RNA (**step 17**) with nuclease-free water in a 1.5-ml tube to a final volume of 150 μl.
2. Denature the RNA by heating the RNA for 5 min at 65 °C. After the incubation is complete, immediately place the tube on ice for 5 min.
3. Proceed with the Vaccina Capping System kit. While the RNA is incubating on ice, prepare a 2 mM solution of SAM by adding 2 μl of 32 mM SAM to 30 μl nuclease-free water. Save the 2 mM SAM on ice.
4. Set up the following reaction on ice in a 1.5-ml tube in the order specified.

Reagent	Volume (μl)
100 μg denatured RNA (from **step 2**)	150
10× capping buffer	20
10 mM GTP	10
2 mM SAM (from **step 3**)	10
Vaccinia capping enzyme	10

5. Mix by setting a pipettor to 50 μl and pipetting up and down several times.
6. Aliquot the reaction into five 0.2-ml PCR tubes with 50 μl in each tube. Incubate the tubes at 37 °C in a thermocycler for 30 min.
7. Combine the reactions into one 1.5-ml tube. Add 700 μl of nuclease-free water, and 100 μl of 3 M sodium acetate, pH 5.2 to each tube. Briefly vortex.
8. Divide the reaction into two 1.5-ml tubes by adding 500 μl of the reaction to each tube.

9. To each tube, add 500 μl of acid-phenol:chloroform pH 4.5 (with IAA, 125:24:1). Vortex on high for 15 s.
10. Centrifuge at 14,000 × *g* for 2 min at room temperature. Remove the supernatants to new 1.5-ml tubes.
11. Add 500 μl of chloroform:isoamyl alcohol 24:1 to the supernatants. Vortex on high for 15 s.
12. Centrifuge at 14,000 × *g* for 2 min at room temperature. Remove each supernatant to one 2.0-ml tube. The volume of the supernatants is typically about 800–900 μl at this step.
13. Add one volume of isopropanol. Briefly vortex to mix and precipitate at −20 °C overnight, or for at least several hours.
14. Centrifuge the RNA at 20,000 × *g* for 30 min at 4 °C.
15. Carefully remove the supernatant as to not disturb the pellet.
16. Wash the pellet by adding 1 ml of 80% ethanol. Briefly vortex and centrifuge the RNA at 20,000 × *g* for 10 min at 4 °C.
17. Carefully remove the supernatant as to not disturb the pellet. Leave the lid open and let the remaining ethanol evaporate (roughly 2–5 min).
18. Resuspend the pellet in 50 μl of nuclease-free water.
19. Measure the concentration of the RNA using a nanodrop. 10 μg of RNA is required for each transfection. It may be necessary to repeat Subheadings 3.2 and 3.3 to obtain enough capped RNA for multiple transfections.
20. Save the RNA at −20 °C overnight or continue on to Subheading 3.4.

3.4 HEK 293 T Cell In Vitro RNA Transfection

1. For each transfection, seed two million cells in a 10-cm tissue culture dish in 10 ml DMEM + 10% FBS. Incubate at 37 °C in an incubator supplemented with 5% CO_2 for 24 h.
2. Prepare 50 ml of Polysome Lysis Buffer (PLB, *see* Subheading 2).
3. For each transfection, add 30 μl of Lipofectamine® Messenger-MAX™ transfection reagent to 500 μl of Opti-MEM Reduced Serum Media in a 1.5-ml tube. Incubate at room temperature for 10 min.
4. In a separate 1.5-ml tube, add 10 μg of RNA (Subheading 3.3, **step 20**) with 500 μl of Opti-MEM Reduced Serum Media.
5. Combine the Lipofectamine complexes (**step 3**) with the RNA (**step 4**) into one tube. Briefly mix with a pipettor set to 500 μl. Incubate the transfection mixture at room temperature for 5 min.
6. Remove the cells from the CO_2 incubator and add the transfection mixture dropwise to the HEK 293T cells while gently

swirling the dish. Place the cells back in the incubator. Incubate for 5.5 h for non-stress conditions, or 5 h for assays under stressed conditions.

7. For the stress condition, after 5 h of incubation, remove the tissue culture dish from the incubator. Add 8 μl of 50 mM sodium arsenite to the dish for a final sodium arsenite concentration of 40 μM. Briefly swirl the media in the dish. Place the tissue culture dish back into the incubator and incubate for 30 min.

8. While the cells are incubating, prepare enough PLB for each transfection (500 μl per transfection) on ice by adding 1 μl of 1 M DTT, 12 μl of TURBO™ DNase, 2 μl of 50 mg/ml cycloheximide, and 1 μl of SUPERase-In™ RNase inhibitor to 1 ml of PLB. Save the PLB on ice.

9. Prepare one 10-ml syringe with a 20 G needle attached per transfection.

10. Prepare 10 ml of ice-cold PBS per transfection. This can be done by aliquoting 10 ml of PBS into a 15-ml conical and burying the conical in ice in an ice bucket.

11. Harvest the cells for each transfection as follows.

    (a) Remove the tissue culture dish containing the cells from the incubator into a biosafety cabinet.

    (b) Remove 0.5 ml of the media from the dish into a 1.5-ml tube.

    (c) Add 20 μl of 50 mg/ml cycloheximide to the 0.5 ml of media and briefly vortex the tube.

    (d) Add the media containing the cycloheximide to the cells in the tissue culture dish. Briefly swirl the media in the dish. Place the cells back into the incubator and incubate for 2 min.

    (e) Remove the cells to a biosafety cabinet. Using a 10-ml pipette, quickly pipette all of the media off of the cells.

    (f) Gently wash the cells by adding 10 ml of ice-cold PBS to the cells by pipetting down the side of the dish, taking care to not disturb the cells.

    (g) Gently swirl the dish and place the dish onto ice.

    (h) Quickly remove all of the PBS using a 10-ml pipette.

    (i) Add 500 μl of the prepared PLB (**step 8**) dropwise to the cells.

    (j) Tilt the dish to about a 20° angle while keeping the dish on ice.

    (k) Using a cell scraper, scrape all of the cells down to the bottom of the dish that is closest to the ice.

(l) Titrate the cells in the dish by passing the cells through a 20 G needle about ten times.

(m) Remove the lysed cells to a 1.5-ml tube on ice. Incubate the lysate on ice for 10 min.

12. Once all of the lysates are collected, centrifuge the lysates for 10 min at 20,000 × *g* at 4 °C.
13. The cell debris should be collected at the bottom of the tube. Remove the lysate without disturbing the cell debris to a new tube. The total volume of the lysate at this point is usually about 0.6 ml.
14. For the total RNA sample, remove 50 μl of lysate to a new 1.5-ml tube. Proceed to Subheading 3.6.
15. Freeze and store the lysates at −80 °C. The lysates can be stored overnight or for up to several months.

3.5 Polysome Fractionation

1. Turn on and cool the ultracentrifuge to 4 °C. Also cool the SW40 rotor to 4 °C (keep in a cold room or refrigerator overnight).
2. Prepare 6.5 ml each of 10% and 50% polysome sucrose-gradient solution (10% PSG and 50% PSG, *see* Subheading 2) per gradient by adding the reagents to the solutions as follows. Save the solutions on ice.

Reagent	Final concentration
DTT	1 mM
Cycloheximide	0.1 mg/ml
SUPERase-In™ RNase inhibitor	20 U/ml

3. Place each ultracentrifuge tube into the marker block and mark the tube along the upper marking edge of the block. Lightly grease the inside of the tube with silicone grease using the provided bolt and piston tip. Label the ultracentrifuge tube with the sample name.
4. Prepare two 10-ml syringes with stainless steel syringe needles. To fill one syringe with approximately 7 ml of 10% PSG, pull the plunger up and down a few times while the syringe is submerged in the 10% PSG. This removes any air bubbles. Gently fill an ultracentrifuge tube with ~6.5 ml of 10% PSG, just so the liquid reaches the marked line.
5. Fill the other syringe with approximately 7 ml of 50% PSG (again, pull the plunger up and down in the PSG to remove the bubbles) and dry the outside of the tip using a lint-free wipe. Gently place the tip of the metal syringe tip at the bottom of the ultracentrifuge and slowly press the syringe plunger down. Observe the heavier, 50% PSG solution, displace the

10% PSG solution as it moves the 10% solution up the tube. Keep pressing the plunger down until the top of the 50% solution is at the marker line. Carefully remove the syringe as to not disrupt the layers. Repeat **steps 4** and **5** for all of the gradients.

6. Place the rubber lids onto the ultracentrifuge tubes so that the small hole in the side of the lid is the last to be placed into the tube.
7. Turn on the gradient station and level the platform and tube holder where the gradients will be placed.
8. Carefully transfer the ultracentrifuge tubes to the gradient station tube holder. Run the 10–50% gradient program on the gradient station (*see* **Note 1**).
9. While the gradient station is running, thaw the lysates on ice if they were previously frozen.
10. Once the gradient station has finished the program, remove the ultracentrifuge tubes from the tube holder and gently place them on ice. Be sure that the tubes are sitting in the ice so that they are upright and not at an angle.
11. Remove the rubber lids. Using a pipette, remove the equivalent volume of lysate plus an additional 100 μl. For example, if the volume of the lysate is 600 μl, remove 700 μl of the sucrose gradient from the top of the tube.
12. Place the gradients into the SW40 buckets. Be sure to record which ultracentrifuge tube was placed into each bucket.
13. Using a pipettor, gently layer the lysate on top of the gradients by pipetting the sample down the side of the tube. This avoids dropping the sample directly on top of the gradient which could disrupt the layers.
14. Weigh each bucket with the gradients in them (along with the caps to the buckets) and balance the buckets that will be across from each other in the rotor using the PLB stock solution.
15. Cap the buckets and place them into the rotor. If there are empty buckets, be sure to load those as well. If there are an odd number of samples, prepare a mock gradient and place it in the bucket opposite of the odd tube.
16. Load the rotor into the centrifuge. Set the acceleration and de-acceleration to 1 and 7, respectively. Centrifuge the gradients at 35,600 rpm (160,000 × *g*) for 2.5 h at 4 °C with the vacuum on.
17. While the centrifuge is running, clean the metal syringe tips and rubber caps by soaking in 10% bleach for 30 min, followed by rinsing with distilled water.
18. Label thirty 1.5-ml microcentrifuge tubes per sample and set aside.

19. *See* **Note 2**. Zero the Triax flow cell using ddH_2O. Leave the fraction collection system on to warm up for at least 30 min before use.
20. Remove the rotor from the ultracentrifuge and save the buckets with the gradients on a stable surface as to not tilt the buckets—in an ice bucket, a cold room, or a fridge.
21. Place thirty 1.5-ml labeled tubes with the lids open, in the tube holder for the fractionator in the correct orientation (*see* fractionator's user manual).
22. Put a silicone-greased piston tip onto the piston and load one sample under the piston.
23. Set up the fractionator to fractionate the sample into 30 fractions using the default settings. Each fraction will be about 400 ml, with the exception of the last fraction which will be less.
24. After the fractionator has finished with fractionating one gradient, rinse the system following the prompts, with ddH_2O. Repeat **steps 21–23** for each gradient.
25. After the last fraction, follow the manufacturer's suggested protocol for cleaning and storing the instrument. We run mock samples filled with warm 0.5% SDS followed by ddH_2O. Then, the instrument lines are rinsed with 70% EtOH and air is pumped through to dry.
26. At this point, the fractions can be saved at −20 °C, or they can be combined as desired (next steps).
27. Examine the polysome traces to determine which fractions are representative of the peaks of interest. Combine the fractions as needed so that the total volume of the combined fractions is 400 μl. For example, if fraction numbers 9 and 10 are where the monosome peak is on the polysome trace, combine 200 μl of fraction 9 and 200 μl of fraction 10 into one 1.5-ml ultracentrifuge tube.
28. Either save the combined fractions at −20 °C or proceed to Subheading 3.6.

3.6 RNA Extraction

Perform **steps 1**, **2**, and **4** in a fume hood. Dispose of TRIzol™ reagent and chloroform as chemical hazardous wastes.

1. Add 800 μl of TRIzol™ reagent to each of the 400 μl of combined fractions (or 50 μl of total unfractionated lysate, Subheading 3.4, **step 14**) and vortex briefly. Incubate at room temperature for 5 min.
2. Add 300 μl of chloroform:isoamyl alcohol 24:1 to the TRIzol™ Reagent/lysate mix and briefly vortex. Incubate at room temperature for 2 min.

3. Centrifuge at 12,000 × *g* for 15 min at 4 °C.
4. Carefully remove the supernatant (roughly 900 μl) to a new 2.0-ml tube.
5. Add one volume of ethanol and briefly vortex.
6. Purify the RNA using a Zymo RNA clean and concentrator-5 kit as per the manufacturer's instructions, and as listed in the subsequent steps. All of the centrifuge spins are at 12,000 × *g* at room temperature.
7. Transfer up to 700 μl of the sample to the spin column and centrifuge for 30 s. Discard flow through. Repeat this step as necessary until all of the sample from **step 5** has been added to the column.
8. Add 400 μl of RNA Prep Buffer and centrifuge for 30 s. Discard the flow through.
9. Add 700 μl of RNA Wash Buffer and centrifuge for 30 s. Discard the flow through.
10. Add 400 μl of RNA Wash Buffer and centrifuge for 30 s. Discard the flow through.
11. Centrifuge the column for an additional 2 min to dry the membrane.
12. Transfer the column to a clean 1.5-ml tube with the cap removed. Add 45 μl of nuclease-free water to the column membrane.
13. Elute the RNA by centrifuging the column at room for 1 min.
14. Transfer the eluate to a fresh 1.5-ml tube.
15. For total RNA (unfractionated) samples, check the quality of the lysate preparation by assaying RNA integrity. Run 1 μl of the total RNA on an RNA ScreenTape using an Agilent TapeStation, following the manufacturer's instructions. The RNA Integrity Value (RIN) should be at least 9.0 for high-quality RNA. The typical range of RIN numbers for this assay ranges between 9.5 and 10.0.
16. Save the RNA at −20 °C or proceed to Subheading 3.7.

3.7 DNase Treatment

1. Set up a DNase treatment reaction in a 0.2-ml PCR tube as follows.

Reagent	Volume (μl)
RNA (Subheading 3.6, **step 14**)	43
10× TURBO™ DNase buffer	5
SUPERase-In™ RNase inhibitor	1
TURBO™ DNase	1

2. Pipette up and down several times with a P200 pipettor set to 45 μl to mix.
3. Incubate the reaction at 37 °C for 30 min.
4. Purify the RNA over an RNA clean and concentrator—5 column following the manufacturer's instructions. Elute the RNA in 15 μl of nuclease-free water.
5. Quantify the RNA using a nanodrop.
6. Either save the RNA at −20 °C or proceed to Subheading 3.8.

3.8 Reverse Transcription

1. In a 0.2-ml tube, set up the reaction on ice as follows.

Reagent	Volume
RNA (Subheading 3.7, **step 8**)	250 ng to 2.5 μg (max 11 μl)
10 mM dNTPs	1 μl
2 μM MGFP-52-72-R primer	1 μl
Nuclease-free water	To 13 μl

2. Incubate the reaction for 5 min at 65 °C in a thermocycler. Place the reaction on ice immediately after the incubation is complete.
3. To the reaction, add the following.

Reagent	Volume (μl)
5× SuperScript™ IV buffer	4
100 mM DTT	1
SUPERase-In™ RNase inhibitor	1
SuperScript™ IV reverse transcriptase	1

4. In a thermocycler, incubate the reaction for 30 min at 50 °C. Heat inactivate at 80 °C for 10 min.
5. Add 30 μl of nuclease-free water to the cDNA.
6. Prepare the bottle of AMPure XP beads by vortexing until the beads are fully resuspended.
7. Add 50 μl of AMPure XP of Agencourt beads to the cDNA. Mix well by setting a pipettor to 50 μl and pipetting up and down.
8. Incubate the cDNA and beads at room temperature for 5 min.
9. Place the tube on a magnetic separation rack. Incubate at room temperature for 2 min. All of the beads should move to one side and the supernatant should be clear.
10. Pipette off the supernatant and discard.

11. While the tubes are still on the magnetic rack, wash the beads by adding 180 μl of 80% EtOH to the side of the tube opposite of the beads. Incubate at room temperature for 30 s.
12. Pipette off the 80% EtOH and discard.
13. While the tube is still on the rack with the lids open, pipette off the ethanol that has been collected at the bottom of the tube and let the remaining ethanol evaporate. The bead pellet should appear to be dry (not shiny) and more matte in appearance.
14. Resuspend the beads by removing the tube from the magnet and adding 20 μl of nuclease-free water to the pellet. Pipette up and down several times so that the beads are fully resuspended. Incubate at room temperature for 2 min.
15. Place the tube back on the magnet and wait for the supernatant to become clear (approximately 30 s to 1 min).
16. Remove the supernatant which now contains the cDNA to a new 1.5-ml tube.
17. Either save the cDNA at −20 °C or continue with Subheading 3.9.

3.9 Sequencing Library Preparation

1. Prepare a 10 μM F/R primer mix by adding 10 μl of 100 μM forward N primer mix and 10 μl of 100 μM reverse N primer mix (*see* Subheading 2), to 80 μl of nuclease-free water. The primer mix can be stored at −20 °C.
2. In a 0.2-ml tube, set up the reaction on ice as follows.

Reagent	Volume
Purified cDNA (Subheading 3.8, **step17**)	3
Nuclease-free water	32
5× Q5 buffer	10
10 mM dNTPs	2
10 μM F/R primer mix (step 1)	2
Q5 polymerase	1

3. Put the reaction in a thermocycler and run the program:
   (a) 98 °C for 30 s.
   (b) 6 cycles of 98 °C for 30 s, 55 °C for 30 s, and 72 °C for 30 s.
4. Prepare the bottle of AMPure XP beads by vortexing until the beads are fully resuspended.
5. Add 50 μl of AMPure XP of Agencourt beads to the amplification reaction. Mix well by setting a pipettor to 50 μl and pipetting up and down.

6. Incubate amplification reaction with the beads at room temperature for 5 min.
7. Place the tube on a magnetic separation rack. Incubate at room temperature for 2 min. All of the beads should move to one side, and the supernatant should be clear.
8. Pipette off the supernatant and discard.
9. While the tubes are still on the magnetic rack, wash the beads by adding 180 μl of 80% EtOH to the side of the tube opposite of the beads. Incubate at room temperature for 30 s.
10. Pipette off the 80% EtOH and discard.
11. While the tube is still on the rack with the lids open, pipette off the ethanol that has been collected at the bottom of the tube and let the remaining ethanol evaporate. The bead pellet should appear to be dry (not shiny) and more matte in appearance.
12. Resuspend the beads by removing the tube from the magnet and adding 50 μl of nuclease-free water to the pellet. Pipette up and down several times so that the beads are fully resuspended. Incubate at room temperature for 2 min.
13. Place the tube back on the magnet and wait for the supernatant to become clear (approximately 30 s to 1 min).
14. Remove the supernatant which now contains the library to a new 1.5-ml tube.
15. Either save the library at −20 °C or continue with the subsequent steps.
16. Dilute the NEBNext® Multiplex Oligos for Illumina to 2 μM by adding 5 μl of undiluted primer to 20 μl of nuclease-free water. This can be done in 0.2-ml PCR strip tubes or 1.5-ml tubes. Take care to record which well in the plate in the undiluted primers originated from.
17. In a 0.2-ml tube, set up the reaction on ice as follows. Be sure to add a different primer pair to each sample, and record which sample received which primer pair.

Reagent	Volume (μl)
Purified library PCR 1 (from **step 15**)	2
Nuclease-free water	11.5
5× Q5 buffer	5
10 mM dNTPs	1
2 μM NEB primer (**step 16**)	5
Q5 polymerase	0.5

18. Put the reaction in a thermocycler and run the program:
    (a) 98 °C for 30 s.
    (b) 15 cycles of 98 °C for 10 s, 64 °C for 10 s, and 72 °C for 30 s.
19. Check for amplification and library quality using a DNA ScreenTape (D1000 or similar). Load 1 μl of the PCR product onto the TapeStation (Fig. 4). If the PCR products are the

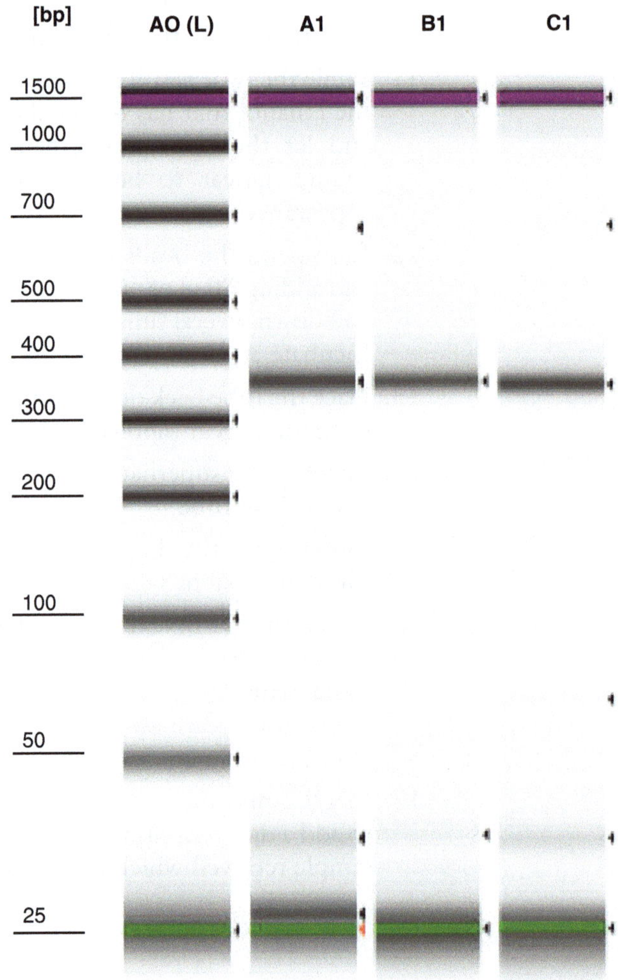

Fig. 4 TapeStation gel image of sequencing library final PCR products (Lanes A1, B1, C1). A standard DNA ladder is shown in lane A0, and molecular size standards are at 1500 bp (purple) and 25 bp (green). The libraries range from 355 to 368 bp due to the incorporation of up to 13 random bases during the first PCR step. The PCR product will sometimes contain a small amount of non-specific product (~700 bp) which does not affect the quality of the sequencing library. Non-specific products and excess primers less than 100 bp are removed during the subsequent bead clean up step

correct size and the no RT control is negative for a band, proceed to with the next steps.

20. Add 25 μl of AMPure XP of Agencourt beads to the amplification reaction. Mix well by setting a pipettor to 50 μl and pipetting up and down.
21. Continue with the bead cleanup procedure (**steps 6–11** above).
22. Resuspend the beads by removing the tube from the magnet and adding 15 μl of nuclease-free water to the pellet. Pipette up and down several times so that the beads are fully resuspended. Incubate at room temperature for 2 min.
23. Place the tube back on the magnet and wait for the supernatant to become clear (approximately 30 s to 1 min).
24. Remove the supernatant which now contains the library to a new 1.5-ml tube.
25. Either save the library at −20 °C or continue with the subsequent steps.
26. Quantify the concentration using the Qubit™ dsDNA HS Assay Kit according to the manufacturer's instructions. Typically, 1–2 μl of library is sufficient for accurate quantification. At this point, the libraries can be pooled and run on an Illumina sequencer.

4 Notes

1. We use the gradient (10–50%) program as described in Aboulhouda et al. [17] (described below). Depending on the gradient platform being used, other preset programs for 10–50% sucrose gradients can be used.
   (a) 05/85/35.
   (b) 01/77/0.
   (c) 04/86/35.
   (d) 03/86.5/35.
   (e) 20/81/14.
   (f) 07/86/20.
   Sequence = abcbdbabcbdbef.
2. The protocol for fractionation may be different depending on the fractionator being used. The steps in this procedure are for the Gilson fraction collection system. Refer to the user's manual for the Gilson fraction collection system for more detailed instructions or to your particular fraction collector for alternative instructions.

References

1. Ryoo HD, Vasudevan D (2017) Two distinct nodes of translational inhibition in the integrated stress response. BMB Rep 50: 539–545
2. Wek RC (2018) Role of eIF2α kinases in translational control and adaptation to cellular stress. Cold Spring Harb Perspect Biol 10: 1–16
3. Pakos-Zebrucka K, Koryga I, Mnich K et al (2016) The integrated stress response. EMBO Rep 17:1374–1395
4. Vattem KM, Wek RC (2004) Reinitiation involving upstream ORFs regulates ATF4 mRNA translation in mammalian cells. Proc Natl Acad Sci U S A 101:11269–11274
5. Zhang H, Wang Y, Lu J (2019) Function and evolution of upstream ORFs in eukaryotes. Trends Biochem Sci 44:782–794
6. McGillivray P, Ault R, Pawashe M et al (2018) A comprehensive catalog of predicted functional upstream open reading frames in humans. Nucleic Acids Res 46:3326–3338
7. Melnikov A, Murugan A, Zhang X et al (2012) Systematic dissection and optimization of inducible enhancers in human cells using a massively parallel reporter assay. Nat Biotechnol 30:271–277
8. Sharon E, Kalma Y, Sharp A et al (2012) Inferring gene regulatory logic from high-throughput measurements of thousands of systematically designed promoters. Nat Biotechnol 30:521–530
9. Soemedi R, Cygan KJ, Rhine CL et al (2017) Pathogenic variants that alter protein code often disrupt splicing. Nat Genet 49:848–855
10. Adamson SI, Zhan L, Graveley BR (2018) Vex-seq: high-throughput identification of the impact of genetic variation on pre-mRNA splicing efficiency. Genome Biol 19:1–12
11. Sample PJ, Wang B, Reid DW et al (2019) Human 5′ UTR design and variant effect prediction from a massively parallel translation assay. Nat Biotechnol 37:803–809
12. Dvir S, Velten L, Sharon E et al (2013) Deciphering the rules by which 5′-UTR sequences affect protein expression in yeast. Proc Natl Acad Sci U S A 110:E2792–E2801
13. Lin Y, May GE, Kready H et al (2019) Impacts of uORF codon identity and position on translation regulation. Nucleic Acids Res 2:1–10
14. Jia L, Mao Y, Ji Q et al (2020) Decoding mRNA translatability and stability from the 5′ UTR. Nat Struct Mol Biol 27:814–821
15. Noderer WL, Flockhart RJ, Bhaduri A et al (2014) Quantitative analysis of mammalian translation initiation sites by FACS-seq. Mol Syst Biol 10:748
16. May GE, McManus CJ (2022) High-throughput quantitation of yeast uORF regulatory impacts using FACS-uORF. Methods Mol Biol 2404:331–351. https://doi.org/10.1007/978-1-0716-1851-6_18. PMID: 34694618
17. Aboulhouda S, Di Santo R, Therizols G, Weinber D (2017) Accurate, Streamlined Analysis of mRNA Translation by Sucrose Gradient Fractionation. Bio Protoc 7(19):e2573. https://doi.org/10.21769/BioProtoc.2573. PMID: 29170751. PMCID: PMC5697790

Chapter 4

Measuring Bulk Translation Activity in Single Mammalian Cells During the Integrated Stress Response

Alyssa M. English and Stephanie L. Moon

Abstract

The attenuation of global translation is a critical outcome of the integrated stress response (ISR). Consequently, it is important to effectively detect and measure protein synthesis in studies seeking to evaluate the ISR. This chapter details two methods, surface sensing of translation (SUnSET) and fluorescent noncanonical amino acid tagging (FUNCAT), to measure global translation activity in individual cells using fluorescence microscopy as a read-out. Detecting bulk translation activity in single cells is advantageous for the concurrent observation of newly synthesized proteins and other cellular structures and to identify differences in translation activity among individuals within a population of cells.

Key words Translation, Puromycin, Puromycylation, Surface sensing of translation, SUnSET, Fluorescent noncanonical amino acid tagging, FUNCAT, Immunofluorescence, Click chemistry, Integrated stress response

1 Introduction

The suppression of bulk translation activity is one of the first events to occur during the integrated stress response (ISR). Stress-sensing kinases phosphorylate the major translation initiation factor eukaryotic initiation factor 2α (eIF2α), which reduces ternary complex formation and suppresses translation activity [1, 2]. Thus, protein synthesis is a critical process to monitor as a central readout of ISR activity. Several approaches are available to measure bulk translation activity at the single-cell and population levels. This chapter focuses on two separate methods, surface sensing of translation (SUnSET) [3] and fluorescent noncanonical amino acid tagging (FUNCAT) [4, 5], to quantify the translation activity of single cells by fluorescence microscopy.

Classically, protein synthesis is measured by the incorporation of radioactive amino acids, typically methionine and cysteine, into newly synthesized proteins during translation elongation. This

Daniel Matějů and Jeffrey A. Chao (eds.), *The Integrated Stress Response: Methods and Protocols*,
Methods in Molecular Biology, vol. 2428, https://doi.org/10.1007/978-1-0716-1975-9_4,

powerful technique is typically used to quantify the translation activity of a population of cells, as nascent proteins from cell lysates are commonly visualized following their separation on SDS-PAGE gels [6–8] . More recently, additional methods to evaluate protein synthesis have been developed including SUnSET [3] and FUNCAT [4, 5] that permit facile detection of bulk translation activity in single cells. SUnSET and FUNCAT identify newly synthesized proteins by fluorescence-based techniques and do not require the use of radioactive materials. A primary advantage of these methods is that they detect the translation activity of single cells. This not only allows for the assessment of the subcellular location of newly synthesized proteins but also permits the simultaneous observation of newly synthesized proteins and other fluorescently labeled biomolecules or subcellular structures such as organelles. Additionally, it enables visualization of the range of translation activity within a population of cells. Thus, SUnSET and FUNCAT are advantageous methods that are highly suitable for investigating pathways that impact translation, including the ISR, in individual cells, and we detail these methods below.

SUnSET is an easily accessible, rapid, and sensitive method to measure bulk translation activity in situ [3, 7]. SUnSET utilizes puromycin, a protein synthesis inhibitor that resembles the 3′ end of an aminoacylated tyrosyl-tRNA (Fig. 1a) [9, 10]. In a process known as puromycylation, puromycin binds to the ribosomal A-site and the nascent polypeptide chain is transferred to it (Fig. 1a). This results in translation termination and release of the polypeptide chain from the ribosome (Fig. 1a) [9–12]. Puromycylated proteins can be detected by anti-puromycin antibodies and secondary antibodies labeled with fluorophores using immunofluorescence microscopy (Fig. 1a). Alternatively, newly synthesized proteins can be labeled with O-propargyl puromycin (OPP) and detected by click chemistry instead of immunofluorescence [13]; however, this approach will not be discussed in this article. The benefits of using SUnSET are that it is easy, sensitive, and fast—nascent polypeptide labeling occurs within 5 min [3]. Further, SUnSET can be combined with other methods such as immunofluorescence microscopy or fluorescent fusion protein detection to concurrently visualize bulk translation activity and subcellular structures [7, 14] or markers of specific cell or tissue types within a population [15]. However, three limitations are associated with the use of this assay to quantify protein synthesis activity. Namely, (1) puromycin disrupts translation machinery by reacting with nascent polypeptide chains, causing them to be ejected from ribosomes along with mRNA [16–18], (2) two steps are needed to detect puromycylated polypeptides, and (3) puromycylation generates truncated proteins [9, 12] that may activate protein quality control pathways or otherwise perturb cellular protein metabolism. Thus,

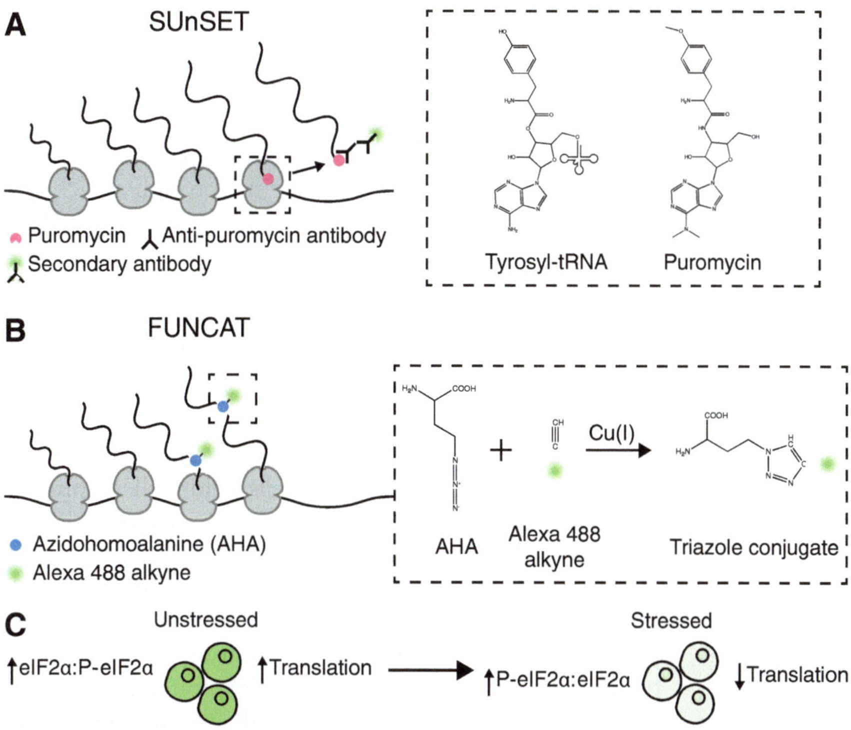

Fig. 1 Surface sensing of translation (SUnSET) and fluorescent noncanonical amino acid tagging (FUNCAT) enable visualization of nascent protein translation in situ in mammalian cells. (**a**) Puromycin is incorporated into nascent polypeptide chains and causes them to be ejected from the ribosome (left). At right, the structure of puromycin resembles the structure of the tyrosyl-tRNA, (adapted from [9, 16]). (**b**) The bioorthogonal methionine analog azidohomoalanine is incorporated into nascent polypeptide chains (left). At right, the click reaction between azidohomoalanine and Alexa 488 alkyne results in a stable triazole conjugate. (**c**) Both assays will result in a read-out of global translation activity in individual cells (green), which is reduced during the integrated stress response when the ratio of P-eIF2α:eIF2α is high

care should be taken when designing and interpreting experiments in order to appropriately use SUnSET.

FUNCAT is a second useful method to detect bulk translation activity via metabolic labeling of growing nascent polypeptide chains [4, 5]. This approach relies on the incorporation of bioorthogonal alkyne or azide-containing amino acids into newly synthesized proteins (Fig. 1b) [19]. Upon addition of an alkyne- or azide-containing fluorophore, the fluorophore covalently attaches to its complementary alkyne- or azide-containing amino acid ortholog by click chemistry (Fig. 1b). The click reaction is a simple, high-yield linking reaction that occurs rapidly at room temperature in conditions amenable to labeling and purifying biomolecules [20–22]. Nascent proteins labeled via FUNCAT are detected by fluorescence microscopy, and this method can be combined with immunofluorescence microscopy or the detection of fluorescent fusion proteins to simultaneously visualize other cellular structures

[5, 23], specific cell types [5, 24], and cells expressing certain proteins [4]. In contrast to SUnSET, FUNCAT does not interfere with the translation machinery and, as a result, does not create truncated, aberrant proteins. Three disadvantages of the FUNCAT method are: (1) it is costlier than SUnSET, (2) it requires two steps to detect newly synthesized proteins through a click reaction, and (3) similar to radiolabeling nascent proteins, FUNCAT is typically performed following a short-term amino acid deprivation period, which can increase the sensitivity of the assay at the expense of potentially altering cell metabolism. Therefore, it is important to consider the limitations of FUNCAT when designing and interpreting experiments.

Together, the SUnSET and FUNCAT fluorescence-based approaches permit a wide range of scientific questions regarding translation activity in single cells and cell populations to be explored. These methods are extensively used to investigate protein synthesis in a variety of cell types [4, 7, 13, 24–26], tissues [13, 26–29], and whole organisms [5, 16, 30]. Specifically, SUnSET and FUNCAT can be used to explore translation activity during the ISR (Fig. 1c) [7, 23]. Additionally, because puromycylation produces truncated proteins, SUnSET can be exploited to study the outcome of defective ribosomal products (DRiPs) [31, 32]. It is important to note that, although these techniques identify nascent proteins, they do not necessarily label sites of active translation [16]. This chapter will focus on the use of SUnSET and FUNCAT with a fluorescence microscopy read-out in mammalian cell culture systems to study translation activity regulation during the ISR in the context of arsenite stress.

2 Materials

2.1 SUrface SEnsing of Translation (SUnSET) Assay

1. Adherent tissue culture cells, such as U-2 OS cells.
2. Humidified cell culture incubator with 5% CO_2.
3. Biosafety cabinet.
4. Hemocytometer.
5. Light microscope with 10× objective for cell counting.
6. Trypsin–EDTA solution (0.25%).
7. Phosphate-buffered saline (PBS).
8. Cell culture maintenance medium: Dulbecco's Modified Eagle Medium, 10% fetal bovine serum, 1% streptomycin/penicillin, 2 mM glutamine (glutaMAX).
9. 12-well tissue culture plate(s).
10. Glass coverslips (circular #1, 18 mm diameter).
11. Sodium arsenite solution (100×): 50 mM in sterile water.

12. Puromycin (1000×): 10 μg/μL in sterile water.
13. Paraformaldehyde: 4% in PBS.
14. Triton X-100: 0.5% in PBS.
15. Bovine serum albumin (BSA): 3% in PBS.
16. Bovine serum albumin (BSA): 0.3% in PBS.
17. 12-well uncoated plate(s) (sterile).
18. Primary mouse anti-puromycin antibody (Millipore Sigma, MABE343) (1:1000).
19. Secondary anti-mouse antibody (e.g., goat anti-mouse IgG FITC, Abcam ab97022, 1:1000).
20. Glass slides.
21. VECTASHIELD® Antifade Mounting Medium with DAPI (Vector laboratories).
22. Wide-field fluorescence microscope with deconvolution software.

2.2 Fluorescent Noncanonical Amino Acid Tagging (FUNCAT) Assay

1. Adherent tissue culture cells, such as U-2 OS cells.
2. Humidified cell culture incubator with 5% CO_2.
3. Biosafety cabinet.
4. Hemocytometer.
5. Light microscope with 10× objective for cell counting.
6. Trypsin–EDTA solution (0.25%).
7. Phosphate-buffered saline (PBS).
8. Cell culture maintenance medium: Dulbecco's Modified Eagle Medium, 10% fetal bovine serum, 1% streptomycin/penicillin, 2 mM glutamine (glutaMAX).
9. Labeling medium: Dulbecco's Modified Eagle Medium lacking methionine, 10% dialyzed fetal bovine serum, 1% streptomycin/penicillin, 2 mM glutamine (glutaMAX).
10. 12-well tissue culture plate(s).
11. Glass coverslips (circular #1, 18 mm diameter).
12. Sodium arsenite solution (100×): 50 mM in sterile water.
13. Methionine (1000×): 50 mM in dimethyl sulfoxide.
14. Azidohomoalanine (Click-iT AHA C10102) (1000×): 50 mM in dimethyl sulfoxide.
15. Deionized sterile water.
16. Paraformaldehyde: 4% in PBS.
17. Triton X-100: 0.5% in PBS.
18. Bovine serum albumin (BSA): 3% in PBS.
19. Click-iT™ Cell Reaction Buffer Kit (Invitrogen, cat. C10269).

20. Alexa fluor 488 alkyne (1000×): 2 mM in dimethyl formamide.
21. Glass slides.
22. VECTASHIELD® Antifade Mounting Medium with DAPI (Vector laboratories).
23. Wide-field fluorescence microscope with deconvolution software.

3 Methods

All reagents and solutions should be sterile and protease-free when possible.

3.1 SUnSET Assay

1. In a biosafety cabinet, place one glass coverslip per condition (for unstressed and stressed conditions, plate two coverslips) into each well of a 12-well plate. Wash twice with 70% ethanol and once with cell culture maintenance medium.
2. Split and count cells using a hemocytometer. Resuspend cells in cell culture maintenance medium to plate them in a 1-mL volume on the glass coverslips to achieve ~50% confluency the next day. Remove any trapped air beneath the cover slip by pressing them down gently with a sterile pipette tip.
3. The next day, remove the medium and replace it with 1 mL of warm (37 °C) fresh medium without or with the stressor (e.g., 0.5 mM sodium arsenite). Incubate in the humidified cell culture incubator (5% CO_2, 37 °C) for at least 20 min to evaluate translation at 30 min post-stress, or 50 min to evaluate translation for 60 min post-stress.
4. Ten minutes prior to the desired post-stress time-point, add enough puromycin to reach 10 μg/mL (1 uL) to each well. Incubate for 10 min in the cell culture incubator (*see* **Note 1**).
5. In a fume hood, rinse the cells twice with 1 mL PBS to remove free puromycin. Add 250 μL of 4% paraformaldehyde (*see* **Note 1**) to each well and incubate for 10 min (*see* **Note 2**).
6. Rinse the cells with 1 mL PBS and permeabilize the cells with 0.5% Triton X-100 for 5 min.
7. Wash the cells with 1 mL PBS for 5 min. Incubate the cells in 3% BSA for 1 h or at 4 °C overnight (*see* **Note 3**).
8. Wash cells thrice with 1 mL PBS for 10 min each.
9. Incubate with primary antibody against puromycin (1:1000) in 0.3% BSA for 1 h at room temperature by placing coverslips upside down onto a ~100 μL droplet of antibody solution in a petri dish humidified chamber (*see* **Note 4**).

10. Place coverslips into the wells of an uncoated 12-well dish, cell-side up. Wash cells thrice with 1 mL PBS for 10 min each.
11. Incubate with secondary anti-mouse antibody (e.g., goat anti-mouse IgG FITC) at 1:1000 in 0.3% BSA for 1 h at room temperature (*see* **Note 5**).
12. Place coverslips into new wells of an uncoated 12-well dish, cell-side up. Wash cells thrice with 1 mL PBS for 10 min each.
13. Remove coverslips from the 12-well plate, and briefly dry by turning them upside down onto a fresh kimwipe. Place the dry coverslip on top of ~15 μL mounting medium (e.g., VECTA SHIELD®) containing DAPI nuclear stain on a glass slide, cell side down.
14. Allow to cure overnight in the dark.
15. Proceed to imaging and image analysis of cells, Subheading 3.3.

3.2 FUNCAT Assay

1. In a biosafety cabinet, place one glass coverslip per condition (for one unstressed and one stressed condition, plate three coverslips to include a methionine labeling unstressed control, **Note 6**) into each well of a 12-well plate. Wash coverslips twice with 70% ethanol and once with maintenance medium.
2. Split and count cells using a hemocytometer. Resuspend cells in cell culture maintenance medium to plate them in a 1 mL volume on the glass coverslips to achieve ~50% confluency the next day. Remove any trapped air beneath the cover slip by pressing them down gently with a sterile pipette tip.
3. Prepare and store Click-iT™ Cell Reaction Buffer and Click-iT™ Cell Buffer Additive according to the Click-iT™ Cell Reaction Buffer Kit manufacturer's instructions.
4. The next day, replace the cell culture maintenance medium with 1 mL pre-warmed (37 °C) labeling medium. Incubate for 30 min in the cell culture incubator.
5. Replace labeling medium with at least 0.5 mL fresh pre-warmed (37 °C) labeling medium containing either azido-homoalanine (0.5 μL of 1000× stock per dish) or methionine (0.5 μL of 1000× stock per dish) in the presence or absence of stressor (e.g., 0.5 mM sodium arsenite) (*see* **Note 1**). Incubate in the cell culture incubator for 30 min to 1 h (*see* **Note 7**).
6. In a fume hood, remove the medium and replace with ~250 μL paraformaldehyde (*see* **Note 1**). Incubate for 10 min.
7. Wash cells once with PBS.
8. Permeabilize cells with Triton X-100 (0.5% in PBS) for 5 min.
9. Incubate cells with 3% BSA in PBS for 10 min.

10. While cells are incubating (**step 9**), prepare a 1.5 mL master mix of Click-iT™ cell reaction cocktail by combining 1.32 mL 1× Click-iT™ reaction buffer, 30 μL copper sulfate, 150 μL cell buffer additive, and 1.5 μL of Alexa Fluor 488 alkyne.
11. Incubate the cells under 0.5 mL each of Click-iT™ reaction cocktail for 30 min protected from light with rocking (*see* **Note 8**).
12. Incubate cells for 10 min with 3% BSA in PBS (*see* **Note 9**).
13. Mount coverslips by first drying by gently placing upside down on a kimwipe, and placing onto a droplet (~15 μL) of mounting medium (e.g., VECTASHIELD® plus DAPI nuclear stain).
14. Allow samples to cure overnight in the dark.
15. Proceed to imaging and image analysis, *see* Subheading 3.3.

3.3 Imaging and Image Analysis

Samples can be visualized using widefield fluorescence microscopy such as with a DeltaVision Elite microscope equipped with a 100× objective and a PCO Edge sCMOS camera. Choose a system with an appropriate wavelength of light for excitation and mirrors/filters to detect emission of selected fluors, e.g., in the FITC channel.

1. Determine imaging parameters (i.e., exposure time and intensity) empirically to obtain fluorescence read-outs of all samples within the dynamic range, and avoid over-saturated pixels in micrographs of cells with high FUNCAT or SUnSET labeling.
2. Determine the number of Z sections and step sizes empirically. For U2-OS cells, 15 Z sections at 0.2 μm intervals are adequate to span the entire cell volume.
3. Acquire photomicrographs using the same imaging parameters for each sample.
4. Deconvolve image data using a non-destructive method, such as the softWoRx DeltaVision software.
5. The open-source software Fiji/ImageJ [33, 34] can be used to create maximum intensity projections. Use the Fiji/ImageJ Analyze and Set Measurements tools to obtain pixel intensities in cells demarcated as regions of interest (ROI) to obtain a read-out of relative translation activity in individual cells. Useful resources for Fiji/ImageJ analysis are the online forum https://forum.image.sc/ and the website https://imagej.net/ImageJ.

4 Notes

1. Puromycin-containing medium should be inactivated by autoclaving or properly disposed of as a toxic chemical product

[35]. Arsenite and paraformaldehyde solutions must be properly disposed of as hazardous chemical waste.

2. Incubations are performed at room temperature unless otherwise noted.
3. Incubations can be either at 1 h room temperature or overnight at 4 °C. Overnight incubations should be done by sealing the coverslips in a humidified chamber to reduce evaporation. Samples stored in 12-well plates can be humidified by wrapping them in wet paper towels and covering in plastic wrap. Alternatively, a petri dish incubator can be assembled to reduce the amount of antibody solution needed (*see* **Note 4**).
4. Assemble a petri dish humidifier by placing a layer of 3–5 paper towels cut to fit inside the bottom dish. Wet the paper towels with sterile water, and place a piece of parafilm cut such that it is slightly smaller than the paper towel on top. Mark the parafilm at the edges with a marker to identify samples. Gently place the coverslip onto a droplet of antibody solution (100 μL). Close the petri dish and seal with parafilm if incubating overnight.
5. It is not advised to incubate samples in secondary antibody overnight.
6. A methionine control is highly recommended. This control condition is fed methionine instead of azidohomoalanine and should be used to determine the non-specific fluorescence contribution to observed fluorescence intensities in FUNCAT conditions.
7. For brevity, we describe FUNCAT using azidohomoalanine and the fluor-labeled alkyne Alexa Fluor 488 alkyne. However, other bioorthogonal amino acids such as L-homopropargylglycine, and other fluor-labeled alkynes such as Alexa Fluor 594 alkyne, are available.
8. Copper sulfate must be properly disposed of as a toxic chemical product.
9. To perform FUNCAT labeling and immunofluorescence microscopy, proceed after Subheading 3.2.11 to the primary antibody incubation step using a standard immunofluorescence labeling protocol.

References

1. Costa-Mattioli M, Walter P (2020) The integrated stress response: from mechanism to disease. Science 368:eaat5314. https://doi.org/10.1126/science.aat5314
2. Pakos-Zebrucka K, Koryga I, Mnich K et al (2016) The integrated stress response. EMBO Rep 17:1374–1395
3. Schmidt EK, Clavarino G, Ceppi M, Pierre P (2009) SUnSET, a nonradioactive method to monitor protein synthesis. Nat Methods 6: 275–277
4. Dieterich DC, Hodas JJL, Gouzer G et al (2010) In situ visualization and dynamics of newly synthesized proteins in rat hippocampal neurons. Nat Neurosci 13:897–905

5. Tom Dieck S, Müller A, Nehring A et al (2012) Metabolic labeling with noncanonical amino acids and visualization by chemoselective fluorescent tagging. Curr Protoc Cell Biol. Chapter 7:Unit7.11
6. Moon SL, Parker R (2018) EIF2B2 mutations in vanishing white matter disease hypersuppress translation and delay recovery during the integrated stress response. RNA 24: 841–852
7. Burke JM, Moon SL, Matheny T, Parker R (2019) RNase L reprograms translation by widespread mRNA turnover escaped by antiviral mRNAs. Mol Cell 75:1203–1217.e5
8. Bonifacino JS, Dasso M, Harford JB et al (2001) Detection and quantitation of radiolabeled proteins in gels and blots. In: Current protocols in cell biology. John Wiley & Sons, Inc., Hoboken, NJ, USA, p 132
9. Aviner R (2020) The science of puromycin: from studies of ribosome function to applications in biotechnology. Comput Struct Biotechnol J 18:1074–1083
10. Yarmolinsky MB, Haba GL (1959) Inhibition by puromycin of amino acid incorporation into protein. Proc Natl Acad Sci U S A 45: 1721–1729
11. Semenkov YP, Shapkina TG, Kirillov SV (1992) Puromycin reaction of the A-site bound peptidyl-tRNA. Biochimie 74:411–417
12. Nathans D (1964) Puromycin inhibition of protein synthesis: incorporation of puromycin into peptide chains. Proc Natl Acad Sci U S A 51:585–592
13. Liu J, Xu Y, Stoleru D, Salic A (2012) Imaging protein synthesis in cells and tissues with an alkyne analog of puromycin. Proc Natl Acad Sci U S A 109:413–418
14. Kamelgarn M, Chen J, Kuang L et al (2018) ALS mutations of FUS suppress protein translation and disrupt the regulation of nonsense-mediated decay. Proc Natl Acad Sci U S A 115: E11904–E11913
15. Goodman CA, Hornberger TA (2013) Measuring protein synthesis with SUnSET: a valid alternative to traditional techniques? Exerc Sport Sci Rev 41:107–115
16. Enam SU, Zinshteyn B, Goldman DH et al (2020) Puromycin reactivity does not accurately localize translation at the subcellular level. Elife 9:e60303. https://doi.org/10.7554/eLife.60303
17. Kedersha N, Cho MR, Li W et al (2000) Dynamic shuttling of TIA-1 accompanies the recruitment of mRNA to mammalian stress granules. J Cell Biol 151:1257–1268
18. Khong A, Parker R (2018) mRNP architecture in translating and stress conditions reveals an ordered pathway of mRNP compaction. J Cell Biol 217:4124–4140
19. Dieterich DC, Link AJ, Graumann J et al (2006) Selective identification of newly synthesized proteins in mammalian cells using bioorthogonal noncanonical amino acid tagging (BONCAT). Proc Natl Acad Sci U S A 103:9482–9487
20. Hein CD, Liu X-M, Wang D (2008) Click chemistry, a powerful tool for pharmaceutical sciences. Pharm Res 25:2216–2230
21. Rostovtsev VV, Green LG, Fokin VV, Sharpless KB (2002) A stepwise huisgen cycloaddition process: copper(I)-catalyzed regioselective "ligation" of azides and terminal alkynes. Angew Chem Int Ed Engl 41:2596–2599
22. Tornøe CW, Christensen C, Meldal M (2002) Peptidotriazoles on solid phase: [1,2,3]-triazoles by regiospecific copper(i)-catalyzed 1,3-dipolar cycloadditions of terminal alkynes to azides. J Org Chem 67:3057–3064
23. Ruggieri A, Dazert E, Metz P et al (2012) Dynamic oscillation of translation and stress granule formation mark the cellular response to virus infection. Cell Host Microbe 12:71–85
24. Erdmann I, Marter K, Kobler O et al (2015) Cell-selective labelling of proteomes in Drosophila melanogaster. Nat Commun 6:7521
25. Signer RAJ, Magee JA, Salic A, Morrison SJ (2014) Haematopoietic stem cells require a highly regulated protein synthesis rate. Nature 509:49–54
26. Alvarez-Castelao B, Schanzenbächer CT, Hanus C et al (2017) Cell-type-specific metabolic labeling of nascent proteomes in vivo. Nat Biotechnol 35:1196–1201
27. Seedhom MO, Hickman HD, Wei J et al (2016) Protein translation activity: a new measure of host immune cell activation. J Immunol 197:1498–1506
28. Deliu LP, Ghosh A, Grewal SS (2017) Investigation of protein synthesis in drosophila larvae using puromycin labelling. Biol Open 6: 1229–1234
29. Liang V, Ullrich M, Lam H et al (2014) Altered proteostasis in aging and heat shock response in C. elegans revealed by analysis of the global and de novo synthesized proteome. Cell Mol Life Sci 71:3339–3361
30. Hinz FI, Dieterich DC, Tirrell DA, Schuman EM (2012) Non-canonical amino acid labeling in vivo to visualize and affinity purify newly synthesized proteins in larval zebrafish. ACS Chem Neurosci 3:40–49

31. Lelouard H, Ferrand V, Marguet D et al (2004) Dendritic cell aggresome-like induced structures are dedicated areas for ubiquitination and storage of newly synthesized defective proteins. J Cell Biol 164:667–675
32. Turakhiya A, Meyer SR, Marincola G et al (2018) ZFAND1 recruits p97 and the 26S proteasome to promote the clearance of Arsenite-induced stress granules. Mol Cell 70: 906–919.e7
33. Schindelin J, Arganda-Carreras I, Frise E et al (2012) Fiji: an open-source platform for biological-image analysis. Nat Methods 9: 676–682
34. Schindelin J, Rueden CT, Hiner MC, Eliceiri KW (2015) The ImageJ ecosystem: an open platform for biomedical image analysis. Mol Reprod Dev 82:518–529
35. Meyer EL, Golston G, Thomaston S et al (2017) Is your institution disposing of culture media containing antibiotics? Appl Biosaf 22: 164–167

Chapter 5

Quantitative Translation Proteomics Using mePROD

Kevin Klann and Christian Münch

Abstract

Multiplexed enhanced protein dynamic mass spectrometry (mePROD MS) enables robust quantification of translation in cell culture. Tandem mass tags (TMT) are combined with pulsed stable isotope labeling in cell culture (pSILAC) to monitor newly synthesized proteins on a proteome wide scale. While approaches combining pSILAC and TMT typically require long labeling times to reach sufficient intensity of the newly synthesized peptides in the mass spectrometer, mePROD uses a carrier signal that boosts the survey scan intensity and strongly increases identification rates. Hence, this protocol provides an easy and cost-efficient method to profile proteome-wide translatome changes at a temporal resolution of minutes.

Key words Proteomics, Mass spectrometry, SILAC, TMT, Translation, Protein dynamics, pSILAC, Bottom-up proteomics

1 Introduction

The cellular proteome is highly dynamic, representing a fine-tuned balance between protein synthesis, modification, and degradation. When the cellular environment changes, cells need to adapt their proteome to the altered conditions. As an example, under stress conditions, such as a viral infection or protein misfolding, the translation landscape changes tremendously. General translation efficiency drops, while specialized mRNAs, coding for specific stress response proteins, are translated to counteract the stressor [1–6]. These changes are often highly timed events and might not be visible directly on the level of total protein abundance.

To directly monitor changes in protein expression, several methods are available: First, next-generation sequencing (NGS) approaches, such as ribosome profiling [7, 8], exhibit superior coverage. During ribosome profiling experiments, mRNA fragments (so called footprints) protected by the ribosome from nuclease digestion are sequenced, giving information on ribosome position on the mRNA. NGS approaches rely on an indirect read-out (ribosomes on a mRNA does not directly correspond to protein

Daniel Matějů and Jeffrey A. Chao (eds.), *The Integrated Stress Response: Methods and Protocols*, Methods in Molecular Biology, vol. 2428, https://doi.org/10.1007/978-1-0716-1975-9_5,

synthesis) and suffer from an inherent normalization bias, challenging measurements in conditions where global translation levels change extensively [9]. Second, metabolic labeling and subsequent proteomics measurements offer an alternative to measure changes in protein dynamics. Stable isotope-labeled amino acids (or other labeled metabolites) are incorporated into newly synthesized proteins. However, these proteomics approaches require long labeling times or enrichment steps to reach sufficient signal intensity and depth of analysis [10–12]. Long labeling times prevent the high temporal resolution needed to profile time-dependent effects, such as the integrated stress response (ISR), while enrichment compatible labeling agents can change translation itself [13, 14].

We recently developed multiplexed enhanced protein dynamic (mePROD) proteomics to overcome these limitations (Fig. 1). Our approach utilizes a carrier sample, comprised of a fully SILAC-labeled whole-cell digest, that improves survey scan intensity of newly synthesized (i.e., heavy labeled) peptides. TMT multiplexing enables sample coding by using isobaric tags that modify peptides with identical masses, thereby summing all identical peptides from different samples into one peptide peak [15]. Upon fragmentation, the tags dissociate into up to 16 different reporter ions [16], enabling quantification of individual sample. By adding a booster (also called carrier) sample to the multiplexed pulse-labeled samples, the overall peptide intensity of the pulse-labeled peptides can be elevated [17]. This allows shortening of the labeling time to the time scale of minutes to achieve high temporal resolution in an unbiased system. Since no extensive cell culture is needed to perform the metabolic labeling, mePROD is well suited for the use with primary cell samples, as long as the cells can be cultured for the required pulse time in the presence of the isotopic amino acids. Furthermore, combination of mePROD experiments, i.e., applying a carrier/booster channel, with specialized acquisition methods [18] minimizes the required sample input to less than 100,000 cells. Thus, mePROD proteomics is an ideal method to monitor translation changes, as observed upon activation of the ISR.

2 Materials

For cell culture, all used solutions need to be appropriately sterilized to avoid contamination. After cell lysis, all solutions should be prepared with HPLC grade water to avoid chemical contamination that would perturb the MS measurement (e.g., PEG, detergents). Solutions for MS sample preparation should not be stored in glassware that were cleaned in a dishwasher, since remains of detergents might transfer to the samples. For disposal, follow the official rules of your institution.

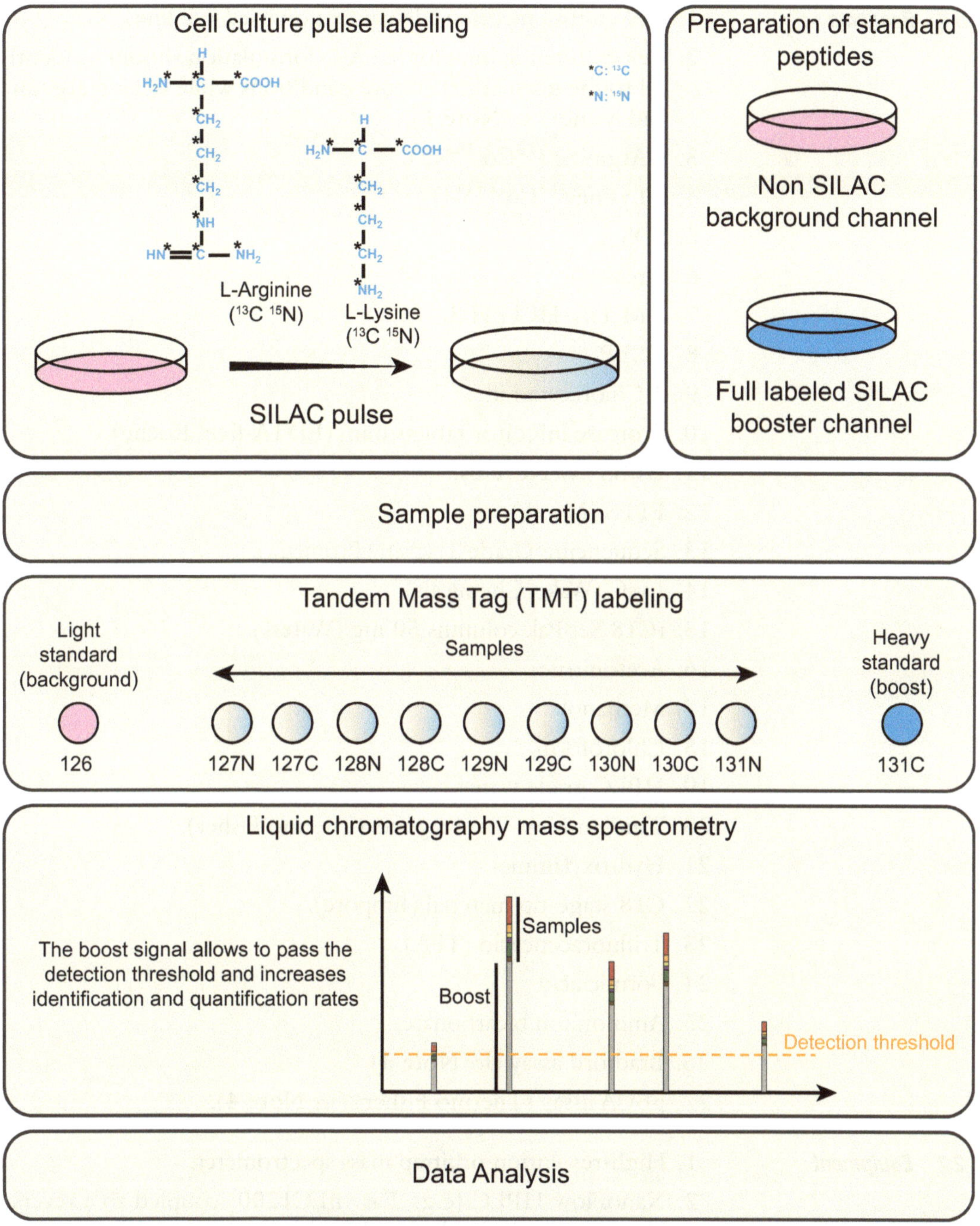

Fig. 1 Workflow scheme of mePROD experiments. The cultured cells are pulse labeled with heavy isotope containing L-arginine and L-lysine for the desired time. To enable the mePROD workflow, standard peptides from cell lysates without SILAC amino acids (for the background signal) and fully heavy labeled cells (for signal boosting) are collected. After the desired MS sample preparation protocol, the samples are labeled with TMT reagents together with the light and heavy standard peptides. During MS measurement, the boost channel will alleviate the overall peptide intensity, making detection of the peptide and subsequent quantification more likely. The background channel will allow for the calculation of co-isolation derived signal intensity and increases the dynamic range of the measurements. MS data is then processed to obtain ratios of newly synthesized protein for each individual sample

2.1 Reagents

1. Cell culture medium (dependent on used cell line).
2. Cell culture medium for SILAC (formulation should be identical to the normal cell culture conditions while lacking arginine and lysine) (*see* **Note 1**).
3. L-Arginine (^{13}C6 ^{15}N2).
4. L-Lysine (^{13}C6 ^{15}N4)
5. SDS.
6. Urea.
7. 1 M Tris–HCl pH 8.
8. TCEP.
9. 2-Chloracetamide.
10. Protease inhibitor tablets mini (EDTA-free, Roche).
11. DPBS (*see* **Note 2**).
12. EPPS (Merck).
13. Sequencing Grade Trypsin (Promega).
14. LysC (Wako Chemicals).
15. tC18 SepPak columns 50 mg (Waters).
16. Acetonitrile.
17. Methanol.
18. Chloroform.
19. HPLC grade water.
20. Tandem Mass Tag reagents (Thermo Fisher).
21. Hydroxylamine.
22. C18 stage-tip material (Empore).
23. Trifluoraceticacid (TFA).
24. Formic acid.
25. Ammonium bicarbonate.
26. Bradford assay (*see* **Note 3**).
27. μBCA assay (Thermo Fisher) (*see* **Note 4**).

2.2 Equipment

1. High-resolution orbitrap mass spectrometer.
2. NanoFlow HPLC (e.g., Easy nLC1200) coupled to a reverse phase column suitable for MS.
3. Probe sonication device.
4. SepPak vacuum manifold (Waters).
5. Benchtop microcentrifuge.
6. Speedvac.
7. HPLC for peptide fractionation.

8. Software for MS analysis (e.g., Proteome Discoverer) (*see* **Note 5**).
9. Protein low bind reaction tubes (e.g., Eppendorf).

2.3 SILAC Amino Acid Stock Solutions

Prepare stock solutions of amino acids at a suitable concentration, depending on the used cell culture medium. Since DMEM is routinely used, all concentrations given here correspond to DMEM, please modify if needed. Final amino acid concentrations should be the same as in the original medium formulation (the additional mass of heavy amino acids has to be considered). SILAC amino acids are dissolved in PBS to concentrations of 87.9 g/L (arginine) and 151.2 g/L (lysine). Amino acid stock solutions need to be filtered with a 0.22-μM filter syringe to be sterile and can be stored at −20 °C after aliquoting. The noted concentrations represent 1000× stock solutions when using DMEM. To create labeling medium, dilute amino acids 1:1000 in DMEM. Store medium at 4 °C (*see* **Note 6**).

2.4 Buffers

1. Lysis buffer: 2% SDS, 100 mM Tris–HCl pH 8, 150 mM NaCl, 10 mM TCEP, 40 mM 2-chloracetamide, protease inhibitor cocktail tablets.
2. Resuspension buffer: 8 M Urea, 10 mM EPPS pH 8.2 in water.
3. Buffer A: 0.1% TFA in water.
4. Buffer B: 70% acetonitrile, 0.1% TFA in water.
5. TMT wash buffer: 5% acetonitrile, 0.1% TFA in water.
6. TMT labeling buffer: 200 mM EPPS pH 8.2, 10% acetonitrile in water.
7. Quenching solution: 5% hydroxylamine in water.
8. Fractionation buffer: 5% acetonitrile, 10 mM ammonium bicarbonate.

2.5 Standard Peptides

1. mePROD uses a carrier signal that is comprised of a fully heavy labeled cell digest. To create this digest, culture the cell line of interest with heavy SILAC medium for a minimum of 2 weeks (*see* **Note 7**). Process these cells like regular samples, as described below (*see* **Note 8**). Carrier samples can be aliquoted and used for multiple experiments, without the need of preparing them freshly.
2. To improve data analysis and overcome ratio compression, mePROD also contains a non-SILAC-labeled sample that serves as a background proxy for the quantification calculation. To create this sample, harvest cells cultured in light medium and process as described below.

3 Methods

After pulse labeling of cells, the samples can be processed using any common protocol for whole-cell proteomics (e.g., SP3, iST). We routinely use a SDS/urea-based protocol [17] that is outlined here, but may be substituted with other protocols.

3.1 Pulse Labeling of Cells and Harvest

1. Culture cells according to the experimental design.
2. Wash cells twice with warm (37 °C) DPBS (or equivalent buffer) to remove remaining non-heavy amino acids.
3. Add heavy SILAC cell culture medium to cells.
4. Incubate for desired pulse time (e.g., 1 h) (*see* **Note 9**).
5. Pre-heat lysis buffer to 95 °C.
6. Aspirate medium and wash cells 3× with warm DPBS (or PBS) to remove medium and remaining serum.
7. Lyse cells on plate with hot lysis buffer (six-well plate; 200 μL per well).
8. Scrape lysate and transfer to 2 mL Eppendorf tube (*see* **Note 10**).
9. Incubate lysate at 95 °C for 5 min.
10. Sonicate lysate with probe sonicator to shear genomic DNA.
11. Incubate lysate at 95 °C for 5 min.

    Pause point: Lysates can be stored at −20 °C for short periods (several days). Long-term storage at −80 °C.

3.2 Sample Preparation for Mass Spectrometry

As stated before, this part can be substituted with any suitable whole-cell proteomics protocol if needed. 1 volume (vol.) refers to original sample volume in lysis buffer (e.g., 200 μL).

1. Add 3 vol. ice-cold methanol and mix.
2. Add 1 vol. chloroform and mix.
3. Add 2.5 vol. water and mix.
4. Spin samples in a microcentrifuge at 4 °C with 16,000 × *g* for 10 min (*see* **Note 11**).
5. Aspirate upper aqueous phase. Be careful not to disturb the interface layer.
6. Add 3 vol. ice-cold methanol and mix.
7. Spin samples in a microcentrifuge at 4 °C with 20,000 × *g* for 5 min.
8. Aspirate liquid.
9. Add 3 vol. ice-cold methanol and mix.

10. Spin samples in a microcentrifuge at 4 °C with 20,000 × *g* for 5 min.
11. Aspirate all liquid.
12. Dry pellets at 60 °C with open lids to evaporate remaining methanol. Be careful not to overly dry pellets, since these will be hard to resuspend.
    Pause point: Dried proteins can be stored at −20 °C.
13. Resuspend proteins in 1 vol. Resuspension buffer.
14. Measure protein concentration by Bradford assay.
15. Transfer 100 μg of proteins to new 2 mL Eppendorf tube (*see* **Note 12**).
16. Adjust all samples to same volume with resuspension buffer.
17. Dilute samples to 1 M Urea with 10 mM EPPS pH 8.2.
18. Add 1:50 (w/w) LysC and 1:100 (w/w) Trypsin.
19. Incubate at 37 °C overnight.
    Pause point: Digests can be frozen for short-term storage at −20 °C.
20. Acidify samples using TFA to a pH between 2 and 3.
21. Prepare SepPak columns: Volume for each wash corresponds to one column fill, depending on column size. Be careful that columns do not run dry, as these will not retain peptides properly. If dried, start again with the first methanol wash.
22. Wash SepPak column with methanol.
23. Wash SepPak with Buffer B.
24. Wash SepPak with Buffer A.
25. Load acidified peptide solution.
26. Wash peptides with Buffer A.
27. Elute into Eppendorf tubes with 1 mL Buffer B.
28. Dry peptides in speedvac.
    Pause point: Dried peptides can be stored at −20 °C.
29. Resuspend peptides in TMT labeling buffer.
30. Measure peptide concentration using μBCA assay.
31. Transfer 25 μg of peptides to new 1.5-mL tube (*see* **Note 13**).
32. Adjust all samples to equal volume using TMT-labeling buffer.
33. Set up suitable TMT-labeling scheme (e.g., TMT126 = Light standard; TMT127N = Condition 1; and so on).
34. Add 2.5 μL of TMT reagents to tube according to labeling scheme.
35. Prepare 50 μg of carrier peptides in TMT-labeling buffer (*see* **Note 14**).

36. Add 5 μL TMT to carrier peptides.
37. Prepare 25 μg light standard peptides in TMT-labeling buffer.
38. Add 2.5 μL TMT to light standard peptides.
39. Incubate all labeled samples at room temperature for 1 h.
40. Quench reaction by addition of 1:10 (v/v) 5% hydroxylamine solution.
41. Incubate at room temperature for 15 min.

    Pause point: Labeled peptides can be stored at −20 °C.
42. To ensure equal sample loading, pool 0.5 μg of each sample for test run.
43. Purify using C18 stage-tips.
44. Test-run on LC-MS/MS (*see* **Note 15**).
45. Pool remaining sample and perform SepPak purification as before (replace Buffer A with TMT wash buffer).
46. Dry eluates in speedvac.

    Pause point: Dried samples can be stored at −20 °C.

3.3 High pH Reverse Phase Fractionation

To reach sufficient depth and a comprehensive dataset, we recommend fractionation of samples across a second dimension. For this step, any peptide fractionation method compatible with LC-MS workflows is suitable. We recommend 12–24 fractions for translation measurements of up to 8000 proteins. In this protocol, we describe a high-pH reverse phase fractionation [17, 19] using a Dionex Ultimate 3000 system.

1. Resuspend multiplexed sample in 700 μL of fractionation buffer.
2. Place sample in autosampler.
3. Fractionate sample into 96 fractions using a multistep gradient from 100% Solvent A (5% acetonitrile, 10 mM ammonium bicarbonate) to 60% Solvent B (90% acetonitrile, 10 mM ammonium bicarbonate) over 70 min. Eluting peptides are collected every 45 s in a total of 96 fractions.
4. Cross-concatenate fractions into 24 fractions. (Mix fraction 1 with fractions 25, 49, and 73; Mix fraction 2 with fractions 26, 50, and 74, and so on.) Use 200 μL per fraction for concatenation. Remaining fractions can be stored at −20 °C.
5. Dry concatenated samples in speedvac.

    Pause point: Dried peptides can be stored at −20 °C.

3.4 LC-MS^2

To enable high-depth measurements, we recently established the usage of a targeted mass difference (TMD) filter for the acquisition of pSILAC/TMT data [18]. This acquisition filter is available on all newer ThermoFisher Orbitrap mass spectrometers since the introduction of the Orbitrap Fusion. TMD searches for isotope pairs (i.e., labeled and non-labeled peptide) during the survey scan, triggering the dependent scan(s) for both ions. This enables higher identification rates since heavy labeled peptides are measured once their higher intensity "light" counterpart is identified. TMD can boost identification rates of labeled peptides ~2.5-fold in a single LC-MS run. However, mePROD experiments can also be measured by conventional data dependent acquisition modes on older instruments [17]. In terms of relative quantification, we did not observe large benefits of using SPS-MS^3 based workflows in mePROD experiments [18]. By using the noise proxy sample (light sample), co-isolation of interfering light ions can be directly accounted for during data analysis. Thus, we routinely use MS^2-based workflows for mePROD experiments.

1. Resuspend samples in 2% ACN, 0.1% TFA.
2. Inject approximately 1 μg of concatenated fractions into nanoflow LC for MS/MS analysis.

3.5 Data Analysis

In this workflow, two separate search nodes are implemented into the data processing in ProteomeDiscoverer. One node searches for light peptides, the other for heavy peptides. In our hands, this approach speeds up data analysis by approximately 200–300%, since only few dynamic modifications are occurring. In any case, a new modification at lysines needs to be created. Currently, the search engines do not allow two modifications at the same residue (i.e., heavy lysine and TMT at lysines). Therefore, light peptides contain the common TMT modification, while the heavy peptides contain a custom TMT modification with the additional isotope mass. Quantification of TMT reporter ions is unaffected by the modification. Since data analysis of proteomics data is a wide field, we only state special settings for mePROD in this protocol. All other settings can be handled as for any other proteomics experiment. If proteomics is performed for the first time, we recommend the default workflows provided with ProteomeDiscoverer, including the modifications below.

1. Open ProteomeDiscoverer software.
2. Create new chemical modification. TMTK8 or TMTproK8 by adding the modification masses of TMT/TMTpro and the isotopic lysine modification together (TMTK8: $\Delta M = 237.1771$, Δ Average Mass = 237.2061; TMTproK8: $\Delta M = 312.2213$, Δ Average Mass = 312.2554).
3. Create a new study and load RAW files generated by the MS instrument.
4. Create an MS2 reporter ion-based workflow.

5. Set up two search engine nodes for peptide identification: Use appropriate sequence database (e.g., current Human SwissProt) in both nodes. Set common modifications as desired (e.g. methionine oxidation). Set carbamidomethyl as static cysteine modification in both nodes.
6. First node (light peptides): Set static modification at the N-terminus to TMT. Set TMT as static modification at lysines.
7. Second node (heavy peptides): Set static modification at the N-terminus to TMT. Set TMTK8 as static modification at lysines and Arginine10 as static modification at arginines.
8. Validate identification FDRs using common methodology (e.g., target-decoy approach).
9. Run analysis using default settings (*see* **Note 16**).
10. Export peptide data in the preferred format.
11. Normalize samples using total intensity normalization. The sum of all heavy and light peptides over the whole proteome is assumed to be equal between all samples.
12. Filter for peptides containing heavy SILAC modifications.
13. For each labeled peptide, the background signal will be subtracted. The background signal is derived from the TMT intensity in the non-SILAC-labeled sample that should not contain the peptide that was measured. Through co-isolation/ fragmentation of contaminating light ions however, this intensity will not be zero. Therefore, the value of the non-SILAC-labeled sample is subtracted from all other channels.
14. To calculate protein translation, all corresponding noise subtracted peptide intensities are summed for each channel (*see* **Note 17**).

4 Notes

1. If the general formulation of the "normal" and "SILAC" media are not identical, the medium change may cause biological side effects visible during the proteomics experiment (e.g., metabolic pathway changes, due to different nutrient concentrations). If no equivalent SILAC medium is available, a non-labeled medium can be obtained by substituting a suitable SILAC medium with light amino acids to be used for cell culture before the heavy SILAC pulse. We recommend culturing cells for at least 24 h in new media before performing the experiment to avoid media-driven effects.
2. Any other cell culture compatible buffer may be used instead of DPBS.

3. May be replaced by alternative protein quantitation assays.
4. The μBCA assay may be substituted by other high-sensitivity peptide quantitation assays.
5. In our hands, MaxQuant was not able to perform searches with TMT quantification and SILAC amino acids as dynamic modifications. This may change with future updates.
6. For short SILAC pulses as used during translation proteomics with mePROD, no dialyzed FBS is necessary.
7. We recommend using the same cell line for standard (carrier or background) peptides as used in the final experiment. However, if this is not possible (e.g., for primary cells), we suggest using an established cell line from the same tissue to approximate the proteome composition of the analyzed cells.
8. To evaluate suitable labeling times for the experimental setup, the percentage of newly synthesized proteins can be calculated in relation to the fully labeled standard peptides at a given timepoint.
9. We validated mePROD with labeling times as short as 15 min. However, longer labeling times further increase quantification robustness and identification rates. We recommend using 1–2 h pulses, unless higher temporal resolution is necessary.
10. We recommend using protein low bind tubes to avoid sample loss during preparation.
11. If the upper phase still contains white precipitate, spin until it is collected at the interphase.
12. 100 μg represents an easy-to-handle amount and accounts already for possible sample loss during the following steps. However, for extensive fractionation 10–25 μg peptide per sample are sufficient (depending on the column size on the fractionating HPLC; for very small columns, the required amount might be even lower). Thus, the protein amount for digestion can be adjusted accordingly.
13. On an Aeris Peptide XB-C18 (4.6 mm ID, 2.6 μM particle size), we commonly use 250 μg total multiplexed peptides (with a TMT 10plex 25 μg per sample). The needed amount to achieve good peptide separation can differ if smaller column sizes are used. Adjust peptide amount for labeling accordingly so that the resulting total multiplex fits the fractionation column. If unfractionated measurements are sufficient, 1–5 μg per sample are sufficient for labeling.
14. Usage of more carrier peptides will result in more identifications of peptides; however, using more than twice the amount of the regular samples will weaken the sample quantifications, due to the constrained nature of TMT measurements

[17, 20]. We recommend double-molar usage of the carrier signal.

15. If sample loading is not equal, use median TMT abundance per channel as an approximation and calculate normalization factors for pooling the samples. Equal sample loading will enhance quantification robustness and avoids artifacts caused by extensive post-acquistion normalization.
16. The default settings using ProteomeDiscoverer workflow nodes are sufficient for most applications and may be kept if the experimenter is new to the field of proteomics.
17. To ensure robust quantification, we recommend using only peptides that have enough remaining intensity value after noise subtraction. The best value has to be determined empirically, depending on the used instrumentation and data analysis.

Acknowledgments

We thank the Quantitative Proteomics Unit (IBC2, Goethe University Frankfurt). C.M. was supported by the European Research Council under the European Union's Seventh Framework Programme (ERC StG 803565), Collaborative Research Center (CRC) 1177 and the Emmy Noether Programme of the Deutsche Forschungsgemeinschaft (DFG; MU 4216/1-1), the Johanna Quandt Young Academy at Goethe and the Aventis Foundation Bridge Award.

References

1. Holcik M, Sonenberg N (2005) Translational control in stress and apoptosis. Nat Rev Mol Cell Biol 6:318
2. Pakos-Zebrucka K, Koryga I, Mnich K, Ljujic M, Samali A, Gorman AM (2016) The integrated stress response. EMBO Rep 17: 1374–1395
3. Khaperskyy DA, Emara MM, Johnston BP, Anderson P, Hatchette TF, McCormick C (2014) Influenza a virus host shutoff disables antiviral stress-induced translation arrest. PLoS Pathog 10:e1004217
4. Cheng Z, Teo G, Krueger S, Rock TM, Koh HW, Choi H, Vogel C (2016) Differential dynamics of the mammalian mRNA and protein expression response to misfolding stress. Mol Syst Biol 12:855
5. Jiang Z, Yang J, Dai A, Wang Y, Li W, Xie Z (2017) Ribosome profiling reveals translational regulation of mammalian cells in response to hypoxic stress. BMC Genomics 18:638
6. Stern-Ginossar N, Thompson SR, Mathews MB, Mohr I (2019) Translational control in virus-infected cells. Cold Spring Harb Perspect Biol 11:a033001. https://doi.org/10.1101/cshperspect.a033001
7. Ingolia NT, Ghaemmaghami S, Newman JRS, Weissman JS (2009) Genome-wide analysis in vivo of translation with nucleotide resolution using ribosome profiling. Science 324: 218–223
8. Ingolia NT, Brar GA, Rouskin S, McGeachy AM, Weissman JS (2012) The ribosome profiling strategy for monitoring translation in vivo by deep sequencing of ribosome-protected mRNA fragments. Nat Protoc 7: 1534–1550
9. McGlincy NJ, Ingolia NT (2017) Transcriptome-wide measurement of translation by ribosome profiling. Methods 126: 112–129

10. Welle KA, Zhang T, Hryhorenko JR, Shen S, Qu J, Ghaemmaghami S (2016) Time-resolved analysis of proteome dynamics by tandem mass tags and stable isotope labeling in cell culture (TMT-SILAC) Hyperplexing. Mol Cell Proteomics 15:3551–3563
11. Schwanhäusser B, Gossen M, Dittmar G, Selbach M (2009) Global analysis of cellular protein translation by pulsed SILAC. Proteomics 9:205–209
12. Rothenberg DA, Taliaferro JM, Huber SM, Begley TJ, Dedon PC, White FM (2018) A Proteomics approach to profiling the temporal translational response to stress and growth. iScience 9:367–381
13. Kiick KL, Saxon E, Tirrell DA, Bertozzi CR (2002) Incorporation of azides into recombinant proteins for chemoselective modification by the Staudinger ligation. Proc Natl Acad Sci U S A 99:19–24
14. Klann K, Münch C (2020) Unbiased translation proteomics upon cell stress. Mol Cell Oncol 7:1763150
15. Thompson A, Schäfer J, Kuhn K, Kienle S, Schwarz J, Schmidt G, Neumann T, Hamon C (2003) Tandem mass tags: a novel quantification strategy for comparative analysis of complex protein mixtures by MS/MS. Anal Chem 75:1895–1904
16. Thompson A, Wölmer N, Koncarevic S, Selzer S, Böhm G, Legner H, Schmid P, Kienle S, Penning P, Höhle C, Berfelde A, Martinez-Pinna R, Farztdinov V, Jung S, Kuhn K, Pike I (2019) TMTpro: design, synthesis, and initial evaluation of a proline-based isobaric 16-Plex tandem mass tag reagent set. Anal Chem 91:15941–15950
17. Klann K, Tascher G, Münch C (2020) Functional translatome proteomics reveal converging and dose-dependent regulation by mTORC1 and eIF2α. Mol. Cell 77: 913–925.e4
18. Klann K, Münch C (2020) Instrument logic increases identifications during Mutliplexed Translatome measurements. Anal Chem 92: 8041–8045
19. Batth TS, Francavilla C, Olsen JV (2014) Off-line high-pH reversed-phase fractionation for in-depth phosphoproteomics. J Proteome Res 13:6176–6186
20. O'Brien JJ, O'Connell JD, Paulo JA, Thakurta S, Rose CM, Weekes MP, Huttlin EL, Gygi SP (2018) Compositional proteomics: effects of spatial constraints on protein quantification utilizing isobaric tags. J Proteome Res 17:590–599

Chapter 6

Quantifying the Binding of Fluorescently Labeled Guanine Nucleotides and Initiator tRNA to Eukaryotic Translation Initiation Factor 2

Martin D. Jennings and Graham D. Pavitt

Abstract

The translation initiation factor eIF2 is critical for protein synthesis initiation, and its regulation is central to the integrated stress response (ISR). eIF2 is a G protein, and the activity is regulated by its GDP or GTP-binding status, such that only GTP-bound eIF2 has high affinity for initiator methionyl tRNA. In the ISR, regulatory signaling reduces the availability of eIF2-GTP and so downregulates protein synthesis initiation in cells. Fluorescence spectroscopy can be used as an analytical tool to study protein–ligand interactions in vitro. Here we describe methods to purify eIF2 and assays of its activity, employing analogs of GDP, GTP, and methionyl initiator tRNA ligands to accurately measure their binding affinities.

Key words Translation initiation, eIF2, Fluorescence spectroscopy, GDP, GTP, Met–tRNAi

1 Introduction

In protein synthesis, the translation factor eukaryotic initiation factor 2 (eIF2) delivers initiator methionyl tRNA (Met–tRNAi) to initiating ribosomes and aids accurate AUG start-codon recognition [1, 2]. As eIF2 is a G-protein, its nucleotide-bound status eIF2 is tightly coupled to its Met–tRNAi binding ability, and this makes eIF2 a target for global translational control via regulation of GDP/GTP binding. eIF2 binds Met–tRNAi when it is bound to GTP with an affinity of 1–2 nM, but eIF2-GDP complexes have 50- to 100-fold lower affinity for Met–tRNAi. However, eIF2 has a higher affinity for GDP than GTP, such that it requires the action of eIF2B to perform guanine-nucleotide exchange and promote formation of the eIF2-GTP-Met–tRNAi ternary complex (TC) [2]. eIF2 also interacts with other translation initiation factors and 40S ribosomes to facilitate protein synthesis initiation. One factor that is directly relevant is eIF5. It acts as a GTPase activating factor for TC [3] and as a GDP-dissociation inhibitor following

Daniel Matějů and Jeffrey A. Chao (eds.), *The Integrated Stress Response: Methods and Protocols*,
Methods in Molecular Biology, vol. 2428, https://doi.org/10.1007/978-1-0716-1975-9_6,

eIF2-GDP release from ribosomes [4, 5]. In addition, and central to global and gene-specific control of protein synthesis, there is a family of protein kinases that each phosphorylate eIF2 on its alpha subunit at the same site, commonly called serine 51. The mammalian kinases are HRI, PKR, Perk, and Gcn2. The generated eIF2 (αP) forms a non-productive complex with eIF2B that reduces its GEF activity and thereby lowers cellular TC levels. This can have a large impact on cellular protein synthesis altering translation of individual mRNAs either up and down to ensure an integrated stress response (ISR) [6].

Elaborate purification schemes for fractionating cell lysates and obtaining purified eukaryotic translation factors were initially developed in the 1970s (e.g., [7, 8]). Although variations of the traditional purification schemes remain in use (e.g., [9]), modern recombinant expression methods and affinity chromatography methods have enabled more streamlined purification schemes. In our laboratory, yeast eIF2 can be purified in relatively high quantities from yeast cells overexpressing epitope-tagged eIF2 [10–12]. We have also successfully "humanized" yeast cells that overexpress human eIF2 cDNAs in place of the yeast *eIF2* genes and developed a similar scheme for its purification [13]. Others have developed mammalian cell lines expressing tagged versions of eIF2α and used these as a source of mammalian eIF2 (e.g., [14]). Below we describe our optimized and updated method for purifying yeast eIF2 (Fig. 1). One advantage of the yeast

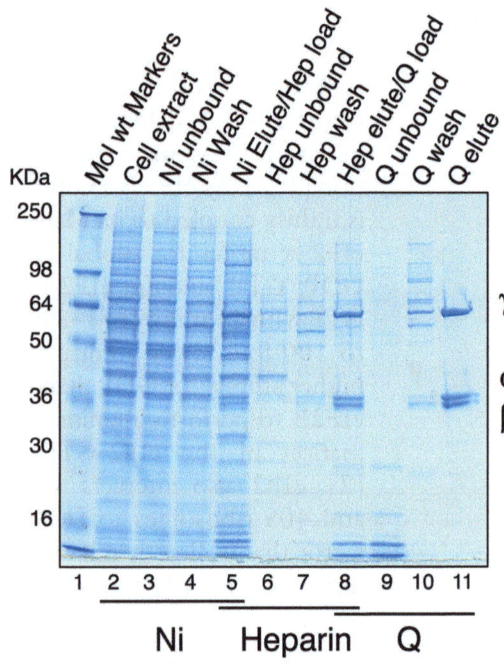

Fig. 1 Overview of eIF2 purification. Coomassie-stained SDS-PAGE gel showing samples from successive steps outlined in Subheading 3.1

expression system is that by using cells deleted for the sole yeast eIF2α kinase Gcn2, the resulting eIF2 is entirely free of phosphorylated eIF2α.

Radio-isotopically labeled nucleotides and methionine were used initially to monitor ligand binding to purified eIF2. Mixed eIF2/nucleotide/Met–tRNAi solutions were sampled and bound to nitrocellulose filters. As only protein-bound radioactivity was retained on the filters, unbound ligands were able to be washed away by filtration under vacuum [15–18]. Gel-shift assays monitoring the inclusion of Met–tRNAi into larger complexes with eIF2 ± 40S ribosomes and additional purified initiation factors have also been developed (e.g., [12]). The development of commercially available fluorescent analogs has enabled the use of fluorescence spectroscopy to accurately monitor the binding of nucleotides and Met–tRNAi to eIF2 [14, 19, 20]. Upon binding to eIF2, the fluorescence of these labeled molecules increases. The assays are set up, so that eIF2 is titrated into a fluorescent solution, and the change in fluorescence is recorded. This increase in fluorescence anisotropy as a function of protein concentration can then be used to fit the dissociation constant (K_d). Importantly, the assay is in solution and, unlike earlier radioisotope/filter binding assays, does not require removing samples from reactions at defined time points. By performing such assays, we have measured the binding of BODIPY™ FL-labeled GTP and GDP nucleotides to eIF2 [19, 20]. We have also confirmed that this assay works with MANT-labeled GTP, and possibly other ribose ring-labeled fluorescent analogs could also be used in this manner. Using these analogs, we were able to measure the affinity of BODIPY™ FL-GDP and BODIPY™ FL-GTP to be 19.1 nM and 35.2 nM, respectively (Fig. 2a, b).

In a similar manner, we have used BODIPY-labeled Met–tRNAi (amino labeled on the methionine) to monitor initiator tRNA binding to eIF2-GDP or eIF2-GTP, termed ternary complex

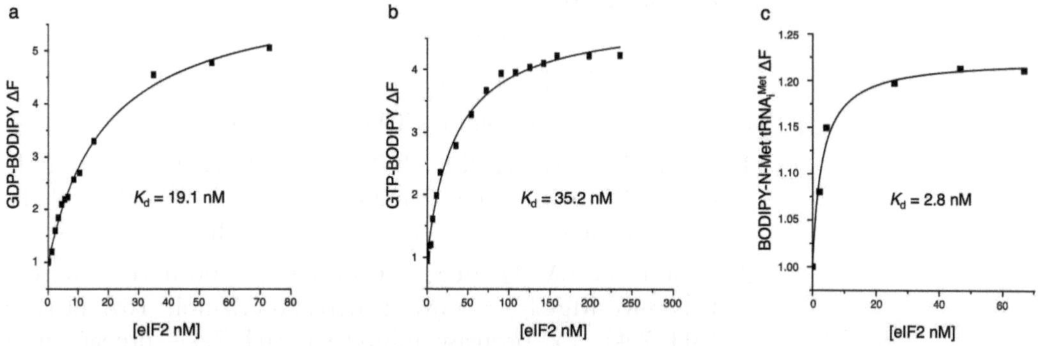

Fig. 2 Example fluorescence binding curve fittings and nucleotide/Met–tRNA dissociation constant determination. Fluorescence binding of BODIPY-GDP (**a**) and BODIPY-GTP (**b**) to apo-eIF2 and BODIPY-N-Met–tRNA$_i$ to eIF2 with 1 mM GTP (**c**). Inserts show the calculated binding dissociation constant (K_d)

formation [19, 20]. Using this assay, we have been able to monitor ternary complex formation with GTP or GDP-bound eIF2 (dissociation constants of 1.1 nM and 55 nM, respectively) (Fig. 2c) [19]. These affinities are comparable to rates demonstrated using ^{35}S-Methionine-labeled tRNA (e.g., [18]).

We and others have also developed expression and purification procedures for FLAG-tagged eIF2B (the guanine-nucleotide exchange factor for eIF2) that can be used to stimulate nucleotide release from eIF2, as well as the eIF2α kinase PKR, which can be used to phosphorylate purified eIF2 and inhibit the activity of eIF2B as measured in variations of the assays described herein [19, 20]. We have also shown that Phos-tag™ SDS-PAGE gels can be used to evaluate the extent of phosphorylation [19, 21]. Readers are referred to the cited sources for further information on how to adapt these assays. For a recent review of the thermodynamics of eIF2–ligand interactions, *see* [22], and for computational modeling of the control mechanism, *see* [23].

Here we describe the purification of eIF2 and the use of fluorescent spectroscopy to monitor non-protein eIF2 ligand binding.

2 Materials

2.1 Purification of eIF2

1. Yeast eIF2 overexpressing strain, e.g., GP3511 (*MATα leu2-3 leu2-112 ura3-52 ino1 gcn2Δ pep4::LEU2 sui2Δ HIS4-lacZ* p [*SUI2 SUI3* His$_6$-*GCD11 URA3* 2 μm]) (*see* **Note 1**).
2. Yeast extract peptone dextrose (YPD) medium.
3. Cryogenic mill (Spex 6870D SamplePrep Freezer Mill, or similar).
4. Nickel-NTA agarose resin (Qiagen).
5. SnakeSkin Dialysis tubing, 10k MWCO (Thermo Fisher).
6. HiTrap Heparin HP column (Cytiva).
7. HiTrap Q HP column (Cytiva).
8. ÅKTA pure system (Cytiva).
9. cOmplete EDTA-free protease inhibitor cocktail (Roche).
10. Pierce centrifuge columns 10 mL (Thermo Fisher).
11. Ni10: 30 mM HEPES, 800 mM KCl, 10 mM Imidazole, 0.1 mM $MgCl_2$, 5 mM 2-mercaptoethanol, 10% glycerol, 1× protease inhibitors (pH 7.4)—prepare fresh and chill to 4 °C.
12. Ni500: 30 mM HEPES, 500 mM KCl, 500 mM Imidazole, 0.1 mM $MgCl_2$, 5 mM 2-mercaptoethanol, 10% glycerol (pH 7.4), 1× protease inhibitors (pH 7.4)—prepare fresh and chill to 4 °C.

13. HS100: 30 mM HEPES, 100 mM KCl, 0.1 mM $MgCl_2$, 10% glycerol (pH 7.4)—prepare fresh, vacuum filter, and chill to 4 °C.
14. HS150: 30 mM HEPES, 150 mM KCl, 0.1 mM $MgCl_2$, 10% glycerol (pH 7.4)—prepare fresh, vacuum filter, and chill to 4 °C.
15. HSA: 30 mM HEPES, 0.1 mM $MgCl_2$, 10% glycerol (pH 7.4)—prepare fresh, vacuum filter, and chill to 4 °C.
16. HSB: 30 mM HEPES, 1 M KCl 0.1 mM $MgCl_2$, 10% glycerol (pH 7.4)—prepare fresh, vacuum filter, and chill to 4 °C.
17. eIF2 sample buffer: 30 mM HEPES, 100 mM KCl, 0.1 mM $MgCl_2$, 1 mM DTT, 10% glycerol (pH 7.4)—prepare fresh and chill to 4 °C.
18. Bio-Rad Protein Assay Dye Reagent Concentrate.
19. Eppendorf Protein LoBind tubes (Merck).

2.2 Preparation of Apo-eIF2

1. EDTA dialysis buffer: 30 mM HEPES, 100 mM KCl, 1 mM DTT, 1 mM EDTA, 10% glycerol, pH 7.4.
2. Slide-A-Lyzer™ MINI Dialysis Device 10 K MWCO (Thermo Fisher).
3. Eppendorf Protein LoBind tubes (Merck).

2.3 Fluorescence Spectroscopy Analysis

1. BODIPY™ FL-GDP (Thermo Fisher).
2. BODIPY™ FL-GTP (Thermo Fisher).
3. Assay buffer: 30 mM HEPES, 100 mM KCl, 10 mM $MgCl_2$, pH 7.4.
4. Spectrofluorometer (Fluoromax 4 or similar—Horiba).
5. Fluorometer cuvettes (*see* **Note 2**).
6. Purified apo eIF2.
7. 1 mM GDP in Assay buffer.
8. 1 mM GTP in Assay buffer.
9. BODIPY™ Bop Met-tRNAi (tRNAProbes).
10. Curve fitting software, e.g., Prism (GraphPad).

3 Methods

3.1 eIF2 Purification

This method describes the growth, harvest, lysis, and step-by step purification of eIF2 from yeast cells bearing a high-copy plasmid overexpressing all three subunits of eIF2. In our adapted strain, *GCD11* encoding eIF2γ is 6xhistidine tagged. However, Nickel affinity only partially purifies eIF2, so additional chromatography steps are required to further purify and polish eIF2 (Fig. 1).

3.1.1 Cell Growth, Harvest, and Lysis

1. Streak eIF2 overexpressing strain onto selective media and incubate at 30 °C for 2–3 days.
2. From a single colony, grow approximately 12 L of a yeast strain to a high culture density (OD_{600} = 4–6) at 30 °C, 180 rpm in the same selective media (*see* **Note 3**).
3. Harvest by centrifugation at 5000 × *g*, 10 min, 4 °C.
4. Resuspend cells in 2 mL per gram of cells of Ni10 buffer, then freeze in liquid nitrogen.
5. Grind cells under liquid nitrogen in cryogenic mill (we use 3 × 2 min runs at 15 cycles per second with 2 min cool time between runs in cryogenic mill).

   At this point, cells can be frozen at −80 °C until start of purification (*see* **Note 4**).

3.1.2 Lysate Clarification and Nickel-Affinity Chromatography

1. Thaw cell lysate on ice.
2. Clear cell lysate from insoluble debris by centrifugation at 5000 × *g*, 5 min, 4 °C.
3. Pass lysate through Pierce centrifuge column.
4. Keep supernatant and further clarify by centrifugation at 20,000 × *g*, 20 min, 4 °C.
5. Repeat the above **step 1** and **2** more times, if necessary, until lysate is fully clear (*see* **Note 5**).
6. Prepare 20 mL of Ni-NTA resin by washing twice with 20 mL of Ni10 buffer in batch—harvesting resin by centrifugation at 500 × *g*, 5 min, 4 °C between washes.
7. Bind lysate to Ni-NTA resin for 2 h at 4 °C with rotation.
8. Collect resin by centrifugation at 500 × *g*, 5 min, 4 °C.
9. Wash four times with 10× resin bed volume of Ni10 (collecting resin by centrifugation each time as above) (*see* **Note 6**).
10. Mix with 15 mL of Ni500 for 1 h at 4 °C, centrifuge, then keep liquid (elution).
11. Repeat above for a second elution and pool both eluates.
12. Dialyze using SnakeSkin tubing at 4 °C into 1 L of HS150 with mixing for at least 2 h or overnight.

3.1.3 Heparin HP Chromatography

1. Clarify sample at 20,000 × *g*, 20 min, 4 °C (*see* **Note 7**).
2. Prepare the ÅKTA pure system, connecting a 5 mL HiTrap Heparin HP column and the HSA and HSB buffers (*see* **Note 8**).
3. Equilibrate the column in 150 mM KCl (*see* **Note 9**).
4. Apply sample to column.

5. Wash column sequentially with 40 mL of 150 mM, 250 mM, and 350 mM KCl.
6. Elute column with 650 mM KCl in 2 mL fractions.
7. Pool fractions with a large peak of A_{254}.
8. Dialyze using SnakeSkin tubing at 4 °C into 1 L of HS100 with mixing for at least 2 h or overnight.

3.1.4 HiTrap Q Chromatography and Final Steps

1. Prepare the ÅKTA pure system, connecting a 1 mL HiTrap Q HP column and the HSA and HSB buffers.
2. Equilibrate the column at 10% HSB (100 mM KCl).
3. Wash column sequentially with 10 mL of 100 mM and 200 mM KCl.
4. Elute column 600 mM KCl in 0.3 mL fractions.
5. Pool fractions with a large peak of A_{254}.
6. eIF2 concentration can be quantified using Bio-Rad Protein Assay Dye Reagent and an appropriate standard (*see* **Note 10**).
7. If Apo-eIF2 is required, proceed to Subheading 3.2.
8. Dialyze sample into eIF2 sample buffer using a Slide-A-Lyzer™ with mixing on a stirrer at 4 °C for at least 2 h, or overnight.
9. Aliquot sample into small tubes and store at −80 °C in LoBind tubes.

3.2 Preparation of Apo-eIF2

Removing co-purifying nucleotide (GDP) from eIF2 to generate apo-eIF2 is necessary before monitoring nucleotide binding to eIF2 and to control the nucleotide status of eIF2 in other assays. As nucleotide binding is stabilized by Mg^{2+}, both can be removed by dialysis in buffer containing EDTA, which chelates Mg^{2+}.

1. Prepare 1 L of EDTA dialysis buffer and 1 L of eIF2 sample buffer and chill to 4 °C in a cold room in a glass beaker.
2. Dialyze eIF2 for 2 h in EDTA dialysis buffer using a Slide-A-Lyzer™ with mixing on a stirrer at 4 °C.
3. Dialyze eIF2 for 2 h or overnight in eIF2 sample buffer.
4. Aliquot sample into small tubes and store at −80 °C in LoBind tubes.

3.3 Fluorescence Spectroscopy Analysis of Nucleotide Binding to eIF2

Here purified eIF2 is titrated into a solution containing a fluorescently labeled nucleotide. Interaction between the protein and the nucleotide alter the local environment and increase fluorescent signal.

1. Prepare 100 nM of BODIPY-labeled nucleotide in 180 μL of assay buffer in a cuvette (*see* **Note 11**).
2. Also prepare 180 μL of assay buffer in a cuvette as a control.

3. Leave to equilibrate at room temperature for 5 min (*see* **Note 12**).
4. Measure the fluorescence of the labeled nucleotide and control samples using an excitation wavelength of 490 nM and emission wavelength of 509 nM, with a slit setting of 5 nm on both emission and excitation.
5. Titrate in increasing amounts of eIF2 (approximately in the range of 1–100+ nM final concentration) into both the labeled nucleotide and control cuvettes (*see* **Note 13**).
6. At each titration, pipette up and down steadily to mix, without introducing any bubbles, then leave samples for 5 min at room temperature to equilibrate, then repeat the fluorescence measurement.
7. Continue titration until change in wavelength is saturated (fluorescence readings plateau, then start to decrease due to dilution of the fluorescent nucleotide).

3.4 Fluorescence Spectroscopy Analysis of Met–tRNAi Binding to eIF2

The assay setup is similar to the assay described in Subheading 3.3, but with unlabeled nucleotides and instead BODIPY™-labeled methionine bound to initiator tRNA. We obtained BOP-N-Met–tRNAi from tRNA Probes Inc. Alternatively, it can be generated in house, by purifying bulk tRNA, or via in vitro transcription using a bespoke ribozyme-linked tRNAi sequence, as described in [12]. Met–tRNAi is amino-acylated with methionine, using purified *E. coli* MetRS. This enzyme only amino-acylates tRNAi, not eukaryotic elongator tRNAs [12]. Finally, BODIPY-FL or another fluorophore is coupled to methionine [24], and the resulting tRNA is purified before use.

1. Prepare 20 nM of BODIPY-labeled Met-tRNAi in 180 μL of assay buffer with 1 mM of GDP or GTP in separate reactions.
2. Also prepare 180 μL of assay buffer in a cuvette as a control.
3. Leave to equilibrate at room temperature for 5 min (*see* **Note 12**).
4. Measure the fluorescence of the labeled nucleotide and control samples using an excitation wavelength of 490 nM and emission wavelength of 509 nM, with a slit setting of 5 nm on both emission and excitation.
5. Titrate in increasing amounts of eIF2 (approximately in the range of 1–70 nM final concentration for GTP) into both the labeled nucleotide and control cuvettes. With GDP reactions, eIF2 concentration needs to be significantly higher to saturate (*see* **Note 13**).
6. At each titration, pipette up and down steadily to mix, without introducing any bubbles, then leave samples for 5 min at room temperature to equilibrate, then repeat the fluorescence measurement.

7. Continue titration until change in wavelength is saturated (fluorescence readings plateau, then start to decrease due to dilution of the fluorescent nucleotide).

3.5 Fluorescence Data Analysis

1. Correct measurements for any effects of eIF2 by subtracting the control sample at each titration point. The control sample should be negligible in terms of fluorescence.
2. Adjust each value for the dilution effects of eIF2 titration (corrected value = fluorescence × (current volume/starting volume). Starting volume in this case is 180 μL.
3. Calculate the relative change in fluorescence (ΔF) at each point by dividing by the starting fluorescence.
4. Calculate the eIF2 concentration in the cuvette at each titration point (taking into account the change in volumes).
5. Plot eIF2 concentration (x) against ΔF (y) (Fig. 2).
6. Fit against a single site binding model $y = 1 + [(\Delta F_{max} - 1) \times (x/(x + K_d))]$ which calculates the maximum change in fluorescence (ΔF_{max}) and the dissociation constant K_d (*see* **Note 14**).

4 Notes

1. This particular strain overexpresses all three eIF2 subunits (α, β, γ) with a His_6 tag at the gamma subunit N-terminus. It has the vacuolar protease Pep4 deleted to reduce degradation of the overexpressed protein during cell lysis and early purification steps. In addition, the eIF2 kinase, Gcn2, is deleted to ensure that the eIF2 produced is not the inhibitory phosphorylated version.
2. Cuvette needed is dependent on the machine used. In this case, a 100 μL volume with 10 mm pathlength, z height of 15 mm and 2 mm × 5 mm window.
3. The particular strain referenced can be grown in YPD to allow maximum growth rate.
4. We have found the eIF2 purification to be easily achieved over 3 days. Day 1 involves nickel affinity purification and then dialysis overnight into HS150. Day 2 involves heparin purification, dialysis for 1 h into HS100, Mono Q purification and dialysis overnight (buffer depending on whether Apo-eIF2 is needed). Day 3 involves any final dialysis, quantification, aliquoting samples, running gels, and machine cleaning.
5. The high efficiency of cell lysis achieved by nitrogen grinding results in a large degree of lipid content that interferes with nickel resin binding. Repeat centrifugation and centrifuge column clearing until lysate is clear to ensure a high degree of

binding. Careful extraction of the lysate from centrifuge bottles when lipid can be observed near the top also helps with this.

6. The sample often goes cloudy during dialysis. The white precipitate obtained following centrifugation does not contain eIF2.
7. The high KCl concentration in buffers Ni10 and Ni500 ensures that eIF2 is purified away from its partner protein eIF2B.
8. Typically, a flow rate of 2 mL per min is used for the larger heparin column and 0.5 mL per min is used for the smaller Q column.
9. HSA is 0 M KCl and HSB is 1 M KCl, so the % B used in the system represents the % of 1 M KCl. For example, to achieve 150 mM KCl, 15% HSB is used (85% HSA).
10. Typically, a concentration of 2 mg/mL is achieved with a total yield of ~4 mg.
11. Magnesium concentration can be optimized at this point. We observe variability from purification to purification.
12. Kinetics of protein–ligand interactions are temperature-dependent. Ideally ensure replicates are done at the same known temperature to reduce variability in results.
13. Note that calculation of the eIF2 concentration at each point must take into the dilution effects of repeated additions of eIF2 which increase the assay volume.
14. This is a single site–binding model similar to that found within GraphPad Prism (https://www.graphpad.com/) with an adjustment to go through 1.

Acknowledgments

This work was supported by grant BB/S014667/1 from the UK Biotechnology and Biological Sciences Research Council (BBSRC).

References

1. Dever TE, Kinzy TG, Pavitt GD (2016) Mechanism and regulation of protein synthesis in Saccharomyces cerevisiae. Genetics 203(1): 65–107. https://doi.org/10.1534/genetics.115.186221
2. Merrick WC, Pavitt GD (2018) Protein synthesis initiation in eukaryotic cells. Cold Spring Harb Perspect Biol 10:a033092. https://doi.org/10.1101/cshperspect.a033092
3. Algire MA, Maag D, Lorsch JR (2005) Pi release from eIF2, not GTP hydrolysis, is the step controlled by start-site selection during eukaryotic translation initiation. Mol Cell 20(2):251–262
4. Jennings MD, Pavitt GD (2010) eIF5 is a dual function GAP and GDI for eukaryotic translational control. Small GTPases 1(2):118–123. https://doi.org/10.4161/sgtp.1.2.13783
5. Jennings MD, Pavitt GD (2010) eIF5 has GDI activity necessary for translational control by eIF2 phosphorylation. Nature 465(7296):

378–381. https://doi.org/10.1038/nature09003

6. Pavitt GD (2018) Regulation of translation initiation factor eIF2B at the hub of the integrated stress response. Wiley Interdiscip Rev RNA 9(6):e1491. https://doi.org/10.1002/wrna.1491
7. Schreier MH, Erni B, Staehelin T (1977) Initiation of mammalian protein synthesis. I. Purification and characterization of seven initiation factors. J Mol Biol 116(4):727–753. https://doi.org/10.1016/0022-2836(77)90268-6
8. Safer B, Anderson WF, Merrick WC (1975) Purification and physical properties of homogeneous initiation factor MP from rabbit reticulocytes. J Biol Chem 250(23):9067–9075
9. Sidrauski C, Tsai JC, Kampmann M, Hearn BR, Vedantham P, Jaishankar P, Sokabe M, Mendez AS, Newton BW, Tang EL, Verschueren E, Johnson JR, Krogan NJ, Fraser CS, Weissman JS, Renslo AR, Walter P (2015) Pharmacological dimerization and activation of the exchange factor eIF2B antagonizes the integrated stress response. eLife 4:e07314. https://doi.org/10.7554/eLife.07314
10. Pavitt GD, Ramaiah KV, Kimball SR, Hinnebusch AG (1998) eIF2 independently binds two distinct eIF2B subcomplexes that catalyze and regulate guanine-nucleotide exchange. Genes Dev 12(4):514–526. https://doi.org/10.1101/gad.12.4.514
11. Asano K, Phan L, Krishnamoorthy T, Pavitt GD, Gomez E, Hannig EM, Nika J, Donahue TF, Huang HK, Hinnebusch AG (2002) Analysis and reconstitution of translation initiation in vitro. Methods Enzymol 351:221–247. https://doi.org/10.1016/s0076-6879(02)51850-4
12. Acker MG, Kolitz SE, Mitchell SF, Nanda JS, Lorsch JR (2007) Reconstitution of yeast translation initiation. Methods Enzymol 430:111–145. https://doi.org/10.1016/S0076-6879(07)30006-2
13. de Almeida RA, Fogli A, Gaillard M, Scheper GC, Boesflug-Tanguy O, Pavitt GD (2013) A yeast purification system for human translation initiation factors eIF2 and eIF2Bepsilon and their use in the diagnosis of CACH/VWM disease. PLoS One 8(1):e53958. https://doi.org/10.1371/journal.pone.0053958
14. Sekine Y, Zyryanova A, Crespillo-Casado A, Fischer PM, Harding HP, Ron D (2015) Stress responses. Mutations in a translation initiation factor identify the target of a memory-enhancing compound. Science 348(6238):1027–1030. https://doi.org/10.1126/science.aaa6986
15. Safer B, Adams SL, Anderson WF, Merrick WC (1975) Binding of MET-TRNAf and GTP to homogeneous initiation factor MP. J Biol Chem 250(23):9076–9082
16. Merrick WC (1979) Assays for eukaryotic protein synthesis. Methods Enzymol 60:108–123. https://doi.org/10.1016/s0076-6879(79)60011-3
17. Panniers R, Rowlands AG, Henshaw EC (1988) The effect of Mg2+ and guanine nucleotide exchange factor on the binding of guanine nucleotides to eukaryotic initiation factor 2. J Biol Chem 263:5519–5525
18. Kapp LD, Lorsch JR (2004) GTP-dependent recognition of the methionine moiety on initiator tRNA by translation factor eIF2. J Mol Biol 335(4):923–936
19. Jennings MD, Kershaw CJ, Adomavicius T, Pavitt GD (2017) Fail-safe control of translation initiation by dissociation of eIF2alpha phosphorylated ternary complexes. eLife 6:e24542. https://doi.org/10.7554/eLife.24542
20. Jennings MD, Kershaw CJ, White C, Hoyle D, Richardson JP, Costello JL, Donaldson IJ, Zhou Y, Pavitt GD (2016) eIF2beta is critical for eIF5-mediated GDP-dissociation inhibitor activity and translational control. Nucleic Acids Res 44(20):9698–9709. https://doi.org/10.1093/nar/gkw657
21. Adomavicius T, Guaita M, Zhou Y, Jennings MD, Latif Z, Roseman AM, Pavitt GD (2019) The structural basis of translational control by eIF2 phosphorylation. Nat Commun 10(1):2136. https://doi.org/10.1038/s41467-019-10167-3
22. Bogorad AM, Lin KY, Marintchev A (2018) eIF2B mechanisms of action and regulation: a thermodynamic view. Biochemistry 57(9):1426–1435. https://doi.org/10.1021/acs.biochem.7b00957
23. Khan MF, Spurgeon S, von der Haar T (2018) Origins of robustness in translational control via eukaryotic translation initiation factor (eIF) 2. J Theor Biol 445:92–102. https://doi.org/10.1016/j.jtbi.2018.02.020
24. Gite S, Mamaev S, Olejnik J, Rothschild K (2000) Ultrasensitive fluorescence-based detection of nascent proteins in gels. Anal Biochem 279(2):218–225. https://doi.org/10.1006/abio.1999.4472

Chapter 7

Mammalian In Vitro Translation Systems

Yulia Gonskikh, Valentina Pecoraro, and Norbert Polacek

Abstract

Under cellular stress, tight and coordinated regulation of the gene expression allows to minimize cellular damage, maintains cellular homeostasis, and ensures cell survival. Among stress-induced cellular responses, alteration of translation rates represents one of the most effective and rapid regulatory mechanisms available for cells. Here we report on detailed protocols of mammalian in vitro translation systems. While most of the available in vitro translation methods are based on bacterial or yeast components, tailor-made and robust mammalian systems are sparse. Our protocols allow measuring global translation of the total mRNA pool as well as translation of one specific reporter mRNA. Furthermore, it provides access to measuring translational activity of isolated ribosomes combined with non-ribosomal cytosolic fractions using reduced amounts of biological starting material. The herein described method can be applied to (1) investigate the effects of stress-dependent soluble factors regulating translation (such as tRNA fragments or ribosome-associated ncRNAs), (2) compare translational activity and translational fidelity of different ribosomes supplemented with the same non-ribosomal fractions, and (3) to investigate protein biosynthesis in various mammalian cell lines as well as tissue samples.

Key words Protein synthesis, Ribosomes, Translation control, Stress response, In vitro translation

1 Introduction

Numerous studies show that mRNA and protein levels have poor correlation especially in mammalian cells, thus emphasizing the importance of posttranscriptional regulation of gene expression [1]. This regulation becomes particularly crucial when cells encounter stress and thus need to adapt their proteome accordingly. Cellular stress response includes radical reprograming of protein synthesis. Regulation of translation, as the final step of gene expression, allows immediate response to physiological changes that is essential for the stress adaptation and the cell survival [2]. Protein synthesis is one of the most energy-consuming processes in the cell, and it consumes about ~75% of the cellular energy [3]. Activation of the stress-induced pathways typically causes phosphorylation of several translational factors resulting in

Daniel Mátějů and Jeffrey A. Chao (eds.), *The Integrated Stress Response: Methods and Protocols*,
Methods in Molecular Biology, vol. 2428, https://doi.org/10.1007/978-1-0716-1975-9_7,

inhibition of global translation [4]. Translational slow-down allows conserving energy and re-directs it toward the cellular needs. Stress-induced reduction of overall protein translation is accompanied by the selective translation of specific mRNAs involved in cellular stress-response. Besides canonical translational factors, trans-acting factors like ncRNA have been shown to contribute to translation regulation in all domains of life [5]. In the recent past, tRNA-derived fragments (tRFs) have been uncovered as additional regulatory molecules in translation control and beyond [6]. A subclass of tRFs has been shown to directly bind to and regulate the ribosome. These so-called ribosome-associated ncRNAs (rancRNAs) represent an emerging group of ribo-regulators of protein biosynthesis [5, 7], and they also seem to be at work in mammalian cells [8]. Even though translation regulation during stress is a widespread phenomenon and found in all cells and organisms investigated, its molecular mechanisms are far from being understood in molecular terms.

A molecular understanding of the mechanism of translation and translational control benefits from the functional cell-free translational system. It provides a fast, inexpensive tool to study protein biosynthesis under controllable in vitro conditions. Here, we present a robust protocol to address global translation in vitro using mammalian cell lines and tissues (Fig. 1). Moreover, we present a reconstituted in vitro translation system that allows minimizing the required cellular material and monitoring the translational competence of either isolated ribosomes themselves or the soluble translation factors separately (Fig. 2). Furthermore, by combining isolated crude mammalian ribosomes with commercially available cytosolic extracts from rabbit reticulocytes allows the in vitro translation of single mRNA reporters (Fig. 3). We think the combinatory nature of the presented in vitro translation protocols represents a valuable tool for scientists interested in assessing the functional competence of mammalian ribosomes or studying the roles of soluble cytosolic regulators in translational control.

2 Materials

All solutions should be prepared using ultrapure water. If not otherwise indicated all steps should be performed on ice.

2.1 Translation with the Crude Cell Lysate

1. 10× translation buffer: 300 mM HEPES/KOH (pH 7.6), 1.5 M KOAc, 39 mM $MgOAc_2$. Prepare 10 ml, store at 4 °C (*see* **Note 1**).
2. Wash buffer: 30 mM HEPES/KOH (pH 7.6), 150 mM KOAc, 3.9 mM $MgOAc_2$, 4 mM DTT, 1 mM PMSF (*see* **Note 1**).

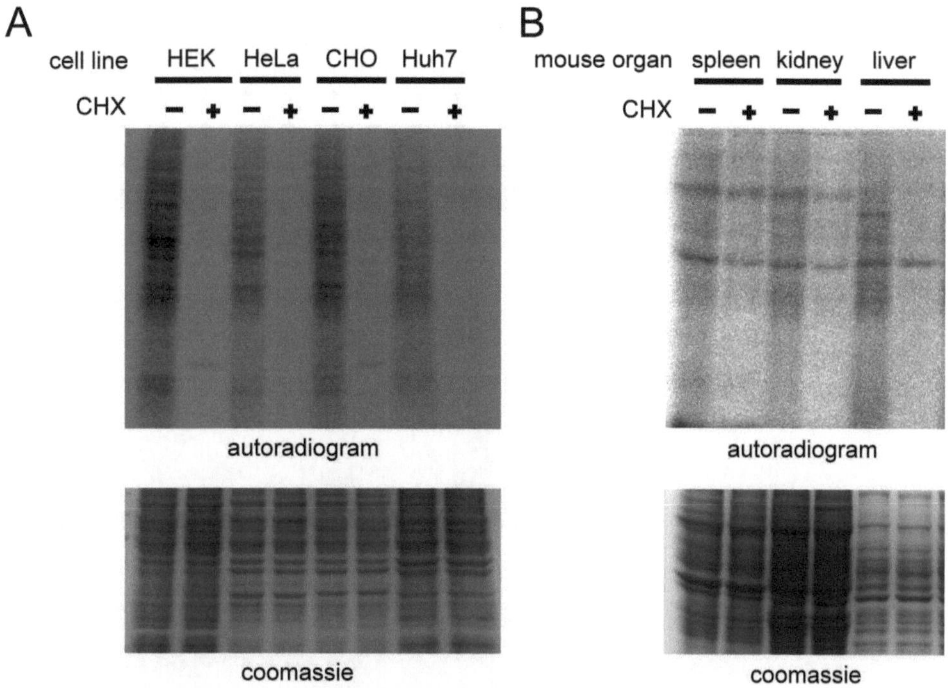

Fig. 1 In vitro translation with the crude cell lysates originating from mammalian cell lines (**a**) or mouse organs (**b**). Autoradiograms of dried SDS polyacrylamide gels represent ^{35}S-methionine-containing newly in vitro synthesized proteins. Cycloheximide (+CHX) at a final concentration of 7.5 mg/ml was used to inhibit in vitro protein synthesis and ensure that the radioactive signals obtained are indeed genuine in vitro translation products. The origins of immortalized mammalian cell lines (**a**) or mouse organs (**b**) are indicated. Coomassie-stained gels shown at the bottom visualize proteins of the cell lysate and serve as loading controls

3. Lysis buffer: 30 mM HEPES/KOH (pH 7.6), 150 mM KOAc, 3.9 mM $MgOAc_2$, 4 mM DTT, 1% Triton, protease inhibitor Tm complete (Roche), RNase inhibitor RNasin (Promega) (*see* **Note 1**). Prepare 1 ml aliquots, store it at −20 °C.
4. Pre-cooled minicentrifuge for cell debris removal.
5. Bradford assay reagent to measure protein concentration of the cell lysate.
6. Translational cocktail: 150 mM HEPES/KOH (pH 7.6), 750 mM KOAc, 19.5 mM $MgOAc_2$, 4 mM GTP, 17.5 mM ATP, 500 μM 19 amino acid mix except methionine (*see* **Note 1**).
7. 3 M creatine phosphate prepared in water, stored at −20 °C.
8. 20 mg/ml creatine phosphokinase prepared in water with 40% glycerol, stored at −20 °C.
9. ^{35}S-methionine 10 μCi/μl (Hartman Analytic).
10. 4× Laemmli buffer.

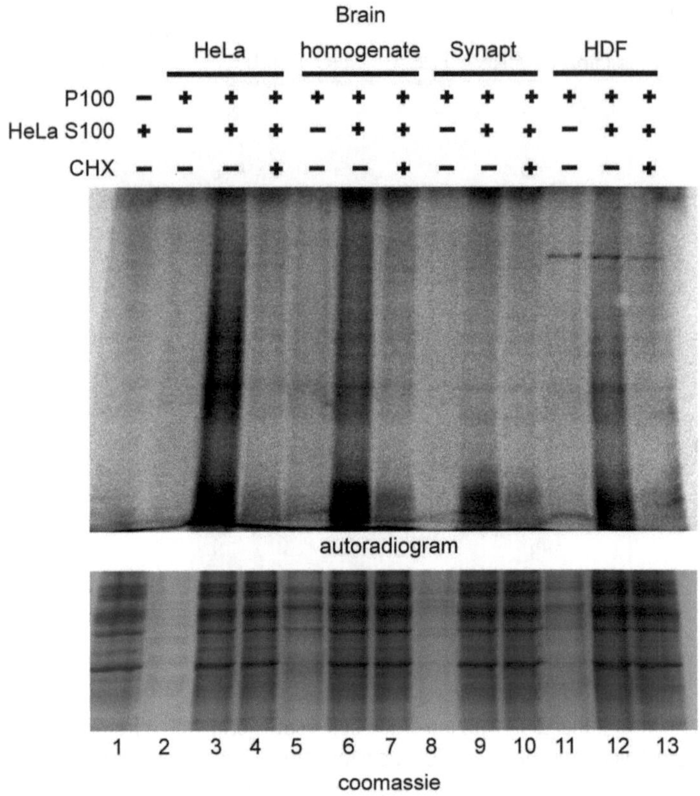

Fig. 2 Reconstituted in vitro translation system utilizing the endogenous mRNA pool. Crude ribosome pellets (P100) originating from HeLa cells, mouse brain homogenate, mouse synaptoneurosomes (Synapt), or from primary human cells Human Dermal Fibroblasts (HDF) were combined with the HeLa non-ribosome-containing cytosolic fraction (S100). Autoradiogram represents ^{35}S-methionine incorporated into newly synthesized proteins. Cycloheximide (CHX) at a final concentration of 4.5 mg/ml was used to inhibit in vitro protein synthesis (lanes 4, 7, 10, 13). Only upon combination of the P100 and S100 preparations, in vitro translation was apparent (lanes 3, 6, 9, 12). Reactions in the absence of P100 (lane 1) or S100 (lanes 2, 5, 8, 11) control for the purity of the preparations and show putative minimal background translational activities of the S100 or P100 preps alone. The Coomassie-stained gel shows proteins present in S100 and P100 preparations and serves as loading control

11. 10–12% SDS polyacrylamide gel and SDS-PAGE running buffer.
12. Destaining solution: 100 ml acetic acid, 400 ml methanol, water up to 1 l.
13. Coomassie staining solution: dissolve 8 g Coomassie blue in 1 l destaining solution.
14. Vacuum gel dryer.

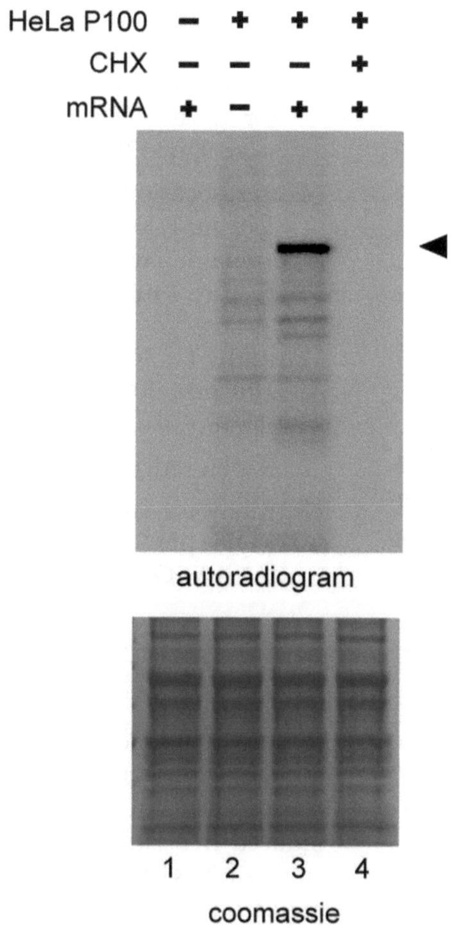

Fig. 3 Reconstituted in vitro translation system using a single reporter mRNA. Crude ribosome pellet (P100) originating from HeLa cells was combined with the non-ribosome-containing cytosolic fraction (S100) obtained from a commercial rabbit reticulocyte lysate and 100 ng of uncapped luciferase mRNA (lane 3). The 62 kDa reaction product is indicated by an arrow head. Smaller bands visible in the autoradiogram either represent products from minor amounts of endogenous mRNAs co-purified with the P100 preparation (lane 2) or products originating from luciferase mRNA fragments (lane 3). Cycloheximide (CHX) at a final concentration of 4.5 mg/ml was used to inhibit in vitro protein synthesis (lane 4). In the absence of P100, no in vitro translation reaction products are apparent (lane 1) thus controlling for the absence of active ribosomal particles in the S100 prep used. The Coomassie-stained gel shows proteins present in the S100 and P100 preparations and serves as loading control

2.2 Global In Vitro Translation with Isolated Ribosomes and Non-ribosome-Containing Cytosolic Fractions

1. Pre-cooled miniultracentrifuge with a fixed angle rotor S140AT (Beckman).
2. Sucrose solution: 1.1 M Sucrose, 30 mM HEPES/KOH (pH 7.6), 150 mM KOAc, 3.9 mM $MgOAc_2$, 4 mM DTT, 1 mM PMSF (*see* **Note 1**). Prepare 10 ml solution, store it at 4 °C for up to 2 days.

3. Ribosome storage buffer A: 20% glycerol, 30 mM HEPES/KOH (pH 7.6), 150 mM KOAc, 3.9 mM $MgOAc_2$, 4 mM DTT, protease inhibitor Tm complete (Roche), RNase inhibitor RNasin (Promega) (*see* **Note 1**). Prepare 1 ml aliquots, store it at −20 °C.
4. Resuspension buffer: 30 mM HEPES/KOH (pH 7.6), 150 mM KOAc, 3.9 mM $MgOAc_2$, 4 mM DTT, protease inhibitor Tm complete, RNase inhibitor RNasin (*see* **Note 1**). Prepare 1 ml aliquots, store it at −20 °C.

2.3 In Vitro Translation of One Specific Reporter mRNA with Isolated Ribosomes and Non-ribosome-Containing Fractions

1. Buffer C: 30 mM HEPES/KOH (pH 7.6), 2 mM $MgOAc_2$, 100 mM KOAc, 1 mM ATP, 0.2 mM GTP. Prepare 1 ml aliquots, store it at −20 °C.
2. Rabbit reticulocyte cell lysate (Promega).
3. Ribosome storage buffer C: 20% glycerol, 30 mM HEPES/KOH (pH 7.6), 2 mM $MgOAc_2$, 100 mM KOAc, 1 mM ATP, 0.2 mM GTP. Prepare 1 ml aliquots, store it at −20 °C.

3 Methods

3.1 In Vitro Translation with the Crude Cell Lysate

In vitro translation system with crude cell lysates is designed to monitor global translation using extracts of mammalian cells and organs (Fig. 1). This method allows to study effects of stress-induced translational factors and other soluble regulatory molecules, such as tRFs, on protein synthesis. The potential translational regulators of interest can be either depleted before cell harvesting (e.g., via RNAi or knock-out strategies) or can be added externally during the in vitro translation reaction. This method was successfully applied for studying tRFs regulating translation by directly binding to ribosomes (rancRNAs for ribosome-associated ncRNAs) in different immortalized mammalian cell lines (CHO, HeLa, HEK) [8, 9].

1. Mammalian cell lines: Harvest cells at 90–95% confluency. For a large-scale experiment, pellet cells from several 150 cm^2 plates. Alternatively, if the material is limited, pellet cells from one or two wells of a six-well plate.

   Mouse organs and tissues: after dissection rinse mouse organs with wash buffer twice.

   Optional: snap freeze cell pellet/organs and store at −80 °C.
2. Mammalian cell lines: Resuspend cells in lysis buffer. Use 400 μl of the lysis buffer for the cell pellet obtained from one 150-cm^2 plate, use 50 μl lysis buffer for the cell pellet obtained from one well of the six-well plate. To ensure sufficient cell opening pass the cell suspension 30 times through a 25G

needle. For the small-scale experiment, replace this step by intensive vortexing.

Mouse organs and tissues: Homogenize mouse organ in lysis buffer. Ideally add one volume lysis buffer to one volume of tissue/organ. In order not to dilute the sample too much, try to add as little lysis butter as possible.

3. Remove cell debris via centrifugation at 18,000 × *g* for 15 min at 4 °C in a pre-cooled centrifuge. For preparation of the mouse organ lysate, this step should be repeated several times until the lysate becomes clear.
4. Estimate protein concentration of the cell lysate using the Bradford assay. Dilute the lysate to 10–15 mg/ml with lysis buffer (*see* **Note 2**). Prepare small aliquots of the cell lysate to reduce the number of freezing–thawing cycles. Snap freeze aliquots and store at −80 °C.
5. To set up a translation reaction, pipette 6 μl of the cell lysate and add water up to 9.9 μl. To study the effect of the translational regulators, water can be replaced by the addition of the translational regulator (e.g., tRF, rancRNA) dissolved in water. As a control, translational inhibitors, such as harringtonine or cycloheximide, can be added at this step (*see* **Note 3**).
6. Pre-incubate the reaction at 37 °C for 10 min (*see* **Notes 4** and **5**).
7. During the pre-incubation time, prepare the translational master mix. For one reaction pipette 1.2 μl translation cocktail, 0.17 μl water, 0.08 μl 3 M Creatine phosphate, 0.06 μl 20 mg/ml creatine phosphokinase, 0.625 μl S^{35}-Methionine.
8. After the pre-incubation step, add 2.1 μl of the freshly prepared translational mix to each reaction.
9. Incubate reactions for 30 min at 37 °C (*see* **Note 6**).
10. Add 12 μl of 2× Laemmli buffer. Optional: store samples at −20 °C.
11. Boil samples for 5–10 min.
12. Run samples on a 10–12% SDS polyacrylamide gel.
13. Stain the gel for 10–15 min in the Coomassie staining solution. Coomassie staining visualizes mainly proteins originating from the cell lysate and serves as a loading control.
14. Distain the gel from a few hours to overnight.
15. Vacuum-dry the gel at 70 °C. For small (10 × 10 cm) gels, 45–60 min is sufficient.
16. Expose the dried gel to a phosphor imager screen from overnight up to several days.

3.2 Global In Vitro Translation with Isolated Ribosomes and the Non-ribosome-Containing Cytosolic Fraction

This method presents a reconstituted system consisting of the crude ribosomal pellet fraction after 100,000 × *g* centrifugation (P100) and the corresponding soluble fraction containing not-pelleted translation factors (S100) (Fig. 2). In this method, ribosomes (P100) isolated from mammalian cell lines, primary cells, or mouse organs are supplemented with the S100 fraction extracted from HEK, HeLa, or any other mammalian cell lines with an active translational status. This experimental setup is especially suited for studies focusing on the translational competence of the ribosome itself, e.g., after certain stress situations or for ribosomes isolated from different cell lines, or from tissues/organs at different developmental stages.

3.2.1 Isolation of P100 from Mammalian Cell Lines, Primary Cells, or Mouse Tissues

1. Prepare cell lysate as described in Subheading 3.1 (**steps 1–4**).
2. Pipette 0.5 ml of pre-cooled sucrose solution prepared in resuspension buffer in a pre-cooled 2 ml miniultracentrifuge tube.
3. Layer 200–500 μl cell lysate on top of the sucrose cushion.
4. Perform ultracentrifugation at 200,000 × *g* at 4 °C for 2 h in a fixed angel rotor S140AT.
5. Obtained pellet is P100. Resuspend pellet in ribosome storage buffer A.
6. Prepare 1:1000 or 1:100 dilution in water and measure OD_{260}. Estimate ribosome concentration using 1 OD_{260} unit equals to 18 pmol of crude 80S ribosomes.
7. Make 0.5 μM dilution with the ribosome storage buffer A, aliquot it, and snap freeze. Aliquots can be snap frozen and thawed several times.

3.2.2 Preparation of Concentrated S100

1. Prepare cell lysate from a HeLa or HEK cell pellet originating from two 150 cm^2 plates as described in Subheading 3.1 (**steps 1–4**).
2. Add the cell lysate to a miniultracentrifuge tube. Ensure that the volume of the lysate is large enough for the centrifugation step (usually should not be below 500 μl).
3. Perform ultracentrifugation at 200,000 × *g* at 4 °C for 2 h.
4. Collect the supernatant (S100). Aliquot it to avoid multiple freezing-thawing cycles. Snap freeze the aliquots and store it at −80 °C.

3.2.3 Setting up Translation Reaction

1. Pipette 0.5 pmol of P100 into a pre-cooled tube. As a control, translational inhibitors can be added at this step (*see* **Notes 3** and 7).
2. Add 1× resuspension buffer to bring the volume up to 10 μl.

3. Pre-incubate the reaction at 37 °C for 10 min (*see* **Notes 4** and **5**).
4. During the pre-incubation time prepare the translational master mix. For one reaction pipette 3.75 μl of concentrated S100 (prepared in Subheading 3.2.2), 1.2 μl translation cocktail, 0.08 μl of 3 M Creatine phosphate, 0.06 μl of 1 mg/ml Creatine phosphokinase, 0.625 μl of ^{35}S-Methionine, resuspension buffer up to 10 μl.
5. After the pre-incubation step, add 10 μl of freshly prepared master mix to each reaction.
6. Incubate the reaction at 37 °C for 1 h (*see* **Note 6**).
7. Add 7 μl of 4× Laemmli buffer. Optional: store samples at −20 °C.
8. Follow **steps 11–16** in Subheading 3.1.

3.3 Translation of One Specific mRNA Reporter with Isolated Ribosomes and the Non-ribosome-Containing Cytosolic Fraction

In this system, mammalian crude 80S ribosomes (P100) isolated from cell lines (Fig. 3), primary cells, or mouse organs are supplemented with the S100 obtained from the commercial Rabbit Reticulocyte Lysate (RRL) system. This reconstitution enables to perform in vitro translation of one mRNA reporter with crude ribosomes isolated from the cells of interest.

3.3.1 Isolation of P100 from Mammalian Cell Lines, Primary Cells or Mouse Tissues

1. Isolate ribosomes from the cells of interest following the steps in Subheading 3.2.1.
2. Resuspend P100 pellet in ribosome storage buffer C.
3. Estimate ribosome concentration based on OD_{260} using 1 OD_{260} unit equals to 18 pmol of crude 80S (P100).
4. Make a 0.2 μM dilution of P100 with the ribosome storage buffer C, aliquot, and snap freeze. It can be snap frozen and thawed several times.

3.3.2 Prepare RR S100

1. Combine two aliquots of the RRL. Prepare S100 fraction from RRL following the steps in Subheading 3.2.2.

3.3.3 Set up in Vitro Translation Reaction

1. Pipette 1 μl of 0.2 μM P100. As a control, translational inhibitors can be added at this step (*see* **Notes 3** and 7).
2. Prepare translational master mix. For one reaction, pipette 2.5 μl of RRL S100, 0.625 μl S^{35}-Methionine, RNase inhibitor RNasin (Promega), 7 μM Amino Acid Mixture Minus Methionine (Promega), 2 ng of capped mRNA or 100 ng of uncapped mRNA, buffer C up to 19 μl (*see* **Note 8**).
3. Add 19 μl of freshly prepared master mix to each reaction.
4. Incubate the reaction at 37 °C for 1 h (*see* **Note 6**).

5. Add 7 μl of 4× Laemmli buffer. Optional: store samples at −20 °C.
6. Follow **steps 11–16** in Subheading 3.1.

4 Notes

1. Prepare 10× translation buffer in 10 ml volume, store at 4 °C. Use it to prepare the wash buffer, lysis buffer, sucrose cushion buffer, and ribosome storage buffer A, resuspension buffer by making 1:10 dilution, and translational cocktail by making 1: 2 dilution.
2. The measured protein concentration of the lysate does not necessarily directly correspond to its translational activity. Even different batches of the lysate deriving from the same kind of cells can have different translational activities. Titration of a new batch of a cell lysate can be informative to address the translation status of the lysate.
3. To include translation inhibition control, use cycloheximide or harringtonine dissolved in ethanol. First, pipette 0.9 μl of 100 mg/ml cycloheximide or 3 μl 10 mg/ml of harringtonine to the empty reaction tube. Evaporate ethanol by running a SpeedVac vacuum concentrator for 10 min at room temperature. Pipette all remaining components to set up the translation reaction.
4. During the pre-incubation, (1) translational inhibitor and translational regulators pre-bind the ribosome; (2) ribosomes that have already initiated on endogenous mRNAs present in the P100 preparation will finish their elongation.
5. To avoid formation of water condensates on the lid of the tube use a 37 °C incubator instead of water bath or heat block.
6. Depending on the translational activity of the cell lysate or P100/S100, the incubation time recommended in the protocol (30 or 60 min) can already correspond to the end point of the reaction. If translational activities of different cell lysates are compared or if the effects of translational factors or other soluble translation regulators are under investigation, time-course experiments should be performed.
7. To visualize the background signal originating from possible contaminations of the P100 and S100 fractions, set up one control reactions without P100 and one control reaction without S100 (Figs. 2 and 3). Substitute P100 with the ribosome storage buffer, S100—with the resuspension buffer in Subheading 3.2 and Buffer C in Subheading 3.3.

8. During P100 preparation, ribosomes are pelleted together with a small fraction of endogenous mRNAs that are also translated during the subsequent in vitro translation reaction. To ensure that the detected translational product is specific to the exogenously added mRNA, set up control reaction without the mRNA reporter (Fig. 3).

Acknowledgments

We thank Steve Brown and Sarah Bernardez (Institute of Pharmacology and Toxicology, University of Zurich) for providing mouse brain and synaptoneurosome samples as well as Britta Engelhardt and Sara Barcos (Theodor-Kocher-Institute, University of Bern) for mouse organs. We are grateful to Iolanda Ferro and Fabian Nagelreiter for providing Huh7 and HDF cells, respectively. This work was supported by the D-A-CH grant 310030E-162559/1 and by the grant 310030-188969 both funded by the Swiss National Science Foundation.

References

1. Schwanhäusser B, Busse D, Li N, Dittmat G, Schuchardt J, Wolf J, Chen W, Selbach M (2011) Global quantification of mammalian gene expression control. Nature 473:337–342
2. Ed N, Ammerer G, Posas F (2011) Controlling gene expression in response to stress. Nat Rev Genet 12:833–845
3. Lane N, Martin W (2010) The energetics of genome complexity. Nature 467:929–934
4. Gonskikh Y, Polacek N (2017) Alterations of the translational apparatus during aging and stress response. Mech Ageing Dev 168:30–36
5. Barbosa CC, Calhoun SH, Wieder H-J (2020) Non-coding RNAs: what are we missing? Biochem Cell Biol 98:23–30
6. Polacek N, Ivanov P (2020) The regulatory world of tRNA fragments beyond canonical tRNA biology. RNA Biol 17:1057–1059
7. Pircher A, Gebetsberger J, Polacek N (2014) Ribosome-associated ncRNAs: an emerging class of translation regulators. RNA Biol 11: 1335–1339
8. GonskikhY GM, Kos M, Borth N, Schosserer M, Grillari J, Polacek N (2020) Modulation of mammalian translation by a ribosome-associated tRNA half. RNA Biol 30:1125–1136
9. Fricker R, Brogli R, Luidalepp H, Wyss L, Fasnacht M, Joss O, Zywicki M, Helm M, Schneider A, Cristodero M, Polacek N (2019) A tRNA half modulates translation as stress response in *Trypanosoma brucei*. Nat Commun 10:118

During P100 preparation, ribosomes are pelleted together with a small fraction of endogenous mRNAs that are also translated during the subsequent in vitro translation reactions. To ensure that the detected translational product is specific to the exogenously added mRNA, set up control reaction without the mRNA reporter (Fig. 3).

Acknowledgments

[illegible]

References

[illegible]

Chapter 8

Measuring Repeat-Associated Non-AUG (RAN) Translation

Shaopeng Wang and Shuying Sun

Abstract

Expansions of short nucleotide repeats account for more than 50 neurological or neuromuscular diseases. Many repeat expansion-containing RNAs can generate toxic repeat proteins through repeat-associated non-AUG (RAN) translation in all the reading frames. Understanding how RAN translation occurs and what cellular factors regulate this process will help decipher the basic mechanism of the molecular process and disease pathogenesis. Using reporter systems to quantitatively measure RAN translation provides a platform to examine candidate genes/pathways and screen for modifiers of this non-canonical pathway. In this chapter, we describe the dual-luciferase reporter system to measure RAN translation using *C9ORF72* GGGGCC^{exp} as an example, which is the most common genetic cause of amyotrophic lateral sclerosis (ALS) and frontotemporal dementia (FTD).

Key words Repeat expansion, RAN translation, Cap-independent, Integrated stress response, eIF2α phosphorylation

1 Introduction

Short tandem repeats (STRs), also known as microsatellites, are a set of short nucleotide sequences (usually 2–6 base pairs) repeated consecutively. Approximately 3% of the human genome consists of STRs, which may have regulatory roles in gene expression and function [1]. However, these loci are prone to DNA replication and repair errors. The abnormal expansions of STRs account for more than 50 neurological or neuromuscular diseases [2, 3]. Most of the disorders are dominantly inherited, but the disease mechanisms vary depending on the repeat sequences and gene context. For the repeats located in the protein-coding open reading frames (ORFs), the peptide repeat expansion and the disruption of the host protein function can cause toxicity related to proteostasis perturbation. The non-ORF located repeats could lead to either altered expression of the host gene or the RNA transcripts containing the repeat expansion can elicit toxicities by sequestering specific RNA-binding proteins (RBPs) and affect their functions.

Daniel Mátějů and Jeffrey A. Chao (eds.), *The Integrated Stress Response: Methods and Protocols*,
Methods in Molecular Biology, vol. 2428, https://doi.org/10.1007/978-1-0716-1975-9_8,

The discovery of repeat-associated non-AUG (RAN) translation adds another layer of complexity in pathogenic mechanisms. This non-canonical translation initiation process occurs in all possible reading frames independent of AUG start codon and produces multiple poly-peptide repeat proteins. Since it was first described in CAG^{exp} and CUG^{exp} associated spinocerebellar ataxia type 8 (SCA8) and myotonic dystrophy type 1 (DM1) [4], RAN translation has also been found in Fragile X-associated tremor/ataxia syndrome (CGG^{exp} and antisense CCG^{exp}) [5, 6], DM2 ($CCUG^{exp}$ and $CAGG^{exp}$) [7], Huntington's disease (CAG^{exp}) [8], spinocerebellar ataxia type 36 (SCA36) ($UGGGCC^{exp}$) [9], and C9ORF72-linked amyotrophic lateral sclerosis (ALS) and frontotemporal dementia (FTD) ($GGGGCC^{exp}$ and $CCCCGG^{exp}$) [10–12]. The repeat expansions subjected to RAN translation are located in a surprising variety of RNA contexts, including untranslated regions (UTRs), protein-coding ORFs, and even introns. The potential toxicities of the RAN-translated poly-peptides have been increasingly studied in various neurodegenerative diseases. Understanding the basic mechanism and regulation of the RAN translation pathway will likely provide insights on disease pathogenesis and identify modifiers to inhibit the abnormal accumulation of the poly-peptides as novel therapeutic strategy. We now describe a reporter system, which allows quantitative measurement of RAN translation, to decipher the mechanisms and test or screen candidate modifiers of this non-canonical molecular pathway.

The hexanucleotide GGGGCC repeat expansion in the intron 1 region of the *C9ORF72* gene is recognized as the most frequent genetic cause of both ALS and FTD [13–15]. The bidirectional transcription produces both sense GGGGCC and antisense CCCCGG repeat-containing RNAs [10–12]. RAN translation from all the six reading frames of both strands produces five different dipeptide repeat (DPR) proteins [10, 16–18]: poly-glycine-alanine (poly-GA) and poly-glycine-arginine (poly-GR) from sense repeats, poly-proline-alanine (poly-PA) and poly-proline-arginine (poly-PR) from antisense repeats, and poly-glycine-proline (poly-GP) generated from both strands (Fig. 1).

One unique feature of the $GGGGCC^{exp}$ is its location in the intron region. After transcription, pre-mRNA will be rapidly spliced and intron will be excised into a lariat structure, subsequently debranched and degraded [19]. The only two possible RAN translation templates, the unspliced pre-mRNA and spliced intron, are not expected to be exported to the cytoplasm where translation happens [20, 21]. Additionally, the spliced intron will not have the $5'$-m^7G cap that is required for canonical translation initiation [22]. It is important to take those unique features of the intronic $GGGGCC^{exp}$ into consideration when studying the RAN translation mechanisms.

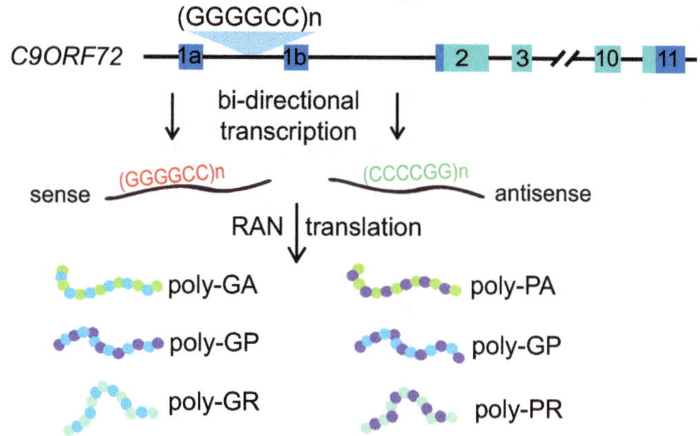

Fig. 1 Diagram of the *C9ORF72* gene and RAN translation of sense and antisense repeats. The GGGGCC^{exp} is located in the first intron. There are sense and antisense RNA generated through bidirectional transcription. Both sense GGGGCC^{exp}- and antisense CCCCGG^{exp}-containing RNAs undergo RAN translation, which produces five DPRs. We focus on the translation of sense repeats in this chapter

Integrated stress response (ISR) is a common adaptive pathway that is activated in eukaryotic cells upon diverse stress stimuli [23]. Upon ISR activation, the phosphorylation of the eukaryotic translation initiation factor 2 alpha (eIF2a) is upregulated which leads to global attenuation of cap-dependent translation [24]. However, the translation of selective cellular RNAs is resistant to the inhibition, or even upregulated by ISR, which helps restore cellular homeostasis under stress. Those transcripts usually have non-canonical translation initiation mechanism. As stress responses and stress granule alteration have been increasingly associated with adult-onset progressive neurodegenerative diseases [25], investigation of how RAN translation is modulated under ISR will advance our understanding of disease mechanisms and provide insights on the development of new therapeutic strategy.

From our previous study [26], we have demonstrated that RAN translation of GGGGCC^{exp} can happen without the 5′ Cap using various reporter systems. We identified that the GGGGCC^{exp}-containing spliced intron can be used as the template for RAN translation. The cap-independent RAN translation of GGGGCC repeats is upregulated by various stress stimuli through eIF2a phosphorylation. In this chapter, we describe the main steps involved in constructing luciferase-based reporter systems with various gene contexts to study RAN translation mechanisms and its response to stress stimuli. The protocol presented here pertains to the study of GGGGCC^{exp} in C9ORF72-ALS/FTD, but can be easily adapted to other repeat expansions.

2 Materials

2.1 Cloning of Plasmid Constructs

1. Standard thermal cycler.
2. Pfu DNA polymerase and buffer (Agilent Technologies).
3. Phusion high-fidelity DNA polymerase and buffer (Thermo Fisher).
4. PCR purification kit (Thermo Fisher).
5. Restriction enzymes and enzyme buffer (NEB).
6. Agarose.
7. Loading dye.
8. TAE buffer: 40 mM Tris, 20 mM acetic acid, 1 mM EDTA.
9. Agarose gel electrophoresis system.
10. Gel extraction kit (Thermo Fisher).
11. T4 DNA ligase and T4 DNA ligase buffer (NEB).
12. *E. coli* competent cells (top 10 for regular cloning and NEB stable for repeat-containing sequence cloning).
13. Water bath (42 °C).
14. LB medium: 10 g/L NaCl, 10 g/L tryptone, 5 g/L yeast extract.
15. LB agar plate: LB medium with 15 g/L bacto-agar.
16. 2× YT medium: 5 g/L NaCl, 16 g/L tryptone, 10 g/L yeast extract.
17. Antibiotics (1000× stock solutions): 100 mg/mL ampicillin.
18. 30 °C and 37 °C shaker and incubator.
19. Plasmid isolation and purification kits (Miniprep and Midiprep, Thermo Fisher).

2.2 Stable Cell Line Generation

1. Hela Flp-In cells and 293 phoenix cells.
2. Cell growth medium: DMEM supplied with 10% fetal bovine serum (FBS), antibiotics (100 U/mL penicillin and 100 μg/mL streptomycin).
3. Cell culture plate (10 cm dishes and six-well plates).
4. Trypsin 0.05%.
5. Phosphate-buffered saline (1× PBS).
6. Cell culture incubator (37 °C in humidified environment containing 5% CO_2).
7. Opti-MEM reduced-serum medium.
8. Transfection reagent TransIT-LT1 (Mirus Bio).
9. Retrovirus packing vectors: pCMV-VSV-G (Addgene 8454) and pCL-Eco (Addgene 12371) plasmid.

10. pOG44 Flp-recombinase expression vector (Thermo Fisher).
11. Hygromycin B (50 mg/mL).
12. Polybrene Infection/Transfection Reagent (Sigma).
13. Puromycin dihydrochloride (2 mg/mL).
14. Dynabeads Protein G (Thermo Fisher).
15. Magnetic stand.
16. PBST buffer: 137 mM NaCl, 2.7 mM KCl, 10 mM Na_2HPO_4, 1.8 mM KH_2PO_4, 0.02% Tween-20.
17. Doxycycline 1000× stock solution (2 mg/mL).
18. Cell lysis buffer: 0.3% (v/v) NP-40, 200 mM NaCl, 50 mM Tris (pH 7.4), 1 mM 1,4-Dithiothreitol (DTT)), 0.1 mM EDTA, and protease-inhibitor cocktail.
19. Syringe (1 mL) and needles (18 gauge, 22 gauge, and 26 gauge).
20. Syringe (50 mL) and 0.45-μm sterile syringe filter.
21. Laemmli sample buffer (Bio-Rad).
22. NuPAGE Bis-Tris protein gels and running buffer (Thermo Fisher).
23. Prestained Protein Ladders (Bio-Rad).
24. 0.2 μm Nitrocellulose membrane.
25. Trans-Blot Turbo Transfer System.
26. Transfer buffer: 200 mL 5× TransBlot Turbo transfer buffer, 200 mL ethanol, and 600 mL H_2O.
27. TBST buffer: 20 mM Tris, 150 mM NaCl, and 0.1% Tween 20 (pH 8).
28. Non-fat milk.
29. Bovine serum albumin (BSA).
30. Primary antibody: MYC antibody (Sigma, 05-724), poly-GA (Rb4334), poly-GP (Rb4336).
31. Secondary antibody: goat anti-mouse or anti-rabbit IgG horseradish peroxidase-conjugated antibody (GE Healthcare).
32. Chemiluminescent detection reagents (Thermo Scientific).
33. ChemiDoc Imaging Systems (Bio-Rad).

2.3 Luciferase Activity Assay

1. 5× passive lysis buffer (Promega).
2. Nano-Glo dual-luciferase reporter assay system (Promega).
3. 96-well white microplate.
4. 96-well transparent microplate.
5. BCA protein assay kit (Pierce).
6. Plate reader (TECAN, Infinite 200 PRO).

2.4 Assessing the Cap-Independent RAN Translation

1. Lipofectamine RNAiMAX (Invitrogen).
2. siRNAs: siGENOME eIF4E siRNA and siGENOME Non-Targeting siRNA (GE Dharmacon).
3. Primary antibody: eIF4E (Bethyl, A301-154A), GAPDH (Cell Signaling, 97166).
4. Secondary antibody: goat anti-mouse or anti-rabbit IgG horseradish peroxidase-conjugated antibody (GE Healthcare).

2.5 Investigating the RNA Template of RAN Translation

1. Purification lysis buffer: 20 mM HEPES KOH (pH 7.4), 10 mM KCl, 3 mM $MgCl_2$, 0.3%(v/v) NP-40, 0.1 mM EDTA, 1 mM DTT, 100 μg/mL cycloheximide, protease-inhibitor cocktail, and 200 U/mL RNase inhibitor.
2. Dynabeads Protein G (Thermo Fisher).
3. GFP antibody (Memorial Sloan Kettering Cancer Center Monoclonal Antibody Core Facility, Htz-GFP19C8).
4. High-salt polysome wash buffer: 20 mM HEPES (pH 7.4), 350 mM KCl, 5 mM $MgCl_2$, 1 mM DTT, 1% NP-40, and 100 μg/mL cycloheximide.
5. Trizol reagent.
6. Chloroform.
7. Isopropanol.
8. Ethanol.
9. Nuclease-free water.
10. RQ1 DNase I and buffer (Promega).
11. DNase I stop buffer (Promega).
12. High-capacity cDNA reverse transcription kit (Applied Biosystems).
13. IQ SYBR green supermix (Bio-Rad).
14. 96-Well PCR plates.
15. CFX96 real-time PCR detection system (Bio-Rad).
16. Primers.

2.6 Modulation of RAN Translation by Integrated Stress Response

1. Sodium arsenite.
2. MG132.
3. Primary antibody: G3BP (BD, 611126), phospho-eIF2α (Cell Signaling, 9721), eIF2α (Cell Signaling, 9722), GAPDH (Cell Signaling, 97166).
4. Secondary antibody: Alexa Fluor 546-conjugated secondary antibodies (Thermo Fisher), goat anti-mouse or anti-rabbit IgG horseradish peroxidase-conjugated antibody (GE Healthcare).

5. 4% paraformaldehyde: 4 g PFA dissolved in 100 mL PBS.
6. Triton X-100.
7. Bovine serum albumin.
8. Goat serum.
9. 4′,6-Diamidino-2-phenylindole (DAPI).
10. Mount solution (Thermo Fisher).
11. Fluorescence microscope (Zeiss Axiophot).
12. ISRIB (Sigma).
13. GSK260641 (PERKi, Sigma).

3 Methods

To monitor RAN translation in a timely manner in vivo, we build stable cell lines with tetracycline-inducible dual-luciferase reporters. Each reporter is cloned in the pcDNA5-FRT-TO vector and engineered in a single genomic locus in HeLa Flp-In cells using a site-directed recombinase (Flip) [27]. We use Nanoluc Luciferase (NLuc) fused with GGGGCC^{exp} to monitor RAN translation and Firefly Luciferase (FLuc) with AUG start codon to represent canonical translation as internal control. We show here the design of three translation reporter systems. In the monocistronic reporter, NLuc and FLuc are in separate transcripts, both of which will have the m^{7}G cap and poly(A) tail (Fig. 2a). The bicistronic reporter contains two cistrons in one transcript. The translation of RAN-NLuc requires translation initiation in the middle of the transcript independent of the 5′ cap (Fig. 2b). The bicistronic splicing reporter includes exons and introns of the endogenous gene context (Fig. 2c), which helps assess the RAN translation of repeat expansions located in the intronic region.

3.1 Construct Design and Cloning

3.1.1 Construct Design Strategy

1. *Monocistronic reporter:* Clone the RAN-NLuc reporter into the pcDNA5-FRT-TO vector for inducible expression and clone the AUG-FLuc reporter into the pBABE (retrovirus vector) for constitutive expression in the cells. The RAN-NLuc reporter is designed with three principles taken into consideration: (1) prevent any leakage from AUG-translation; (2) keep the endogenous 5′ end context of the C9ORF72 repeats which might influence the RAN translation initiation; (3) include a C-terminal tag to validate the RAN translation product.

   First, introduce multiple stop codons in all three reading frames after the CMV promoter and before the multiple cloning sites of the pcDNA5-FRT-TO vector by QuickChange site-directed mutagenesis. Next, clone the repeats into the modified pcDNA5-FRT-TO vector by HindIII and NotI. At last, PCR

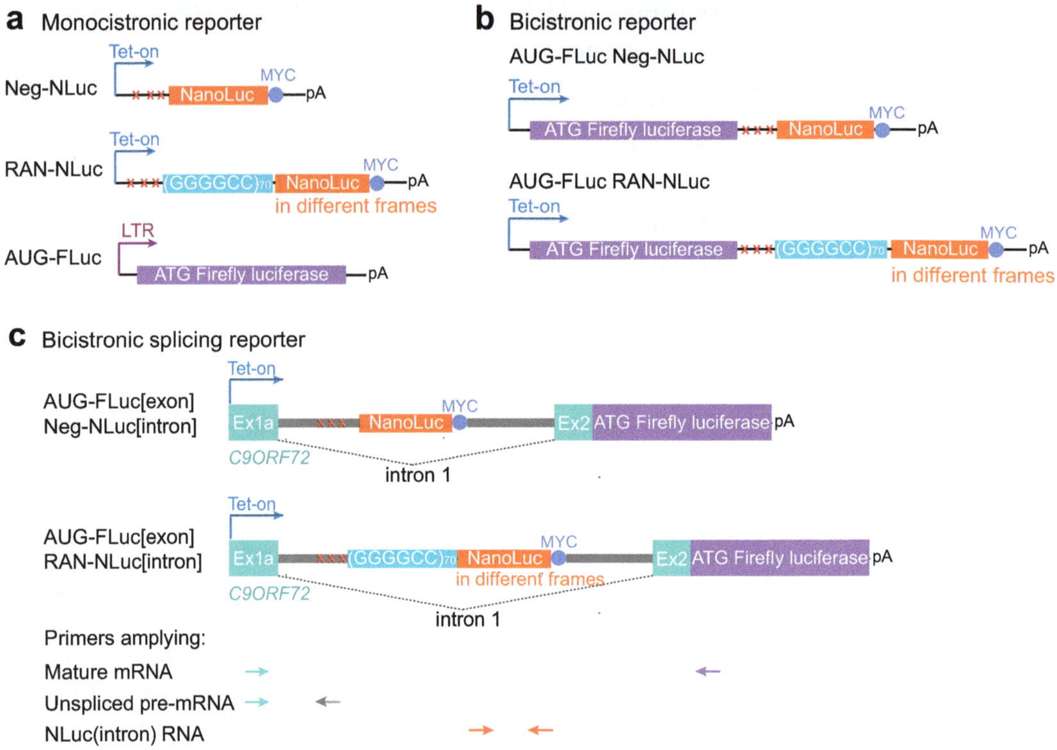

Fig. 2 Schematic of dual-luciferase reporters for RAN translation. We use NLuc to monitor RAN translation and FLuc to represent canonical translation. (**a**) Monocistronic reporter: The RAN-NLuc transgene is cloned in the pcDNA5-FRT-TO vector under tetracycline-inducible promoter. Around 70× GGGGCC repeats are fused with the NLuc coding sequence lacking the AUG start codon. A nucleotide shift before the NLuc is introduced to tag RAN translation products in different reading frames. Stop codons in all reading frames are included before the repeats to avoid the canonical translation. The same construct lacking the GGGGCC repeats serves as the negative control. The AUG-FLuc transgene is cloned in the pBABE retrovirus vector for constitutive expression as internal control. (**b**) Bicistronic reporter: Both NLuc and FLuc are in one transcript under the tetracycline-inducible promoter. The AUG-FLuc is at the 5′ end with canonical translation. The RAN-NLuc cassette (from the monocistronic construct) is placed downstream of FLuc after three stop codons. (**c**) Bicistronic splicing reporter: The transgene contains *C9ORF72* exon 1a, exon 2, and ~200 bp intron sequences adjacent to the splice sites. RAN-NLuc is located in intron 1 and AUG-FLuc is in exon 2 fused in frame with the original *C9ORF72* start codon. The positions of primers amplifying different RNA species are indicated below the diagram

amplifies the NLuc-MYC without AUG and clone after the repeats by NotI and XhoI (*see* **Note 1**). The fusion with different reading frames can be achieved by inserting one nucleotide shift during PCR amplification of the NLuc gene. AUG-FLuc coding sequence is cloned into pBABE (retrovirus vector) via BamHI and SalI.

2. *Bicistronic reporter:* The bicistronic reporter is designed to test mechanism of the cap-independent RAN translation. In this construct, the first cistron (AUG-FLuc) is translated from the

5′ end by canonical cap- and AUG-dependent initiation and is terminated by stop codons placed after FLuc in all the reading frames. This is followed by the second cistron encoding RAN-NLuc, which is translated only if the repeats can recruit ribosomes by a cap-independent mechanism (*see* **Note 2**).

First, PCR amplifies the FLuc coding sequence containing multiple stop codons at the 3′ end in three reading frames and clone into the pcDNA5-FRT-TO vector via KpnI and EcoRV. Second, cut the RAN-NLuc out from the monocistronic reporter by HindIII and XhoI, blunt the HindIII end, and clone after FLuc via EcoRV and XhoI sites.

3. *Bicistronic splicing reporter*: The GGGGCC repeat expansion is located in the first intron of the *C9ORF72* gene. The bicistronic splicing reporter is designed to mimic this endogenous context and dissect the molecular mechanism of intronic repeat translation. The splicing reporter contains the exon 1a and exon 2 of *C9ORF72*, as well as the intron sequences adjacent to the 5′ and 3′ splice sites. The RAN-NLuc is inserted in the middle of the intron and the AUG-FLuc is placed in exon 2 in frame with the C9ORF72 AUG start codon to monitor the total transgene expression level.

   First, synthesize the sequences including *C9ORF72* exon 1a, exon 2, and around 200 nt intron 1 sequences from each exon–intron junction and clone into the pcDNA5-FRT-TO vector via NheI and XhoI sites. Include HindIII and BamHI sites close to the repeat expansion location in the synthesized sequence. Second, cut the RAN-NLuc out from the monocistronic reporter by HindIII and BamHI and clone into the intron region. At last, clone the AUG-FLuc coding sequence after exon2 via XhoI and PmeI.

3.1.2 Molecular Cloning

1. For site-directed mutagenesis, set up 50 μL PCR using Pfu polymerase with 20 ng template DNA. After 18–20 cycles in a standard thermal cycler, add 1 μL DpnI to digest the template DNA and incubate at 37 °C overnight. Proceed to transformation.
2. Use a high-fidelity polymerase to PCR amplify the fragment for molecular cloning. Purify the product by PCR purification kit or gel extraction kit.
3. Perform restriction enzyme digestion of PCR product and plasmid vector at appropriate condition according to the manufacturer's instruction of each enzyme. Conduct agarose gel electrophoresis and purify the digested products by gel extraction kit.
4. Ligate the purified insert and the vector using T4 DNA ligase in a 10 μL reaction.

5. Transform competent *E. coli* cells (*see* **Note 3**) with the ligation mix or mutagenesis mix and plate on petri dishes with antibiotic selection (100 μg/mL ampicillin for pcDNA5-FRT-TO and pBABE). Incubate the plates at 37 °C overnight (*see* **Note 4**).
6. Pick 4–6 colonies and propagate in 15-mL falcon tubes containing 4 mL liquid LB medium with 100 μg/mL ampicillin at 37 °C with shaking overnight (*see* **Note 5**).
7. Harvest bacteria by centrifugation and extract plasmid DNA using a miniprep kit according to the manufacturer's instruction.
8. Verify the presence and correct size of the insert by restriction enzyme digestion followed by agarose gel electrophoresis, and further validate by Sanger sequencing.

3.2 Stable Cell Line Generation

3.2.1 Making Stable Lines by Flp-In

1. Day 1, seed healthy and logarithmically dividing Hela Flp-In cells in six-well plate at a density of 2×10^5 cells/well in 2.5 mL growth medium and allow them to grow overnight at 37 °C in humidified environment containing 5% CO_2.
2. Day 2, add 2.25 μg pOG44 plasmid and 0.25 μg Flp-In constructs into 250 μL Opti-MEM medium, mix gently by pipetting (*see* **Note 6**).
3. Add 7.5 μL TransIT-LT1 reagent into the mixture from **step 2** and mix gently by pipetting (*see* **Note** 7). Incubate at room temperature for 15 min.
4. Add the mixture from **step 3** dropwise to the seeded cells from **step 1** and shake the plate gently.
5. Twenty four hours after transfection, split the cells to a 10-cm dish.
6. Start selection by adding hygromycin B with a final concentration of 200 μg/mL (*see* **Note 8**). Change with fresh medium containing hygromycin B every 3 days until there are visible colonies (*see* **Note 9**).
7. Pool all the colonies together and expand the culture for future use.

3.2.2 Making Stable Lines by Retrovirus Transduction

Stably express AUG-FLuc in monocistronic reporter cells and EGFP-RPL10a in bicistronic splicing reporter cells via retrovirus transduction.

1. Day 1, trypsinize healthy and logarithmically dividing 293 phoenix cells and plate 3–5 $\times 10^6$ cells in a 10-cm plate in 10 mL medium.
2. Day 2, change the cell culture medium to 5 mL pre-warmed Opti-MEM 1 h before transfection.

3. Add 6 μg pBABE, 2 μg pCMV-VSV-G, and 2 μg pCL-Eco plasmids into 1 mL Opti-MEM medium, mix gently by pipetting.
4. Add 28 μL TransIT-LT1 reagent into the mixture from **step 3** and mix gently by pipetting. Incubate at room temperature for 15 min.
5. Add the mixture from **step 4** to the plate of **step 2** dropwise and shake the plate gently.
6. After 6 h, change to 10 mL regular DMEM growth medium (*see* **Note 10**).
7. Day 3 evening, change to 6 mL fresh growth medium.
8. Days 4–6, collect the virus-containing medium twice a day, and replace with 6 mL fresh growth medium. Store the virus at 4 °C for short term.
9. Day 6, pass the collected virus-containing medium through a 0.45-μm filter to remove cell debris. Aliquot and store at −80 °C.
10. Day 5, seed HeLa Flp-In cells to be infected at appropriate density (40–50% confluency) in a 10-cm dish.
11. Day 6, remove the cell growth medium, mix 10 mL virus-containing medium from **step 9** with polybrene (final concentration 8 μg/mL) and add to the cells.
12. Day 7, change to fresh growth medium.
13. Day 8, split cells. After the cells settle and attach, add puromycin at 1 μg/mL for selection (*see* **Note 8**).

3.2.3 Confirmation of RAN Translation Products

To validate the RAN translation products, perform the immunoprecipitation experiment with MYC antibody and western blotting with the corresponding DPR antibody.

1. Vortex the Protein G Dynabeads briefly. Take out 0.3 mg beads (*see* **Note 11**) and put into a 1.5-mL Eppendorf (EP) tube. Place on the magnetic stand for 2 min. Remove the supernatant.
2. Wash once with 500 μL PBST buffer. Resuspend the beads in 40 μL PBST.
3. Dilute 1 μg MYC antibody in the 40 μL beads. Rotate for 1 h at room temperature.
4. Wash two times with 500 μL PBST buffer.
5. Place the 10 cm cell culture dish (90% confluence of cells with 24-h doxycycline induction) on ice. Remove medium and wash the cells with ice-cold PBS twice. Remove PBS completely.
6. Add 400 μL cell lysis buffer directly to the 10-cm plate and scrape the adherent cells off the dish using a cold plastic cell

scraper. Gently transfer the cell suspension into a pre-cooled EP tube.

7. Homogenize cells using syringe and needle. Pass cells through 18 gauge needle three times, 22 gauge needle three times, and 26 gauge needle three times. Incubate the cell lysate on ice for 5 min.
8. Centrifuge the lysate at 15,000 × *g* for 20 min at 4 °C.
9. Transfer the supernatant to the beads from **step 4**. Rotate in the cold room overnight (*see* **Note 12**).
10. Put the tube on the magnetic stand for 5 min and remove the supernatant. Wash the beads with 1 mL cell lysis buffer for five times.
11. Resuspend the beads in 30 μL Laemmli sample buffer, heat at 95 °C for 5 min. Cool the tube down, place back on the magnetic stand and transfer the sample to a new EP tube. Directly proceed for western blot or save at −20 °C for later use.
12. Run the protein samples on the NuPAGE Bis-Tris protein gel. Perform protein transfer to nitrocellulose membrane using the Trans-Blot Turbo Transfer System.
13. Rinse the membrane in TBST buffer briefly and block with 10 mL TBST contains 5% non-fat milk at room temperature for 1 h on a slow shaker. Rinse once with 20 mL TBST.
14. Dilute the primary antibody in 10 mL TBST with 5% non-fat milk or 1% BSA and incubate with the membrane at room temperature for 1–3 h or at 4 °C overnight on a slow shaker.
15. Wash the membrane with 20 mL TBST for 10 min with gentle shaking. Repeat this step three times.
16. Dilute 1 μL secondary antibody in 10 mL TBST buffer with 5% non-fat milk, incubate with the membrane at room temperature for 1 h.
17. Wash the membrane with 20 mL TBST for 10 min. Repeat this step three times.
18. Add the chemiluminescent detection reagents and develop the membrane on the ChemiDoc imaging systems.

3.3 Luciferase Assay to Measure RAN Translation

Measure NLuc activity and normalize to FLuc activity or total protein concentration to compare RAN translation levels among different reporters and conditions.

1. Seed different reporter cells in 24-well plate (triplicate) at appropriate density (40–50%) and allow the cells to grow overnight.
2. Induce the reporter gene expression by adding Doxycycline with a final concentration of 2 μg/mL to the medium. Return the cells to the CO_2 incubator and culture for another 24 h.

3. Wash the cells with 1× PBS three times. Remove PBS as much as possible after the last wash.
4. Add 70 μL 1× passive lysis buffer to each well and shake the plate at room temperature for 5 min.
5. Resuspend the cell lysate and transfer 60 μL to 96-well white microplate.
6. Add 60 μL One-Glo™ EX luciferase assay reagent to each well, incubate at room temperature with shaking for 3 min (avoid light) and measure the firefly luciferase activity by Tecan plate reader (*see* **Note 13**).
7. Add 60 μL NanoDLR™ Stop & Glo reagent to each well of the plate, incubate at room temperature with shacking for 10 min (avoid light) and measure the NanoLuc luciferase activity by Tecan plate reader.
8. Add 10 μL cell lysate from **step 4** into a 96-well transparent plate.
9. Add 10 μL albumin (BSA) standards (0, 25, 125, 250, 500, 750, 1000, 1500, 2000 μg/mL) in a new row.
10. Add 200 μL BCA working reagent (reagent A:reagent B = 50: 1) to each well. Incubate the plate at 37 °C for 30 min. Measure the absorbance of all the samples at 562 nm by Tecan plate reader.
11. Determine the concentration of each sample based on the BSA standard curve.
12. Compare RAN translation activity under different conditions by normalizing RAN-NLuc to (a) AUG-FLuc or (b) total protein concentration as internal control, using the following equations for each sample. Present the data as mean ± standard deviations from at least three biological replicates. Calculate the statistical significance between groups by two-tailed Student's *t* test.
    (a) Relative RAN translation activity = the activity of NLuc/ the activity of FLuc.
    (b) Relative RAN translation activity = the activity of NLuc/ protein concentration.

3.4 Assessing the Cap-Independent RAN Translation

Knock down the eukaryotic translation initiation factor 4E (eIF4E) by siRNA transfection and determine whether it influences RAN translation levels. eIF4E facilitates initiation complex assembly and mRNA scanning [28]. If the RAN-NLuc is not affected, it indicates RAN translation can initiate independent of the 5′-cap.

1. Day 1, seed cells in a 24-well plate (triplicate) and allow the cells to grow overnight.
2. Day 2, change the medium to 210 μL pre-warmed Opti-MEM about 1 h before transfection.

3. For one well, dilute 0.9 μL Lipofectamine RNAiMAX in 15 μL Opti-MEM and add 0.6 μL 10 μM siRNA in another 15 μL Opti-MEM (final siRNA concentration is 25 nM). Combine together and mix gently by pipetting up and down a few times. Incubate at room temperature for 15 min.
4. Add the mixture from **step 3** into each well dropwise and return the cells to the incubator. Six hours after transfection, change to regular growth medium.
5. Day 4, induce reporter gene expression 48 h after transfection by adding Doxycycline with a final concentration of 2 μg/mL to the medium.
6. Day 5, analyze luciferase activity 24 h after induction as described in Subheading 3.3 or protein expression by Western blot as described in Subheading 3.2.3.

3.5 Investigating the RNA Template of RAN Translation

When the repeat expansion is located in the intron, both the unspliced pre-mRNA and the spliced intron RNA contain the repeats. To examine which RNA species could be the template for RAN translation, we quantify the relative enrichment of different RNA species (Fig. 2c) associated with poly-ribosomes isolated by the translating ribosome affinity purification (TRAP) method [29, 30].

3.5.1 Translating Ribosome Affinity Purification

1. Seed the bicistronic splicing reporter cells stably expressing EGFP-RPL10a in a 10-cm dish and let the cells grow overnight.
2. Induce the reporter gene expression by adding Doxycycline with a final concentration of 2 μg/mL to the medium.
3. After 24 h induction, wash the cells with PBS twice. Add 500 μL ice-cold PBS and scrape the cells off the plate. Transfer the cell suspension into a pre-cooled EP tube. Centrifuge at 200 × *g* for 5 min at 4 °C, remove the supernatant.
4. Add 400 μL purification lysis buffer to each tube. Disperse the pellet by pipetting up and down gently. Incubate the lysate on ice for 5 min.
5. Centrifuge the lysate at 2300 × *g* for 5 min at 4 °C. Carefully transfer the supernatant (cytoplasmic fraction) without disturbing the pellet to a new 1.5-mL EP tube. Adjust the KCl concentration to 150 mM.
6. Centrifuge the lysate from **step 5** at 13,000 × *g* for 20 min at 4 °C.
7. Coat Protein G Dynabeads with GFP antibody as described in Subheading 3.2.3.
8. Incubate the supernatant from **step 6** with the GFP antibody-coated beads from **step 7** at 4 °C overnight with rotation.

9. Wash the beads five times with high-salt polysome wash buffer. Remove the buffer and directly add 1 mL Trizol to the beads for RNA extraction.

3.5.2 RNA Extraction

1. Add 200 μL chloroform to each tube, mix vigorously, and let it sit at room temperature for 15 min.
2. Centrifuge at 12,000 × *g* for 15 min at 4 °C. Carefully transfer the upper aqueous phase containing RNA into a new 1.5-mL RNase-free EP tube.
3. Precipitate the RNA by adding 1 volume of isopropanol to each tube, mix by invert the tube several times, put at −20 °C for 15 min.
4. Centrifuge at 12,000 × *g* for 15 min at 4 °C. Discard the supernatant without disturbing the RNA pellet. Wash the RNA pellet with 1 mL ice-cold 75% ethanol.
5. Centrifuge at 12,000 × *g* for 5 min at 4 °C. Discard the supernatant and dry the RNA at room temperature.
6. Dissolve the RNA in nuclease-free water. Put the RNA in −80 °C for long-term storage.

3.5.3 First Strand cDNA Synthesis

1. Treat 1 μg RNA by RQ1 DNase I in 10 μL reaction at 37 °C for 30 min.
2. Inactive DNase I by adding 1 μL DNase I stop buffer and incubate at 65 °C for 10 min.
3. Add 10 μL reverse transcription reaction mix (1 μL reverse transcriptase, 1 μL RNase inhibitor, 2 μL random hexamer, 2 μL 10× buffer, 0.8 μL 100 mM dNTP, 3.2 μL nuclease-free water) to each reaction from **step 2**. Incubate in a standard thermal cycler at 25 °C for 10 min followed by 37 °C for 2 h, and a final step at 85 °C for 10 min.

3.5.4 Real-Time PCR Quantification

1. Dilute the cDNA ten times with nuclease-free water and add 8 μL into each well of the qPCR plate.
2. Add 12 μL of qPCR master mix (10 μL SYBR green supermix, 0.3 μL of each forward and reverse primer, 1.4 μL water) to each well. Seal the plate with transparent film. Vortex to mix reaction components thoroughly. Spin the plate briefly to collect the reaction mixture to the bottom.
3. Run the reaction on the real-time PCR thermocycler. The cycling parameters are: 50 °C for 2 min, 95 °C for 3 min, 40 cycles of 95 °C for 10 s and 57 °C for 30 s, and finally 55–95 °C with 0.5 °C/5 s for melt curve. All reactions should be performed with at least two technical replicates and three biological replicates with internal standards for normalization (*see* **Note 14**).

4. Calculate the polysome-associated RNA values as the $2^{-\Delta\Delta Ct}$ relative to the values of input control samples and compare the relative enrichment of different RNA species. Present data as mean ± standard deviations from three biological replicates with statistical analysis using the two-tailed Student's *t* test.

3.6 Modulation of RAN Translation by Integrated Stress Response

The integrated stress response (ISR) is a conserved cell signaling network that helps to restore cellular homeostasis in response to variable environmental and pathological conditions [31]. The core event of this pathway is the phosphorylation of eIF2α, which can lead to global protein synthesis reduction, paradoxically coupled with elevated translation of a subset of cellular RNAs. This includes mRNAs containing short upstream open reading frames (uORFs) in the 5′ UTRs with translation reinitiation at downstream coding sequences, and IRES-containing mRNAs with cap-independent translation [23]. As stress factors play important roles in neurodegenerative diseases, it is important to examine how ISR modulates the non-canonical RAN translation.

3.6.1 Measuring RAN Translation Under Stress Stimuli

The ISR can be activated by multiple stimuli. Two examples are shown here. We examine the ISR activation by measuring the eIF2α phosphorylation and stress granule formation. We induce the expression of the RAN-NLuc reporter upon stress in order to measure the translational changes in response to stress stimuli excluding the influence from preexisting RAN proteins.

1. Seed cells in a 24-well plate (triplicate) and allow the cells to grow overnight.
2. For stress stimuli, treat the cells with sodium arsenite at 200 μM or MG132 at 10 μM. Treat cells with the same volume of DMSO as negative control. At the same time, induce the reporter gene expression by adding 2 μg/mL Doxycycline to the medium.
3. After 6 h, collect the cells for luciferase assay and protein quantification as described in Subheading 3.3.
4. Measure eIF2α and phospho-eIF2α protein levels by Western blot as described in Subheading 3.2.3.

3.6.2 Immunofluorescence of Stress Granules

1. Seed an aliquot of cells from **step 1** in Subheading 3.6.1 on coverslips in a 24-well plate and continue with the same treatment.
2. When harvesting, wash the cells with 1× PBS and fix the cells with 500 μL 4% paraformaldehyde in PBS at room temperature for 15 min. Wash with PBS three times, 5 min each.
3. Permeabilize the cells with 0.2% Triton X-100 in PBS at room temperature for 10 min. Wash with PBS three times, 5 min each.

4. Adding 500 μL blocking solution (1% BSA, 2% goat serum in PBS) to each well and incubate for 30 min.
5. Incubate the cells with G3BP antibody diluted in blocking solution (1:300) at room temperature for 1 h. Wash with 500 μL PBS three times, 10 min each.
6. Incubate the cells with Alexa Fluor 546-conjugated secondary antibody diluted in blocking solution (1:1000) at room temperature for 1 h. Wash with 500 μL PBS three times, 10 min each. In the second wash, use PBS containing 1 μg/mL DAPI for nuclear staining.
7. Mount the coverslips onto slides with mounting solution and dry overnight at room temperature.
8. Image the slides by fluorescence microscope.

3.6.3 Inhibition of the Integrated Stress Response Pathway

Treat cells with small molecule inhibitors of the eIF2α pathway. ISRIB blocks downstream signaling of phospho-eIF2α without changing the level of eIF2α phosphorylation [32]. GSK2606414 (PERKi) inhibits PRKR-like ER kinase (PERK) [33], one of the kinases activated by unfolded protein response and phosphorylating eIF2α [34].

1. Seed cells in a 24-well plate (triplicate) and allow the cells to grow overnight.
2. Treat the cells with ISRIB at a final concentration of 0.5 μM or GSK260641 at a final concentration of 1 μM.
3. Eighteen hours later, induce stress by treating the cells with sodium arsenite at 200 μM or MG132 at 10 μM. At the same time, induce reporter gene expression by adding 2 μg/mL Doxycycline.
4. After 6 h, perform luciferase assay as described in Subheading 3.3. Examine the eIF2α phosphorylation by Western blot as described in Subheading 3.2.3, and stress granule formation by immunofluorescence as described in Subheading 3.6.2.

4 Notes

1. The restriction enzyme sites, MYC tag, or other sequences are included in primers and introduced into the insert fragments by PCR.
2. The cap-independent translation should also be validated by knockdown of cap-binding proteins and in vitro translation assay.
3. Use Top10 or DH5α bacteria strains for regular constructs. Use NEB Stable (NEB, C3040) or SURE competent cells

(Agilent, 200227) for constructs with GGGGCC repeat expansion.

4. Grow the plates at 30 °C to improve the stability of repeat elements in the plasmids.
5. For constructs with repeat expansion, grow the bacteria in liquid LB or 2×YT medium at 30 °C overnight.
6. Use 10:1 ratio of pOG44 plasmid to the Flp-In construct to ensure that the recombination occurs in the desired site rather than random insertion.
7. The ratio of plasmid and transfection reagent needs to be optimized to get the best transfection efficiency. To optimize the transfection efficiency, vary the amount of transfection reagent from 2 to 6 μL per 1 μg plasmid.
8. Titrate the drug dosage on each cell type before selection.
9. The colonies should be visible in about 2 weeks. It is important to include a negative control in which only pOG44 is transfected. The negative control cells should be killed within 1 week after drug selection.
10. Experiments using recombinant retrovirus require approval from the biosafety committee. Make sure to follow the safety guidelines.
11. The amount of beads needs to be optimized to achieve optimal immunoprecipitation efficiency. For cells from one 10-cm dish, we found that 0.3 mg beads are usually enough.
12. The incubation time varies depending on the affinity of the antibody and the abundance of the target protein.
13. Luminescence is measured with 1 s integration time, attenuation "automatic."
14. Detection of RNA transcripts of "housekeeping" genes as internal control, such as *GAPDH*.

Acknowledgments

This work was supported by the grants to S.S from the National Institutes of Health (R00NS091538, R01NS107347, and RF1NS113820), Target ALS, and the Robert Packard Center.

References

1. Genomes Project C, Abecasis GR, Altshuler D, Auton A, Brooks LD, Durbin RM et al (2010) A map of human genome variation from population-scale sequencing. Nature 467(7319):1061–1073. https://doi.org/10.1038/nature09534
2. Paulson H (2018) Repeat expansion diseases. Handb Clin Neurol 147:105–123. https://

doi.org/10.1016/B978-0-444-63233-3.00009-9

3. Sznajder LJ, Swanson MS (2019) Short tandem repeat expansions and RNA-mediated pathogenesis in myotonic dystrophy. Int J Mol Sci 20(13):3365. https://doi.org/10.3390/ijms20133365
4. Zu T, Gibbens B, Doty NS, Gomes-Pereira M, Huguet A, Stone MD et al (2011) Non-ATG-initiated translation directed by microsatellite expansions. Proc Natl Acad Sci U S A 108(1): 260–265. https://doi.org/10.1073/pnas.1013343108
5. Krans A, Kearse MG, Todd PK (2016) Repeat-associated non-AUG translation from antisense CCG repeats in fragile X tremor/ataxia syndrome. Ann Neurol 80(6):871–881. https://doi.org/10.1002/ana.24800
6. Todd PK, Oh SY, Krans A, He F, Sellier C, Frazer M et al (2013) CGG repeat-associated translation mediates neurodegeneration in fragile X tremor ataxia syndrome. Neuron 78(3):440–455. https://doi.org/10.1016/j.neuron.2013.03.026
7. Zu T, Cleary JD, Liu Y, Banez-Coronel M, Bubenik JL, Ayhan F et al (2017) RAN translation regulated by Muscleblind proteins in myotonic dystrophy type 2. Neuron 95(6): 1292–1305. e5. https://doi.org/10.1016/j.neuron.2017.08.039
8. Banez-Coronel M, Ayhan F, Tarabochia AD, Zu T, Perez BA, Tusi SK et al (2015) RAN translation in Huntington disease. Neuron 88(4):667–677. https://doi.org/10.1016/j.neuron.2015.10.038
9. McEachin ZT, Gendron TF, Raj N, Garcia-Murias M, Banerjee A, Purcell RH et al (2020) Chimeric peptide species contribute to divergent dipeptide repeat pathology in c9ALS/FTD and SCA36. Neuron 107(2): 292–305. e6. https://doi.org/10.1016/j.neuron.2020.04.011
10. Gendron TF, Bieniek KF, Zhang YJ, Jansen-West K, Ash PEA, Caulfield T et al (2013) Antisense transcripts of the expanded C9ORF72 hexanucleotide repeat form nuclear RNA foci and undergo repeat-associated non--ATG translation in c9FTD/ALS. Acta Neuropathol 126(6):829–844. https://doi.org/10.1007/s00401-013-1192-8
11. Mizielinska S, Lashley T, Norona FE, Clayton EL, Ridler CE, Fratta P et al (2013) C9orf72 frontotemporal lobar degeneration is characterised by frequent neuronal sense and antisense RNA foci. Acta Neuropathol 126(6): 845–857. https://doi.org/10.1007/s00401-013-1200-z
12. Lagier-Tourenne C, Baughn M, Rigo F, Sun S, Liu P, Li HR et al (2013) Targeted degradation of sense and antisense C9orf72 RNA foci as therapy for ALS and frontotemporal degeneration. Proc Natl Acad Sci U S A 110(47): E4530–E4539. https://doi.org/10.1073/pnas.1318835110
13. DeJesus-Hernandez M, Mackenzie IR, Boeve BF, Boxer AL, Baker M, Rutherford NJ et al (2011) Expanded GGGGCC Hexanucleotide repeat in noncoding region of C9ORF72 causes chromosome 9p-linked FTD and ALS. Neuron 72(2):245–256. https://doi.org/10.1016/j.neuron.2011.09.011
14. Renton AE, Majounie E, Waite A, Simon-Sanchez J, Rollinson S, Gibbs JR et al (2011) A Hexanucleotide repeat expansion in C9ORF72 is the cause of chromosome 9p21-linked ALS-FTD. Neuron 72(2):257–268. https://doi.org/10.1016/j.neuron.2011.09.010
15. Renton AE, Chio A, Traynor BJ (2014) State of play in amyotrophic lateral sclerosis genetics. Nat Neurosci 17(1):17–23. https://doi.org/10.1038/nn.3584
16. Mori K, Arzberger T, Grasser FA, Gijselinck I, May S, Rentzsch K et al (2013) Bidirectional transcripts of the expanded C9orf72 hexanucleotide repeat are translated into aggregating dipeptide repeat proteins. Acta Neuropathol 126(6):881–893. https://doi.org/10.1007/s00401-013-1189-3
17. Zu T, Liu YJ, Baez-Coronel M, Reid T, Pletnikova O, Lewis J et al (2013) RAN proteins and RNA foci from antisense transcripts in C9ORF72 ALS and frontotemporal dementia. Proc Natl Acad Sci U S A 110(51): E4968–E4E77. https://doi.org/10.1073/pnas.1315438110
18. Mori K, Weng SM, Arzberger T, May S, Rentzsch K, Kremmer E et al (2013) The C9orf72 GGGGCC repeat is translated into aggregating dipeptide-repeat proteins in FTLD/ALS. Science 339(6125):1335–1338. https://doi.org/10.1126/science.1232927
19. Wilkinson ME, Charenton C, Nagai K (2020) RNA splicing by the spliceosome. Annu Rev Biochem 89:359–388. https://doi.org/10.1146/annurev-biochem-091719-064225
20. Katahira J (2015) Nuclear export of messenger RNA. Genes 6(2):163–184. https://doi.org/10.3390/genes6020163
21. Tutucci E, Stutz F (2011) Keeping mRNPs in check during assembly and nuclear export. Nat Rev Mol Cell Biol 12(6):377–384. https://doi.org/10.1038/nrm3119

22. Galloway A, Cowling VH (2019) mRNA cap regulation in mammalian cell function and fate. Biochim Biophys Acta Gene Regul Mech 1862(3):270–279. https://doi.org/10.1016/j.bbagrm.2018.09.011
23. Pakos-Zebrucka K, Koryga I, Mnich K, Ljujic M, Samali A, Gorman AM (2016) The integrated stress response. EMBO Rep 17(10): 1374–1395. https://doi.org/10.15252/embr.201642195
24. Ryoo HD, Vasudevan D (2017) Two distinct nodes of translational inhibition in the integrated stress response. BMB Rep 50(11): 539–545. https://doi.org/10.5483/BMBRep.2017.50.11.157
25. Li YR, King OD, Shorter J, Gitler AD (2013) Stress granules as crucibles of ALS pathogenesis. J Cell Biol 201(3):361–372. https://doi.org/10.1083/jcb.201302044
26. Cheng W, Wang S, Mestre AA, Fu C, Makarem A, Xian F et al (2018) C9ORF72 GGGGCC repeat-associated non-AUG translation is upregulated by stress through eIF2α phosphorylation. Nat Commun 9(1):1–12. https://doi.org/10.1038/s41467-017-02495-z
27. Sun S, Ling SC, Qiu J, Albuquerque CP, Zhou Y, Tokunaga S et al (2015) ALS-causative mutations in FUS/TLS confer gain and loss of function by altered association with SMN and U1-snRNP. Nat Commun 6: 6171. https://doi.org/10.1038/ncomms7171
28. Sonenberg N, Hinnebusch AG (2009) Regulation of translation initiation in eukaryotes: mechanisms and biological targets. Cell 136(4):731–745. https://doi.org/10.1016/j.cell.2009.01.042
29. Heiman M, Schaefer A, Gong S, Peterson JD, Day M, Ramsey KE et al (2008) A translational profiling approach for the molecular characterization of CNS cell types. Cell 135(4): 738–748. https://doi.org/10.1016/j.cell.2008.10.028
30. Sun S, Sun Y, Ling SC, Ferraiuolo L, McAlonis-Downes M, Zou Y et al (2015) Translational profiling identifies a cascade of damage initiated in motor neurons and spreading to glia in mutant SOD1-mediated ALS. Proc Natl Acad Sci U S A 112(50): E6993–E7002. https://doi.org/10.1073/pnas.1520639112
31. Costa-Mattioli M, Walter P (2020) The integrated stress response: from mechanism to disease. Science 368(6489):eaat5314. https://doi.org/10.1126/science.aat5314
32. Sidrauski C, McGeachy AM, Ingolia NT, Walter P (2015) The small molecule ISRIB reverses the effects of eIF2alpha phosphorylation on translation and stress granule assembly. eLife 4:e05033. https://doi.org/10.7554/eLife.05033
33. Axten JM, Medina JR, Feng Y, Shu A, Romeril SP, Grant SW et al (2012) Discovery of 7-methyl-5-(1-{[3-(trifluoromethyl)phenyl]acetyl}-2,3-dihydro-1H-indol-5-yl)-7H-p yrrolo[2,3-d]pyrimidin-4-amine (GSK2606414), a potent and selective first-in-class inhibitor of protein kinase R (PKR)-like endoplasmic reticulum kinase (PERK). J Med Chem 55(16):7193–7207. https://doi.org/10.1021/jm300713s
34. Walter P, Ron D (2011) The unfolded protein response: from stress pathway to homeostatic regulation. Science 334(6059):1081–1086. https://doi.org/10.1126/science.1209038

Chapter 9

Analysis of Ribosome Profiling Data

Carine Legrand, Khanh Dao Duc, and Francesca Tuorto

Abstract

Ribosome profiling methods are based on high-throughput sequencing of ribosome-protected mRNA footprints and allow to study in detail translational changes. Bioinformatic and statistical tools are necessary to analyze sequencing data. Here, we describe our developed methods for a fast and reliable quality control of ribosome profiling data, to efficiently visualize ribosome positions and to estimate ribosome speed in an unbiased way. The methodology described here is applicable to several genetic and environmental conditions including stress and are based on the R package RiboVIEW and calculation of quantitative estimates of local and global translation speed, based on a biophysical model of translation dynamics.

Key words Ribosome profiling, Bioinformatic analysis, RiboVIEW, TASEP, Ribo-Seq

1 Introduction

Protein synthesis is a highly regulated and fine-tuned process, which enables a fast response to metabolic and environmental changes including stress. Proteins are synthesized by ribosomes that translate mRNA codons into amino acids while moving along transcripts. Ribosome profiling is a method which allows to directly measure protein synthesis by detecting the position of ribosomes on mRNAs [1]. The method is based on the deep sequencing of the short mRNA fragments (footprints) protected by the ribosomes upon RNase digestion (Fig. 1). Bioinformatic alignment of these footprints allows to determine the position of translating ribosomes on mRNAs, even at single-codon resolution. Quality tools assessing the structure of the data, detection of contaminants, but also artifacts and batch effects, should initially be used to verify the quality of ribosome profiling datasets. Later, a more global analysis can be performed to assess codon enrichment (unbiased codon occupancy) and translation efficiency (TE) (*see* **Note 1**). Here we show how to use RiboVIEW R package [2] to obtain relevant quality properties of sequenced datasets. Furthermore, RiboVIEW provides unbiased estimates of codon enrichment, which can be

Daniel Matějů and Jeffrey A. Chao (eds.), *The Integrated Stress Response: Methods and Protocols*,
Methods in Molecular Biology, vol. 2428, https://doi.org/10.1007/978-1-0716-1975-9_9,

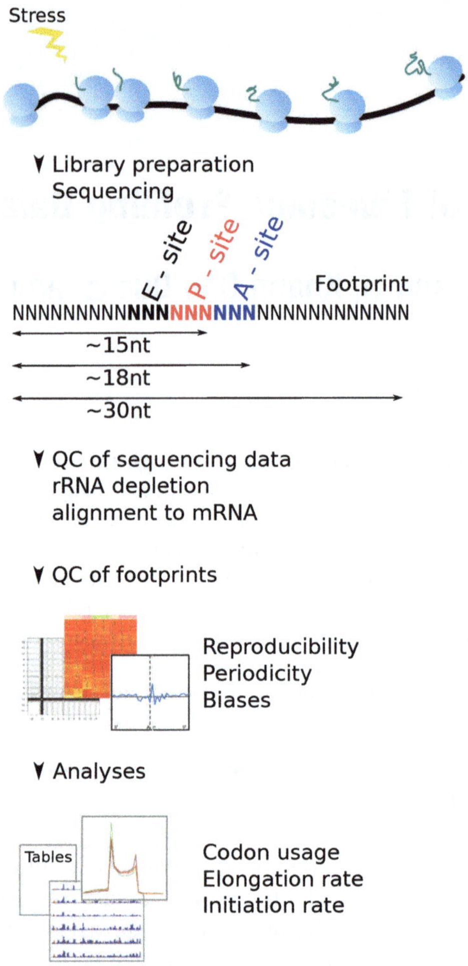

Fig. 1 Workflow for the analysis of ribosome profiling data from stress stimulus to quality control and specific analyses

used to detect codon specific effects in different genetic or metabolic conditions, including amino acid deprivation or lack of RNA modifications [3–5]. Variation in ribosome density dependent on specific codons can be interpreted as variation in translational speed or codon stalling mainly in dependence on the size effect exerted on the translation machinery.

The totally asymmetric simple exclusion process (TASEP) is a stochastic model for the movement of interacting particles, such as ribosomes moving along the mRNA during translation. Various factors can influence the movement of ribosomes, and the observed heterogeneity of ribosome density along transcript sequences remains only partially explained. In this context, we recently developed analytical and computational methods to analyze and simulate the TASEP model, with direct applications for studying mRNA translation and interpreting experimental ribosome profiling data

[6–8]. As a result of our theoretical study [7], one can infer, under the TASEP model, the elongation rates associated with a given ribosome density profile. Conversely, we developed a software called EGGTART (Extensive GUI gives TASEP-realization in real time) to visualize the ribosome profile and quantify the translation speed for a given input sequence of elongation and initiation rates [8].

Here we describe methods which require basic bioinformatics and programming knowledge to carry out quality control checks and determine codon enrichment (occupancy) using RiboVIEW, and necessitate mathematics and bioinformatics knowledge to apply the TASEP model to infer translation rates from a density profile, and visualize the ribosome traffic using EGGTART software.

2 Materials

Ribosome profiling datasets are generally presented in the form of BAM-formatted files, one per sample. Reads should be present in a sufficient extent. Solid analyses can be expected with at least tens of millions of sequenced reads per sample, as an input. The analysis generally stays feasible for amounts as low as a few millions of reads per sample, albeit with results of a diminished quality.

The following system characteristics allow smooth analyses for most ribosome profiling data.

1. CPU: 64 bits, $\geq$ 3.4GHz
2. RAM memory: $\geq$ 32 GB
3. Cores: several (2 to 4) cores are advisable to analyse several samples in parallel. More cores can be used advantageously, provided the user keeps some capacity for other tasks, and provided the cores do not interfere with each other (overheating or too frequent switching between cores could impair calculations).
4. Memory: depending on number of samples and coverage, a set of input and results files commonly weighs at least hundreds of megaoctets.

2.1 Quality Control and Codon Occupancy Estimates

The analysis presented here necessitate reads aligned after a quality check and depletion of non-mRNA reads from demultiplexed sequence reads. This step is well explained in Riboview "Supplementary Template Workflow" [2]. The analysis relies on R package RiboVIEW, which necessitates that R (version 3.4.4 or newer) and Python (2.7.6 - 2.7.9) are installed on the working computer. An access to Internet is necessary to download packages and reference sequences.

RiboVIEW makes use of the following R CRAN packages (https://cran.r-project.org/) which should be available and working: ggplot2 (version 3.0.0 or newer), gplots (3.0.1 or newer), gridExtra (2.3 or newer), latex2exp (0.4.0 or newer), MASS (7.3-50 or newer), png (0.1-7 or newer), RColorBrewer (1.1-2 or newer), PythonInR (0.1-7 or newer), R.devices (2.16.1 or newer), tseriesChaos (0.1-13 or newer), Rtsne (0.15 or newer), and VennDiagram (1.6.20 or newer).

Similarly, RiboVIEW makes use of the following Python packages which should be installed: Biopython (version 1.72 or newer), Numpy (1.8.2 or newer), and pysam (0.15.0 or newer).

2.2 Inference and Visualization of Translation Dynamics Using the TASEP

For a ribosome profile associated with a coding region of n codons, the inference method (Subheading 3.2) requires a corresponding vector of size n, such that the i-th coordinate gives the number of footprints with ribosome A-site assigned at position i. In addition, it also requires the transcript abundance (obtained, e.g., from matched RNA-Seq data) to normalize the data (*see* **Note 2**). The visualization and perturbation analysis (Subheading 3.3) require an input of elongation rates, written as a single-column comma-separated values file (.csv). Such an input file can be obtained from a ribosome profile following the method presented in Subheading 3.2.

The methods presented for inference of elongation rates (Subheading 3.2) are based on explicit mathematical formulae, and therefore require minimal computational power. The visualization software EGGTART is provided as an executable version and thus poses almost no requirements on neither computational resource nor preinstalled software. To launch EGGTART, simply download the version developed for your operating system at https://github.com/songlab-cal/EGGTART, and double-click the corresponding executable file. We provide distributions of EGGTART for Mac OS X, Linux (Ubuntu 18.04) and Windows, with most extensive testing conducted on Mac.

3 Methods

In this paragraph we present how to perform quality control and generation of codon enrichment (codon occupancy) using RiboVIEW, and calculation of quantitative estimates of elongation using the TASEP. RiboVIEW visualizes translation elongation at codon level and provides relevant quality properties. Furthermore, RiboVIEW provides unbiased estimates of codon enrichment and detects some causal or confounding covariates [2]. In addition, we provide the visualization of elongation rates based on the TASEP model using EGGTART software [8].

3.1 Quality Control

In this step, several quality properties are examined. First of all, it is necessary that ribosome footprints have 3-nucleotide periodicity. Further, replicates should be consistent with each other, and there should be no bias at the level of footprints selection. Finally, potential biases due to cycloheximide or any other drug or molecule used during the experiment are inspected.

3.1.1 Installation of RiboVIEW

1. Open R from the command line by typing:
   > R
   or, open Rstudio or a similar application for R.
2. From R, install the devtools package if not already installed, and import the devtools library, by typing the following commands.
   > install.packages("devtools")
   > library(devtools)
3. Download RiboVIEW from zenodo: https://doi.org/10.5281/zenodo.4401399
4. Install RiboVIEW:
   > install_local("path-to-downloaded-RiboVIEW-folder/RiboVIEW_2.0.tar.gz")
5. Import the RiboVIEW library:
   > library(RiboVIEW)

3.1.2 Fetching Reference Sequences and Reference Annotation

RiboVIEW needs reference mRNA sequences for checking purposes, for unbiased estimates of codon enrichment, and for mRNA tracks. In combination with these sequences in FASTA format, an annotation of the coding sequence start and stop positions in tabular format is necessary.

1. Fetching a FASTA-formatted file of mRNA reference sequences: the reference sequences for mRNA should be identical, as much as possible, to those used during the alignment. Here, we use sequences from Ensembl FTP [9] (*see* **Notes 3** and **4**); FTP addresses are found at: http://www.ensembl.org/info/data/ftp/. For mRNAs, the FASTA file in the category "cDNA" should be selected and downloaded.
2. Build the mRNA reference: sequences corresponding to pseudogenes or to mitochondrial mRNAs should be identified and removed from the downloaded FASTA. This clean FASTA file constitutes the mRNA reference sequences file. This step can be done for example in bash or perl (*see* **Note 5**). The resulting file's address and name should be provided to RiboVIEW as follows:

```
> refFASTA <- "home/username/ref/organism_mRNAsequences.fas-
ta"
```

3. Fetching mRNA annotation: Similarly, the annotation for the coding sequence should correspond exactly to the mRNA sequences, in particular mRNA identifiers should be strictly the same. Continuing with Ensembl, the GTF-formatted file in the category “gene sets” should be selected and downloaded.
4. Here we present the expected format of mRNA annotation: The annotation table expected in input by RiboVIEW is tab-separated and contains one line per mRNA, and columns as in the following example.

ID	localStart	localEnd	strandDir	exonLength
mRNA1	0	450	+	483
mRNA2	51	843	+	1041
mRNA3	40	1240	+	1642
(etc.)	(...)	(...)	(...)	(...)

In this table, *ID* is the identifier or name of the mRNA; *localStart* is the position of the start codon (most often the “A” of “AUG”), in nucleotides, and one uses 0 as the convention for the first position of the mRNA; *localEnd* is the position of the first nucleotide after the coding sequence, that is to say just after the stop codon. As a check, the length from start to stop defined by localStart and localEnd should correspond to an integer number of codons, and therefore be a multiple of 3: in the example above mRNA1 has $450/3 = 150$ codons (including start and stop), and mRNA2 has $(843 - 51)/3 = 264$ codons; *strandDir* is the direction of the mRNA along the strand: + or −; and *exonLength* is the length of the full mRNA, including the 5′ UTR (if present), the CDS, and 3′ UTR (if present).

5. Building the CDS table using python script ’gtf2table’ [Ribo-VIEW paper]: this can be made by first defining the name and address of the downloaded GTF file, and of the wanted CDS table,

```
> refGTF <- "home/username/ref/organism_genesets.gtf"
> refCDS <- "home/username/ref/organism_CDStable.tsv"
```

and second, by using the following command to generate the CDS table (*see* **Note 6**).

```
> gtf2table(refGTF, refCDS)
```

3.1.3 Setting Up Parameters

A template for the following commands can be obtained by typing "help(RiboVIEW)".

1. Define the address and name of aligned sequences (in BAM format) of each sample, for instance for condition 1 and replicate 1, here denoted "c1" and "r1":

   > reads_c1_r1 <- "/home/username/data/myXP/aligned-reads-condition1-rep1.bam"

2. Define the list of files:

   > list.bam <- list(reads_c1_r1, reads_c1_r2, reads_c1_r3, reads_c2_r1, reads_c2_r2, reads_c2_r3)

   This example corresponds to two different conditions and three replicates. This can be expanded as appropriate for three or more conditions, and for more replicates.

3. Define names for the conditions of your experiment. These names will be used by RiboVIEW in tables and graphics. They should hence be clear and preferably short. Here we use the generic terms "Stress" and "Control" for the 2 conditions, repeated for each replicate:

   > XP.conditions <- c("Stress", "Stress", "Stress", "Control", "Control", "Control")

   Further, the conditions also have to be defined as a numeral indicator 1, 2, and so on:

   > XP.conditions.i <- c(2,2,2,1,1,1)

   Finally, each replicate name should be defined (*see* **Note 7**):

   > XP.names <- c("Stress.R1", "Stress.R2", "Stress.R3", "Ctrl.R1", "Ctrl.R2", "Ctrl.R3")

4. Now that the experiment is well defined, one still needs to type the address of mRNA annotation and sequences, if not done previously:

   > refCDS<- "home/username/ref/organism_CDStable.tsv"

   > refFASTA <- "home/username/ref/organism_mRNA-sequences.fasta"

   The mRNA sequences should be given in FASTA format.

5. Finally, one needs to define and create the output directory where the results will be saved to:

   > pathout <- "home/username/Analyses/myXP/resu/"

   > mkdir(pathout).

3.1.4 Preliminary Calculation and Check (Periodicity)

As for the previous steps, a template for the following commands can be obtained by typing "help(RiboVIEW)".

1. Load the package RiboVIEW by typing:

   > library("RiboVIEW")

2. Run the preliminary calculations for periodicity:

   > periodicity(list.bam, refCDS, refFASTA, pathout, XP.names, versionStrip = FALSE,

python.messages = FALSE, mitochondrion = FALSE)

If needed, set one the options to TRUE:

- versionStrip=TRUE is a convenience option in case one of the BAM, CDS or FASTA files contains a version number, separated by a point, but not the other files. In this case, this version number is trimmed in order to make the two references identical. For instance mRNA1.1 and mRNA1 would be considered identical if versionStrip=TRUE but not identical if versionStrip=FALSE.

- python.messages=TRUE triggers display of informative messages from the Python script which is performing the core calculations.

- mitochondrion=TRUE allows start codons "AUA" and "AUU", in addition to "AUG".

3. Review and select adequate footprint length via the following command.

> attach(listminmax <- select.FPlen(list.bam, pathout, XP.names))

This command will display one panel per sample, in our example "Stress.R1", "Ctrl.R1", etc. Each panel is stratified by footprint length from 25nt to 32nt. For each of these lengths, a recurrence plot is displayed (Fig. 2). One expects a first black column at -12nt, indicating the presence of footprints at

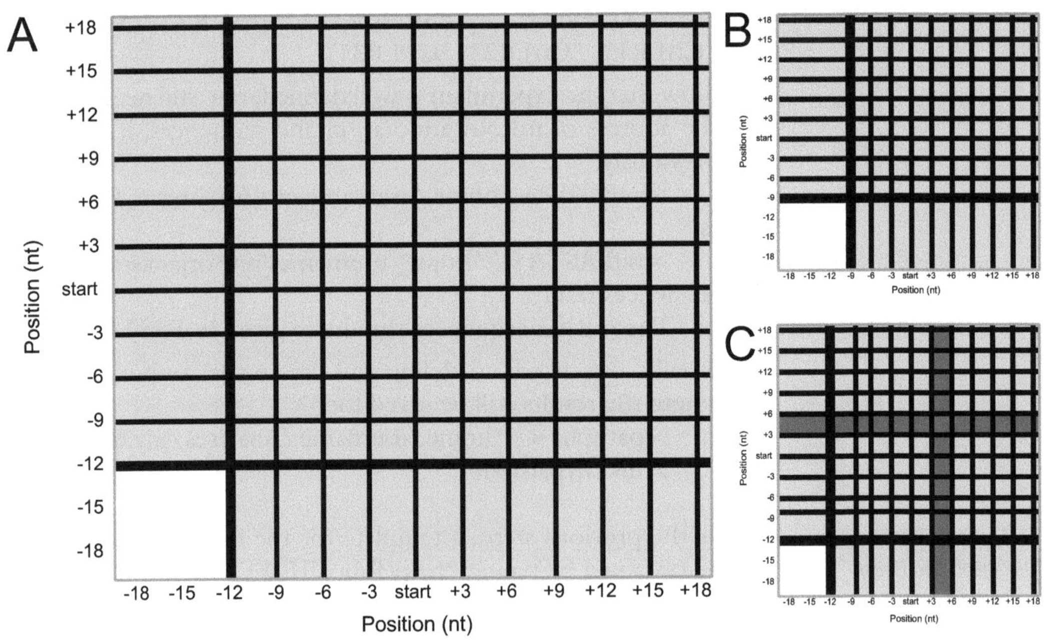

Fig. 2 Recurrence plot schematic examples. Axis: nucleotide (nt) position, 0: first nt of start codon (**a**) ideal recurrence plot, (**b**) recurrence plot with correct 3nt-periodicity, but initiation in P-site is not at -12nt, indicating that the rationale for codon position would have to be adapted, (**c**) recurrence plot with incorrect or insufficient periodicity

initiation (P-site in this case), and then one black column every 3 nucleotides, ideally separated by two white positions, meaning that all footprints are in the 3nt-periodicity frame. Light greyish tones can be tolerated. If a footprint length is meeting these expectations, it can be selected for further analyses. If the peak is not at -12nt, or if there is no separation between successive peaks, the footprint length has to be rejected. These review and selection steps have to be carefully repeated for each sample.

After all samples have been reviewed, the user will be prompted to type the minimum and maximum footprint lengths which are appropriate in all samples. The "attach" part of the command, "attach(listminmax" allows the variables containing the selected minimum/maximum footprint length, *mini* and *maxi*, to be available in the workspace, for use in subsequent commands.

3.1.5 Run Calculations for Quality Control and Visualization

The following list of commands generates results and plots for quality control and visualization. In these commands, options versionStrip, python.messages, and mitochondrion should be set to TRUE if appropriate (*see* description of these options above and *see* **Note 8** for customization of mRNA tracks). Otherwise, these commands can be typed exactly as follows:

```
> enrichmentNoccupancy(list.bam, refCDS, refFASTA, mini, maxi,
XP.names,
 pathout, versionStrip = FALSE, python.messages=FALSE,
 mitochondrion = FALSE)
> generate.m.s(XP.conditions, XP.names, pathout, B=1000)
> visu.m.s.enrichmnt.res <- visu.m.s.enrichmnt(XP.conditions,
XP.names, pathout)
> visu.tracks.res <- visu.tracks(XP.conditions, XP.names,
pathout, refCDS, mRNA="random",
> codon.labels=FALSE, codon.col="darkslateblue")
> Venn.all.res <- Venn.all(XP.names, XP.conditions, pathout)
> enricht.aroundA.res <- enricht.aroundA(XP.conditions, XP.
names, pathout)
> repl.correl.counts.Venn.res <- repl.correl.counts.Venn(XP.
conditions, XP.names, pathout)
> repl.correl.gene.res <- repl.correl.gene(XP.conditions, XP.
names, pathout)
> repl.correl.codon.res <- repl.correl.codon(list.bam,
refCDS, refFASTA,
 mini, maxi,XP.names, XP.conditions, pathout)
> repl.correl.heatmap.res <- repl.correl.heatmap(XP.condi-
tions.i, XP.names, pathout)
> chx.artefacts.res <- chx.artefacts(XP.conditions, XP.names,
pathout)
```

```
> ntcodon.freq.nt.res <- ntcodon.freq.nt(XP.conditions, XP.
names, pathout)
> ntcodon.freq.cod.res <- ntcodon.freq.cod(XP.conditions, XP.
names, pathout)
> batch.effects.lm.e.res <- batch.effects.lm.e(XP.conditions,
XP.names, pathout)
> batch.effects.pca.res <- batch.effects.pca(XP.conditions,
XP.names, pathout)
> metagene.res <- metagene.all(XP.conditions, XP.names, path-
out)
> outputQc(pathout, XP.conditions)
> outputMine(pathout, XP.conditions)
```

Of note, default option $B = 1000$, in command *generate.m.s*, corresponds to the number of bootstrap resampling, for calculation of mean and standard error of ratios. This default value is sufficient for a small number of replicates per condition. This number can be increased if there are numerous replicates.

The above set of commands creates tables and pictures (*see* **Note 9**), as well as browser-readable files "Results-Qc.html" and "Results-Mine.html". These two files respectively show a dashboard of quality properties and display an overview of results.

3.1.6 Reviewing Quality Controls

In the previous paragraph, underlying settings and calculations have been completed. As a result, the file "Results-Qc.html" was created. In the following we use this file to review quality properties. There are also alternative and sometimes complementary quality controls (*see* **Note 10**).

1. Open Results-Qc.html with a browser, such as Firefox, Chrome, Brave, etc. The browser will display a page named "RiboQC" with four clickable buttons "Periodicity" (shown by default), "Replicates," "Footprints," and "Drugs," which allow to navigate between categories.
2. Remain in category "Periodicity" (or click the corresponding button if you changed to another category). A panel with two tabs, "Recurrence" and "Coverage" is shown. Stay (or click the tab) "Recurrence." This reports the recurrence plots for the first sample and the footprint lengths selected. This control has been carried out previously, but one gets here the opportunity to double check the chosen footprint lengths.
3. Click the tab "Coverage. This shows the footprints stratified by length, from 25nt to 32nt. Inside each length, coverage is shown at each position between AUG -18nt to AUG+18nt. Are the high-covered lengths among the selected footprint lengths? Normally, the selected footprint lengths also encompass the lengths with the highest coverage. If it is not the case,

this corresponds to a notable loss of experimental data, possibly due to insufficient or too aggressive read digestion.

4. Click the category "Replicates." This panel entails three tabs which allow to verify the consistency between replicates. Note that replicates consistency can be poor due to a high level of biological variability, even if the experiment has been conducted properly. In this case, sample size calculation, and an accordingly large number of replicates, might be necessary to detect significant differences between conditions.
5. Stay (or click) on the first tab "heatmap and clustering." The heatmap displayed corresponds to codon enrichment values. Sample names are written on the bottom, and correspond to color-coded conditions displayed on the top. An important feature is the hierarchical clustering tree displayed at the top. This tree assembles samples based on their similarity, in an unsupervised manner. Do replicates of the same condition cluster in effect together? Is the Spearman correlation, written below the heatmap large enough (*see* **Note 11**)?
6. Then, click the tab "Correlation between codons." This panel shows codon-level plots and the corresponding Spearman correlation for each pair of replicates. Do the plotted points follow a "y=x" line? Are the Spearman correlations comparable to similar studies [10]? A guidance is provided by the text below the figure: this text indicates which level of averaging is necessary to obtain a >0.6 Spearman correlation. A high correlation combined to a low level of averaging indicates that codon-level analysis is possible.
7. To finish with the analysis of replicates consistency, click the tab "Correlation between genes." This displays the RPKM (Reads Per Kilobase transcript per Million reads) of genes between pairs of replicates. The Spearman correlation of RPKM is always very large, and should be >0.9, if possible >0.95.
8. Next, click the category "Footprints." This panel allows to examine biases at the level of footprint reads, for example a ligation bias, if not using UMIs (Unique Molecular Identifiers) in the experiment. This category contains three tabs (nt logo, codon logo, and metagene).
9. Click the tab "Ligation bias (nt logo)", which displays the frequency, in bits, of nucleotides A, C, G, U at the 5′ end (left) and 3′ end (right). There are as many rows as there are samples in the analysis. A high value in bits signals an over- or underrepresented nucleotide, and triggers a warning written below the figure. A bias here could be due to defects in the library preparation, for example in the adapter ligation steps, but it could also be due to trimming in the bioinformatics pipeline.

10. Click the tab "Ligation bias (codon logo)". Similarly, codon frequency at 5′ (left) and 3′ (right) is displayed, with one sample per row. A high value in bits signals over or under representation and is signalled by a message.
11. Click the tab "Metagene (Monosome selection, drop-off)" to display the footprint density in metagene coordinates. In this figure, density is plotted in the UTRs as well as in the CDS. Proper monosome selection corresponds to actually translating, fully built ribosomes, which should therefore lay predominantly in the CDS. There might be some density in the 5′UTR due to alternative start sites, but to a small extent, and only in certain organisms.
12. Select the last category of quality controls: "Drugs". This shows potential biases due to incomplete ribosome arrest when using cycloheximide in yeast (*S. cerevisiae*), or due to other drug-related artifacts [11].
13. Stay (or click) tab "Cycloheximide." This shows enrichment in a window of −90nt to +90nt around the A site, for arginine codons, which are known to be particularly affected (*see* graphs enrichment-all_*.eps or enrichment-all_*.png, in your output folder, to see enrichment for all codons). Is there a distinct wave pattern 5 to 20 codons downstream of the ribosome? This is expected on the 5′ side and might be easier to see on the magnification plots (enrichment-all-zoom_*.eps or enrichment-all-zoom_*.png, *see* **Note 9**). Such a pattern would point to a cycloheximide bias. There might also be large noise revealed by this plot, which triggers a message below the plot.
14. Next, click the tab "Inflation at start codon," which shows coverage (not normalized) in the CDS, in metagene coordinates. Is there an unusually high number of reads located near the CDS, suggesting continued initiation? A message shows the percentage of reads near CDS and indicates if it corresponds to standards for this kind of experiments.
15. Finally, click the tab "Leakage." This shows footprint density around the START codon (respectively, STOP codon) in the top panel (resp. bottom). Is the density increasing after START, pointing to ribosome leakage, and continued elongation? A message provides here again some guidance: it tells if there is a positive significant slope or not after START. Similarly, is there a nonzero footprint density after STOP, indicating leakage of the STOP codon? In both cases, leakage might be due to a drug bias, but could also be due to the translation characteristics of the organism studied, and should therefore be interpreted with that in mind.

This concludes the quality controls section. If all quality controls are satisfactory, the data may be used without constraint. Otherwise, it might be necessary to generate more samples, adapt the protocol or adapt the analysis, in order to exclude or limit defects due to the biases.

3.1.7 Codon Enrichment (Occupancy) and Further Results

In this section, we go through the overview of results given in RiboVIEW's output file "Results-Mine.html". Similarly, there are alternatives described elsewhere (*see* **Note 10**). The categories presented here encompass results in individual conditions, comparisons between conditions, and codon enrichment in the vicinity of the A site.

1. Select category "Within conditions," which describes results for individual samples, and click the first tab "Enrichment by sample." This shows unbiased codon enrichment for each codon identity, in alphabetic order. The dots indicate the mean, and the error bars correspond to +/− standard deviation. A value larger than 1 (respectively, lower than 1) indicates that a codon is more often found than expected, meaning that it is paused (resp., accelerated). These values are also available in files Enrichmnt-per-condition_weighted-* (*see* **Note 9**).
2. Select the second tab, "Nucleobases," which shows the contribution of each base A, C, G, or U to codon enrichment, sample by sample. This is often independent, but if this contribution is significant, it will be indicated by a message below this plot, as well as the name of samples involved.
3. Select the third track in this category, "Tracks." This shows the tracks for an individual mRNA, given in the *x*-axis name. This mRNA is picked at random, but the user can select another mRNA and other options (*see* **Note 8**). The track is given for each sample (one sample per row). The *y*-axis gives the raw coverage in the A-site. This plot can be reconstructed and adapted from output files (single codon occupancy, *see* **Note 9**).
4. Select then the "Venn diagram" tab. This shows mRNA common to at most five samples. If there are more samples, you might consider using an upset plot instead of a Venn diagram.
5. Continuing with the results per experimental condition, select category "Between conditions," and the first tab "Comparisons between conditions." This tab shows the codon enrichment in one condition relative to (that is to say, divided by) another condition. This is obtained by sampling all possible ratios, which yields the mean and standard error of the ratio. The values displayed correspond to the mean and standard error, and codons are ordered from the most slowed to the most accelerated (for corresponding files *see* **Note 9**).
6. Next, the "PCA" tab is used to detect separation or structure in the samples. A guiding *p*-value for significant structure (or batch effect) is given in the text below the figure.

7. The "tSNE" tab also serves the purpose of detecting clusters and special structures in the data, for the case of a large number of samples. This plot has limited usefulness when there is a small number of samples.
8. Next, select category "Codon enrichment" and tab "A-site +/-90". This shows enrichment in the window A-site +/- 90 nt for a subset of codons. This allows to detect possible enrichments in the P or E site (or, potentially, biases, if there is enrichment deviating from unity, distantly from the A-site).
9. Finally, a magnification is given in tab "A-site +/-15", which shows the window of +/-15nt around the A-site. This is a more convenient view to examine codon enrichment in the P or E sites. Plots and files for all codons are provided in RiboVIEW output folder (*see* **Note 9**).

3.2 Inference of Rates Under the TASEP Model

The following procedure describes how to estimate various translation rates from ribosome profiles, under the TASEP model. For a given quality-controlled ribosome profile R, **steps 1–4** (Subheading 3.3) lead to estimates of rates which are scaled by an unknown parameter, since multiplying all the rates by a same constant in the TASEP model does not modify the observed density profile. While one can use these rates to simulate a density profile and test the impact of the different parameters, as described in Subheading 3.3, **steps 5–7** further allow to determine the scaling constants for comparison between genes. However, determining these constants relies on some assumptions detailed in the notes, and we invite the reader to carefully consider them before running these additional steps.

1. *Profile normalization*: Divide each entry of the profile R by the total number of transcripts M, to obtain a normalized profile ρ of densities of ribosome footprints ($\rho = R/M$, *see* **Note 12**).
2. *Smoothing*: Apply a moving average of length 10 (codons) and stride 1 to ρ, to form the smoothed density vector $\bar{\rho}$ with xth entry (*see* **Notes 12** and **13**).

$$\bar{\rho}(x) = \frac{1}{10}\sum\nolimits_{k=x}^{x+9}\rho(k)$$

3. *Scaled smoothed elongation rates*: Estimate the scaled vector of smoothed elongation rate $\bar{\lambda}$ with x-th entry $\bar{\lambda}(x) = \frac{1-9\bar{\rho}(x)}{\bar{\rho}(x)(1-\bar{\rho}(x))}$ (*see* **Note 14**).
4. *Scaled initiation rate and termination rate*: Estimate the scaled initiation rate $\alpha' = 1/(1-10\bar{\rho}(1))$, and the termination rate $\beta' = 1/\rho(n)$, where n is the gene length in codons (that is the length of the vector ρ, *see* also **Note 15**).
5. *Deconvolution*: Deconvolve the scaled smoothed elongation rates λ' into scaled codon-specific elongation rates. For a set of unique codons $\{c_i\}$ present in the gene sequence, solve the

scaled codon-specific elongation rates $\{\varepsilon'(c_i)\}$ as the least-squares solution (*see* **Note 16**) of the linear system

$$\left\{\sum_{k=x}^{x+9} \varepsilon'(c(k)) = 10\tilde{n}\lambda'(x)\right\}_{Lxn-9},$$

where L is a fixed integer smaller than $n-9$ (*see* **Note 17**), $c(k)$ is the codon at position k, and x varies from L to $n-9$ (yielding a system of $n-8-L$ equations).

6. *Scaling constant and production rate*: Determine the scaling constant τ as the mean of $\{\varepsilon'(c_i)\}$, and the production rate as $\frac{1}{\tau}$.
7. *Unscaled rates*: Compute the unscaled elongation, initiation and termination rates (*see* **Note 18**), respectively as

$$\lambda = \lambda'/\tau, \alpha = \alpha'/\tau, \beta = \beta'$$

3.3 Visualization and Perturbation Analysis

For a given sequence of elongation rates, which can be inferred from a ribosome profile with procedure Subheading 3.2, one can conversely predict the profile and other quantities of interest from a mathematical analysis of the TASEP model of translation [7]. The software presented in Erdmann-Pham et al. [8] was built upon these theoretical results to provide a graphical user interface for visualizing the current and local densities of ribosomes along the mRNA. The following steps describe how to use this software, called EGGTART, to visualize the predicted profile and translation rates, and explore the parameter landscape of the TASEP model to test the impact of modifying the initiation rate and/or the codon sequence).

1. Load the elongation rate input file (*see* Subheading 2) into the software EGGTART.
2. Once loaded, EGGTART will display, as shown in Fig. 3, (*see* also **Note 19** six panels detailing: in the left column (i) the elongation profile itself (bottom) and (ii) the resulting ribosome density profile (up), *in middle column*, (iii) the currently chosen initiation and termination rates α and β together with a ribosome size l (bottom), and (iv) the phase diagram (up) describing when changes in initiation rates will produce possibly discontinuous modifications in densities [7] and in right column graphs depicting the dependence of protein production rates on both initiation and termination.
3. Adjust panels (i), (iii), and (iv) to interactively explore in the other panels, the effects of local elongation rate perturbations and/or fluctuations in initiation and termination rates on both ribosome densities and protein production rates (*see* **Note 20**).
4. Export the results by clicking on the File Tab then Export command. After naming the export folder (and choosing where to save it), EGGTART will create separate pdf figures

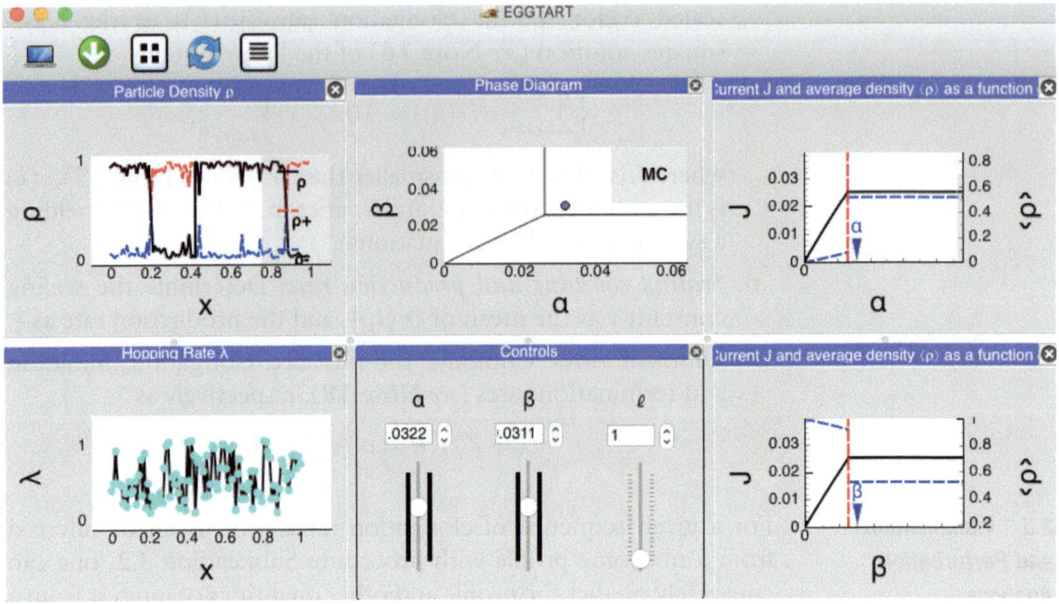

Fig. 3 EGGTART default interface. The default interface generated after loading the input file is divided into six panels. Top row, left to right: Smoothed density profile, phase diagram (*see* ref. 8), and production rate as a function of the initiation rate. Bottom row, left to right: elongation rates, parameters control panel, production rate as a function of the termination rate

of the different plots displayed and csv files detailing the input and output values.

4 Notes

1. Translation efficiency (TE) is the rate of mRNA translation into proteins. It is normally calculated by dividing normalized ribosome footprint reads by the normalized RNA sequencing reads. The concept is based on the assumption that translation interferences are absent, and the more ribosomes are on a transcript, the more efficient is protein synthesis.
2. The precise procedure to estimate total transcript abundances from RNA-Seq reads may depend on the protocol used for the RNA-Seq experiment. For example, if cDNAs are used, remember to only average reads near the 3′-end and discard all remaining locations.
3. Choice of reference sequences: one could alternatively retrieve reference sequences from the UCSC table browser [12] (http://www.genome.ucsc.edu/cgi-bin/hgTables). In the main text we use Ensembl database because it offers data in a

systematic way and for a vast array of organisms. However, some kind of data are only found from the UCSC browser [13]. This regards essentially RefSeq which offers a non-redundant set of reference sequences.

4. For the depletion and alignment step described in [2], reference sequences can also be retrieved, for instance, from Ensembl FTP [9]. In this case, one should select the category non-coding RNA (ncRNA) [9], which contains transfer RNA, ribosomal RNA and the category coding RNA (cDNA) for the mRNA reference and download the corresponding FASTA files. To build the depletion reference, it is not sufficient to only use the ncRNA FASTA. One should also add the sequences corresponding to mitochondrial mRNA or to pseudogenes, which can be extracted from the cDNA reference. To build the mRNA reference, in a complementary way, this set of sequences (mitochondrial mRNA and pseudogenes) should be removed from the file. The depletion reference can be further refined by adding tRNA sequences from the GtRNAdb [14] (http://gtrnadb.ucsc.edu) and rRNA sequences from arb-silva [15] (https://www.arb-silva.de/). In this case, identical or almost identical sequences should be identified and removed. Identical sequences can be found using for instance a bash or perl script. Identical and almost-identical sequences can be found using the sequence similarity matrix outputted by a multiple alignment tool such as Clustal Omega (https://www.ebi.ac.uk/Tools/msa/clustalo/) [16].

5. The following bash commands can be adapted and used to clean an mRNA reference from unwanted sequences. Here, pseudogenes and mitochondrially translated mRNA sequences are removed.

```
filecDNA=myref.fasta.gz
filemRNA=myref_mRNA.fasta
# Unzip, and convert the fasta to a fasta-like format (one
line per sequence).
gzip -dc $chemin/$filecDNA | awk '
   !/^>/ { printf "%s", $0; n = "\n" }
   /^>/ { print n $0"<"; n = "" }
   END { printf "%s", n }
   ' | tr '\n' ' ' | sed 's/ >/\n>/g' > $filecDNA.fastalike2
# Remove pseudogenes and MT-encoded mRNAs, convert back to
fasta
cat $filecDNA.fastalike2 |\
   grep -v "gene_biotype:pseudogene" |\
   grep -v ":Mito:" |\
   sed 's/< /\n/g' | sed 's/\([acgtuACGTUN]\{80\}\)/\1\n/g'
```

```
$filemRNA
# Check number of sequences at start, intermediate and final
steps, mRNA reference
gzip -dc $chemin/$filecDNA | grep -c "^>"
grep -c "^>" $filecDNA.fastalike2
grep -c "^>" $filemRNA
grep -c "protein_coding" $filemRNA
less $filemRNA
```

6. Command "gtf2table" might fail if the GTF file selected in input is not exactly as expected by the script. This can be solved by using script options in the command help, obtained by typing "help(gtf2table)" in R. One should also additionally use option "verbose=1" to understand the cause of the problem. If the common options are not sufficient, the user might want to adapt the script to the available GTF file. This is more convenient by using the standalone Python script which contains command "gtf2table." This script is named "gtf2table_standalone.py" and it can be found in the sources of RiboVIEW on zenodo (https://doi.org/10.5281/zenodo.4401399). This script can then be modified and run for debug, as follows:

```
> python gtf2table_standalone.py \
  --refGTF $filegtf \
  --out $fileCDS \
  --endinside1 1 \
  --stopoutside1 1 \
  --stopminusoutside1 1 \
  --verbose 1
```

   In this command, endindside1 and so on are the default settings, which should be modified depending on the GTF file's characteristics. Be cautious if using verbose=2 since this setting prints very detailed information, which is useful for debug, but leads to a slow execution and a large volume of messages printed to screen.

7. XP.names: it is wise to use a clear but short name for each sample defined in the list "XP.names", since these names are used in tables or graphs. Short names will prevent these from being overcrowded with long names.

8. mRNA tracks can be customized, by using the options of RiboVIEW command "visu.tracks":

   mRNA="random" picks an mRNA with a large number of footprints aligned to it and plots this mRNA track. mRNA="mRNA1" selects mRNA1 for plotting.

codon.labels=TRUE displays codon labels on the mRNA tracks plot.

codon.col="darkslateblue" by default but can be set to any other R-compatible color.

Additionally, codon.col can be set as a list of colors for each codon:

```
> codon.colors <- rbind( c('aaa','tomato'), c('aac','or-
ange'), ..., c('uuu','forestgreen') )
```

9. The full list of results outputted by RiboVIEW is summed up in Table 1.
10. This note provides alternatives or complementary analyses to results and quality control done by RiboVIEW. For complementary quality controls, *see* the following.
    (a) Read density profiles for detection of sequence artifacts, in particular secondary structure artifacts: see RUST (Ribo-seq Unit Step Transformation)
    (b) 3-nt periodicity at STOP codon: RiboViz [17].
    (c) 3nt-periodicity on full CDS: RiboSeqR (https://rdrr.io/bioc/riboSeqR), etc. (*see* ref. 2 for a review).
    (d) biases on reads distribution: RiboViz.

    For supplemental results, *see* the following.
    (a) Translation efficiency: TripsViz (https://trips.ucc.ie), RiboGalaxy (tool riboSeqR).
    (b) Genome tracks: GwipsViz/RUST (https://gwips.ucc.ie).
    (c) Galaxy implementation: etc. (*see* ref. 2 for a review).

    For alternative presentations of quality controls or results, see in particular the set of tools RiboViz, GwipsViz/RUST, TripsViz (*see* also a list in RiboVIEW [2]).
11. The Spearman correlation written below the heatmap in RiboQC is taken between the vector of conditions "XP.conditions" and the vector of groups automatically inferred from the hierarchical clustering tree. In particular, the tree is cut to form as many groups as there are conditions. If there are many conditions and few replicates, this grouping might not work perfectly, and the user should then trust the tree, rather than the correlation value.
12. After normalizing by transcript abundances (**step 1** of Subheading 3.3), the ribosomal density vectors ρ should feature entries between 0 and 1 (only one ribosome can occupy a given site at any time), and smoothed density profiles (**step 2** of Subheading 3.3) $\bar{\rho}$ should lie between 0 and 1/10 (ribosome footprints span approximately ten codons, so they cannot

Table 1
List of results outputted by RiboVIEW

Description	File names
Codon enrichment by sample, here condition 1, replicate 1 (mean, standard deviation over mRNA transcripts)	Sample-Stress.R1_Codon-enrichment-unbiased_mean Sample-Stress.R1_Codon-enrichment-unbiased_sd
Codon enrichment by condition (mean, standard deviation and standard error over samples)	Enrichmnt-per-condition_weighted-mean.tsv Enrichmnt-per-condition_weighted-sdev.tsv Enrichmnt-per-condition_weighted-serr.tsv
Comparison between conditions, here condition 1 relative to condition 2	Condition_Stress_relative-to_Control_mean-and-stderr.tsv
Bulk codon occupancy by sample. For unbiased codon occupancy *see* "Codon enrichment by sample," "Codon enrichment by condition," and "Comparison between conditions"	Sample-Stress.R1_BCO
Hussmann's codon enrichment (mean and standard deviation)	Sample-Stress.R1_Hussmann_mean Sample-Stress.R1_Hussmann_sd
Number of reads per kilobase of transcript per million mapped reads	Sample-Stress.R1 .RPKM
List of paused sites (mRNA, position, codon and context sequence, relative peak height)	Sample-Stress.R1 .paused
Single codon occupancy (mRNA, codon, coverage, occupancy at codon resolution, % mRNA sequence covered)	Sample-Stress.R1 .SCO
Dashboard of quality control parameters, and overview of results. These files are readable in standard browsers such as Firefox, Chrome, Brave, Safari, or Edge.	Results-Qc.html Results-Mine.html
Metagene coordinates and coverage for mRNA transcripts	Sample-Stress.R1 .metagene
Raw read count per mRNA transcript	Sample-Stress.R1 .nbreads
Coverage at footprint boundaries (at codon resolution and at nucleotide resolution)	Sample-Stress.R1 .limCod Sample-Stress.R1 .limNt
Coverage around AUG, stratified by footprint lengths (25nt to 32nt)	Sample-Stress.R1_OL.txt
Pictures corresponding to Results-QC.html and Results-Mine.html in higher quality, in EPS format	*.eps
Pictures corresponding to Results-QC.html and Results-Mine.html in, in PNG format	*.png
Data underlying the generation of Results-Qc.html and Results-Mine.html. These files are readable in R	*.RData

occur with a frequency of more than once per ten codons per transcript). If either ρ or $\bar{\rho}$ violate these constraints, the corresponding gene should be filtered out. Also note that if the profile contains a region with zero reads, the inference procedure can only assign an infinite local elongation rate in this region, while large density peak values yield small values for the density profiles in other sites, which would potentially create numerical instabilities for the method. The user can consider inputing some default values or applying smoothing or thresholding on the profiles to mitigate the effects caused by these extreme values.

13. $\bar{\rho}$ is a vector of size $n - 9$. While inferring initiation and termination rates, pay particular attention on whether ρ or $\bar{\rho}$ are manipulated; accidentally mistaking one for the other may lead to cascading misinference downstream.
14. $\bar{\lambda}$ is a vector of size $n - 9$. While the smoothed elongation rates λ provide slightly more coarse-grained information than the unsmoothed jump rates that give rise to them, they are sufficient for assessing translation efficiency within the inhomogeneous l-TASEP framework.
15. Given that we solve the TASEP model in the so-called hydrodynamic limit [7, 8], the formulas used here become theoretically exact in the limit of large transcript lengths n, and otherwise provides predictions of accuracy proportional to n^{-1}. In practice, for genes of at least 100 codons ($n \geq 100$), theoretical predictions become indistinguishable from empirical observations, and so provide a reliable means to accurately infer elongation rates, initiation rates, termination rates and translation efficiencies from high-quality Ribo-Seq experiments.
16. The least square solution of a linear system can for example be directly computed using the function lsqr in MATLAB. In the Deconvolution step of Subheading 3.3, run $lsqr(A, b)$, where b is an array of size $n - 8 - L$ containing the terms from $10 \cdot \lambda'(x)$ $(L \leq x \leq n - 9)$, and A is a matrix of size $n - 8 - L$, such that entries of the row labeled by x are 1 from x to $x + 9$, and 0 else. Also note that while we detail how to apply the method for one gene sequence, the user can also jointly find the least square solution for all the genes in the dataset, to get an estimate of the codon-specific average elongation rate across all genes.
17. In **step 5** of Subheading 3.2, the user specifies a parameter L, which can be interpreted as a value set under the assumption that after the first L codons, the elongation rate is not biased by the codon position. Such a bias was indeed found toward the 5′-end in yeast ribosome profiling data (also known as the 5′

translational ramp) [18] and affects all codons [6]. This led us to introduce this parameter, which we recommend to set to 150 when analyzing Ribo-Seq data from Yeast. If there is no evidence of such a bias, the user can simply set L to 1, while being careful of any bias of density due to the position, which might also affect the estimates obtained in **step 5**.

18. The normalization procedure of **step 7** works best if every gene contains every (coding) codon, which evidently is not true in most genomes. However, statistically speaking, the scheme continues to produce high-quality estimates if the set of missing codons is sufficiently close to a randomly sampled subset of codons, which is true for most organisms. Care must be taken when comparing results across different genes with a sizable number (15) of missing codons, as their statistical inference will lack power.
19. In **step 2** of Subheading 3.3 and Fig. 3, we describe the default interface of EGGTART. The user can also adjust the panels by moving them or remove some of them. The elongation rates are plotted such that the values on x-axis go from 0 to 1, as we normalize the codon position by the sequence length (in codons). Finally, it is important to note that the density profile displayed by EGGTART is obtained under the hydrodynamic limit, and thus represent the smooth profile obtained after smoothing the input rates as in **step 2** of Subheading 3.2.
20. Single modifications of the elongation rate sequence can be done by selecting each dot on panel (i) and move the dot to the desired value. The control panel (iii) allows to adjust the initiation and termination rates. In addition, the user can adjust a parameter l that corresponds to the size of the particle in the TASEP model. In the context of simulating ribosome profile data, we recommend to leave it at 10 (i.e., the size occupied by the ribosomes).

Acknowledgments

We thank Đan Erdmann-Pham for early discussion and feedback on the manuscript.

References

1. Ingolia NT, Ghaemmaghami S, Newman JR, Weissman JS (2009) Genome-wide analysis in vivo of translation with nucleotide resolution using ribosome profiling. Science 324(5924): 218–223. https://doi.org/10.1126/science.1168978
2. Legrand C, Tuorto F (2020) RiboVIEW: a computational framework for visualization, quality control and statistical analysis of ribosome profiling data. Nucleic Acids Res 48(2): e7. https://doi.org/10.1093/nar/gkz1074
3. Tuorto F, Legrand C, Cirzi C, Federico G, Liebers R, Muller M, Ehrenhofer-Murray AE,

Dittmar G, Grone HJ, Lyko F (2018) Queuosine-modified tRNAs confer nutritional control of protein translation. EMBO J 37(18). https://doi.org/10.15252/embj.201899777

4. Tuorto F, Herbst F, Alerasool N, Bender S, Popp O, Federico G, Reitter S, Liebers R, Stoecklin G, Grone HJ, Dittmar G, Glimm H, Lyko F (2015) The tRNA methyltransferase Dnmt2 is required for accurate polypeptide synthesis during haematopoiesis. EMBO J 34(18):2350–2362. https://doi.org/10.15252/embj.201591382
5. Muller M, Legrand C, Tuorto F, Kelly VP, Atlasi Y, Lyko F, Ehrenhofer-Murray AE (2019) Queuine links translational control in eukaryotes to a micronutrient from bacteria. Nucleic Acids Res 47(7):3711–3727. https://doi.org/10.1093/nar/gkz063
6. Dao Duc K, Song YS (2018) The impact of ribosomal interference, codon usage, and exit tunnel interactions on translation elongation rate variation. PLoS Genet 14(8):e1007620. https://doi.org/10.1371/journal.pgen.1007620
7. Erdmann-Pham DD, Dao Duc K, Song YS (2020) The key parameters that govern translation efficiency. Cell Syst 10(2):183–192.e186. https://doi.org/10.1016/j.cels.2019.12.003
8. Erdmann-Pham DD, Son W, Dao Duc K, Song YS (2021). EGGTART: A tool to visualize the dynamics of biophysical transport under the inhomogeneous ℓ-TASEP. Biophysical Journal, 120(8), 1309-1313. https://doi.org/10.1016/j.bpj.2021.02.004
9. Yates AD, Achuthan P, Akanni W, Allen J, Allen J, Alvarez-Jarreta J, Amode MR, Armean IM, Azov AG, Bennett R, Bhai J, Billis K, Boddu S, Marugan JC, Cummins C, Davidson C, Dodiya K, Fatima R, Gall A, Giron CG, Gil L, Grego T, Haggerty L, Haskell E, Hourlier T, Izuogu OG, Janacek SH, Juettemann T, Kay M, Lavidas I, Le T, Lemos D, Martinez JG, Maurel T, McDowall M, McMahon A, Mohanan S, Moore B, Nuhn M, Oheh DN, Parker A, Parton A, Patricio M, Sakthivel MP, Abdul Salam AI, Schmitt BM, Schuilenburg H, Sheppard D, Sycheva M, Szuba M, Taylor K, Thormann A, Threadgold G, Vullo A, Walts B, Winterbottom A, Zadissa A, Chakiachvili M, Flint B, Frankish A, Hunt SE, IIsley G, Kostadima M, Langridge N, Loveland JE, Martin FJ, Morales J, Mudge JM, Muffato M, Perry E, Ruffier M, Trevanion SJ, Cunningham F, Howe KL, Zerbino DR, Flicek P (2020) Ensembl 2020. Nucleic Acids Res 48 (D1):D682–D688. https://doi.org/10.1093/nar/gkz966
10. Diament A, Tuller T (2016) Estimation of ribosome profiling performance and reproducibility at various levels of resolution. Biol Direct 11:24. https://doi.org/10.1186/s13062-016-0127-4
11. Hussmann JA, Patchett S, Johnson A, Sawyer S, Press WH (2015) Understanding biases in ribosome profiling experiments reveals signatures of translation dynamics in yeast. PLoS Genet 11(12):e1005732. https://doi.org/10.1371/journal.pgen.1005732
12. Karolchik D, Hinrichs AS, Furey TS, Roskin KM, Sugnet CW, Haussler D, Kent WJ (2004) The UCSC Table Browser data retrieval tool. Nucleic Acids Res 32(Database issue):D493–D496. https://doi.org/10.1093/nar/gkh103
13. O'Leary NA, Wright MW, Brister JR, Ciufo S, Haddad D, McVeigh R, Rajput B, Robbertse B, Smith-White B, Ako-Adjei D, Astashyn A, Badretdin A, Bao Y, Blinkova O, Brover V, Chetvernin V, Choi J, Cox E, Ermolaeva O, Farrell CM, Goldfarb T, Gupta T, Haft D, Hatcher E, Hlavina W, Joardar VS, Kodali VK, Li W, Maglott D, Masterson P, McGarvey KM, Murphy MR, O'Neill K, Pujar S, Rangwala SH, Rausch D, Riddick LD, Schoch C, Shkeda A, Storz SS, Sun H, Thibaud-Nissen F, Tolstoy I, Tully RE, Vatsan AR, Wallin C, Webb D, Wu W, Landrum MJ, Kimchi A, Tatusova T, DiCuccio M, Kitts P, Murphy TD, Pruitt KD (2016) Reference sequence (RefSeq) database at NCBI: current status, taxonomic expansion, and functional annotation. Nucleic Acids Res 44(D1):D733–D745. https://doi.org/10.1093/nar/gkv1189
14. Chan PP, Lowe TM (2016) GtRNAdb 2.0: an expanded database of transfer RNA genes identified in complete and draft genomes. Nucleic Acids Res 44(D1):D184–D189. https://doi.org/10.1093/nar/gkv1309
15. Quast C, Pruesse E, Yilmaz P, Gerken J, Schweer T, Yarza P, Peplies J, Glockner FO (2013) The SILVA ribosomal RNA gene database project: improved data processing and web-based tools. Nucleic Acids Res 41(Database issue):D590–D596. https://doi.org/10.1093/nar/gks1219
16. Madeira F, Park YM, Lee J, Buso N, Gur T, Madhusoodanan N, Basutkar P, Tivey ARN, Potter SC, Finn RD, Lopez R (2019) The EMBL-EBI search and sequence analysis tools APIs in 2019. Nucleic Acids Res 47(W1):W636–W641. https://doi.org/10.1093/nar/gkz268

17. Carja O, Xing T, Wallace EWJ, Plotkin JB, Shah P (2017) riboviz: analysis and visualization of ribosome profiling datasets. BMC Bioinformatics 18(1):461. https://doi.org/10.1186/s12859-017-1873-8
18. Tuller T, Carmi A, Vestsigian K, Navon S, Dorfan Y, Zaborske J, Pan T, Dahan O, Furman I, Pilpel Y (2010) An evolutionarily conserved mechanism for controlling the efficiency of protein translation. Cell 141(2): 344–354. https://doi.org/10.1016/j.cell.2010.03.031

Chapter 10

Analysis of Translational Control in the Integrated Stress Response by Polysome Profiling

Michael J. Holmes, Jagannath Misra, and Ronald C. Wek

Abstract

Translational control provides a strategy for rapid optimization of gene expression and restoration of protein homeostasis in response to cellular stresses. An important mechanism for translational control involves phosphorylation of eIF2, which invokes the integrated stress response (ISR). In the ISR, initiation of bulk protein synthesis is lowered coincident with enhanced translation efficiency of select gene transcripts that serve critical functions in stress adaptation. In this chapter, we focus on polysome profiling as a tool for establishing and characterizing translation control induced by eIF2 phosphorylation during environmental stresses. We describe in detail the experimental strategies of polysome profiling for detecting bulk repression of the translational machinery and quantifying translational control of key stress-induced gene transcripts. These experimental strategies can be adjusted to measure individual gene transcripts or genome-wide analyses and can be adapted to measure changes in the levels of ribosome subunits and associated factors invoked by various cellular cues in the ISR.

Key words Translational control, Integrated stress response, Polysome profiling, mRNA translation, ATF4

1 Introduction

The integrated stress response (ISR) is a widely conserved mechanism that adjusts the rate of bulk translational initiation to environmental stimuli [1]. In mammals, four different protein kinases sense distinct stress insults and phosphorylate the alpha subunit of eukaryotic initiation factor 2 (eIF2α) at serine 51. These eIF2α kinases include HRI (EIF2AK1) that is induced by heme and iron depletion in erythroid cells and by mitochondrial stress [2–4], PKR (EIF2AK2), which is activated by viral infection [5], the endoplasmic reticulum (ER)-resident PERK (EIF2K3) that responds to ER stresses such as the accumulation of unfolded proteins within its

Michael J. Holmes and Jagannath Misra contributed equally.

Daniel Mateju and Jeffrey A. Chao (eds.), *The Integrated Stress Response: Methods and Protocols*, Methods in Molecular Biology, vol. 2428, https://doi.org/10.1007/978-1-0716-1975-9_10, lumen [1, 6], and GCN2 (EIF2AK4), which is activated by amino acid depletion and UV irradiation [7–10].

The eIF2 bound to GTP delivers initiating Met-tRNA$_i^{Met}$ to 40S ribosomal subunit and facilitates recognition of the mRNA start codon by the scanning ribosome. During the initiation process, GTP associated with eIF2 is hydrolyzed to GDP and eIF2 is released from the translation machinery. To participate in subsequent rounds of translation initiation, eIF2 requires the exchange of its bound GDP for a new GTP by the guanine exchange factor eIF2B. Whereas the nucleotide exchange occurs efficiently under optimal growth conditions, stress-dependent phosphorylation of eIF2α (eIF2α ~ P) increases its affinity for eIF2B thereby limiting eIF2 recycling. Lowered eIF2-GTP levels dampen the overall rate of translational initiation [1]. Importantly, eIF2α ~ P also selectively enhances the translation of specific stress-related mRNAs, such as the transcription factor *ATF4* [11–13]. ATF4 serves to program the mRNA expression of the ISR that is critical for restoration of protein homeostasis and feedback resumption of translation [1, 14, 15].

Polysome profiling is a well-established method for investigating translational control. In a polysome profiling experiment, cell lysate is generated and applied to the top of a continuous sucrose gradient followed by high-speed ultracentrifugation. The sucrose gradient centrifugation separates the free 40S and 60S ribosomal subunits, mRNAs bound to a single ribosome (monosomes), and the mRNAs that are being translated by many ribosomes (polysomes). Total RNA content of the gradient is analyzed by continuous absorbance measurements at 254 nm as a piston gradient fractionator traverses the sucrose gradient column. In nonstressed cells, large polysomes are observed relative to monosomes which is indicative of efficient translation occurring in growing cells; however, lowered bulk translation seen during the ISR is characterized by the reduction of large polysomes with a concomitant increase in monosomes [1]. A significant benefit of this method is that it also enables the simultaneous collection of fractions from the gradient that can be used to characterize shifts in the distribution of translation machinery as well as gene transcripts during cellular stress.

Here, we utilize two stress agents, tunicamycin and sodium arsenite ($NaAsO_2$), which induce PERK [16] and HRI [17], respectively, to activate the ISR in mammalian cells. We outline the steps involved in generating high-quality cell lysate and using it for polysome profiling. We also describe how to assess changes in bulk translation capacity by calculating the ratio of monosomes to polysomes, how to investigate cases of transcript-specific translational control by determining changes in mRNA sedimentation by reverse transcription and real-time PCR (RT-qPCR), and how to measure changes in the sedimentation profile of translational machinery by western blotting that occur during the ISR.

2 Materials

2.1 Cell Culture and Chemical Treatment

1. CO_2 cell culture incubator.
2. DMEM supplemented with 10% heat-inactivated fetal bovine serum.
3. 10 cm tissue culture dishes.
4. Cultured mammalian cells, such as mouse embryonic fibroblasts (MEF) or human foreskin fibroblasts (HFF).
5. 2 mM stock tunicamycin (Calbiochem) dissolved in DMSO (*see* **Note 1**).
6. 1 M stock $NaAsO_2$ (Sigma-Aldrich) dissolved in H_2O (*see* **Note 1**).
7. 100 mg/ml stock solution of cycloheximide (CHX, Sigma-Aldrich) prepared in DMSO (*see* **Note 1**).

2.2 Making Sucrose Gradients and Other Solutions

1. 50 ml of 10% sucrose solution: 10% sucrose (w/v), 20 mM Tris–HCl, pH 7.4, 100 mM NaCl, 10 mM $MgCl_2$, 50 μg/ml CHX (*see* **Note 2**)
2. 50 ml of 50% sucrose solution: 50% sucrose (w/v), 20 mM Tris–HCl, pH 7.4, 100 mM NaCl, 10 mM $MgCl_2$, 50 μg/ml CHX
3. 120 ml of 1× PBS freshly supplemented with 50 μg/ml CHX
4. 5 ml of polysome lysis solution: 20 mM Tris–HCl, pH 7.4, 100 mM NaCl, 10 mM $MgCl_2$, 50 μg/ml CHX, 0.4% NP-40 (v/v) (*see* **Note 3**), 1× EDTA-free protease and phosphatase inhibitor (*see* **Note 4**).
5. Six Seton Scientific 9/16″ × 3-1/2″ polyclear ultracentrifuge tubes (Fisher Scientific) (*see* **Note 5**)
6. 4-in. 14-gauge Luer lock cannula
7. 10 ml syringe
8. Biocomp gradient making station and gradient-making accessories, including rubberized caps and marker block, for Beckman Coulter SW41-Ti rotor (Biocomp Instruments; Fig. 1) (*see* **Note 7**)
9. Fine point indelible marker.

2.3 Collecting Lysate and Ultracentrifugation

1. 1× PBS freshly supplemented with 50 μg/ml CHX
2. Large flat tray with raised edges filled with crushed ice
3. Polysome lysis solution
4. Cell scraper
5. Prechilled 1.5 ml microcentrifuge tubes
6. 23-gauge needle and syringe

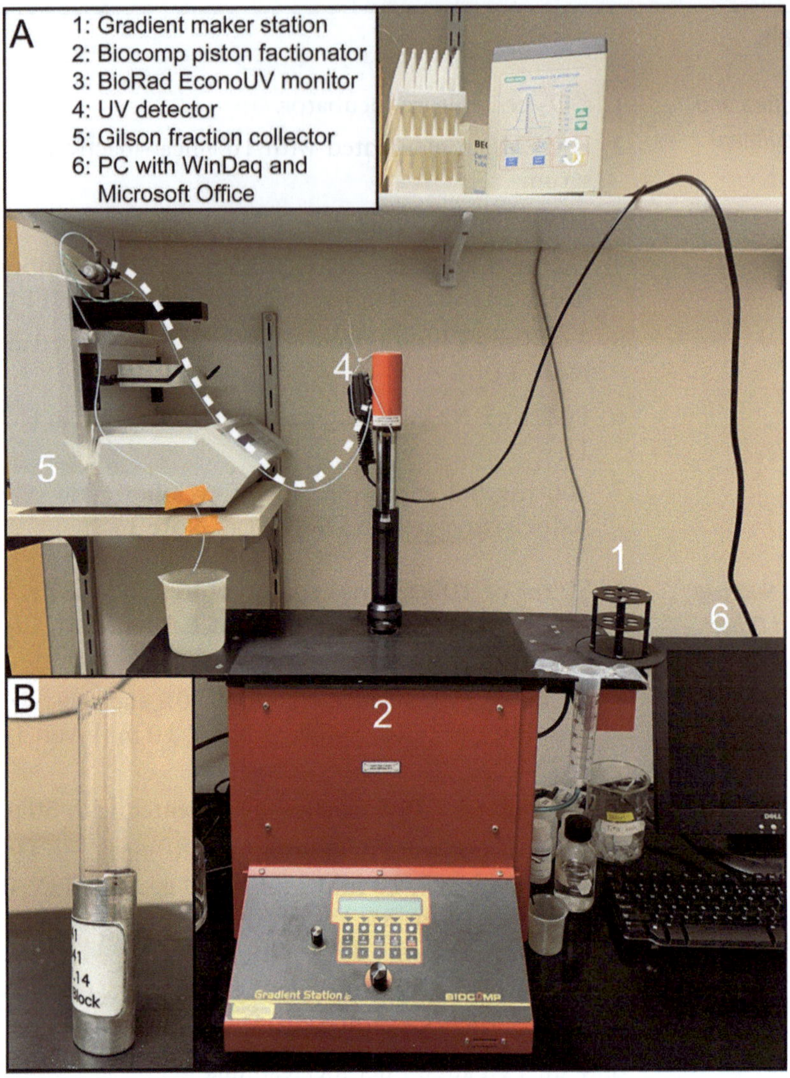

Fig. 1 Depiction of Biocomp piston fractionator and associated instrumentation. (**a**) All instrumentation is numbered corresponding to the inset. The dotted white line corresponds to the path that the eluate travels between the UV detector and the fraction collector. In order to accurately compare fractions to the polysome profile, the time it takes for the eluate to travel this distance must be determined for each particular setup (*see* **Note 6**). (**b**) Image of marked ultracentrifuge tube inside marker block. The black line approximately halfway up the tube shows the quantity of 10% sucrose to add in the first step when forming sucrose gradients

7. Refrigerated tabletop centrifuge prechilled to 4 °C
8. Laboratory scale
9. Tube rack that supports ultracentrifuge tubes
10. SW41-Ti rotor prechilled to 4 °C
11. Beckman Coulter Optima L-series ultracentrifuge prechilled to 4 °C

2.4 Polysome Analysis and Fractionation

1. Biocomp piston fractionator system and accessories equipped with Bio-Rad Econo UV monitor and Gilson fraction collector (Fig. 1) (*see* **Note 7**).
2. One spare Seton Scientific polyclear ultracentrifugation tube.
3. RNase-free water.
4. WinDaq Data Acquisition and Analysis software (DATAQ, Ohio).
5. 2 ml free-standing screw cap tubes (Fisher).
6. −80 °C freezer.
7. Microsoft Office software.

2.5 Analysis of Transcript Sedimentation

1. Luciferase Control RNA (Promega) diluted to 1 μg/ml stock solution with RNase-free water.
2. TRIzol™ LS (Ambion).
3. Chloroform.
4. Refrigerated tabletop centrifuge prechilled to 4 °C.
5. Isopropanol.
6. GlycoBlue™ coprecipitant (15 mg/ml) (Invitrogen).
7. Ice-cold 75% EtOH made with RNase-free water.
8. RNase-free water.
9. Reverse transcription reagents (*see* **Note 8**).
10. RNase inhibitor.
11. SYBR Green real-time PCR (qPCR) reagents (*see* **Note 8**).
12. Primers for firefly luciferase spike-in (forward: CCAGGGATTT CAGTCGATGT and reverse: AATCTCACGCAGG CAGTTCT) and other genes of interest (e.g., mouse *ATF4*; forward: GCCGGTTTAAGTTGTGTGCT and reverse: CTG GATTCGAGGAATGTGCT) (*see* **Note 9**).
13. Microsoft Office software.

2.6 Analysis of Translational Machinery Sedimentation

1. 1.5 ml microcentrifuge tubes.
2. Trichloroacetic acid, 100% (w/v) solution, prechilled to 4 °C.
3. Refrigerated tabletop centrifuge prechilled to 4 °C.
4. Ice-cold acetone.
5. 2× SDS-PAGE loading dye.
6. Heat block set to 95 °C.
7. Standard equipment for running SDS-PAGE.
8. Standard equipment for western blot transfer and detection.
9. Rabbit polyclonal anti-RPS17 IgG (Abcam).
10. Graphics software, such as Adobe Illustrator.

3 Methods

3.1 Cell Culture and Chemical Treatment

1. Subculture mammalian cell line into 10 cm tissue culture dishes using appropriate techniques for your experiment and cell line (*see* **Note 10**).
2. When cells reach about 70% confluency (*see* **Note 11**), replace with fresh media and add tunicamycin to 2 μM for 6 h or $NaAsO_2$ to 500 μM for 1 h (*see* **Note 12**).
3. Replace cells in CO_2 incubator for allotted time.

3.2 Making Sucrose Gradients and Other Solutions

1. Prepare PBS, 10% and 50% sucrose, and polysome lysis buffer between 2 and 12 h in advance of intended cell lysate collection time. Store chilled at 4 °C.
2. Using the upper edge of the marker block, trace a line on the exterior of six polyclear centrifuge tubes with a fine point indelible marker (Fig. 1b).
3. Using a 10 ml syringe affixed with the 4-in. cannula, add the 10% sucrose solution up to the marked line. Wipe off excess sucrose solution from the cannula between each tube. For consistency, be sure to avoid getting sucrose solution on the walls of the centrifuge tube (*see* **Note 13**). After the last tube, rinse the cannula with RNase-free water.
4. Using a 10 ml syringe affixed with the 4-in. cannula, add the 50% sucrose solution to the bottom of the centrifuge tube to layer it under the 10% sucrose solution. Add enough 50% sucrose to form a convex meniscus at top of the centrifuge tube. Carefully withdraw the cannula while being careful to minimize disturbances at the interface between the 10% and 50% sucrose solutions and without letting the cannula touch the walls of the centrifuge tube. A sharp interface should be visible between the two layers. Wipe off excess sucrose solution from the cannula between each tube (*see* **Note 14**).
5. Place rubberized caps on each centrifuge tube and remove excess sucrose from caps and outside of centrifuge tube.
6. Transfer centrifuge tubes to Biocomp gradient making station and run its preset program for making 10–50% sucrose (w/v) gradients for the SW41-Ti rotor. Once the program is complete, return tubes to 4 °C until they are ready to be loaded with cell lysates.

3.3 Collecting Lysate and Ultracentrifugation

1. 5–10 min prior to intended harvest time, add CHX to the media to reach 50 μg/ml and return the sample to incubator (*see* **Note 15**).

2. Remove the media, place plate on ice, and rinse twice with 8 ml prechilled PBS supplemented with CHX (*see* **Note 16**).
3. Remove excess PBS and add 500 μl polysome lysis buffer to plate. Scrape cells vigorously. Transfer lysate to an RNase-free 1.5 ml microcentrifuge tube on ice. Maintain lysate on ice until all six samples are collected.
4. Triturate samples 8–10 times using a 23-gauge needle affixed to a syringe.
5. Incubate cell lysates on ice for 10 min.
6. Centrifuge samples for 10 min at 21,000 × *g* in prechilled tabletop centrifuge to clarify lysate. Transfer supernatant to new prechilled 1.5 ml microcentrifuge tube.
7. Remove the rubber stoppers from each ultracentrifuge tube and remove 300 μl from each of the top of the sucrose gradients.
8. Add 400 μl of precleared lysate to the top of each gradient in a dropwise manner while being careful not to disturb the top of the gradient.
9. Record the weight of each tube. Add polysome lysis buffer to lighter tubes so that the weight of each tube is equivalent prior to ultracentrifugation (*see* **Note 17**).
10. Transfer tubes to prechilled SW41-Ti swinging rotor.
11. Centrifuge the samples at 40,000 rpm for 2 h in a Beckman Coulter L-series ultracentrifuge prechilled to 4 °C.

3.4 Polysome Analysis and Fractionation

1. Approximately 30 min before the completion of ultracentrifugation, turn on Bio-Rad Econo UV monitor to allow lamp to equilibrate (*see* **Note 7**).
2. During the 30-min warm up time, fill a spare ultracentrifugation tube with RNase-free water and use it to set up the initial position of the piston on the fractionator. The position of the piston should be calibrated to the point just before the water reaches the UV detector. Also set up the piston movement rate to 0.3 and the distance the piston travels so that it will reach the bottom of the tube.
3. Upon completion of the piston setup, blank the monitor system with RNase-free water.
4. Process the samples using the Biocomp piston fractionator as instructed in the user manual while using the WinDaq software to record the polysome profiles.
5. As the piston fractionator is running, collect the eluate in equal fractions of at least 500 μl into screw-cap tubes using a Gilson fraction collector.

6. Store fractions at −80 °C until they are ready to be processed for RNA extraction or protein concentration.
7. When finished processing all samples, run warm RNase free water through both the piston fractionator and the fraction collector (*see* **Note 18**).
8. Export all data to Microsoft Excel file in order to plot and compare polysome profiles (*see* **Note 6**).
9. To obtain the monosome–polysome ratio, use Microsoft Excel to calculate the area under the curve of the monosome peak and the heavy polysome region (*see* **Note 19**).

3.5 Analysis of Transcript Sedimentation

1. Allow fractions to thaw slowly on ice and transfer 250 μl from each to sample to a new 1.5 ml microcentrifuge tube (*see* **Note 20**). Make sure to mix each fraction well prior to transfer since the sucrose will sediment during thawing.
2. Add 3 μl from a 1 μg/ml stock of firefly luciferase RNA as a spike-in to each sample (*see* **Note 21**).
3. Add 750 μl TRIzol LS reagent to each sample and mix well.
4. Isolate the RNA according to the manufacturer's protocol with the following modifications:
   (a) Add 1 μl GlycoBlue coprecipitant along with the addition of isopropanol (*see* **Note 22**).
   (b) Resuspend the RNA pellet in 20 μl RNase-free water (*see* **Note 23**).
5. Perform the reverse transcription reaction according to the manufacturer's protocol in a 50 μl reaction while supplementing with RNase inhibitor and using 19.5 μl input of each RNA sample (*see* **Note 24**). For short term storage, keep the cDNA at −20 °C until ready to perform real-time PCR.
6. Dilute the cDNA 1/10 using RNase-free water.
7. Perform real-time PCR as per the manufacturer's instructions in technical triplicates for *ATF4* (or other genes of interest) and the firefly luciferase spike-in control.
8. Transfer the RT-qPCR data to Microsoft Excel.
9. Taking the firefly luciferase as an internal standard, use the $2^{-\Delta\Delta CT}$ method to measure the relative levels of *ATF4* mRNA throughout the fractions (Fig. 2a) and calculate the percentage shift in its distribution along the gradient (Fig. 2b; *see* **Notes 25** and **26**).

3.6 Analysis of Translational Machinery Sedimentation

1. Allow fractions to thaw slowly on ice and transfer 500 μl to a new 1.5 ml microcentrifuge tube. Make sure to mix the fraction prior to transfer since the sucrose will sediment during thawing.

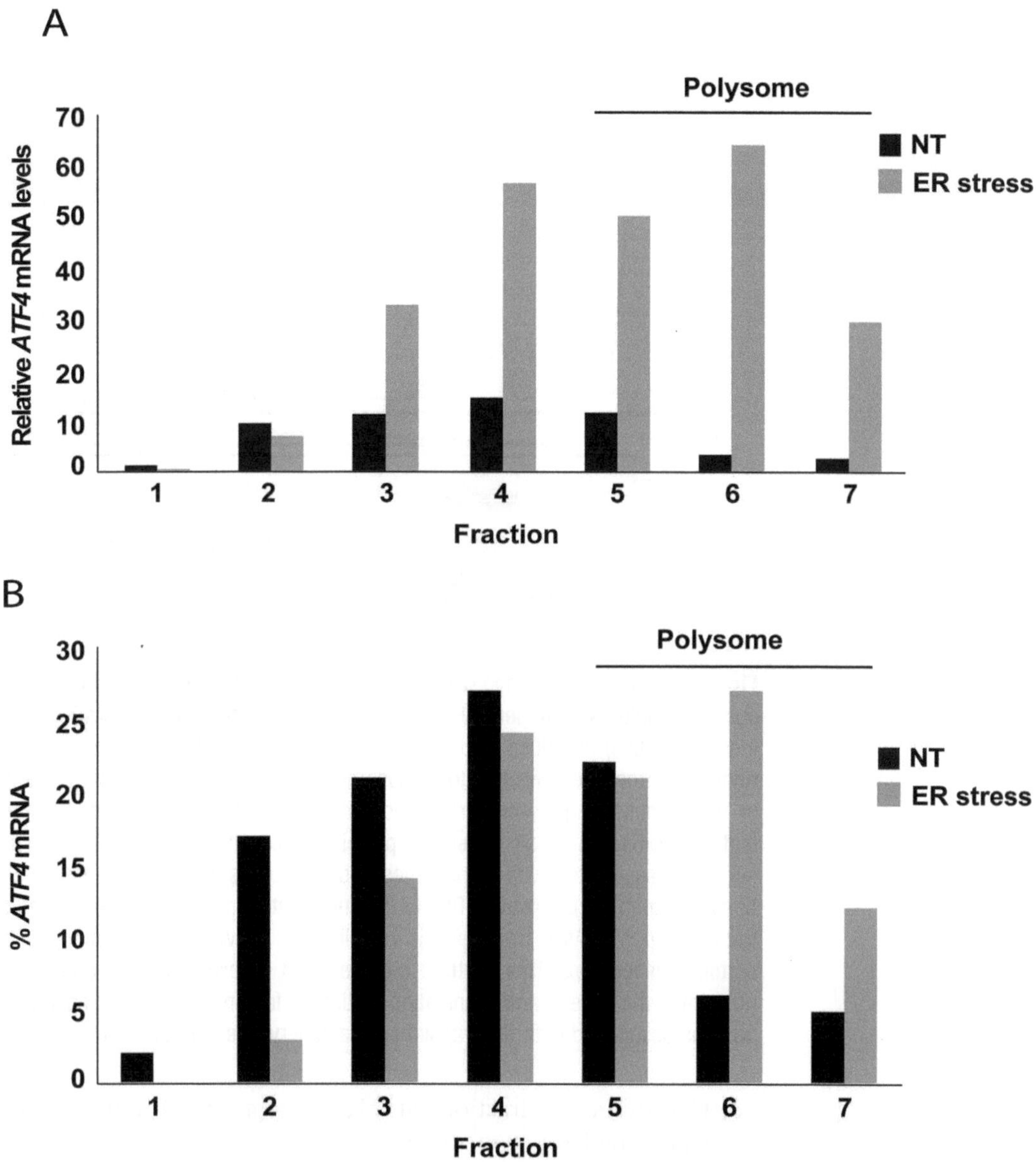

Fig. 2 ER stress triggers preferential translation of *ATF4* mRNA. Fourteen fractions were collected by sucrose gradient analyses of lysates prepared from MEF cells treated with tunicamycin for 6 h or left untreated (NT). Every two fractions were pooled together (*see* **Note 20**). (**a**) Relative *ATF4* mRNA levels present in each fraction were determined by RT-qPCR. This figure shows the combination of the transcriptional and translational control responses. (**b**) The distribution of *ATF4* mRNA in each fraction was normalized across each sample to show its shift towards heavier polysome fractions in order to highlight the translational control aspect of *ATF4* regulation upon ISR induction. The bar graph data presents the average of two biological replicates

2. Add 125 μl 100% trichloroacetic acid solution and mix well. Transfer fractions to −20 °C for 1 h to aid in protein precipitation.

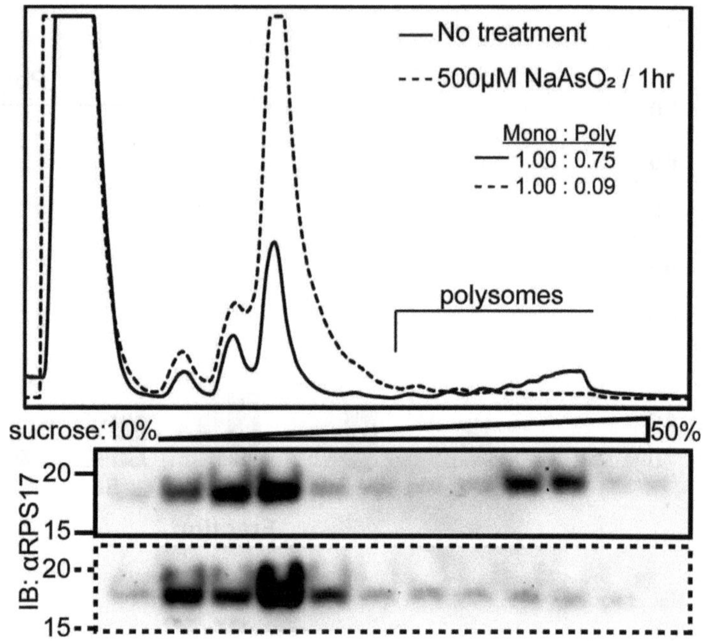

Fig. 3 Polysome profiling and western blot analyses of proteins in the fractions of sucrose gradients. Confluent HFF cells were treated with 500 μM $NaAsO_2$ for 1 h (dashed outline) to induce oxidative stress. Polysome profiling shows a shift of ribosomes from the polysome to monosome fractions and, consequently, a shift in the monosome–polysome ratio indicative of reduced translation. Western blots of RPS17, a small ribosomal protein, shows that the small ribosomal subunit is associated with monosomes and polysomes during nonstress conditions (top solid box). However, RPS17 is concentrated in low- and nontranslating fractions upon $NaAsO_2$ treatment as visualized in the western blot panel in the bottom dashed box. This method can be used to examine how other core ribosomal machinery and translation initiation factors, and other accessory proteins associated with the translation machinery, respond to ISR induction

3. Centrifuge the fractions at 21,000 × *g* for 30 min in a prechilled tabletop centrifuge.
4. Remove the supernatant. The pellet contains precipitated protein.
5. Rinse the tubes with 500 μl ice-cold acetone and incubate at room temperature for 10 min.
6. Quickly centrifuge the fractions and remove the acetone. Allow samples to air-dry for 10 min.
7. Resuspend the pellets in 2× SDS-PAGE loading dye and heat samples for 10 min at 95 °C.
8. Process the fractions for SDS-PAGE and western blotting using standard procedures.

9. Once western blot results are complete, use graphic software to align the images of the blots to the original polysome profiles in order to determine changes in sedimentation patterns (Fig. 3; *see* **Note 6**).

4 Notes

1. This chemical is a hazardous material and should be disposed of using appropriate precautions.
2. All solutions should be prepared with RNase-free Milli-Q water. We have found DEPC treatment to be unnecessary.
3. Alternatively, 1% Triton X-100 can be used with similar results.
4. $MgCl_2$ is required for the assembly of ribosomal subunits, and chelation of divalent cations (Mg^{2+}) by EDTA will cause dissociation of the ribosomal subunits from mRNAs. Intentional chelation of Mg^{2+} can be useful to run gradients in which both ribosomal subunits dissociate to allow for relative quantitation of one subunit to the other or to quantitate ribosomal content between treated and untreated samples.
5. The SW41-Ti rotor and the gradient maker hold six tubes. Since small variations in gradient formation can disproportionally impact the polysome profile, we limit direct comparisons between samples that were run at the same time.
6. It is important to be able to relate the captured fractions to their position in the polysome profile. First, determine the time it takes for liquid to pass from the UV detector to the fraction collector which corresponds to the delay between sample detection and sample collection. This path is depicted by the dotted line in Fig. 1a. It is also useful to record the time that the fraction collector takes to capture each fraction. Use this timing information to accurately compare profiles to each other and position RNA and/or protein measurements to their corresponding position in the profile.
7. We use an automatic gradient making station to generate sucrose gradients. For polysome profiling, we use a piston gradient fractionator paired with accessory machinery for UV monitoring and fraction collection. Since individual equipment will vary, precise set up instructions should be tailored to individual setups as per the manufacturer's manual or guidelines for analogous equipment.
8. We have used a variety of reagents and vendors for reverse transcription and qPCR steps with success.
9. When designing primers for RT-qPCR experiments that will be performed using SYBR green chemistry, we recommend using

software to verify that the primers will not form hairpins or homo/heterodimers that will create nonproductive by-products. We typically use online tools, such as OligoAnalyzer by IDT.

10. Most experiments examining induction of the ISR are best performed when cells are approximately 70% confluent. For MEFs, 1.5 million cells are grown in 10 cm dishes overnight. Growth rates and detailed subculture method will vary between cell lines.

11. Higher cell density can reduce cell proliferation and protein synthesis which can complicate assay interpretation. Confluency is suggested to activate one or more eIF2 kinases, triggering eIF2 phosphorylation that reduces bulk translation initiation.

12. The final concentration of compounds needed to illicit the ISR requires optimization for individual cell types and experiments. We have also seen variation in tunicamycin activity between vendors and individual product lots. ISR induction can be verified by western blotting for eIF2α ~ P compared to total eIF2α amounts under a variety of concentrations and treatment times as appropriate.

13. In order to accurately compare between different samples within the same experiment, consistency in gradient preparation is of prime importance. Videos from Biocomp demonstrating ideal techniques that are available on youtube.com are excellent resources.

14. Assuring a clean exterior of the centrifuge tubes reduces their sticking in the SW41 tube holders and ensures more accurate weight measurement of the loaded tubes prior to ultracentrifugation.

15. Prolonged CHX exposure can cause alterations in polysome profiles due to blockage and accumulation of monosomes at early codons soon after the start codon. Consequently, it is advised to maintain a consistent CHX exposure time for each sample by adding CHX to each sample prior to the collection of the previous (e.g., add CHX to sample 3 just prior to collection of sample 2) since it takes approximately 5 min to process each sample.

16. From this point on, all steps should be performed on ice or at 4 °C in a cold room. If the samples are not cold, ribosomes may detach from mRNAs.

17. Many protocols contain instructions to measure the A260 absorbance of the lysate prior to loading it on the sucrose gradient that it can be normalized to the absorbance units. However, it is worth noting that the detergents used for lysis

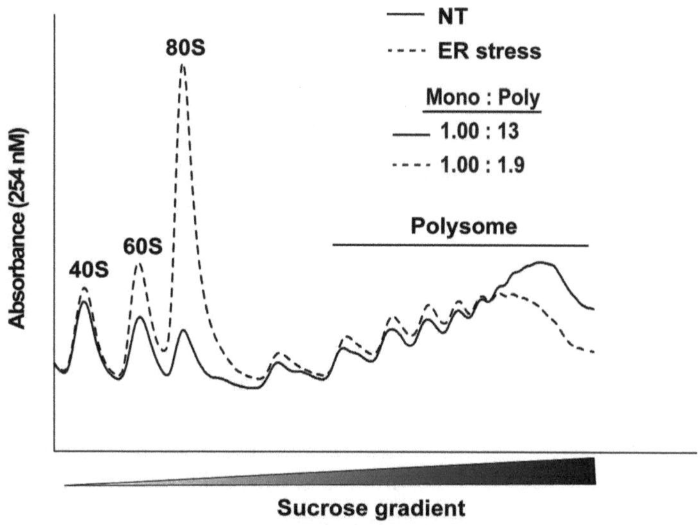

Fig. 4 ER stress triggers the induction of the ISR which lowers bulk mRNA translation initiation as measured by sucrose gradient centrifugation. MEF cells were treated with ER stress agent tunicamycin 6 h or left untreated (NT). Lysates were then analyzed by centrifugation in a 10–50% sucrose gradient and the profiles were monitored by absorbance at 254 nm. The reduction in heavy polysomes and concomitant increase in 80S monosomes are indicative of repressed translation initiation by eIF2α ~ P during ER stress

(NP-40 or Triton X-100) strongly absorb at this wavelength, rendering A260 normalization of little use. We have found that normalizing to cell input, such as outlined in this protocol here provides more consistent results.

18. Running warm water through all piping prevents the buildup of dried sucrose which can cause problems over time. Additionally, we periodically run 70% ethanol followed by water through our machine to prevent the buildup of mold.
19. Comparison of the monosome–polysome ratio is frequently used to demonstrate changes in the bulk translational capabilities during ISR induction. For the polysome value, we calculate the area under the curve from the trisome (three ribosomes) until the heavier polysomes at the right-hand end of the curve (Fig. 4).
20. One can combine every two fractions together (e.g., 125 μl from fraction 1 mixed with 125 μl from fraction 2, etc.) at this step to save on PCR reagents. While this setup decreases resolution, it is typically sufficient to see changes in transcript sedimentation during ISR induction.
21. The addition of exogenous RNA spike-in serves as an internal control for normalizing transcript abundance between fractions during RT-qPCR. The firefly RNA contains a poly-A

tail, making it compatible with an oligo-dT–primed reverse transcription reaction.

22. The glycogen will act as a carrier to help precipitate the RNA. GlycoBlue makes it easier to visualize the RNA pellets.
23. At this stage, the RNA fractions can be pooled into monosome and polysome fractions and processed for high-throughput sequencing if whole transcriptome analysis is desired.
24. Since the RNA content of each fraction has been diluted during polysome profiling, we prefer to include as much RNA as possible during reverse transcription. An optimal approach is to directly resuspend the RNA pellet from the previous step into 1× reverse transcription mix. Also, depending on the experiment and the target genes, we have used random hexamer and oligo-dT priming for generating cDNA.
25. ER stress causes not only preferential translation upregulation but also a transcriptional increase in *ATF4* mRNA levels. Relative transcript abundance was measured by normalizing the amount of ATF4 mRNA present in each fraction by luciferase mRNA present in the respective fraction (Fig. 2a). In order to suppress the contribution of the transcriptional response and highlight the translational control of *ATF4*, the amount of mRNA present in each fraction is expressed in terms of percentage of total ATF4 mRNA for a treatment group. To do so, the sum of all *ATF4* mRNA present in each of the seven fractions should be calculated and the *ATF4* signal should be calculated as the percent of total for that treatment group (Fig. 2b). This method can be applied to examine other gene transcripts that are subjected to translational control mechanisms, such as selective preferential translation and translational repression.
26. The length of coding sequence determines the maximum number of ribosomes that can be accommodated on an mRNA. Therefore, transcripts with long coding sequences that are well translated will typically appear predominantly in the heavier polysome fractions and can show greater migration between fractions during ISR induction than mRNAs with shorter coding sequences.

Acknowledgments

The authors would like to thank Sheree Wek for technical support. This work was supported by National Institutes of Health Grants R35GM136331, R01GM049164, and R01AI24723 (to R.C.W.). M.J.H. is supported by an American Heart Association postdoctoral fellowship (20POST35150019). R.C.W. has received grant support from Eli Lilly and Company and is a scientific advisor to

HiberCell; the other authors declare no conflicts of interest. The content is solely the responsibility of the authors and does not necessarily represent the official views of the American Heart Association and National Institutes of Health.

References

1. Wek RC (2018) Role of eIF2α kinases in translational control and adaptation to cellular stress. Cold Spring Harb Perspect Biol 10(7): pii:a032870. https://doi.org/10.1101/cshperspect.a032870
2. Chen JJ, Zhang S (2019) Heme-regulated eIF2α kinase in erythropoiesis and hemoglobinopathies. Blood 134(20):1697–1707. https://doi.org/10.1182/blood.2019001915
3. Guo X, Aviles G, Liu Y, Tian R, Unger BA, Lin YT, Wiita AP, Xu K, Correia MA, Kampmann M (2020) Mitochondrial stress is relayed to the cytosol by an OMA1-DELE1-HRI pathway. Nature 579(7799):427–432. https://doi.org/10.1038/s41586-020-2078-2
4. Fessler E, Eckl EM, Schmitt S, Mancilla IA, Meyer-Bender MF, Hanf M, Philippou-Massier J, Krebs S, Zischka H, Jae LT (2020) A pathway coordinated by DELE1 relays mitochondrial stress to the cytosol. Nature 579(7799):433–437. https://doi.org/10.1038/s41586-020-2076-4
5. Pindel A, Sadler A (2011) The role of protein kinase R in the interferon response. J Interf Cytokine Res 31(1):59–70. https://doi.org/10.1089/jir.2010.0099
6. Walter P, Ron D (2011) The unfolded protein response: from stress pathway to homeostatic regulation. Science 334(6059):1081–1086. https://doi.org/10.1126/science.1209038
7. Baird TD, Wek RC (2012) Eukaryotic initiation factor 2 phosphorylation and translational control in metabolism. Adv Nutr 3(3): 307–321. https://doi.org/10.3945/an.112.002113
8. Hinnebusch AG (2005) Translational regulation of Gcn4 and the general amino acid control of yeast. Annu Rev Microbiol 59:407–450. https://doi.org/10.1146/annurev.micro.59.031805.133833
9. Jiang HY, Wek RC (2005) GCN2 phosphorylation of eIF2alpha activates NF-kappaB in response to UV irradiation. Biochem J 385 (Pt 2):371–380. https://doi.org/10.1042/BJ20041164
10. Deng J, Harding HP, Raught B, Gingras AC, Berlanga JJ, Scheuner D, Kaufman RJ, Ron D, Sonenberg N (2002) Activation of GCN2 in UV-irradiated cells inhibits translation. Curr Biol 12(15):1279–1286. https://doi.org/10.1016/s0960-9822(02)01037-0
11. Vattem KM, Wek RC (2004) Reinitiation involving upstream ORFs regulates ATF4 mRNA translation in mammalian cells. Proc Natl Acad Sci U S A 101(31):11269–11274. https://doi.org/10.1073/pnas.0400541101
12. Lu PD, Harding HP, Ron D (2004) Translation reinitiation at alternative open reading frames regulates gene expression in an integrated stress response. J Cell Biol 167(1): 27–33. https://doi.org/10.1083/jcb.200408003
13. Harding HP, Novoa I, Zhang Y, Zeng H, Wek R, Schapira M, Ron D (2000) Regulated translation initiation controls stress-induced gene expression in mammalian cells. Mol Cell 6(5):1099–1108. https://doi.org/10.1016/s1097-2765(00)00108-8
14. Marciniak SJ, Ron D (2006) Endoplasmic reticulum stress signaling in disease. Physiol Rev 86(4):1133–1149. https://doi.org/10.1152/physrev.00015.2006
15. Harding HP, Zhang Y, Zeng H, Novoa I, Lu PD, Calfon M, Sadri N, Yun C, Popko B, Paules R, Stojdl DF, Bell JC, Hettmann T, Leiden JM, Ron D (2003) An integrated stress response regulates amino acid metabolism and resistance to oxidative stress. Mol Cell 11(3): 619–633. https://doi.org/10.1016/s1097-2765(03)00105-9
16. Teske BF, Wek SA, Bunpo P, Cundiff JK, McClintick JN, Anthony TG, Wek RC (2011) The eIF2 kinase PERK and the integrated stress response facilitate activation of ATF6 during endoplasmic reticulum stress. Mol Biol Cell 22(22):4390–4405. https://doi.org/10.1091/mbc.E11-06-0510
17. McEwen E, Kedersha N, Song B, Scheuner D, Gilks N, Han A, Chen JJ, Anderson P, Kaufman RJ (2005) Heme-regulated inhibitor kinase-mediated phosphorylation of eukaryotic translation initiation factor 2 inhibits translation, induces stress granule formation, and mediates survival upon arsenite exposure. J Biol Chem 280(17):16925–16933. https://doi.org/10.1074/jbc.M412882200

Chapter 11

High-Resolution Ribosome Profiling for Determining Ribosome Functional States During Translation Elongation

Mohammad Shafieinouri, Britnie Santiago Membreno, and Colin Chih-Chien Wu

Abstract

Translation elongation is a highly choreographed process that involves substantial conformational changes of the ribosome to accommodate aminoacyl-tRNAs and traverse along the mRNA template. To capture distinct functional states of the ribosome, a high-resolution ribosome profiling-based approach has been developed. By deep-sequencing differently sized ribosome-protected mRNA fragments, this approach captures not only ribosome positions but also their functional states in vivo across the *Saccharomyces cerevisiae* transcriptome with codon resolution. This chapter presents a condensed and step-by-step protocol for preserving ribosomes in their functional states using a cocktail of antibiotics that traps distinct steps of elongating ribosomes and for constructing a cDNA library derived from the ribosome-protected mRNA fragments for deep sequencing.

Key words High-resolution ribosome profiling, Ribosome functional states, Elongation regulation

1 Introduction

During translation elongation, ribosomes undergo substantial conformational changes to complete iterative cycles of three major reactions: aminoacyl-tRNA (aa-tRNA) decoding, peptide-bond formation, and translocation, which are catalyzed by eEF1A, the large ribosomal subunit, and eEF2, respectively (Fig. 1). Accordingly, there are three ribosome functional states corresponding to the three major reactions (PreAcc, PrePT, and PreTrans in Fig. 1). Over the past few years, studies have shown that elongation can be monitored and regulated at various steps, including aa-tRNA decoding and translocation [1]. It requires the ability not only to determine ribosome positioning but also to distinguish ribosome functional states in the cell to better understand translational regulation at the phase of elongation. Ribosome profiling provides an unprecedented transcriptome-wide view of ribosome positions by

Daniel Matějů and Jeffrey A. Chao (eds.), *The Integrated Stress Response: Methods and Protocols*,
Methods in Molecular Biology, vol. 2428, https://doi.org/10.1007/978-1-0716-1975-9_11,

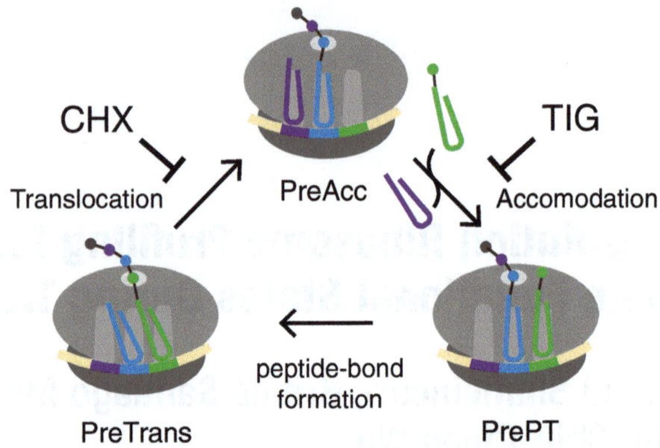

Fig. 1 Eukaryotic elongation cycle. A schematic representation of the eukaryotic elongation cycle. Ribosome functional states are abbreviated: *PreAcc* preaccommodation, *PrePT* pre–peptide bond formation, *PreTrans* pretranslocation. Cycloheximide (CHX) and tigecycline (TIG) are used to inhibit translocation and accommodation, respectively

deep sequencing ribosome-protected mRNA fragments (RPFs) [2, 3]. While elongating ribosomes occupy two distinct RPFs that are different in length (21 nt and 28 nt) [4, 5], 21 nt RPFs correspond to ribosomes with an empty A site and 28 nt RPFs correspond to ribosomes with a full A site [6]. To trap the distribution of ribosome functional states in vivo, we have devised a systematic approach using a cocktail of cycloheximide and tigecycline, which inhibit translocation and aa-tRNA accommodation, respectively [7, 8] (Fig. 1). This methodology provides means to isolate 21 nt RPFs that represent ribosomes in a preaccommodation state (i.e., decoding ribosomes) and 28 nt RPFs that predominantly correspond to ribosomes in a pretranslocation state. In addition, 21 nt RPFs exhibit a superior correlation with cellular tRNA abundance compared to 28 nt RPFs [6]. This characteristic of 21 nt RPFs further enables us to perform quality control on samples prepared with this improved method.

In a high-resolution ribosome profiling experiment, cycloheximide and tigecycline are used during cell lysis to trap ribosomes in their functional state. After cell lysis, polysome cushioning, and RNase digestion, monosomes are isolated and RPFs are purified. Following linker ligation and reverse transcription, RPFs are then converted to a cDNA library. By deep sequencing the cDNA library, transcriptome-wide ribosome functional states can be determined based on RPF lengths. The aim of this chapter is to present a concise protocol for an experimentalist to construct RPF sequencing libraries and to perform quality control on the prepared libraries prior to further tailored analysis.

2 Materials

Prepare all solution using RNase-free water.

1. Ambion ™ RNase I (ThermoFisher).
2. SUPERase*In (ThermoFisher).
3. T4 Polynucleotide kinase (New England Biolabs).
4. T4 RNA ligase 2, truncated KQ (New England Biolabs).
5. Superscript™ III reverse transcriptase (ThermoFisher).
6. CircLigase ssDNA ligase I (Lucigen).
7. Phusion DNA polymerase (ThermoFisher).
8. High sensitivity DNA kit and reagents (Agilent).
9. Cycloheximide (Sigma-Aldrich).
10. Tigecycline (Sigma-Aldrich).
11. SYBR Gold (ThermoFisher).
12. 10 bp DNA ladder (ThermoFisher).
13. GlycoBlue (ThermoFisher).
14. Oligo clean and concentrate kit (Zymo).
15. Acid phenol (VWR).
16. Betaine (Sigma-Aldrich).
17. riboPOOL yeast ribosomal RNA depletion kit (siTOOLs BIOTECH).
18. 15% TBE-urea polyacrylamide gel (Bio-Rad Laboratories).
19. 10% TBE-urea polyacrylamide gel (Bio-Rad Laboratories).
20. 8% TBE polyacrylamide gel.
21. Corning Costar spin-X column (Sigma-Aldrich).
22. Whatman cellulose nitrate 0.45 μm filter membrane (Cytiva).
23. Welch W direct vacuum pump (Welch).
24. Freezer mill (SPEX SamplePrep).
25. Small grinding vial set (SPEX SamplePrep).
26. Type 70 Ti rotor (Beckman Coulter).
27. 26.3 mL polycarbonate bottle with cap (Beckman Coulter).
28. SW41 rotor (Beckman Coulter).
29. Open-top SW41 thin-walled tube (Seton).
30. Gradient master (Biocomp).
31. Gradient fractionator (Biocomp).
32. Ultracentrifuge (Beckman Coulter).
33. Typhoon biomolecular imager (Cytiva).

34. Bioanalyzer (Agilent).
35. Footprint lysis buffer A: 20 mM Tris-HCl (pH 8), 140 mM KCl, 1.5 mM $MgCl_2$, 0.1 mg/mL cycloheximide, 0.1 mg/mL tigecycline.
36. Footprint lysis buffer B: 20 mM Tris-HCl (pH 8), 140 mM KCl, 1.5 mM $MgCl_2$.
37. 10× gradient buffer: 200 mM Tris-HCl (pH 8), 1.5 M KCl, 5 mM DTT, 50 mM $MgCl_2$.
38. Sucrose cushion: 20 mM Tris-HCl (pH 8), 150 mM KCl, 0.5 mM DTT, 5 mM $MgCl_2$, 1 M sucrose.
39. Gel staining solution: 5 μL SYBR Gold in 50 mL 1× TBE buffer.
40. RNA loading buffer: TBE urea samples buffer (2×) (VWR).
41. RNA elution buffer: 300 mM NaOAc, 1 mM EDTA (pH 8), 0.1 U/μL SUPERase*In.
42. DNA loading buffer: 50% glycerol in nuclease-free H_2O (xylene cyanol and bromophenol blue can be added at 0.1%).
43. DNA elution buffer: 10 mM Tris-HCl (pH 8), 300 mM NaCl, 1 mM EDTA (pH 8).
44. Dephosphorylation reaction mix per reaction: 0.5 μL of 10× T4 polynucleotide kinase reaction buffer (New England Biolabs), 0.5 μL of SUPERase*In.
45. Ligation reaction mix per reaction: 0.5 μL of 10× T4 RNA ligase reaction buffer (New England Biolabs), 0.2 μL of 50 μM preadenylated 3′ linker, 3.8 μL of 50% PEG 8000.
46. RT reaction mix per reaction: 4 μL of 5× First-Strand buffer, 1 μL of 10 mM dNTPs, 1 μL of 0.1 M DTT, 1 μL of SUPERase*In.
47. DNA elution buffer: 10 mM Tris-HCl (pH 8), 300 mM NaCl, 1 mM EDTA (pH 8).
48. Circularization reaction mix per reaction: 8.8 μL of 10 mM Tris-HCl (pH 8), 3.2 μL of 5 M betaine, 1.6 μL of CircLigase 10× reaction buffer, 0.8 μL of 1 mM ATP, 0.8 μL of 50 mM $MnCl_2$, 0.8 μL of CircLigase I.
49. PCR master mix: 129 μL of nuclease-free H_2O, 40 μL of HF buffer, 4 μL of 10 mM dNTPs, 1 μL of 100 μM PCR universal primer, 2 μL of Phusion DNA polymerase.
50. RNA size marker 15 nt: AUGUACACGGAGUCG.
51. RNA size marker 34 nt: AUGUACACGGAGUCGAG CACCCGCAACGCGAAUG.
52. Preadenylated 3′ linker: /5Phos/ NNNNNNCACTCGGG CACCAAGGA/3ddC/.

53. RT primer: /5Phos/RNNNAGATCGGAAGAGCGTCGTGTAGGGAAAGAGTGTAGATCTCGGTGGTCGC/iSP18/TTCAGACGTGTGCTCTTCCGATCTGTCCTTGGTGCCCGAGTG.
54. PCR universal primer: AATGATACGGCGACCACCGAGATCTACAC.
55. Barcoded PCR primer: CAAGCAGAAGACGGCATACGAGATAGTCGTGTGACTGGAGTTCAGACGTGTGCTCTTCCG (barcode is underlined).

3 Methods

3.1 Yeast Cell Harvesting by Filtration and Flash Freezing

1. Set up filtration apparatus with a Whatman cellulose nitrate 0.45 μm filter membrane and connect to a direct vacuum pump.
2. Place a tube rack in an ice bucket and fill it with liquid nitrogen. Puncture holes into the bottom of a 50 mL conical tube using an 18-G needle and place it on the tube rack. The level of liquid nitrogen needs to be at least 3–4 in. deep to submerge half of a 50 mL conical tube.
3. Pour ~800 mL of the yeast culture into filtration apparatus. While the media is filtering, scrape with a plastic spatula along the surface of membrane till a clump of yeast cells accumulate on the spatula (*see* **Note 1**).
4. Plunge the spatula into the tube submerged in liquid nitrogen and flick off the cells off with a prechilled pipette tip (*see* **Note 2**).
5. Add 1 mL of footprint lysis buffer A into the tube full of liquid nitrogen.
6. Cap the conical tube and store at −80 °C. Liquid nitrogen will drain through the holes on the bottom of the tube.

3.2 Lyse by Freezer Mill and Polysome Isolation

1. Set up a freezer mill. Prechill a sample prep grinder small vial on dry ice.
2. Place frozen yeast sample, lysis buffer, and a grinding rod inside vial.
3. Run freezer mill with settings 1 min on and 1 min off for eight cycles.
4. Take vial out of freezer mill. Remove cap. Immediately pour frozen ground powder into 22 mL of prechilled footprint lysis buffer A in a labeled 50 mL conical tube (*see* **Note 3**).
5. Vortex to thaw frozen lysate powder in the conical tube.

6. Centrifuge at 4 °C (21,300 × *g*, 15 min). Transfer supernatant into a fresh 50 mL conical tube with a serological pipette.
7. Layer supernatant from **step 12** on the top of 3 mL sucrose cushion solution in a polycarbonate bottle (*see* **Note 4**).
8. Centrifuge in a Type 70 Ti rotor at 4 °C (60,000 rpm, 106 min).
9. Remove supernatant from ribosome pellets with a 10 mL serological pipette (*see* **Note 5**).
10. Wash ribosome pellets once gently with 0.5 mL cold footprint lysis buffer B.
11. Resuspend ribosome pellets in 1 mL cold footprint lysis buffer B with gentle pipetting on ice.
12. Determine RNA concentration using absorption.
13. Resuspended ribosome pellets can be flash frozen and kept at −80 °C.

3.3 Monosome Isolation by Sucrose Gradient

1. Digest 350 μg of pelleted polysomes from **step 3.2.13** with 500 U of Ambion™ RNase I in 300 μL footprint lysis buffer B on a thermomixer at 25 °C (500 rpm, 1 h).
2. Stop RNase I reaction by adding 6000 U of SUPERase*In and place digested samples on ice.
3. During RNase I digestion, use a gradient maker to prepare 15–50% sucrose gradient solution in SW41 polyclear tubes.
4. Layer digested solution from **step 3.3.2** onto 15-50% sucrose gradient solution.
5. Centrifuge in a SW41 rotor at 4 °C (40,000 rpm, 2 h).
6. Use a gradient fractionator to isolate monosome fractions. Monosome fractions can be stored at −80 °C for 6 months.
7. Preheat 0.75 mL of acid phenol in 2-mL screw-cap tubes at 65 °C.
8. Add 57 μL of 20% SDS solution per mL of monosome fractions from **step 3.3.6** and heat to 65 °C on a thermomixer to dissolve SDS.

   Add heated monosome fractions to preheated aid phenol and incubate at 65 °C on a thermomixer shaking at 1300 rpm for 5 min (*see* **Note 6**).
9. Chill samples on ice for 3 min. Centrifuge at room temperature (21,300 × *g*, 3 min).
10. Pull aqueous (bottom) layer and add to a 2-mL screwcap tube containing 0.7 mL acid phenol at room temperature.
11. Vortex samples for 3 min at room temperature. Centrifuge at room temperature (21,300 × *g*, 3 min).

12. Pull aqueous (top) layer and add to a 2-mL screw-cap tube containing 0.6 mL chloroform.
13. Vortex samples for 3 min at room temperature. Centrifuge at room temperature (21,300 × *g*, 3 min).
14. Pull aqueous (top) layer. Add 3 M NaOAc (pH 5.5) to 0.3 M. Then add equal volume of isopropanol and 2 μL of GlycoBlue.
15. Precipitate RNA samples on dry ice for 30 min. Centrifuge at 4 °C (21,300 × *g*, 30 min).
16. Discard supernatant and wash blue RNA pellets with 500 μL of 80% ice-cold ethanol, and air-dry for 15 min.
17. Resuspend RNA pellets in 10 -20 μL of 10 mM Tris-HCl (pH 8). Quantify RNA using absorption.

3.4 Construction of a Sequencing Library from Ribosome-Protected mRNA Fragments

3.4.1 Size Select Ribosome-Protected mRNA Fragments

1. Mix 4 μg of RNA from **step 3.3.17** with equal volume of RNA loading buffer.
2. Mix 2 μL of size control ladders (15 nt and 34 nt RNA) with equal volume of RNA loading buffer.
3. Heat samples to 80 °C for 2 min and then immediately place on ice.
4. Load samples on a 15% TBE-urea polyacrylamide gel and run at 340 V for 20 min.
5. Puncture holes into the bottom of a 0.5 mL PCR tube with an 18-G needle. Also prepare a punctured 0.5 mL PCR tube for control ladders.
6. Stain gel with 50 mL gel staining solution for 5 min with shaking at room temperature. Image gel with a Typhoon biomolecular imager and excise footprints between 15 and 34 nt as shown in Fig. 2. Place gel slices into a punctured 0.5 mL PCR tube from **step 3.4.1.5**. Also excise control ladders to keep as a size control (*see* **Note 7**).
7. Place PCR tubes full of gel slices from **step 3.4.1.6** into 1.5 mL RNase-free eppendorfs and centrifuge at room temperature (21,300 × *g*, 3 min) to shred the gel slices into the eppendorfs.
8. Add 450 μL RNA elution buffer to gel slices and elute on a nutator at 4 °C overnight.
9. Transfer gel mixture to a Costar Spin-X column. Centrifuge at 4 °C (21,300 × *g*, 3 min).
10. Transfer eluates from **step 3.4.1.9** into a new RNase-free Eppendorf. Add 2 μL GlycoBlue and 450 μL isopropanol. Mix well and chill on dry ice for 30 min.
11. Centrifuge at 4 °C (21,300 × *g*, 30 min) to pellet isolated RPFs.
12. Remove supernatant, wash pellets with 500 μL of 80% ice-cold ethanol, and air-dry for 15 min.

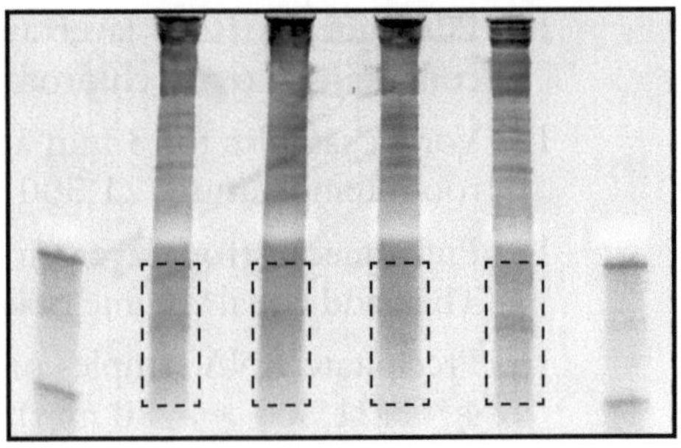

Fig. 2 Ribosome footprint size-selection gel. An image of size selection on a 15% TBE-urea gel from RPFs prepared from *S. cerevisiae* BY4171. Size control ladders (15 nt and 34 nt) are indicated. Areas marked with dashed lines indicate regions to be excised

13. Resuspend pellets in 3.5 μL of 10 mM Tris-HCl (pH 8) and transfer RNA and size control samples to 0.2-mL thin-walled PCR tubes

3.4.2 Dephosphorylation

1. Heat RNA and size control samples to 80 °C for 2 min and place on ice.
2. Add 1 μL of Dephosphorylation reaction mix and 0.5 μL of T4 polynucleotide kinase.
3. Incubate at 37 °C for 1 h in a thermocycler.

3.4.3 Linker Ligation

1. Add 4.5 μL of Ligation reaction mix directly to dephosphorylated sample from **step 52** (*see* **Note 8**).
2. Add 0.5 μL of T4 RNA ligase and incubate at 37 °C for 3 h in a thermocycler.
3. Clean up each reaction with a Zymo oligo clean and concentrate kit (*see* **Note 9**).

3.4.4 Ribosomal RNA Depletion

1. Deplete ribosomal rRNA from samples using a RiboPool yeast rRNA depletion kit according to the manufacturer's instructions (*see* **Notes 10** and **11**).
2. Clean up each reaction with a Zymo oligo clean and concentration kit and elute RNA in 15 μL of nuclease-free water.

3.4.5 Reverse Transcription

1. Mix 10 μL of RNA and size control samples from **step 3.4.4.2** with 1.25 μL of RT primer, heat up to 80 °C for 2 min, and place on ice.
2. Add 7.75 μL of RT reaction mix and then 1 μL of Superscript III.
3. Incubate samples at 55 °C for 30 min in a thermocycler.
4. Add 2 μL of 1 N NaOH to each sample and incubate at 95 °C for 15 min in a thermocycler.

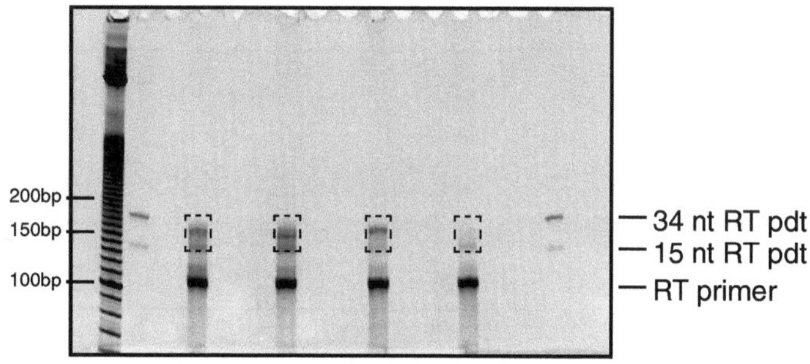

Fig. 3 Reverse transcription gel. A representative gel image of reverse transcription reactions on a 10% TBE-urea gel. 10-bp DNA ladders, 15 nt-RT and 34 nt-RT products are used to guide gel excision. Dashed boxes indicate areas to be excised. Unreacted RT primer is also indicated and should be avoided during gel excision

5. Clean up RNA and size control samples with a Zymo oligo clean and concentration kit and elute in 10 μL of nuclease-free water.
6. Mix the eluate with equal volume of RNA loading buffer.
7. Heat samples to 80 °C for 2 min and then immediately place on ice.
8. Load samples on a 10% TBE-urea polyacrylamide gel and run at 340 V for 1 h.
9. Excise reactions products from gels as described in **steps 3.4.1.6–3.4.1.7**. Use size control samples as a guide to excise reverse transcription products as shown in Fig. 3.
10. Shred gel slices as described in **step 3.4.1.7**.
11. Add 450 μL DNA elution buffer to shredded gel slices and elute on a nutator at room temperature overnight.
12. Transfer gel mixture to a Costar Spin-X column. Centrifuge at room temperature (21,300 × *g*, 3 min).
13. Precipitate reaction products as described in **steps 3.4.1.9–3.4.1.12**.

3.4.6 Circularization

1. Resuspend pellets in 15 μL of Circularization reaction mix and incubate at 60 °C for 2 h on a thermomixer.
2. Heat up samples to 80 °C for 10 min to inactivate the enzyme. Circularized DNA can be stored at −80 °C for 3 months.

3.4.7 Preparative PCR

1. Add 2 μL of circularized DNA sample and 2 μL of barcoded PCR primer to 36 μL of PCR master mix.
2. Aliquot 20 μL sample master mix into two 0.2-mL thin-walled PCR tubes. Each tube will undergo different numbers of PCR cycles.
3. Perform PCR as follows (*see* **Note 12**):

1 cycle:	98 °C	30 s
N cycles:	98 °C	10 s
	65 °C	10 s
	72 °C	5 s
1 cycle:	16 °C	Hold

4. Add 4 μL of DNA loading buffer to each sample.
5. Run an 8% TBE polyacrylamide gel at 340 V for 1 h (Fig. 4).
6. Stain gel with gel staining solution for 5 min with shaking at room temperature. Excise PCR products between 160 and 190 bp. Shred gel slices and precipitate PCR products as described in **steps 3.4.5.9–3.4.5.13**.
7. Resuspend PCR products in 6 μL of 10 mM Tris-HCl (pH 8). Dilute PCR products if necessary.
8. Quantify DNA concentration on a Bioanalyzer with a DNA high-sensitivity chip according to manufacturer's instructions. The size range of PCR products should be around 160–190 bp.
9. Submit quantified PCR products for Illumina sequencing. We recommend 50 bp sequencing on a NextSeq 2000 or NovaSeq 6000 machine.

3.5 Bioinformatic Analysis of RPF Libraries

Downstream analysis of RPF libraries heavily depends on the aims of the experiments. In subsequent steps, we describe our typical procedures in general terms to perform quality control on the prepared RPF libraries and we believe a variety of different approaches would also generate equivalent results (*see* Chapter 9).

1. Remove four random nucleotides included in RT primer from the 5′ end of reads with seqtk (https://github.com/lh3/seqtk) using option "trimfq -b 4" (*see* **Note 13**).
2. Remove 3′ adaptor sequence and discard reads shorter than 15 nt after trimming with skewer [9] using options "-x NNN NNNCACTCGGGCACCAAGGA -l 15".
3. Align trimmed reads to noncoding RNAs (including rRNAs, tRNAs, and snoRNAs) with STAR [10] using option "–outFilterMismatchNoverLmax 0.3".
4. Map unmapped reads against *S. cerevisiae* R64-1-1 S288C reference genome assembly (SacCer3) with STAR using options "–outFilterIntronMotifs RemoveNoncanonicalUnannotated –outFilterMultimapNmax 1 outFilterMismatchNoverLmax 0.1". STAR generates a SAM or BAM file of mapped reads for each input.

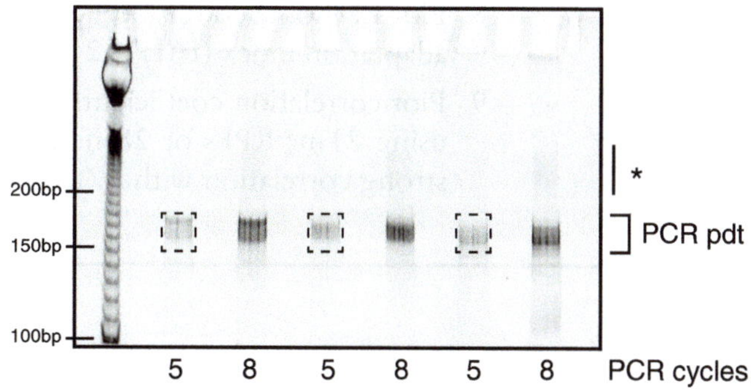

Fig. 4 Preparative PCR gel. A representative gel image of preparative PCR reactions on an 8% polyacrylamide native gel. 10 bp DNA ladders are included to guide gel excision. Dashed boxes indicate gel areas to be excised. Asterisk marks high-molecular weight PCR products. We do not recommend excising PCR products from reactions that have high-molecular weight products

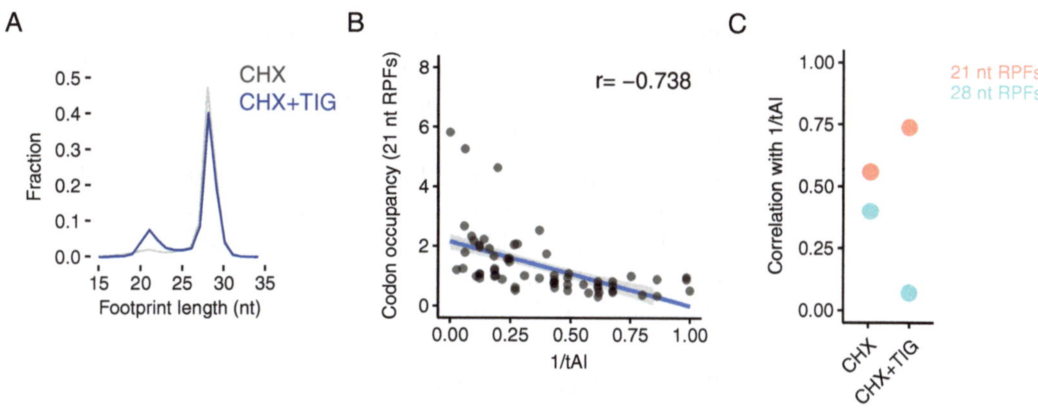

Fig. 5 Characteristics of high-resolution ribosome profiling. (**a**) Length distributions of RPFs comparing libraries prepared with CHX only (gray) and CHX+TIG (blue) in the lysates. (**b**) Correlation plot of codon-specific ribosome occupancies with the inverse of tRNA adaption index (tAI) using 21 nt RPFs prepared from unstressed *S. cerevisiae* BY4741 grown to OD_{600} 0.5 in YPD. *r* spearman rank correlation coefficient. (**c**) Spearman rank correlation coefficients of A-site codon-specific ribosome occupancies with the inverse of tAI from samples prepared with CHX or CHX+TIG

5. Plot size distributions of reads mapped to the reference genome from the SAM or BAM file. Using unstressed *S. cerevisiae* BY4741 grown to OD_{600} 0.5 in YPD as an example, one should see two major populations of RPFs: 21 nt RPFs and 28 nt RPFs (Fig. 5a) (*see* **Note 14**).
6. Calibrate the offset of the A site from the 5′ end of reads using start codons of CDS [11].
7. Compute relative codon occupancies by taking the ratio of ribosome density in a 3-nt window at the codon over the overall density in the CDS (excluding the first and last 15 nt).

8. Plot ribosome codon occupancies against the inverse of tRNA adaptation index (tAI) [12] using 21 nt RPFs (Fig. 5b).
9. Plot correlation coefficients of codon occupancies with 1/tAI using 21 nt RPFs or 28 nt RPFs. 21 nt RPFs should show a strong correlation with 1/tAI compared to 28 nt RPFs.

4 Notes

1. 800 mL of yeast culture at OD_{600} 0.5 in YPD should be able to pass through a filter membrane in 1 min. Prolonged filtering process introduces cellular stress.
2. While filtering media through a filter membrane, liquid should never go dry.
3. Ground powder should not be thawed before diluting in 22 mL footprint lysis buffer A.
4. Pipette 3 mL sucrose cushion solution in a Type 70 polycarbonate bottle first. When layering 22 mL diluted cell lysate, hold the bottle nearly horizontal and dribble the lysate down the side of the bottle.
5. Ribosome pellets should look greenish and glassy on the bottom of the polycarbonate bottle. The polycarbonate bottles should be kept cold on ice during the course of pellet resuspension. A thin layer of lipid is usually present on the top of the supernatant and it should be removed first with a serological pipette.
6. Pipetting phenol and chloroform in a hood is advised.
7. UV transillumination introduces damages to RNA and thus it should not be used for visualization of RPFs.
8. Preadenylated 3′ linker can be prepared enzymatically with a 5′ DNA adenylation kit (New England Biolabs) as described in McGlincy and Ingolia [13] or chemically with adenosine 5′phosphorimidazolide [14].
9. This is a pause point. Size control and RNA samples can be stored at −80 °C for 1 week.
10. Nuclease-based depletion methods are not recommended because they have been shown to introduce severe biases in RPF lengths [15].
11. We usually observe ~60% rRNA-aligning reads in a typical high-resolution ribosome profiling experiment after rRNA depletion. If desired, a second round of rRNA depletion can be performed using custom-designed oligos [3, 16] or depletion of abundant sequences by hybridization (DASH) [17, 18].

12. Numbers of PCR cycles in preparative PCR need to be empirically determined. Using 4 μg of RNA from monosome fractions for size-selection, we found five to eight cycles are good options. Choose PCR products from reactions that do not produce high-molecular weight products (Fig. 4).
13. Our preadenylated 3′ linker and RT primer contain unique molecular identifiers. Prior to read trimming, duplicated reads due to PCR amplification can be removed.
14. If one does not observe 21 nt RPFs in a RPF library prepared from an unstressed growth condition, the most likely cause is inefficient inhibition of accommodation, particularly in more concentrated lysates. We suggest one can dilute lysate with higher volume of footprint lysis buffer A or increase tigecycline concentration in footprint lysis buffer.

Acknowledgments

This protocol was initiated in the Green lab (Johns Hopkins University School of Medicine). The authors would like to thank Rachel Green for her support, and Nicholas R. Guydosh, Boris Zinshteyn, and Danny Nedialkova for helpful tips on developing this protocol. B.S.M. and M.S. are supported by Cancer Research Training Award. The work was support by the Intramural Research Program of the National Cancer Institute, National Institute of Health (C.C.-C.W.).

References

1. Dever TE, Dinman JD, Green R (2018) Translation elongation and recoding in eukaryotes. Cold Spring Harb Perspect Biol. https://doi.org/10.1101/cshperspect.a032649
2. Ingolia NT, Ghaemmaghami S, Newman JR, Weissman JS (2009) Genome-wide analysis in vivo of translation with nucleotide resolution using ribosome profiling. Science 324: 218–223. https://doi.org/10.1126/science.1168978
3. Ingolia NT, Lareau LF, Weissman JS (2011) Ribosome profiling of mouse embryonic stem cells reveals the complexity and dynamics of mammalian proteomes. Cell 147:789–802. https://doi.org/10.1016/j.cell.2011.10.002
4. Wolin SL, Walter P (1988) Ribosome pausing and stacking during translation of a eukaryotic mRNA. EMBO J 7:3559–3569
5. Lareau LF, Hite DH, Hogan GJ, Brown PO (2014) Distinct stages of the translation elongation cycle revealed by sequencing ribosome-protected mRNA fragments. elife 3:e01257. https://doi.org/10.7554/eLife.01257
6. Wu CC-C, Zinshteyn B, Wehner KA, Green R (2019) High-resolution ribosome profiling defines discrete ribosome elongation states and translational regulation during cellular stress. Mol Cell 73:959–970.e5. https://doi.org/10.1016/j.molcel.2018.12.009
7. Budkevich T, Giesebrecht J, Altman RB et al (2011) Structure and dynamics of the mammalian ribosomal pretranslocation complex. Mol Cell 44:214–224. https://doi.org/10.1016/j.molcel.2011.07.040
8. Jenner L, Starosta AL, Terry DS et al (2013) Structural basis for potent inhibitory activity of the antibiotic tigecycline during protein synthesis. Proc Natl Acad Sci U S A 110: 3812–3816. https://doi.org/10.1073/pnas.1216691110
9. Jiang H, Lei R, Ding SW, Zhu S (2014) Skewer: a fast and accurate adapter trimmer for next-generation sequencing paired-end

reads. BMC Bioinformatics 15:182. https://doi.org/10.1186/1471-2105-15-182
10. Dobin A, Davis CA, Schlesinger F et al (2013) STAR: ultrafast universal RNA-seq aligner. Bioinformatics 29:15–21. https://doi.org/10.1093/bioinformatics/bts635
11. Schuller AP, Wu CC, Dever TE et al (2017) eIF5A functions globally in translation elongation and termination. Mol Cell 66:194–205.e5. https://doi.org/10.1016/j.molcel.2017.03.003
12. dos Reis M, Savva R, Wernisch L (2004) Solving the riddle of codon usage preferences: a test for translational selection. Nucleic Acids Res 32:5036–5044. https://doi.org/10.1093/nar/gkh834
13. McGlincy NJ, Ingolia NT (2017) Transcriptome-wide measurement of translation by ribosome profiling. Methods 126: 112–129. https://doi.org/10.1016/j.ymeth.2017.05.028
14. Pfeffer S, Lagos-Quintana M, Tuschl T (2005) Cloning of small RNA molecules. Curr Protoc Mol Biol. https://doi.org/10.1002/0471142727.mb2604s72
15. Zinshteyn B, Wangen JR, Boyang HUA, Green R (2020) Nuclease-mediated depletion biases in ribosome footprint profiling libraries. RNA. https://doi.org/10.1261/rna.075523.120
16. Guydosh NR, Green R (2014) Dom34 rescues ribosomes in 3′ untranslated regions. Cell 156: 950–962. https://doi.org/10.1016/j.cell.2014.02.006
17. Gu W, Crawford ED, O'Donovan BD et al (2016) Depletion of abundant sequences by hybridization (DASH): using Cas9 to remove unwanted high-abundance species in sequencing libraries and molecular counting applications. Genome Biol. https://doi.org/10.1186/s13059-016-0904-5
18. Han P, Shichino Y, Schneider-Poetsch T et al (2020) Genome-wide survey of ribosome collision. Cell Rep 31:107610. https://doi.org/10.1016/j.celrep.2020.107610

Chapter 12

Fluorescence Intensity-Based eIF2B's Guanine Nucleotide-Exchange Factor Activity Assay

Yusuke Sekine, David Ron, and Alisa F. Zyryanova

Abstract

Guanine nucleotide-exchange factors (GEFs) activate the function of guanine nucleotide-binding proteins (G-proteins) by promoting the exchange of GDP for GTP on the latter. Here, we describe a protocol for in vitro measurements of the GEF activity of eukaryotic translation initiation factor 2B, eIF2B, toward its substrate eIF2. This protocol provides a relatively simple method for determining the eIF2B's GEF activity in crude cell extracts. The eIF2 heterotrimeric substrate, with phosphorylated or unphosphorylated eIF2α, is prepared by immunoprecipitation, following subsequent loading of a fluorescent BODIPY-FL dye-attached GDP. The exchange of the bound fluorescent GDP molecule for an unlabeled one on eIF2 promoted by eIF2B is monitored kinetically using a fluorescence microplate reader.

Key words eIF2, eIF2B, Guanine nucleotide exchange factor (GEF), Guanine nucleotide-binding protein (G-protein), Integrated stress response (ISR), Translation, Translation initiation factor

1 Introduction

Guanine nucleotide-exchange factors (GEFs) stimulate the release of GDP from their substrates, guanine nucleotide-binding proteins (G-proteins or GTPases), allowing G-proteins to be activated through binding of GTP. Even though GTP is present at much higher concentration than GDP in cells, the rates of spontaneous nucleotide exchange on some G-proteins are not sufficient to promote their reactivation, thus posing a need for a GEF [1, 2]. In the active GTP-bound form, G-proteins mediate signal to their effectors. Conversely, GTPase activating proteins (GAPs) promote the hydrolysis of GTP by stimulating the intrinsic GTPase activity of G-proteins, thus inactivating the signal [1]. This GDP–GTP exchange cycle on G-proteins serves as a molecular switch for various biological processes.

The eukaryotic translation initiation factor 2B, eIF2B, is a GEF that stimulates the exchange of GDP for GTP on another

Daniel Matějů and Jeffrey A. Chao (eds.), *The Integrated Stress Response: Methods and Protocols*, Methods in Molecular Biology, vol. 2428, https://doi.org/10.1007/978-1-0716-1975-9_12,

eukaryotic translation initiation factor, eIF2 [3, 4]. The GTP-bound eIF2 associates with an initiator transfer RNA ($tRNA_i^{Met}$) forming a ternary complex essential for starting translation. Phosphorylation of the residue serine 51 on the α subunit of eIF2 by stress-activated eIF2α kinases (including HRI, PKR, PERK, and GCN2) converts eIF2 from a substrate to an inhibitor of eIF2B, thereby depleting the number of the ternary complexes and attenuating rates of global translation. However, a selective number of rare mRNAs with upstream open reading frames is induced, marking the characteristic features of the integrated stress response (ISR) [5]. Thus, the relationship between eIF2 and eIF2B is pivotal to the translation regulation, lying at the heart of the ISR. eIF2 is a heterotrimer comprising of α-, β-, and γ-subunits. eIF2B forms a decamer consisting of two sets of heteropentameric units made of α-, β-, γ-, δ-, and ε-subunits. The multimeric composition of these two translational factors, eIF2 and eIF2B, made the analysis of their functional interactions more challenging. Recent structural studies using X-ray crystallography and cryo-electron microscopy combined with single particle analysis uncovered tertiary structures of yeast and human eIF2B decamers associated with either phosphorylated or unphosphorylated eIF2 trimers, providing a significant insight into the regulation of eIF2B's GEF activity toward eIF2 [6–13].

In this chapter, we describe a protocol for an in vitro assay measuring the GEF activity of eIF2B in cell extracts toward purified eIF2 [14]. Historically, tritium (^3H)-labeled GDP has been used for monitoring the loading of GDP on G-proteins and its accelerated dissociation by GEFs in endpoint assays that require a phase-separation step [15, 16]. More recently, it has been demonstrated that fluorescently labeled guanine nucleotides can be used to measure GEF activity for several combinations of GEFs and G-proteins in a continuous assay that does not require phase separation [17–20]. In this protocol, we use a commercially available fluorescent compound, BODIPY-FL-conjugated GDP, which enables us to measure the level of GDP bound to eIF2 by detecting changes in the fluorescence intensity derived from BODIPY-FL-GDP read on a fluorescence microplate leader. The BODIPY's fluorescence, attached on guanine nucleotides, is quenched by photo-induced electron transfer from proximal guanosine bases when it forms a stacked configuration in solution [19, 20]. Loading of BODIPY-FL-GDP onto a G-protein appears to induce a more extended configuration of BODIPY-FL-GDP by burial of the base in the protein's hydrophobic pocket, thus releasing the fluorophore–base quenching interaction and increasing the probe's fluorescence.

The eIF2 substrate trimer is purified through immunoprecipitation of 3×FLAG-tagged eIF2α (3×FLAG-eIF2α) stably expressed in Chinese hamster ovary (CHO) cells [14]. This CHO cell line also stably expresses an artificial eIF2α kinase, Fv2E-PERK, bearing the cytosolic kinase domain of PERK fused to two modified FK506 binding protein (FKBP) domains that dimerize upon treatment with FKBP dimerization molecule, AP20187, thus triggering PERK kinase activity and generating high levels of eIF2α phosphorylation independently of any other upstream stress-inducing signaling [21]. The over-expressed 3×FLAG-eIF2α joins the endogenous eIF2 β- and γ-subunits forming an unphosphorylated eIF2 or phosphorylated eIF2 [eIF2(αP)] trimer. The purified 3×FLAG-tagged eIF2 or eIF2(αP) trimers are subsequently bound to fluorescent BODIPY-FL-GDP. These substrates are mixed with cell extracts prepared from cultured mammalian cells and an excess amount of unlabeled GDP to measure the exchange of the labeled nucleotide for unlabeled one on eIF2 accelerated by eIF2B present in the cell extracts.

The diverse recessive missense mutations identified in all five subunits of eIF2B are causally linked to a human inherited leukoencephalopathy, known as the vanishing white matter (VWM) disease or Childhood Ataxia with Cerebral Hypomyelination (CACH) [3, 22]. By combining the assay described here with the CRISPR-Cas9 genome editing, we could measure the effect of a VWM mutation in eIF2B on its GEF activity [23]. Moreover, eIF2B was identified as an endogenous target of a small molecule-inhibitor of the ISR (ISRIB) [14, 16, 24]. It has been revealed that ISRIB and its analogues directly bind to and stimulate the GEF activity of eIF2B by stabilizing it in an active decameric conformation inimical to the inhibition by phosphorylated eIF2 [7, 8, 13], having salubrious implications for mammalian cellular and organismal models of diverse neurological disorders including VWM disease [24–30]. Therefore, this method is applicable for evaluating the effect of bioactive small compounds, like ISRIB, on the GEF activity of eIF2B, as well as the effect of various eIF2B-associated mutations in vitro.

2 Materials

2.1 Substrate eIF2 Preparation

1. CHO-K1 cells stably expressing Fv2E-PERK and 3×FLAG-eIF2α (*see* **Note 1**), maintained at 37 °C with 5% CO_2.
2. Regular mammalian cell culture medium: Ham's Nutrient Mixture F12, 10% Fetal calf serum, 2 mM L-glutamine, and 1× Penicillin/Streptomycin (final 100 U/ml and 0.1 mg/ml, respectively).
3. 20 μM solution of AP20187 in 100% EtOH.
4. Ice-cold 1× PBS.
5. Ice-cold PBS/EDTA: 1× PBS containing 1 mM EDTA pH 8.0.

6. Lysis buffer: 0.5% Triton X-100, 50 mM Tris-HCl pH 7.4, 150 mM NaCl, 1 mM EDTA, 10% glycerol, and freshly added 1 mM dithiothreitol (DTT), protease inhibitors [2 mM phenylmethylsulfonyl fluoride (PMSF), 10 μg/ml aprotinin, 4 μg/ml pepstatin, 4 μM leupeptin], and phosphatase inhibitors (10 mM β-glycerophosphate, 10 mM tetrasodium pyrophosphate, 100 mM NaF).
7. High salt lysis buffer: the same composition as Lysis buffer except containing 500 mM NaCl instead of 150 mM NaCl.
8. Protein A Sepharose resin (ZYMED).
9. Anti-FLAG M2 affinity gel (Sigma).
10. GDP-mounting buffer: 50 mM Tris-HCl pH 7.4, 150 mM NaCl, 1 mM DTT.
11. 3×FLAG peptide (Sigma): 4 mg/ml solution in 1× TBS.
12. BODIPY-FL-GDP (Invitrogen): 150 nM solution in GDP-mounting buffer.
13. 12 mM $MgCl_2$ in GDP-mounting buffer.
14. G-50 Sephadex column (GE healthcare).
15. GEF assay buffer: 50 mM Tris-HCl pH 7.4, 150 mM NaCl, 1 mM DTT, 2 mM $MgCl_2$, 0.01% Triton X-100.

2.2 Cell Lysate Preparation

1. CHO-K1 cells maintained as described above.
2. Homogenization buffer: 50 mM Tris-HCl pH 7.4, 375 μM magnesium acetate, 75 μM EDTA, 95 mM potassium acetate, 2.5 mg/ml digitonin, 10% glycerol, and freshly added 1 mM DTT, protease and phosphatase inhibitors as listed in the lysis buffer above (*see* **Note 2**).
3. 0.5 ml syringe with 29 G needle.

2.3 GDP-Release Assay

1. 1.5 mM unlabeled GDP pH 7.4 in GEF assay buffer.
2. 384-well round-bottom black polystyrene assay plate (Corning).
3. Fluorescence microplate reader (e.g., Tecan F500 plate reader).

3 Methods

3.1 Substrate eIF2 Preparation

1. Grow CHO-K1 cells stably expressing Fv2E-PERK and 3×FLAG-eIF2α cells in a 10 cm dish until 100% confluent (approximately 1 × 10^7 cells). To prepare enough amount of substrate for multiple experiments, we culture ten dishes of CHO-K1 cells for both unphosphorylated eIF2 and phosphorylated eIF2(αP). To prepare eIF2(αP), treat cells with 20 nM AP20187 for 1 h before harvest.

2. Wash cells with 2 ml of ice-cold PBS twice and collect cells by using cell scraper in 1 ml of ice-cold PBS/EDTA for each dish.
3. Centrifuge at 350–400 × *g* for 5 min at 4 °C and discard the supernatant.
4. Lyse the cell pellets with 4× volumes of ice-cold Lysis buffer for 5–15 min on ice or rotating in a cold room.
5. Transfer the lysates into one 1.5 ml Eppendorf tube and centrifuge at 21,000 × *g* for 15 min at 4 °C. Transfer the supernatant into a new 1.5 ml Eppendorf tube.
6. Add 20 μl of Protein A Sepharose resin to the cleared supernatant and rotate it for 30 min at 4 °C for precleaning. Pellet Protein A Sepharose resin and transfer the supernatant to a new 1.5 ml Eppendorf tube.
7. Add 30 μl of anti-FLAG M2 affinity gel to the precleaned supernatant and rotate it for 30 min at 4 °C to immunoprecipitate 3×FLAG-tagged eIF2 complex.
8. Wash anti-FLAG M2 affinity gel with 0.5–1 ml of high salt lysis buffer three times, and then wash the gel with 50 μl of GDP-mounting buffer twice.
9. Add 40 μl of GDP-mounting buffer containing 125 μg/ml 3×FLAG peptide to the washed anti-FLAG M2 affinity gel and elute the 3×FLAG-tagged eIF2 complex mixing at 1000 rpm in a thermomixer for 30 min at 4 °C. Pellet the gel and transfer the eluate to a new 1.5 ml Eppendorf tube (*see* **Note 3**).
10. Add 20 μl of 150 nM BODIPY-FL-GDP to the eluate and incubate at 25 °C for 20 min to reach equilibrium binding of BODIPY-FL-GDP to eIF2 (*see* **Note 4**).
11. Add 10 μl of 12 mM $MgCl_2$ to the mixture and pass the total amount of 70 μl through a G-50 Sephadex column equilibrated with GEF assay buffer. Collect the flow-through solution and use it as a substrate in the GEF assay. Typical yield is around 100 nM. Aliquot and store the BODIPY-FL-GDP-eIF2 substrate at −80 °C.

3.2 Cell Extract Preparation

1. Grow CHO-K1 cells in a 10 cm dish until 80–90% confluent and collect cell pellets as describe in Subheading 3.1.
2. Add 4× volumes of homogenization buffer to the cell pellet and incubate the lysate mixture on ice for 15 min.
3. Pass the lysate through 0.5 ml syringe with 29G needle five times to shear DNA.
4. Centrifuge at 21,000 × *g* for 15 min at 4 °C and collect the supernatant for measuring eIF2B's GEF activity (*see* **Note 5**).

5. Measure the protein concentration of cell extracts using standard protein quantification methods such as Bradford assay.

3.3 GDP-Exchange Assay on Purified eIF2

1. Mix 2 μl of the BODIPY-FL-GDP-conjugated eIF2 or eIF2 (αP) substrate, 3 μl of GEF assay buffer containing 1.5 mM unlabeled GDP in a 384-well round bottom black polystyrene assay plate (*see* **Note 6**).
2. Add 5 μl of cell extract and immediately start measuring fluorescence intensity of BODIPY-FL-GDP release, using excitation wavelength (485 ± 20) nm and emission wavelength (535 ± 25) nm every 20 s until the fluorescence intensity reaches the plateau using a fluorescence microplate reader (typically takes 0.5–1 h) (*see* **Notes 7** and **8**). A representative plot of the fluorescence intensity is shown in Fig. 1a. One way to present the GEF activity is to calculate the rate of the decline in fluorescence intensity (ΔFI/second, where ΔFI is a difference between FI_0, when time is zero, and FI_n, when time is more than zero) at the initial linear phase of the reactions (typically occurring within the first minutes) (Fig. 1b). Alternatively, you can calculate and compare half-lives of the fluorescence decay using predetermined statistical function in a research software such as GraphPad/Prism. An example of such an equation is a single-phase exponential decay function:

$$Y = (Y0 - \text{Plateau}) * \exp(-K * X) + \text{Plateau}$$

where $Y0$ is the Y value when X (time) is zero, Plateau is the Y value at infinite times, K is the rate constant expressed in reciprocal of the X axis time units.

4 Notes

1. This cell line is available from David Ron's Lab.
2. Dissolve digitonin powder in homogenization buffer just before use.
3. The immunoprecipitated 3×FLAG-eIF2α associated with endogenous eIF2 β- and γ-subunits can be confirmed on a gel stained with Coomassie brilliant blue (CBB) after separation by SDS-PAGE (Fig. 2). Phosphorylation of eIF2α can be detected by immunoblotting using an anti-eIF2α-phospho-Ser51 antibody.
4. Binding of BODIPY-FL-GDP to eIF2 can be confirmed by detecting substantial increase in fluorescence intensity of BODIPY-FL-GDP using the same setting on fluorescence microplate reader as GDP exchange assay (Fig. 3).

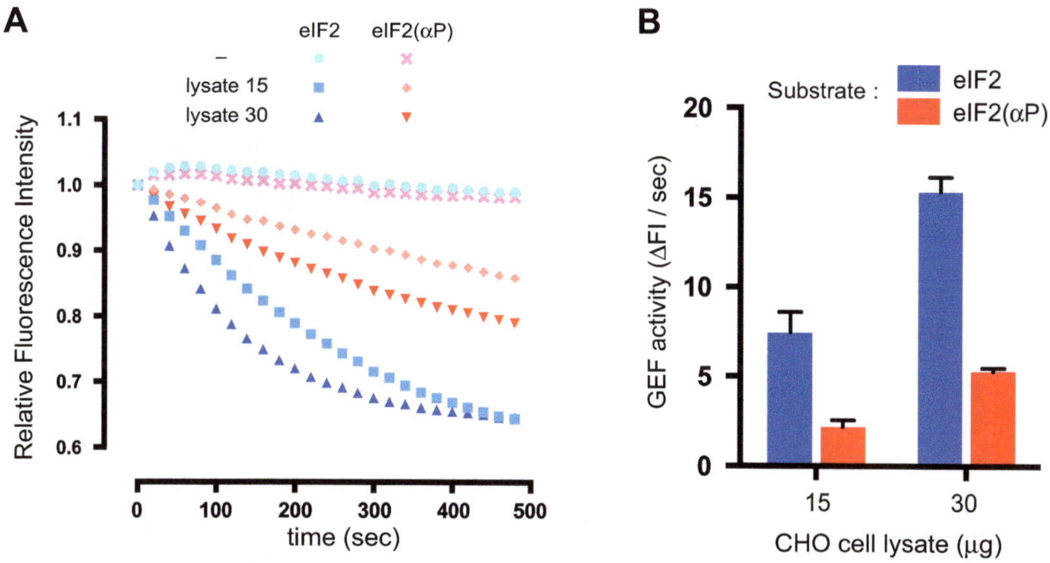

Fig. 1 GEF activity of eIF2B toward eIF2 and eIF2(αP). (**a**) GDP dissociation from eIF2 was reflected in time-dependent decreases in fluorescence of BODIPY-FL-GDP loaded on eIF2 or eIF2(αP) that were incubated with unlabeled GDP in the presence or absence of cell lysate (μg). (**b**) Bar diagram of the GEF activity calculated from the rate of decline of fluorescence intensity (ΔFI/second) at the initial linear phase of the reactions shown in "**a**." Shown are mean ± SEM ($n = 3$). From Sekine et al. (2015) [14]. Reprinted with permission from AAAS

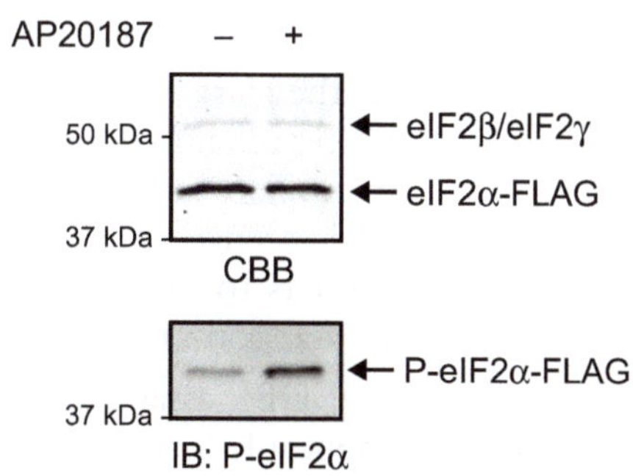

Fig. 2 Immunoaffinity purification of eIF2 complex. 3×FLAG-eIF2α associated with endogenous eIF2β and eIF2γ was immunoprecipitated with anti-FLAG M2 affinity gel and eluted with 3×FLAG peptide. A part of the sample was subjected to SDS-PAGE. The gel was stained with Coomassie brilliant blue (CBB) (upper panel) or subjected to immunoblotting (IB) with an anti-eIF2α-phospho-Ser51 antibody (lower panel). To induce phosphorylation of eIF2α, AP20187 was treated for 1 h before harvest. From Sekine et al. (2015) [14]. Reprinted with permission from AAAS

5. Freeze and thaw cycles dampen the GEF activity of eIF2B in cell extracts. Ideally, freshly prepared cell extracts should be immediately used for GEF assay.

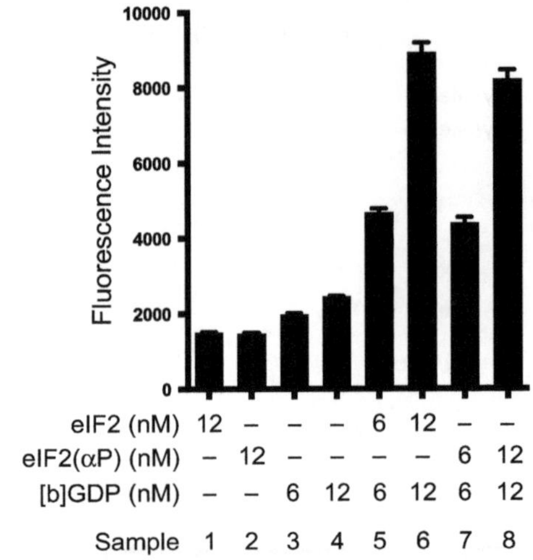

Fig. 3 Loading of BODIPY-FL-GDP on eIF2 increases the fluorescence. Bar diagram of the fluorescence intensity of BODIPY-FL-GDP ([b]GDP) alone and in complex with eIF2 or eIF2(αP). Shown is the mean ± SD ($n = 2$ for sample 1, 2 and $n = 3$ for the other samples). From Sekine et al. (2015) [14]. Reprinted with permission from AAAS

6. When we evaluate an effect of small compounds (like ISRIB), we add a compound and a respective vehicle control within a volume of 0.5–1 μl.
7. At the start of the GDP-release reaction just after adding the GEF-containing cell extract, the settings on fluorescence microplate reader have to be ready in advance to measure the fluorescence intensity decay. Add the GEF-containing cell extracts to the assay plate right next to the microplate reader so that you can start the measurement immediately.
8. The GEF assay using BODIPY-FL-GDP-loaded eIF2 is also applicable for purified eIF2B complexes (*see* refs. 8, 13, 14, 23 for the details).

Acknowledgments

This work was supported by grants from UPMC Competitive Medical Research Fund and The Samuel and Emma Winters Foundation to Y.S. and a Wellcome Trust Principal Research Fellowship to D.R. (Wellcome 200848/Z/16/Z).

References

1. Cherfils J, Zeghouf M (2013) Regulation of small GTPases by GEFs, GAPs, and GDIs. Physiol Rev 93(1):269–309. https://doi.org/10.1152/physrev.00003.2012
2. Maracci C, Rodnina MV (2016) Review: translational GTPases. Biopolymers 105 (8):463–475. https://doi.org/10.1002/bip.22832
3. Pavitt GD (2018) Regulation of translation initiation factor eIF2B at the hub of the integrated stress response. Wiley Interdiscip Rev RNA 9(6):e1491. https://doi.org/10.1002/wrna.1491
4. Marintchev A, Ito T (2020) eIF2B and the integrated stress response: a structural and mechanistic view. Biochemistry 59 (13):1299–1308. https://doi.org/10.1021/acs.biochem.0c00132
5. Pakos-Zebrucka K, Koryga I, Mnich K, Ljujic M, Samali A, Gorman AM (2016) The integrated stress response. EMBO Rep 17 (10):1374–1395. https://doi.org/10.15252/embr.201642195
6. Kashiwagi K, Takahashi M, Nishimoto M, Hiyama TB, Higo T, Umehara T, Sakamoto K, Ito T, Yokoyama S (2016) Crystal structure of eukaryotic translation initiation factor 2B. Nature 531(7592):122–125. https://doi.org/10.1038/nature16991
7. Zyryanova AF, Weis F, Faille A, Alard AA, Crespillo-Casado A, Sekine Y, Harding HP, Allen F, Parts L, Fromont C, Fischer PM, Warren AJ, Ron D (2018) Binding of ISRIB reveals a regulatory site in the nucleotide exchange factor eIF2B. Science 359(6383):1533–1536. https://doi.org/10.1126/science.aar5129
8. Tsai JC, Miller-Vedam LE, Anand AA, Jaishankar P, Nguyen HC, Renslo AR, Frost A, Walter P (2018) Structure of the nucleotide exchange factor eIF2B reveals mechanism of memory-enhancing molecule. Science 359(6383):eaaq0939. https://doi.org/10.1126/science.aaq0939
9. Gordiyenko Y, Llacer JL, Ramakrishnan V (2019) Structural basis for the inhibition of translation through eIF2alpha phosphorylation. Nat Commun 10(1):2640. https://doi.org/10.1038/s41467-019-10606-1
10. Adomavicius T, Guaita M, Zhou Y, Jennings MD, Latif Z, Roseman AM, Pavitt GD (2019) The structural basis of translational control by eIF2 phosphorylation. Nat Commun 10 (1):2136. https://doi.org/10.1038/s41467-019-10167-3
11. Kashiwagi K, Yokoyama T, Nishimoto M, Takahashi M, Sakamoto A, Yonemochi M, Shirouzu M, Ito T (2019) Structural basis for eIF2B inhibition in integrated stress response. Science 364(6439):495–499. https://doi.org/10.1126/science.aaw4104
12. Kenner LR, Anand AA, Nguyen HC, Myasnikov AG, Klose CJ, McGeever LA, Tsai JC, Miller-Vedam LE, Walter P, Frost A (2019) eIF2B-catalyzed nucleotide exchange and phosphoregulation by the integrated stress response. Science 364(6439):491–495. https://doi.org/10.1126/science.aaw2922
13. Zyryanova AF, Kashiwagi K, Rato C, Harding HP, Crespillo-Casado A, Perera LA, Sakamoto A, Nishimoto M, Yonemochi M, Shirouzu M, Ito T, Ron D (2020) ISRIB blunts the integrated stress response by allosterically antagonising the inhibitory effect of phosphorylated eIF2 on eIF2B. Mol Cell. https://doi.org/10.1016/j.molcel.2020.10.031
14. Sekine Y, Zyryanova A, Crespillo-Casado A, Fischer PM, Harding HP, Ron D (2015) Stress responses. Mutations in a translation initiation factor identify the target of a memory-enhancing compound. Science 348 (6238):1027–1030. https://doi.org/10.1126/science.aaa6986
15. Kimball SR, Everson WV, Flaim KE, Jefferson LS (1989) Initiation of protein synthesis in a cell-free system prepared from rat hepatocytes. Am J Phys 256(1 Pt 1):C28–C34. https://doi.org/10.1152/ajpcell.1989.256.1.C28
16. Sidrauski C, Tsai JC, Kampmann M, Hearn BR, Vedantham P, Jaishankar P, Sokabe M, Mendez AS, Newton BW, Tang EL, Verschueren E, Johnson JR, Krogan NJ, Fraser CS, Weissman JS, Renslo AR, Walter P (2015) Pharmacological dimerization and activation of the exchange factor eIF2B antagonizes the integrated stress response. elife 4:e07314. https://doi.org/10.7554/eLife.07314
17. Sajid A, Arora G, Gupta M, Singhal A, Chakraborty K, Nandicoori VK, Singh Y (2011) Interaction of Mycobacterium tuberculosis elongation factor Tu with GTP is regulated by phosphorylation. J Bacteriol 193 (19):5347–5358. https://doi.org/10.1128/JB.05469-11
18. Shang X, Marchioni F, Sipes N, Evelyn CR, Jerabek-Willemsen M, Duhr S, Seibel W, Wortman M, Zheng Y (2012) Rational design of small molecule inhibitors targeting RhoA subfamily Rho GTPases. Chem Biol 19

(6):699–710. https://doi.org/10.1016/j.chembiol.2012.05.009

19. McEwen DP, Gee KR, Kang HC, Neubig RR (2001) Fluorescent BODIPY-GTP analogs: real-time measurement of nucleotide binding to G proteins. Anal Biochem 291(1):109–117. https://doi.org/10.1006/abio.2001.5011
20. Korlach J, Baird DW, Heikal AA, Gee KR, Hoffman GR, Webb WW (2004) Spontaneous nucleotide exchange in low molecular weight GTPases by fluorescently labeled gamma-phosphate-linked GTP analogs. Proc Natl Acad Sci U S A 101(9):2800–2805. https://doi.org/10.1073/pnas.0308579100
21. Lu PD, Jousse C, Marciniak SJ, Zhang Y, Novoa I, Scheuner D, Kaufman RJ, Ron D, Harding HP (2004) Cytoprotection by pre-emptive conditional phosphorylation of translation initiation factor 2. EMBO J 23 (1):169–179. https://doi.org/10.1038/sj.emboj.7600030
22. van der Knaap MS, van Berkel CG, Herms J, van Coster R, Baethmann M, Naidu S, Boltshauser E, Willemsen MA, Plecko B, Hoffmann GF, Proud CG, Scheper GC, Pronk JC (2003) eIF2B-related disorders: antenatal onset and involvement of multiple organs. Am J Hum Genet 73(5):1199–1207. https://doi.org/10.1086/379524
23. Sekine Y, Zyryanova A, Crespillo-Casado A, Amin-Wetzel N, Harding HP, Ron D (2016) Paradoxical sensitivity to an integrated stress response blocking mutation in vanishing white matter cells. PLoS One 11(11): e0166278. https://doi.org/10.1371/journal.pone.0166278
24. Sidrauski C, Acosta-Alvear D, Khoutorsky A, Vedantham P, Hearn BR, Li H, Gamache K, Gallagher CM, Ang KK, Wilson C, Okreglak V, Ashkenazi A, Hann B, Nader K, Arkin MR, Renslo AR, Sonenberg N, Walter P (2013) Pharmacological brake-release of mRNA translation enhances cognitive memory. elife 2: e00498. https://doi.org/10.7554/eLife.00498
25. Halliday M, Radford H, Sekine Y, Moreno J, Verity N, le Quesne J, Ortori CA, Barrett DA, Fromont C, Fischer PM, Harding HP, Ron D, Mallucci GR (2015) Partial restoration of protein synthesis rates by the small molecule ISRIB prevents neurodegeneration without pancreatic toxicity. Cell Death Dis 6:e1672. https://doi.org/10.1038/cddis.2015.49
26. Wong YL, LeBon L, Edalji R, Lim HB, Sun C, Sidrauski C (2018) The small molecule ISRIB rescues the stability and activity of vanishing white matter disease eIF2B mutant complexes. elife 7:e32733. https://doi.org/10.7554/eLife.32733
27. Wong YL, LeBon L, Basso AM, Kohlhaas KL, Nikkel AL, Robb HM, Donnelly-Roberts DL, Prakash J, Swensen AM, Rubinstein ND, Krishnan S, McAllister FE, Haste NV, O'Brien JJ, Roy M, Ireland A, Frost JM, Shi L, Riedmaier S, Martin K, Dart MJ, Sidrauski C (2019) eIF2B activator prevents neurological defects caused by a chronic integrated stress response. Elife 8:e42940. https://doi.org/10.7554/eLife.42940
28. Zhu PJ, Khatiwada S, Cui Y, Reineke LC, Dooling SW, Kim JJ, Li W, Walter P, Costa-Mattioli M (2019) Activation of the ISR mediates the behavioral and neurophysiological abnormalities in Down syndrome. Science 366(6467):843–849. https://doi.org/10.1126/science.aaw5185
29. Chou A, Krukowski K, Jopson T, Zhu PJ, Costa-Mattioli M, Walter P, Rosi S (2017) Inhibition of the integrated stress response reverses cognitive deficits after traumatic brain injury. Proc Natl Acad Sci U S A 114(31): E6420–E6426. https://doi.org/10.1073/pnas.1707661114
30. Abbink TEM, Wisse LE, Jaku E, Thiecke MJ, Voltolini-Gonzalez D, Fritsen H, Bobeldijk S, Ter Braak TJ, Polder E, Postma NL, Bugiani M, Struijs EA, Verheijen M, Straat N, van der Sluis S, Thomas AAM, Molenaar D, van der Knaap MS (2019) Vanishing white matter: deregulated integrated stress response as therapy target. Ann Clin Transl Neurol 6 (8):1407–1422. https://doi.org/10.1002/acn3.50826

Part II

Analysis of Interaction Networks and RNP Granules

Chapter 13

Collective Learnings of Studies of Stress Granule Assembly and Composition

Hadjara Sidibé and Christine Vande Velde

Abstract

Stress granules have gained considerable exposure and interest in recent years. These micron-sized entities, composed of RNA and protein, form following a stress exposure and have been linked to several pathologies. Understanding stress granule function is paramount but has been arduous due to the membraneless nature of these organelles. Several new methodologies have recently been developed to catalogue the protein and RNA composition of stress granules. Collectively, this work has provided important insights to potential stress granule functions as well as molecular mechanisms for their assembly and disassembly. This chapter reviews the latest advancements in the understanding of stress granule dynamics and discusses the various protocols developed to study their composition.

Key words Proteomics Stress granules, Stress response, Stress granule composition, Stress granule assembly, Noncanonical stress granules

1 Introduction

Cells have evolved multiple strategies to maintain cellular homeostasis in response to stress [1–3]. Among these strategies is the formation of stress granules (SGs) [4, 5]. SGs are subcellular nanoscopic membraneless compartments formed in the cytoplasm via a process known as liquid–liquid phase separation (LLPS), in response to noxious stimuli such as oxidative stress, heat shock, viral infection, osmotic stress, and UV irradiation [6–10]. Depending on the stressor, different pathways lead to the formation of either canonical or noncanonical SGs, which seem to differ in function and composition (reviewed in [11]). The membraneless nature of these granules, as well as their dynamic assembly and disassembly, allow for a rapid remodeling of the transcriptome and proteome in order to regulate many different biological processes and effectively determine cell fate/survival due to an external

Daniel Matějů and Jeffrey A. Chao (eds.), *The Integrated Stress Response: Methods and Protocols*,
Methods in Molecular Biology, vol. 2428, https://doi.org/10.1007/978-1-0716-1975-9_13,

stress event [6, 12–14]. However, this dynamism also makes the isolation and study of SGs a laborious enterprise.

Alterations in SG dynamics and composition are associated with several pathologies including neurological disorders, cancer, and viral infections [5, 15–18]. Thus, SGs are of considerable interest to the biomedical research community. To better understand SGs, one of the main questions to be answered is the exact nature of their composition. In recent years, several groups have established protocols to capture the SG proteome and transcriptome. The present chapter will review and synthesize the different protocols and summarize their main findings. Also, we will discuss some of the important variables to consider for future SG studies.

2 Stress Granule Formation

Liquid–liquid phase separation (LLPS) is a process by which a solution of heterogenous composition separates into two or more coexisting phases: a dilute phase and a dense phase. In response to various stressors, protein-laden RNAs dispersed in the cytoplasm coalesce into a concentrated state [19]. As many of these proteins harbor low complexity domains (LCDs), such as RGG domains, these proteins with their bound RNAs, undergo concentration-dependent LLPS to form the membraneless organelles that are SGs. The exact mechanism underlying SG assembly has remained ill defined, although three recent studies have provided new insights on the process [19–21].

Over the years, several models have been proposed to explain SG formation. The first posits that SG assembly requires a solid-like seed composed of the SG nucleating protein G3BP1 [22, 23]. The second model suggests that RNA freed by polysome disassembly attracts RBPs, leading to SG formation [24]. The third model proposes that SGs assemble due to the phase separation of proteins containing LCDs [6, 25]. In contrast, a fourth model invokes the formation of solid core particles that recruit additional RBPs and RNAs via intermolecular base-pairing [26–29]. These models are not mutually exclusive, and all converge on the idea that SG assembly involves RNA and the LCDs of RBPs.

The prevailing challenge is to understand which cues and exact molecular mechanisms drive the assembly of these micron-sized structures [19]. Based on recent studies, several groups have established a common model based on patchy colloid theory in which phosphorylation-induced conformational switching of G3BP1 promotes the formation of condensates [19–21]. This model partly integrates the four hypotheses of SG formation. They refined these models by adding the requirement of a network of multivalent interactions to mediate phase transition [30]. They posit that G3BP proteins are the key proteins for these multivalent interactions, aligning with the first model. According to the patchy colloid theory, the valency (ability to interact with RNA and/or protein) of an SG nucleating protein, such as G3BP1, mediates SG formation [21]. G3BP1, with its large valencies, can mediate a high number of interactions [21].

Under physiological conditions, G3BP1 dimers adopt an auto-inhibited compact conformation that sequesters its RGG domain through electrostatic interactions with the central IDR (intrinsically disordered domain). During stress conditions, the dephosphorylation of residues S149 and S232 induces a conformational change leading to expansion of the RGG domain to permit engagement in multivalent protein-RNA and protein-protein interactions to promote condensate assembly [20, 21]. Once in the open conformation, G3BP1 IDR3 is outcompeted by the freed mRNA from the disassembled polysomes. This aligns with the second model discussed above. G3BP1 binding to RNA enhances its protein valency; thus G3BP1-RNA condensates act as recruiting centers and assembly platforms to enhance the recruitment of other RBPs to G3BP1. Caprin-1 is a RBP known to interact with G3BP1 and contribute to SG assembly [22, 31, 32]. Purified mCherry-Caprin-1 does not phase separate in the presence of RNA [19]. However, Caprin-1 shifts the saturation concentration so that G3BP1-RNA condensates undergo LLPS at a lower concentration and enhances G3BP1 binding to RNA due to polyphasic linkage, a concept defining allosteric and polysteric (aggregating) linkage [33]. On the other hand, overexpression of USP10, another G3BP1 interactor, inhibits SG assembly [22]. USP10 and Caprin-1 compete for overlapping binding sites in the NTF2L domain and thus exhibit mutually exclusive interactions with G3BP1 [22]. It is plausible that USP10, when bound to G3BP1 in opposition to Caprin-1, weakens G3BP1 RNA and protein binding propensity [21]. Very interestingly, without RNA, G3BP1 cannot phase separate even in preformed FUS condensates demonstrating that G3BP1 and FUS do not directly interact. This suggests that FUS recruitment to G3BP1-RNA condensates is a consequence of FUS binding to RNA. The same is expected for other RBPs.

The ability of G3BP1 to interact with other biomolecules in stressed conditions was predicted to drive two distinct layers: a cluster of formation and an intercluster cross-linking [19]. Several studies have highlighted the heterogeneous distribution of RBPs within SGs [29, 34–36]. This led to the hypothesis that SGs contain core-like structures in which specific proteins are concentrated (Model 3). Studies using single molecule fluorescence microscopy following individual proteins, demonstrated that G3BP1 was distributed in immobile hotspots within SGs [35]. This finding was supported by another group using cryo-electron microscopy to show that G3BP1-RNA condensates are indeed not homogeneous [19, 24]. This led to the conclusion that G3BP1 formed clusters that are spaced apart, presumably by RNA-mediated cross-links that are too thin to be resolved. In silico simulations support this claim [19]. Based on the simulations, the concentration of G3BP1 in the condensates is expected to be low as the protein is mainly a cross-linker among RNA molecules. Recent work has highlighted the ability of RNA to aggregate via RNA–RNA interactions and showed the importance of these interactions for SG formation (Model 4) [26, 27], as RNA aggregation is expected during stress due to the sudden release of unfolded and protein-free mRNAs from polysomes. G3BP1 is suggested to inhibit this aggregation by minimizing RNA–RNA interactions as being a physical cross-link to keep RNA apart.

In summary, SGs are formed around the scaffolding proteins G3BP1 and G3BP2. G3BPs are critical to form liquid condensates in response to local changes in RNA concentration. At the heart of this mechanism is a conformational switch that couples RNA binding to a dramatic increase in multivalency [19]. This promotes the formation of G3BP clusters that drive condensation by cross-linking RNA.

3 Canonical Versus Noncanonical SGs

In several instances, stress is accompanied with a general translational repression which can be triggered via either phospho-eIF2α (p-eIF2α)-dependent or -independent mechanisms. In the case of p-eIF2α-dependent translational inhibition, depending on the stress stimulus, one of four kinases (HRI, PERK, GCN2, or PKR) alone or in concert (reviewed in [37]) phosphorylates the initiation factor at residue S51. This leads to stronger binding of GTP within the eIF2α/GTP/tRNAi Met ternary complex and consequently decreased availability of the ternary complex. As this complex is responsible for the loading of the initiator tRNA onto the small ribosomal subunit to begin translation, this phosphorylation event ultimately results in translational arrest, polysome disassembly, and the assembly of canonical SGs [38, 39]. In general, common canonical SG components include stalled 48S preinitiation complexes (PICs) consisting of translationally arrested mRNAs, small ribosomal subunits, translation initiation factors, and RBPs such as G3BP1, Poly A Binding Protein (PABP), and T-cell intracellular antigen-1 (TIA-1) [10, 11, 38].

Although translational inhibition via eIF2α phosphorylation is a major trigger of SG formation, p-eIF2α–independent and translation-independent pathways can also trigger SG assembly (reviewed in [11]). To date, two p-eIF2α–independent pathways leading to SG formation have been identified. The first is via inhibition of the RNA helicase eIF4A, generating canonical SGs; the second is via disruption of the eIF4E-eIF4G interaction (via mTOR), leading to the formation of noncanonical SGs [40]. In both cases, the PIC is incomplete, resulting in translational inhibition and subsequent assembly of SG proteins.

As it can be suggested by the different pathways leading to them, canonical and noncanonical SGs can be triggered by different stressors (Table 1). Moreover, the composition of the two types of SGs is also different. Both p-eIF2α-dependent and -independent canonical SGs are enriched in PICs, 40S ribosomal subunits, initiation factors, and RBPs, but lack eIF2 and eIF5 [11]. These canonical SGs are dynamic and rapidly resolve upon removal of stress to reinitiate translation, therefore playing a pro-survival/cytoprotective role [11, 16]. Inhibitors of eIF4A helicase induce p-eIF2α-independent noncanonical SGs that recruit less polyadenylated mRNA but include eIF2 and eIF5 [16]. mTOR inactivating agents result in noncanonical SGs lacking eIF3 [59, 60]. These noncanonical SGs are less dynamic, pro-apoptotic, and cytotoxic to cells [11, 16, 60]. Therefore, studying both types of granules is essential to understand what can potentially make a granule protective or harmful.

4 Stressors

As previously mentioned, several types of stressors can provoke SG assembly. Heat, arsenite, viruses and pharmacological treatments are the most commonly used stressors (Table 1) [9]. Their popularity is mainly due to their potential physiological relevance and convenience of use. While heat and virus treatments can be used to simulate fever and viral infections, sodium arsenite and diverse pharmacological compounds that have been used are relevant for environmental studies [61]. It is important to consider that different stressors have potentially different impacts on SGs and cellular pathways.

4.1 Sodium Arsenite (SA)

SA is an inorganic sodium salt derivative sodium of arsenic, a naturally occurring element released from the Earth's crust into the environment predominantly by volcanic activity and as the result of industrial and agricultural activities, such as mining, metallurgy, and the excessive use of phosphate fertilizers [62–64]. Although SA contamination is a major public health issue, this element is also an important component of potential anticancer compounds [61]. SA became the most frequently used chemical inducer of SGs because it is inexpensive, easy to handle and induces SGs with high reproducibility. SA stress primarily activates HRI and, to a minor extent, GCN2 [8, 65]. SA treatment is often used interchangeably for "oxidative stress" as it causes an increase of reactive oxygen species (ROS) via its oxidized form and via the inhibition of mitochondrial Complex I. Indeed, this latter aspect

Table 1
Table summurazing common SGs-inducer effects

Stressor	Type of SG	Particularity	References
Heat	Canonical and noncanonical	SG proteins are polyubiquitinated. Induces fewer SGs per cell than SA. Induces high levels of HSC70, HSP70 and HSP27.	[41–43]
Sodium arsenite	Canonical	SG proteins are monoubiquitinated. Induces low levels of HSC70, HSP70, and HSP27.	[41–43]
Hydrogen peroxide	Noncanonical	Promotes 4E-BP hypophosphorylation and leads to 4E-BP1 binding to eIF4E.	[44]
Sorbitol	Canonical	Activates the mitogen-activated protein kinase kinase3/6-p38 signaling pathway.	[45, 46]
NaCl	Noncanonical	Induces the assembly of foci that compositionally resemble canonical SGs but are refractory to drugs that distinguish bona fide SGs from other RNA granules.	[8, 22]
Glucose deprivation	Noncanonical	Inactivates mTOR resulting in nonphosphorylated 4E-BP binding to eIF4E and displacing eIF4G/eIF4A from the mRNA cap–binding eIF4F complex, leading to the inhibition of the cap-dependent translation.	[47, 48]
Pateamine A	Noncanonical	Dissociates eIF4A from eIF4G by inhibiting eIF4A helicase or ATPase activities.	[49, 50]
Hippuristanol	Noncanonical	Inhibits eIF4A and reduces ATPase, helicase, and RNA-binding activities.	[49, 51, 52]
Silvestrol	Noncanonical	Stimulates the RNA-binding and helicase activities of eIF4A, resulting in sequestration of eIF4A from the eIF4F complex.	[53–55]
Rocaglamide A	Noncanonical	Clamps eIF4A onto mRNA and interferes with its helicase activity. SGs contain very little poly (A) mRNA. SGs contain less eIF3B than SGs induced by other stresses.	[8, 56]
Sodium selenite	Noncanonical	Inactivates mTOR to promote the assembly of inhibitory eIF4E:4E-BP1 complexes. SGs lack eIF3B and weakly recruit eIF3E and the small ribosomal protein Rps14. SGs accumulate eIF4E or eIF4G but lack proteins with pro-survival functions such as HDAC6, RACK1 and importin-α1.	[40]
Angiogenin-induced cleavage of mature tRNAs	Noncanonical	Cleavage displaces eIF4G/eIF4A from cap-bound eIF4E.	[55]
Antineoplastic vinca alkaloid	Noncanonical and canonical	Induces eIF4F complex inhibition and phosphorylation of eIF2α-mediated translation inhibition contributes to SG assembly.	[57]

(continued)

Table 1 (continued)

Stressor	Type of SG	Particularity	References
UV	Noncanonical	Triggers eIF2α phosphorylation but this is not required for SG formation. Only partially inhibits translation. SGs lack translation initiation factor eIF4G. SGs contain very little poly(A) mRNA. SGs contain less eIF3B than SGs induced by other stresses.	[8]
MG 132 (proteasome inhibitor)	Canonical	SGs contain less eIF3B than SGs induced by other stresses.	[8, 58]
Thapsigargin	Canonical	Disturbs intracellular calcium homeostasis	[10]

negatively impacts mitochondrial integrity and leads to enormous ROS production [66–68]. However, the impact of SA on cellular metabolism is yet to be fully understood. For example, SA is genotoxic and binds to sulfhydryl groups of many proteins and metabolites, altering the conformation and/or activity of these molecules [69] and disturbs many signaling pathways including glucose uptake, Krebs cycle, fatty acid oxidation, and glutathione production [63, 70, 71]. Most publications examining the connections between SGs and aberrant protein aggregates have used SA for SG induction [72–76]. Variable concentrations of SA and treatment times have been used to produce SA-induced canonical SGs. The typical dosage is 0.5 mM for 30 min, with concentrations ranging from 0.15 to 1 mM [8, 77, 78].

4.2 Heat Stress

Heat stress is another commonly used stressor to induce canonical SGs. Heat shock activates all four eIF2α kinases to induce canonical SGs and also noncanonical SGs via an eIF4E-dependent pathway [8, 37, 79]. Therefore, many molecular pathways are expected to be activated by this stimulus. Heat rapidly induces protein aggregation and misfolding, therefore preventing general translation is essential for the cell to avoid a negative outcome [80, 81]. SGs induced by heat are more likely to sequester misfolded proteins than SA-induced SGs. Indeed, heat-stress is associated with activation of the UPR to alleviate ER stress [82, 83]. In *Drosophila* and yeast, heat-induced SGs sequester model proteins that are prone to misfolding to a significantly greater extent than their wild-type counterparts, while SA-induced SGs do not [84]. This work strongly suggests that in yeast and *Drosophila*, heat shock-induced SGs may be more similar to aberrant aggregates found in neurodegeneration than SA-induced SGs. However, misfolded proteins do not associate with SGs in human cells; this is speculated to be linked to the lack of requirement for eIF2α phosphorylation to mediate translational repression in heat treated yeast, in contrast to mammalian cells [84]. This might be linked to the presence of polyubiquinated proteins [41–43, 84]. Different approaches have been used to induce SG formation by elevated temperature. Most often, heat-induced canonical SG assembly is achieved in cultured cells that have been placed in an incubator at 44 °C for 30 min to 1 h. Other conditions include 43 °C or 45 °C for15 min or 90 min, respectively [8, 85–87].

4.3 Osmotic Stress

The cytosol is filled with several macromolecules including proteins, nucleic acids and polysaccharides [88]. Although the concentration of these macromolecules varies between cell types, their combined concentration can reach up to 400 mg/ml [89, 90]. This high concentration means that the cytosol is densely packed. The osmolarity equilibrium in cells is one of the most tightly controlled physiological parameters, regulated by a balance of hydration and solute concentrations [91]. Osmotic stress is the consequence of cell exposure to an active change of osmotic pressure that can happen during physiologic or pathologic conditions. Intracellular macromolecule concentration is essential to properly adjust the spatial organization within a cell. Hypoosmotic stress swells the cell to a size that is limited by the stretch of the cytoskeleton and adopts a smooth, rounded shape; while hyperosmotic stress shrinks the cell, causing it to assume a convoluted shape. This alteration of the cytoskeleton impacts the entire cell. For example, hypoosmotic stress alters nuclear shape and increases nucleocytoplasmic transport in an attempt to alleviate the stress [92]. In addition, tension introduced to the lamina by swelling of the nucleoplasm can induce histone acetylation and alter genomic architecture [88, 93, 94]. During osmotic stress, cellular activity is altered in an attempt to recover equilibrium and thus prevent these changes. The mechanisms by which cells sense their osmotic environment are not fully understood, but appear to involve a number of ion channel–based transduction pathways [95]. Transient receptor potential (TRP) ion channels is a family of cellular osmosensors that play an important role in osmotic homeostasis [96]. Moreover, osmotic stress gives rise to ROS production, primarily H_2O_2, which can result in DNA damage, disruption of mitochondrial function and apoptosis [97–99]. Osmotic stress activates all four eIF2α kinases but seems to also form SGs via p-eIF2α independent pathways, potentially leading to the formation of both canonical and noncanonical SGs [8, 65, 100–103]. The most commonly used osmotic stress inducer is 0.4 M sorbitol for 1 h (yielding very small SGs in HeLa cells) to 2 h (producing slightly larger SGs, but still smaller than those initiated by SA or heat; unpublished observations); an alternative is 0.2 M NaCl for 1 h.

4.4 ER Stress

The endoplasmic reticulum (ER) is a vital organelle involved in several metabolic processes, such as protein synthesis, calcium ion storage, gluconeogenesis, lipid and cholesterol synthesis and the formation of autophagic vacuoles [104]. The ER receives a constant influx of newly synthesized polypeptides, which undergo various modifications including *N*-linked glycosylation; pairing with chaperones to prevent misfolding and aggregation; formation of intramolecular disulfide bonds; and transportation from the ER to the Golgi for further modification before they reach their final destination [104]. When protein production exceeds the ER's processing capacity, ER stress ensues, leading to activation of the unfolded protein response (UPR) [104]. Once levels of misfolded or unfolded proteins in the ER reach a critical threshold, cells activate a coordinated response mediated by three transmembrane proteins: serine/threonine-protein kinase/endoribonuclease IRE1 (IRE1α), eukaryotic translation initiation factor 2α-kinase 3 (eIF2α kinase, also known as PRKR-like ER kinase (PERK)), and cyclic AMP-dependent transcription factor ATF6α. PERK activation results in translational arrest and subsequent canonical SG formation [8]. To induce ER stress-induced SGs, thapsigargin is often used at 1–5 μM for 30–120 min [8]. Thapsigargin raises cytosolic calcium concentration by blocking ER calcium pumps, leading to depleted ER calcium stores and subsequent UPR activation.

4.5 Eukaryotic Initiation Factor (eIF) 4A Inhibitor

Eukaryotic initiation factor (eIF) 4A is a subunit of the translation initiation complex elF4F. The main function of elF4A during translation initiation is to unwind mRNA secondary structure in the 5′ UTR to facilitate ribosome binding to the mRNA [105]. Interruption of this process results in translational inhibition and potentially to the formation of noncanonical SGs. Hippuristanol, pateamine A and rocaglamide A are three commonly used eIF4A inhibitors [51]. While pateamine A and rocaglamide A induce the dimerization of eIF4A:RNA, hippuristanol and silvestrol prevent both free eIF4A and eIF4F-bound eIF4A from interacting with RNA [49, 51, 106]. Hippuristanol locks eIF4A in a closed conformation preventing its transition from a closed to an open state. This open state is essential to eIF4A helicase activity and RNA binding [107, 108]. Ultimately, both eIF4A stabilization on the RNA and blocking the conformational shift required prevents binding to elF4F and PIC formation. This leads to the formation of noncanonical SGs. Pateamine A is typically used at 0.5–1 μM for 30 min to 1 h, hippuristanol is used at 1 μM for 30 min to 1 h, and rocaglamide A is used at 0.5 μM for 1 h or 2 μM for 2 h.

5 A Critical Appraisal of the Techniques to Study SG Composition

The growing interest to better understand the function, regulation and dynamics of SGs has led to the establishment of new methods to study them (Table 2) [29, 32, 109–112, 115–118]. Traditionally, to identify proteins localized to SGs, researchers have used candidate-based approaches that assess colocalization with previously characterized SG components. Experiments using RNA interference have established certain genes that are essential for SG assembly [77]. Thanks to these methods, in 2009, ~70 different proteins were identified as localized to SGs [119]. However, relying on candidate-based assays to identify SG components is a long and tedious process that new methods have now helped to bypass. Recently, several groups have used mass spectrometry-based and next-generation sequencing approaches to help define SG protein and RNA composition. These methods are either based on cellular fractionation via a series of differential centrifugations, immunoprecipitation or proximity labeling techniques.

5.1 Cellular Fractionation-Based Approaches

Cellular fractionation-based methods exploit the relatively stable nature of SG substructures [29]. Following a gentle lysis, cell lysates are fractionated via sequential centrifugation, followed by immunoprecipitation of an exogenously expressed SG marker. Immunocomplexes are then eluted for analysis by mass spectrometry or next-generation sequencing to identify the SG proteome or transcriptome, respectively.

As an example, Jain and colleagues overexpressed GFP-G3BP1, an SG nucleating protein, in the osteosarcoma cell line U2OS (and yeast) and subjected them to oxidative and heat stress to induce SGs, as defined by colocalization of this exogenous SG marker with PABP1, a known SG marker. In addition, the GFP-G3BP1 foci were observed only following stress, contained polyadenylated mRNA and were inhibited by cycloheximide. In this experiment, differential centrifugation was used to obtain fractions enriched with SGs, as followed by GFP-G3BP1. Superresolution microscopy revealed that G3BP1, PABP1, and polyadenylated mRNA were nonuniformly localized within SGs. Specifically, clusters of concentrated polyadenylated mRNA and G3BP1 signals were observed, and referred to as "cores," with the surrounding less concentrated material designated as the "shell" [29]. A complementary observation was made by Niewidock et al., using single molecule microscopy. Specifically, this work detected a biphasic partitioning within SGs with nanometer-sized nanocores composed of G3BP1 or IMP1 [35]. The super-resolution microscopy suggested that the stable foci obtained via centrifugation were not complete SGs, but instead were suggested to be this stable "core structure" containing G3BP1. In addition, it was noted that the granules obtained using

Table 2
Table summarizing different studies on SG composition

Study	Approach	Targets	Stressors	Cells	Analyses
[109]	Proximity labeling (APEX)	Endogenous G3BP1	SA, heat	HepG2, HEK293T, NPCs, HeLa	Proteomics
[29]	Cell fractionation	G3BP1 overexpressed	SA	U2OS, yeast	Proteomics
[110]	Proximity labeling (APEX)	eIF4E1, eIF4A1 overexpressed	Hippuristanol, heat, SA	HEK293	Transcriptomic
[32]	Proximity labeling (BIOID)	G3BP1, PABPC1 and EIF4A overexpressed	SA	HE 293	Proteomics and transcriptomics
[111]	Immunoprecipitation	Endogenous Caprin-1	SA	HeLa	Proteomics
[112]	Proximity labeling (APEX)	G3BP overexpressed	Thapsigargin, SA, heat	NIH/3T3, HEK293	Transcriptomic
[113]	Proximity labeling (APEX)	G3BP1, FMR1 and FXR1 endogenous	SA	U2OS	Proteomics
[114]	Immunoprecipitation	Ataxin-2 overexpressed	Glucose deprivation	Yeast	Proteomics and transcriptomics

this method are smaller than the expected size for SGs, as determined by light microscopy. However, superresolution microscopy demonstrated that the foci obtained after centrifugation had a size similar to SG cores (U2OS isolated granules: 233.1 ± 18.6 nm; Nanosight Nanoparticle Tracking Analysis system and SG core: 190.8 ± 28.3 using STORM). In any case, this SG enrichment approach identified 411 core mammalian SG proteins. Even though this method enriched for SG cores, it is expected that the identified proteins are indeed SG proteins as the size fractionation step selected granules by size, and thus prevented submicroscopic G3BP1 RNPs from being included.

On the other hand, Namkoong et al. overexpressed RFP-G3BP in NIH/3T3 cells, a spontaneously immortalized fibroblast cell line (Note, it is presumed that G3BP here refers to G3BP1, although this it is not stated explicitly in this paper or in a prior study from which the construct originates.) [112]. In unstressed cells, RFP-G3BP was observed as numerous tiny cytoplasmic puncta, which often occurs when RBPs are overexpressed. However, when cells were treated with the ER stress inducer thapsigargin, RFP-G3BP aggregated into SGs that were inhibited by cycloheximide, confirming their identity as SGs. In this study, SGs were isolated with a series of centrifugation steps, gentler than that used by Jain et al. Known SG markers such as G3BP1, TIA-1, TIAR, eIF4E and YB1 were readily found in the RNP fraction as well as P-body-specific markers (DCP1a, EDC4, and 4E-T) and ribosomal subunits (L4 and S3) but at variable levels. Stress and cycloheximide only affected SG marker levels in the RNP fraction while P body and other granule markers were unchanged. Moreover, some of the nuclear proteins known to associate with SGs, such as HDAC1 and PolII, were also recovered in the RNP fraction, but only in stress conditions. Lastly, markers for other subcellular components, such as the nucleoskeleton, ER, and cytoskeleton were depleted from the RNP fraction. Interestingly, the isolated G3BP structures were of a size comparable to what is observed for SGs by immunofluorescence and confocal microscopy, suggesting that the "core" and "shells" were recovered. However, while SGs may be enriched by this method, proteins from other RNPs were also recovered in this fraction. Thus, proteins that are classified as being in SGs by this approach may actually be associated with other granule types. Of note, however, is that this same study also included comparisons of SG transcriptomes obtained using different stressors (heat and SA), thus permitting the identification of stress-specific transcripts.

5.2 Immunoprecipitation-Based Approaches

As demonstrated by Jain et al., immunoprecipitation-based approaches enrich for SGs using an SG resident protein. However, in this case, no cellular fractionation and selection by size is done. This approach defines SG proteins as those new interactors acquired in the presence of a stressor, compared to unstressed conditions.

Vu et al. isolated SA-induced SGs from HeLa cells by immunoprecipitating endogenous Caprin-1 [111]. Following a 30 min treatment with SA and 60 min recovery, intracellular cross-linking was performed with low concentration formaldehyde, and Caprin-1 immunoprecipitates were recovered and submitted to mass-spectrometry. In order to identify stress-dependent interactions, Caprin-1 bound proteins detected in unstressed condition, which were determined to be somewhat similar to that of G3BP1 in unstressed conditions, were subtracted from the stressed protein interactor list. The authors posit that these prestress interactions may serve as seeds to facilitate rapid nucleation of SGs under stressed conditions, as previously proposed [32, 109]. However, investigation of the overlap between the stress-dependent interactome of Caprin-1, comprising 205 high confidence interactors, and three published G3BP1 proteomics studies showed minimal overlap [29, 32, 109, 111], suggesting that the new interactions gained by Caprin-1 during SA stress are G3BP1-independent. Another explanation could be related to the dynamic nature of the granules. Thus, Caprin-1 may interact with a subpool of G3BP1 and share a basal interactome. As SGs are highly dynamic, it is reasonable to conclude that interactions change during SG assembly and disassembly [109]. Collectively, this work and others suggests that each granule substructure (or "core") may have its own set of interactions and that using several baits will be required to uncover the entirety of the SG proteome.

5.3 Proximity-Labeling Approaches

As an alternative to cell-fractionation methods, proximity-dependent biotinylation techniques have been developed to characterize SGs. Proximity labeling-based proteomic approaches, such as APEX and BioID, use SG components fused with ascorbate peroxidase (APEX) or biotin ligase (BioID) as bait to covalently tag proteins in close proximity following the addition of hydrogen peroxide or biotin-phenol, respectively. As biotinylation induces a covalent liaison, biotinylated proteins can be purified using streptavidin beads, even in harsh conditions. The APEX approach is rapid, labeling proteins within minutes, rather than hours like BioID (will be described in the following paragraph) [120, 121]. Also, APEX-mediated tagging biotinylates proteins at a distance of 150–200 nm, while BioID biotinylates proteins within a 10 nm radius (from 5 nm to 20 nm) [120–123].

Youn et al. used BioID to better understand the overall organization of SA-induced SGs using a tetracycline inducible Flp-In T-REx HEK293 cell system and an abortive biotin ligase (BirA*) fused to three protein residents of SGs (G3BP1, PABPC1, and EIF4A) [32]. The tagging of three SG proteins permits an enhanced capture of tagged proteins as these SGs are approximatively tenfold larger than the range of BioID. This effectively provided a better resolution of SG composition. To limit potential artefacts from overexpression, bait protein expression levels were limited to one- to threefold of the endogenous protein. In the absence of stress, G3BP1, EIF4A, and PABPC1 showed a diffuse cytosolic expression pattern while the proteins localized to SGs during stress. Treatment with biotin led to the biotinylation of structures overlapping the tagged proteins' localization. Streptavidin pulldown followed with mass spectrometry was performed under standard conditions, alongside a set of negative controls.

Even if this method is based on the short range-acting BioID, the utilization of multiple bait proteins amplifies the number of proteins discovered as numerous substructures are expected to be found within an SG. Moreover, via integration of the various datasets, this method permits a fuller realization of the SG proteome compared to using a single bait. However, the choice of bait proteins is very important. If the baits are part of the same interactome in stressed conditions, the same interactome is likely to be identified, and thus minimal additional information will be added.

Markmiller et al. used CRISPR/Cas9-directed genome engineering to insert APEX2-GFP into the endogenous *G3BP1* locus in HEK293T cells [109]. SA exposure of these cells results in G3BP1-

APEX2-GFP fusion protein localization to SGs. The main caveat of this method comes from the fact that not all of the G3BP1 resides in SGs under stress. To specifically recover SG proteins, the group used four experiments. First, they identified G3BP1 stress-dependent interactors (biotinylated proteins in stressed versus unstressed G3BP1-APEX2-GFP cells). Then, compared this list to proteins identified in stressed G3BP1-APEX2-GFP cells incubated with biotin phenol to lysates of identically treated cells for which the biotin phenol substrate was omitted (experiment 2). Third, to control for diffuse cytoplasmic labeling by G3BP1-APEX2-GFP, they compared lysates from stressed G3BP1-APEX2-GFP and the control NES-APEX2-GFP cells which identified nonspecific cytoplasmic binding partners of G3BP1 (experiment 3). Lastly, to define stress-independent as well as stress-sensitive G3BP1 interactors, they compared lysates from unstressed G3BP1-APEX2-GFP and NES-APEX2-GFP cells (experiment 4). Using this elegant approach, this team uncovered several stress and cell-specific SG proteins, as well as the interactome of G3BP1 in physiological conditions. Moreover, this approach revealed the extensive remodeling of the proteome under stress. A total of 123 proteins, including 69 known proteins, were identified.

Lastly, Marmor-Kollet et al. recently reported on SG protein composition in SA-stressed U2OS cells using an APEX2-tag system at 30, 60, 100, and 180 min poststress [113]. Similar to Youn et al., they used multiple APEX-tagged baits in order to obtain a more comprehensive picture of SG composition. The use of several baits also allowed them to not only compare and validate their lists to previous studies [29, 32, 109], but also extend them. As baits, they used G3BP1, FMR1 (FMRP) and FXR1, three extensively studied RBPs that display high SG partition coefficients in U2OS cells [21]. They used the same fused nuclear export signal (NES) to APEX as Markmiller et al., to serve as a cytoplasmic control bait [124]. To avoid overexpression artifacts, Tet-inducible FMR1-APEX and FXR1-APEX constructs were transfected into CRISPR-edited U2OS cells lacking *FMR1/FXR1/FXR2* [125], while the Tet-inducible G3BP1-APEX construct was transfected into U2OS *G3BP1/G3BP2* double knockout cells [22]. Also, induction of APEX expression was carefully titrated so as to approximate endogenous protein levels for all baits. This method, where proteins were bound to at least one of the APEX-tagged baits, identified 489 proteins, including 310 novel proteins. Similarly, 240 were detected with at least two baits, including 109 novel proteins; and 98 were identified by all three APEX-tagged baits, 15 of which were novel proteins. Very interestingly the new SG proteins detected by this approach are previously linked to FMR1 and FXR1, suggesting that focusing on one SG protein to establish the SG proteome is limiting.

6 Insights Gained on SG Composition

6.1 Protein Components

Most of the proteomic studies on SG composition confirmed the presence of the protein families that have been previously associated with SGs, including those involved in RNA metabolism [113]. Comparison between yeast and mammalian SGs confirmed the enrichment of translation factors and RBPs in SGs and helped establish the conservation of SG proteins throughout evolution, reinforcing their importance [29]. tRNA transferases and ribosome biogenesis factors were also found to be enriched in SGs, reflecting the role of SGs in the downregulation of ribosome biogenesis and translational arrest upon stress [29]. Surprisingly the majority of proteins found in SGs were not known to bind mRNA nor predicted to harbor a prion-like domain [29, 110]. Several ATP-dependent proteins, nucleic acid remodeling complexes, protein folding proteins, and proteins involved in posttranslational modifications (polyADP-ribolysation, ubiquitination, or phosphorylation) were associated with SGs [29, 109, 110]. A number of hypotheses could explain the presence of these proteins in SGs. First, it is possible that they are important for SG remodeling, facilitating conformational changes for proteins and nucleic acids to localize to the granules. Also, it is possible that these proteins are localized to SGs in order to prevent their degradation and/or to aid in SG disassembly (will be discussed later) or to facilitate cellular recovery after stress resolution. Interestingly, the proteins identified in SGs form dense protein networks [29]. In mammalian cells, SG-resident proteins on average interact with 8.25 proteins, which is more than what is expected by chance. This is consistent with the patchy colloid theory; that SG nucleating proteins have high valences to favor multiple interactions [21, 29]. This supports the observation that SG resident proteins are promiscuous and may help to explain why these proteins are prone to become immiscible. Indeed, in the dense environment of the cytoplasm, they would be more likely to form complexes and phase separate (and/or aggregate).

Interestingly, among the new SG proteins identified, many are linked to human disorders. Jain et al. noticed that 75 of the proteins on their list were linked to 153 Mendelian-inherited diseases with an enrichment predominantly in neurological, muscular and cardiovascular disorders. For example, the annexin ANXA11 was found to reside in SGs. Mutations in this gene are associated with amyotrophic lateral sclerosis (ALS) and ANXA11 mutant proteins are prone to aggregate [109, 126]. The same observation was made for ATXN2, DPYSL3, DCTN1 and TUBA4A [29, 110, 127, 128], and are supportive of the hypothesis that aberrant SG dynamics are relevant to neurological disorders.

6.2 RNA Components

It is estimated that 80% of an SG is composed of mRNA, as well as several noncoding RNAs [116]. Thus, uncovering the SG transcriptome is also needed to fully understand SG function and dynamics. The preponderance of transcripts in these granules suggests that they are important contributors to SG function [129]. (For more details on the importance of RNA for SG formation, please *see* ref. 26).

Analysis of mRNAs recovered from SGs enriched by cell fractionation revealed that, in general, the transcripts enriched in SGs are more or less the same, regardless of cell type or stressor [112]. The only difference was that the quantity of specific transcripts in the granules varied according to the stressor. For example, *TFRC* is enriched twofold in heat-induced SGs compared to other stressors, while *Polr2a* transcripts are more enriched in SGs induced by heat or sorbitol compared to those induced by SA or thapsigargin [112]. These differences could be due to either the specific translational repression caused by each stressor or to the varied protein composition of SGs. Also, interestingly, depending on the stressor, it is notable that the quantity of cellular mRNA is different. For example, heat increases the percentage of transcripts in the granule fraction and decreases cytoplasmic mRNA levels, while thapsigargin induced a diminution of transcripts in granules without bulk cellular mRNA changes [112]. The authors suggest that this diminution is due to SGs promoting mRNA degradation, a provocative suggestion given that SGs are thought to be implicated in mRNA protection [112]. Finally, these differences were not observed in ER or SA stress conditions.

SG-associated transcripts have specific characteristics. They are generally less abundant, feature long open reading frames (ORFs), low translational efficiency, less GC content and have shorter half-lives. Longer mRNAs are preferentially recruited to SGs following ER and SA stress [28]. The association between mRNA and length could be the consequence of the physical property of RNA itself since longer RNAs more easily form RNPs complexes and complex secondary structures may promote granulation and sedimentation [34]. On average, SG-associated transcripts are 7.1 kb in length while associated noncoding RNAs are 1.9 kb [28]. Interestingly, high folding energy and GC content of the transcripts does not correlate with SG targeting. Moreover, adenylate/uridylate-rich elements (AU-rich elements; AREs) show a positive correlation with SG residence that supersedes that of mRNA length [28]. AREs are *cis*-acting regulatory elements within untranslated regions (UTRs) of short-lived mRNAs which typically function as a signal for rapid mRNA degradation. They are present in 5–8% of human genes, most of which are involved in the regulation of the stress response, cell cycle, inflammation, and apoptosis, and therefore must be tightly regulated [28, 112]. Moreover, SG proteins such as TIA-1, TIAR, and HuR are known to bind and regulate

ARE-containing transcripts. Studies on the RNAs associated with SGs indicate that around 10–15% of the cellular transcriptome is localized to SGs [28, 112]. Curiously, 10–15% of the transcriptome resides in RNPs in basal conditions also [112], suggesting that RNPs, with the transcripts they protect, may be the pre-seeds that coalesce to form SGs in stressed conditions.

Some genes are overrepresented in SGs. Specifically, 185 genes have >50% of their transcripts localized to SGs [28]. This subset of transcripts encode several transcription factors, kinases, as well as proteins involved in cell proliferation, growth and survival such as XIAP, AGO3, CREB1 c-MYC, VEGF-A, FGF13 and RICTOR [28, 110]. Interestingly, transcripts of housekeeping genes such as *GAPDH* and *ACTB* are underrepresented in SGs in both mouse and human cell lines [28, 112]. The high concentration of transcripts encoding protooncogenes and the depletion of those encoding housekeeping functions suggests a role for SGs in limiting cell growth and proliferation, and possibly ensuring basal cell function. However, recent studies have uncoupled translation inhibition and SG formation [130]. For several years, it has been accepted that transcripts residing in SGs were translationally dormant, mostly due to coincidences between the general translation inhibition during stress and SG formation [130]. Single molecule microscopy has recently revealed that some transcripts can be translated while localized within SGs, including the activating transcription factor 4 (ATF4), a protein activated by the UPR governing multiple signaling pathways (autophagy, oxidative stress and inflammation) leading to pro-survival or pro-death pathways depending on the cues [130]. The translation of this factor within SGs suggests that SGs have an active role in pro-survival activities and add another important aspect to SG function.

Other transcripts depleted from SGs are linked to other subcellular sites. For example, transcripts associated with proteins with membrane functions or secreted are depleted from SGs, implying that ER-bound ribosomes are less implicated in SG formation [131]. Moreover, even if some mitochondrial proteins were found in SGs, mitochondrial-encoding transcripts (both nuclear- and mitochondrial-encoded) were depleted from SGs, suggesting that mitochondrial activities are important for the cell during stress, possibly to aid with the high energy demand that overcoming a stress imposes.

7 Lessons that Can Be Learned from These Studies

More than just revealing the RNA and protein composition of SGs, the above cited studies have been essential to advance our understanding of SG dynamics. Indeed, several groups have studied SGs to uncover how their formation or lack of dissolution (disassembly) is linked to human disease. By exposing the composition of these granules, these studies shed light on the proteins and pathways involved in SG physiology.

7.1 Cellular Context Is Important, and Often Overlooked

More than giving general insights on SG composition, the various studies have permitted the confirmation of differential SG composition according to the cell type used. Meta-analyses amongst the different systems demonstrates that, in general, the protein families composing SGs are similar. For example, both yeast and mammalian SGs are enriched in RBPs (TIA1, UBAP2L, Caprin-1, PABPC1, FUS, and ATXN2) and proteins with predicted prion-like domains [29]. SGs from both systems were also found to contain tRNA synthetases and ribosome biogenesis factors (NOP58, FTSJ3, NOP2, and FBL in mammalian cells; Kap123, Rio2, Fap7, and Cam1 in yeast), which might reflect a role for SG formation in the downregulation of ribosome biogenesis during stress [132]. Moreover, a number of ATP-dependent protein and nucleic acid remodeling complexes, including protein chaperones (e.g., Hsp70/Hsp40 and CCT complex) and multiple RNA and DNA helicases (e.g., DEAD-box proteins, MCM, and RVB helicases) are found in yeast and mammalian stress granules [29]. In general, SG proteins are associated with multiple aspects of RNA metabolism [5, 15, 113]. However, proteins associated with DNA or chromatin are also associated with SGs [113]. Also, a number of proteins involved in posttranslational modifications were found to be localized to SGs. Specifically, N6-methyladenosine (m^6A) modification binding proteins, poly-ADP ribosylation, ubiquitination or phosphorylation modifiers are residents of SGs. This supports the critical role of RNA methylation and PTMs regulating SG dynamics and phase separation [133, 134].

Although at a high level there are many similarities in SG component protein families between systems, there are some notable differences. As an example, let us consider G3BP1-APEX2-GFP-expressing human induced pluripotent stem cells (iPSCs) generated using CRISPR/Cas9-mediated gene editing to study SGs in neural progenitor (NPCs) [109]. Interestingly, comparison of SA-induced SGs in NPCs and HEK293FT cells using the same APEX strategy showed that 64% of the proteins were unique to one cell type. This suggests an important diversity in SG composition that is likely due to differences in protein abundance or functions between cell types. Comparison of the proteomes of SA-induced SGs in three cell lines (HepG2, HeLa and NPCs) indicated that among a total of 77 SG-related RBPs examined, only 42 (55%) localized to SGs in all three cell lines while the remaining 35 RBPs exhibited variable patterns among cell types [109]. For example, while the ubiquitin-associated protein 2 like (UBAP2L) localized to SGs in all three cell types examined, the splicing factor SRSF9, the translation initiation factor eIF4G, and the ribonucleoprotein

SRP68 were selectively targeted to SGs in HepG2, HeLa and NPCs, respectively [109]. That 75% of the SG proteins detected in NPCs were not previously associated with SGs further highlights that past studies using commonly available immortalized cell lines have missed potentially important SG proteins with particular neuronal relevance. Moreover, that these neuronal SG proteins have reported functions in protein quality control (PQC) pathways, such as chaperone-assisted protein folding or aggregate clearance, and that these have been implicated in neurodegenerative disorders might help to understand the pathogenesis of several diseases. Indeed, abnormal accumulation of ubiquitin-positive proteins in affected brain regions is a typical pathological hallmark of neurodegenerative diseases such as Alzheimer's disease (AD), Huntington's disease (HD) and Parkinson's disease (PD) and thought to contribute to neurotoxicity. Accumulating evidence suggests that deficiencies in PQC and clearance mechanisms may lead to the abnormal accumulation of proteins in neurodegenerative diseases and/or that an excessive accumulation of misfolded and aggregated proteins may overwhelm the PQC and clearance systems, leading to further protein accumulation, cellular stress, and ultimately to neurodegeneration. Therefore, understanding why neuronal SGs contain PQC proteins may shed light on other affected pathways that could be modulated and perhaps hint at an explanation of why neurons are prone to aggregate formation.

7.2 There Is a Preexisting Network of SG Protein Interactions in Basal Conditions

Studying the interactomes of SG proteins in basal conditions can help advance our understanding of the mechanisms underlying SG formation/dynamics. Most of the proteomic studies to date have demonstrated the steady-state presence of preexisting or submicroscopic RNPs [29, 35]. Indeed, as mentioned previously, less than half of the G3BP1 interactome was stress-dependent [109]. Translation initiation factors such as eIF3 and eIF4 that accumulate in PICs in SGs interact with G3BP1 in basal conditions, as well as Caprin-1 and USP10 [109, 111]. FMR1 and TIA-1 were also found to interact with G3BP1 in basal conditions. Marmor-Kollet et al. confirmed this stress-independent interaction [109, 113]; however, the other SG markers they studied did not show this pattern. Indeed, only 16% of FMR1 and FXR1 interactomes were stress-independent, demonstrating that the majority of the interactions in which FMR1 and FXR1 participate in stressed conditions are specific. This observation is very interesting and reinforces G3BP1 as central to SG formation. It is plausible that G3BP1 interacts with SG proteins in physiological/basal conditions in order to facilitate rapid SG formation upon exposure to a noxious stimulus. G3BP1 would be recruited later to SGs baited with FMR1 as G3BP1 is viewed as the protein that links the RNPs together. This may be in agreement with the observation that G3BP1 cellular depletion leads to small SGs that do not proceed to secondary assembly [135] and the patchy colloid theory [21]. This is also supported by the observation that in contrast to FRX1, G3BP1 is highly mobile in SGs as demonstrated by photobleaching experiments [35, 113]. Moreover, superresolution microscopy revealed that G3BP1 is in a different substructure than FRX1: while the latter is in the periphery, G3BP1 is positioned near the center and moves in and out of the denser part of the SG.

7.3 SG Formation Is an Active Process: The Importance of ATPases

The presence of ATPases in SGs suggests that their activity may actively modulate SG proteins. Cellular depletion of ATP via blocking glycolysis and oxidative phosphorylation with 2-deoxyglucose (2DG) or CCCP, respectively, impairs SG assembly in response to SA. Specifically, when ATP production was blocked 20 min into a SA stress exposure, small granules were observed, indicating the necessity of ATP for SG formation and maturation. It has been suggested that RNP coalescence (and thus SG formation) occurs via ATP-dependent motor-directed movement along the cytoskeleton [136, 137]. Moreover, depletion of ATP after SG formation leads to static SGs and tracks with decreased mobility of the G3BP1 pool and impeded LLPS behavior. Another hypothesis, which is not mutually exclusive, is that ATP-dependent events are necessary to achieve appropriate conformations for proteins and RNAs to assemble with each other into SGs. Indeed, it is plausible that RNPs or preexisting seeds at basal conditions are in conformations that disfavor their self-assembly into larger granules. The presence of ATPases, protein remodeling complexes and helicases in SGs support this hypothesis. The fact that RNPs coalesce to form SGs leads to a series of crucial questions. What does the cell gain from fewer larger granules that is not achieved by numerous smaller granules? What makes these larger granules more cytoprotective than the smaller ones? These questions remain to be answered.

7.4 SG Formation Is a Sequential Process

Examination of the SG proteome at different time points provides very interesting information on SG formation and dynamics. Following heat shock, the eIF4A1 proteome changes rapidly [110]. Over time, different proteins assemble around eIF4A1, indicating a sequential/orderly process of SG formation. Interestingly, Caprin-1 associates with eIF4A1 prior to G3BP1. This is in accordance with data showing that the G3BP1-APEX interactome changes with time [109]. This sequential formation of SGs suggests that G3BP1 is the “glue” that sticks the RNPs together, ultimately leading to the formation of SGs [110]. This is important as it means that SG composition changes with time and the study of these entities must be done with kinetics in mind.

7.5 SG Disassembly May Be Facilitated by a Distinct Set of Interactions

Marmet-Kollet et al. have provided the first glimpse of the SG proteome during SG dissolution. Proteins that are necessary/enriched in SGs during disassembly were named disassembly-engaged proteins (DEPs). A total of 202 proteins were enriched during SG disassembly compared to the stressed phase and gene ontology analyses indicated an enrichment for autophagy and ubiquitin pathways [113]. Moreover, a number of heat shock proteins, RNA helicases, and mitochondrial proteins were observed. Markmiller et al. also reported an enrichment in several factors implicated in autophagy and vesicular transport processes including GABARAPL2, YML2, and ATG8 [109]. Collectively, this suggests a tight surveillance of SG proteins through interactions with ATG8 proteins that may facilitate an important role for autophagy in SG disassembly, as previously suggested [109]. Moreover, the depletion via siRNA of nine proteins involved in the ubiquitin-proteasome, autophagy and SUMOylation machineries revealed that these pathways were necessary for SG disassembly. In addition to autophagy, posttranslational modifications seem to be necessary for SG resolution. Indeed, inhibition of SUMOylation via the chemical inhibition of UBE2L prevented SG formation and disassembly, underlining the importance of this modification for SG dynamics [113]. Therefore, autophagy and PTM seem to be of particular importance for SG disassembly and needs to be considered.

7.6 Canonical and Noncanonical SGs Are Different Entities

Most studies have examined canonical SGs and thus relatively little is known about noncanonical SGs. The few studies focused on noncanonical granules suggest that they are specific entities with functions distinct from canonical SGs and associated transcripts are different between canonical and noncanonical SGs. Specifically, in contrast to heat and SA-induced SGs, transcripts associated with hippuristanol-induced SGs do not demonstrate the same tendencies for length and translational efficiency [110]. This is further supported by a recent study examining SG and PB composition following glucose depletion which induces noncanonical SGs [16, 114]. In this study, Kershaw et al. used an immunoprecipitation approach, similar to Vu et al., in order to isolate SGs, from yeast overexpressing target proteins (Dcp1p, Ataxin 2 ortholog). Proteomic comparison between glucose-deprivation and SA (data from Jain et al.) conditions showed only 25 common proteins, suggesting that the protein composition of these granules is fundamentally different [114]. Similarly, only a modest correlation was found between transcripts from glucose-deprivation induced and SA-induced SGs, highlighting the importance to study both canonical and noncanonical SGs to uncover their functions and potential related dysfunctions associated with disease [114].

8 Conclusion and Future Directions

In recent years, substantial effort has been invested to describe SG RNA and protein compositions. Using cell fractionation to proximity labeling approaches, many insights have been gained which greatly outweigh their respective pitfalls. Perhaps most interesting is that many of the findings corroborate and complement each other to collectively provide a better understanding of SG physiology. This will be crucial as this information is applied to disease contexts where SG dysfunction is suspected. Moreover, it is important to note that the bulk of our knowledge is centered on the composition of canonical SGs, but similar comprehensive studies of noncanonical SGs still remain to be done. This deserves to be underlined, since it is suggested that their composition is different, and may be pro-death [16]. Moreover, relevant to human disease, it remains unknown how chronic stress impacts SG composition and dynamics. Lastly, all the studies mentioned examined SGs in vitro. In order to truly understand SG physiology and their role in pathology, we must begin to rigorously study these entities in vivo. Future work will undoubtedly focus on defining the role of SGs in disease.

Acknowledgments

We thank all members of the Vande Velde lab for helpful discussions, and specifically Asmita Ghosh and Alicia Dubinski for comments on the manuscript. H.S. is supported by an FRQS Doctoral Award. C.V.V. is a FRQS Senior Scholar. Work in the Vande Velde lab is supported by ALS Canada/Brain Canada, Target ALS, and CIHR.

References

1. Galluzzi L, Yamazaki T, Kroemer G (2018) Linking cellular stress responses to systemic homeostasis. Nat Rev Mol Cell Biol 19(11): 731–745. https://doi.org/10.1038/s41580-018-0068-0
2. Kültz D (2005) Molecular and evolutionary basis of the cellular stress response. Annu Rev Physiol 67:225–257. https://doi.org/10.1146/annurev.physiol.67.040403.103635
3. Fulda S, Gorman AM, Hori O, Samali A (2010) Cellular stress responses: cell survival and cell death. Int J Cell Biol 2010: 214074–214074. https://doi.org/10.1155/2010/214074
4. Anderson P, Kedersha N (2002) Stressful initiations. J Cell Sci 115(Pt 16):3227–3234
5. Wolozin B, Ivanov P (2019) Stress granules and neurodegeneration. Nat Rev Neurosci 20(11):649–666. https://doi.org/10.1038/s41583-019-0222-5
6. Molliex A, Temirov J, Lee J, Coughlin M, Kanagaraj AP, Kim HJ, Mittag T, Taylor JP (2015) Phase separation by low complexity domains promotes stress granule assembly and drives pathological fibrillization. Cell 163(1):123–133. https://doi.org/10.1016/j.cell.2015.09.015
7. Kroschwald S, Maharana S, Mateju D, Malinovska L, Nüske E, Poser I, Richter D,

Alberti S (2015) Promiscuous interactions and protein disaggregases determine the material state of stress-inducible RNP granules. elife 4:e06807. https://doi.org/10.7554/eLife.06807

8. Aulas A, Fay MM, Lyons SM, Achorn CA, Kedersha N, Anderson P, Ivanov P (2017) Stress-specific differences in assembly and composition of stress granules and related foci. J Cell Sci 130(5):927–937. https://doi.org/10.1242/jcs.199240
9. Aulas A, Vande Velde C (2015) Alterations in stress granule dynamics driven by TDP-43 and FUS: a link to pathological inclusions in ALS? Front Cell Neurosci 9:423. https://doi.org/10.3389/fncel.2015.00423
10. Kedersha N, Anderson P (2007) Mammalian stress granules and processing bodies. Methods Enzymol 431:61–81
11. Advani VM, Ivanov P (2020) Stress granule subtypes: an emerging link to neurodegeneration. Cell Mol Life Sci 77(23):4827–4845. https://doi.org/10.1007/s00018-020-03565-0
12. Lin Y, Protter DS, Rosen MK, Parker R (2015) Formation and maturation of phase-separated liquid droplets by RNA-binding proteins. Mol Cell 60(2):208–219. https://doi.org/10.1016/j.molcel.2015.08.018
13. Toretsky JA, Wright PE (2014) Assemblages: functional units formed by cellular phase separation. J Cell Biol 206(5):579–588. https://doi.org/10.1083/jcb.201404124
14. Li P, Banjade S, Cheng HC, Kim S, Chen B, Guo L, Llaguno M, Hollingsworth JV, King DS, Banani SF, Russo PS, Jiang QX, Nixon BT, Rosen MK (2012) Phase transitions in the assembly of multivalent signalling proteins. Nature 483(7389):336–340
15. Sidibé H, Vande Velde C (2019) RNA granules and their role in neurodegenerative diseases. In: Oeffinger M, Zenklusen D (eds) The biology of mRNA: structure and function. Springer International Publishing, Cham, pp 195–245. https://doi.org/10.1007/978-3-030-31434-7_8
16. Reineke LC, Cheema SA, Dubrulle J, Neilson JR (2018) Chronic starvation induces noncanonical pro-death stress granules. J Cell Sci 131(19):jcs220244
17. Dhillon P, Durga Rao C (2018) Rotavirus induces formation of remodeled stress granules and P-bodies and their sequestration in viroplasms to promote progeny virus production. J Virol 92(24):e01363-18
18. Humoud MN, Doyle N, Royall E, Willcocks MM, Sorgeloos F, van Kuppeveld F, Roberts LO, Goodfellow IG, Langereis MA, Locker N (2016) Feline calicivirus infection disrupts assembly of cytoplasmic stress granules and induces G3BP1 cleavage. J Virol 90(14):6489–6501. https://doi.org/10.1128/jvi.00647-16
19. Guillén-Boixet J, Kopach A, Holehouse AS, Wittmann S, Jahnel M, Schlüßler R, Kim K, Trussina IREA, Wang J, Mateju D, Poser I, Maharana S, Ruer-Gruß M, Richter D, Zhang X, Chang Y-T, Guck J, Honigmann A, Mahamid J, Hyman AA, Pappu RV, Alberti S, Franzmann TM (2020) RNA-induced conformational switching and clustering of G3BP drive stress granule assembly by condensation. Cell 181(2):346–361.e317. https://doi.org/10.1016/j.cell.2020.03.049
20. Yang P, Mathieu C, Kolaitis R-M, Zhang P, Messing J, Yurtsever U, Yang Z, Wu J, Li Y, Pan Q, Yu J, Martin EW, Mittag T, Kim HJ, Taylor JP (2020) G3BP1 is a tunable switch that triggers phase separation to assemble stress granules. Cell 181(2):325–345.e328. https://doi.org/10.1016/j.cell.2020.03.046
21. Sanders DW, Kedersha N, Lee DSW, Strom AR, Drake V, Riback JA, Bracha D, Eeftens JM, Iwanicki A, Wang A, Wei M-T, Whitney G, Lyons SM, Anderson P, Jacobs WM, Ivanov P, Brangwynne CP (2020) Competing protein-RNA interaction networks control multiphase intracellular organization. Cell 181(2):306–324.e328. https://doi.org/10.1016/j.cell.2020.03.050
22. Kedersha N, Panas MD, Achorn CA, Lyons S, Tisdale S, Hickman T, Thomas M, Lieberman J, McInerney GM, Ivanov P, Anderson P (2016) G3BP-Caprin1-USP10 complexes mediate stress granule condensation and associate with 40S subunits. J Cell Biol 212(7):845–860. https://doi.org/10.1083/jcb.201508028
23. Panas MD, Ivanov P, Anderson P (2016) Mechanistic insights into mammalian stress granule dynamics. J Cell Biol 215(3):313–323. https://doi.org/10.1083/jcb.201609081
24. Bounedjah O, Desforges B, Wu T-D, Pioche-Durieu C, Marco S, Hamon L, Curmi PA, Guerquin-Kern J-L, Piétrement O, Pastré D (2014) Free mRNA in excess upon polysome dissociation is a scaffold for protein multimerization to form stress granules. Nucleic Acids Res 42(13):8678–8691. https://doi.org/10.1093/nar/gku582
25. Patel A, Lee HO, Jawerth L, Maharana S, Jahnel M, Hein MY, Stoynov S, Mahamid J,

Saha S, Franzmann TM, Pozniakovski A, Poser I, Maghelli N, Royer LA, Weigert M, Myers EW, Grill S, Drechsel D, Hyman AA, Alberti S (2015) A liquid-to-solid phase transition of the ALS protein FUS accelerated by disease mutation. Cell 162(5):1066–1077. https://doi.org/10.1016/j.cell.2015.07.047

26. Van Treeck B, Parker R (2018) Emerging roles for intermolecular RNA-RNA interactions in RNP assemblies. Cell 174(4):791–802. https://doi.org/10.1016/j.cell.2018.07.023

27. Van Treeck B, Protter DSW, Matheny T, Khong A, Link CD, Parker R (2018) RNA self-assembly contributes to stress granule formation and defining the stress granule transcriptome. Proc Natl Acad Sci 115(11):2734. https://doi.org/10.1073/pnas.1800038115

28. Khong A, Matheny T, Jain S, Mitchell SF, Wheeler JR, Parker R (2017) The stress granule transcriptome reveals principles of mRNA accumulation in stress granules. Mol Cell 68(4):808–820.e805. https://doi.org/10.1016/j.molcel.2017.10.015

29. Jain S, Wheeler JR, Walters RW, Agrawal A, Barsic A, Parker R (2016) ATPase-modulated stress granules contain a diverse proteome and substructure. Cell 164(3):487–498. https://doi.org/10.1016/j.cell.2015.12.038

30. Banani SF, Lee HO, Hyman AA, Rosen MK (2017) Biomolecular condensates: organizers of cellular biochemistry. Nat Rev Mol Cell Biol 18(5):285–298. https://doi.org/10.1038/nrm.2017.7

31. Reineke LC, Kedersha N, Langereis MA, van Kuppeveld FJ, Lloyd RE (2015) Stress granules regulate double-stranded RNA-dependent protein kinase activation through a complex containing G3BP1 and Caprin1. MBio 6(2):e02486. https://doi.org/10.1128/mBio.02486-14

32. Youn JY, Dunham WH, Hong SJ, Knight JDR, Bashkurov M, Chen GI, Bagci H, Rathod B, MacLeod G, Eng SWM, Angers S, Morris Q, Fabian M, Côté JF, Gingras AC (2018) High-density proximity mapping reveals the subcellular organization of mRNA-associated granules and bodies. Mol Cell 69(3):517–532.e511. https://doi.org/10.1016/j.molcel.2017.12.020

33. Wyman J, Gill SJ (1980) Ligand-linked phase changes in a biological system: applications to sickle cell hemoglobin. Proc Natl Acad Sci 77(9):5239. https://doi.org/10.1073/pnas.77.9.5239

34. Jain A, Vale RD (2017) RNA phase transitions in repeat expansion disorders. Nature 546(7657):243–247. https://doi.org/10.1038/nature22386

35. Niewidok B, Igaev M, Pereira da Graca A, Strassner A, Lenzen C, Richter CP, Piehler J, Kurre R, Brandt R (2018) Single-molecule imaging reveals dynamic biphasic partition of RNA-binding proteins in stress granules. J Cell Biol 217(4):1303–1318. https://doi.org/10.1083/jcb.201709007

36. Wheeler JR, Matheny T, Jain S, Abrisch R, Parker R (2016) Distinct stages in stress granule assembly and disassembly. elife 5:e18413. https://doi.org/10.7554/eLife.18413

37. Taniuchi S, Miyake M, Tsugawa K, Oyadomari M, Oyadomari S (2016) Integrated stress response of vertebrates is regulated by four eIF2α kinases. Sci Rep 6:32886. https://doi.org/10.1038/srep32886

38. Anderson P, Kedersha N (2008) Stress granules: the Tao of RNA triage. Trends Biochem Sci 33(3):141–150

39. Hinnebusch AG (2014) The scanning mechanism of eukaryotic translation initiation. Annu Rev Biochem 83:779–812. https://doi.org/10.1146/annurev-biochem-060713-035802

40. Fujimura K, Sasaki AT, Anderson P (2012) Selenite targets eIF4E-binding protein-1 to inhibit translation initiation and induce the assembly of non-canonical stress granules. Nucleic Acids Res 40(16):8099–8110. https://doi.org/10.1093/nar/gks566

41. Mateju D, Franzmann TM, Patel A, Kopach A, Boczek EE, Maharana S, Lee HO, Carra S, Hyman AA, Alberti S (2017) An aberrant phase transition of stress granules triggered by misfolded protein and prevented by chaperone function. EMBO J 36(12):1669–1687. https://doi.org/10.15252/embj.201695957

42. Markmiller S, Fulzele A, Higgins R, Leonard M, Yeo GW, Bennett EJ (2019) Active protein neddylation or ubiquitylation is dispensable for stress granule dynamics. Cell Rep 27(5):1356–1363.e1353. https://doi.org/10.1016/j.celrep.2019.04.015

43. Xie X, Matsumoto S, Endo A, Fukushima T, Kawahara H, Saeki Y, Komada M (2018) Deubiquitylases USP5 and USP13 are recruited to and regulate heat-induced stress granules through their deubiquitylating activities. J Cell Sci 131(8):jcs210856. https://doi.org/10.1242/jcs.210856

44. Emara MM, Fujimura K, Sciaranghella D, Ivanova V, Ivanov P, Anderson P (2012) Hydrogen peroxide induces stress granule formation independent of eIF2α phosphorylation. Biochem Biophys Res Commun 423(4):763–769. https://doi.org/10.1016/j.bbrc.2012.06.033
45. Allemand E, Guil S, Myers M, Moscat J, Caceres JF, Krainer AR (2005) Regulation of heterogenous nuclear ribonucleoprotein A1 transport by phosphorylation in cells stressed by osmotic shock. Proc Natl Acad Sci U S A 102(10):3605–3610
46. Dewey CM, Cenik B, Sephton CF, Dries DR, Mayer P 3rd, Good SK, Johnson BA, Herz J, Yu G (2011) TDP-43 is directed to stress granules by sorbitol, a novel physiological osmotic and oxidative stressor. Mol Cell Biol 31(5):1098–1108. https://doi.org/10.1128/mcb.01279-10
47. Damgaard CK, Lykke-Andersen J (2011) Translational coregulation of 5'TOP mRNAs by TIA-1 and TIAR. Genes Dev 25(19): 2057–2068
48. Ivanov P, Emara MM, Villen J, Gygi SP, Anderson P (2011) Angiogenin-induced tRNA fragments inhibit translation initiation. Mol Cell 43(4):613–623. https://doi.org/10.1016/j.molcel.2011.06.022
49. Bordeleau ME, Mori A, Oberer M, Lindqvist L, Chard LS, Higa T, Belsham GJ, Wagner G, Tanaka J, Pelletier J (2006) Functional characterization of IRESes by an inhibitor of the RNA helicase eIF4A. Nat Chem Biol 2(4):213–220. https://doi.org/10.1038/nchembio776
50. Dang Y, Kedersha N, Low WK, Romo D, Gorospe M, Kaufman R, Anderson P, Liu JO (2006) Eukaryotic initiation factor 2alpha-independent pathway of stress granule induction by the natural product pateamine A. J Biol Chem 281(43):32870–32878. https://doi.org/10.1074/jbc.M606149200
51. Cencic R, Pelletier J (2016) Hippuristanol—a potent steroid inhibitor of eukaryotic initiation factor 4A. Translation 4(1):e1137381. https://doi.org/10.1080/21690731.2015.1137381
52. Tsumuraya T, Ishikawa C, Machijima Y, Nakachi S, Senba M, Tanaka J, Mori N (2011) Effects of hippuristanol, an inhibitor of eIF4A, on adult T-cell leukemia. Biochem Pharmacol 81(6):713–722. https://doi.org/10.1016/j.bcp.2010.12.025
53. Cencic R, Carrier M, Galicia-Vázquez G, Bordeleau M-E, Sukarieh R, Bourdeau A, Brem B, Teodoro JG, Greger H, Tremblay ML, Porco JA Jr, Pelletier J (2009) Antitumor activity and mechanism of action of the cyclopenta[b]benzofuran, silvestrol. PLoS One 4(4):e5223. https://doi.org/10.1371/journal.pone.0005223
54. Kogure T, Kinghorn AD, Yan I, Bolon B, Lucas DM, Grever MR, Patel T (2013) Therapeutic potential of the translation inhibitor silvestrol in hepatocellular cancer. PLoS One 8(9):e76136. https://doi.org/10.1371/journal.pone.0076136
55. Emara MM, Ivanov P, Hickman T, Dawra N, Tisdale S, Kedersha N, Hu G-F, Anderson P (2010) Angiogenin-induced tRNA-derived stress-induced RNAs promote stress-induced stress granule assembly. J Biol Chem 285(14): 10959–10968. https://doi.org/10.1074/jbc.M109.077560
56. Iwasaki S, Floor SN, Ingolia NT (2016) Rocaglates convert DEAD-box protein eIF4A into a sequence-selective translational repressor. Nature 534(7608):558–561. https://doi.org/10.1038/nature17978
57. Szaflarski W, Fay MM, Kedersha N, Zabel M, Anderson P, Ivanov P (2016) Vinca alkaloid drugs promote stress-induced translational repression and stress granule formation. Oncotarget 7(21):30307–30322. https://doi.org/10.18632/oncotarget.8728
58. Mazroui R, Di Marco S, Kaufman RJ, Gallouzi IE (2007) Inhibition of the ubiquitin-proteasome system induces stress granule formation. Mol Biol Cell 18(7):2603–2618
59. Thedieck K, Holzwarth B, Prentzell Mirja T, Boehlke C, Kläsener K, Ruf S, Sonntag Annika G, Maerz L, Grellscheid S-N, Kremmer E, Nitschke R, Kuehn EW, Jonker Johan W, Groen Albert K, Reth M, Hall Michael N, Baumeister R (2013) Inhibition of mTORC1 by Astrin and stress granules prevents apoptosis in cancer cells. Cell 154(4):859–874. https://doi.org/10.1016/j.cell.2013.07.031
60. Aulas A, Lyons SM, Fay MM, Anderson P, Ivanov P (2018) Nitric oxide triggers the assembly of "type II" stress granules linked to decreased cell viability. Cell Death Dis 9(11):1129. https://doi.org/10.1038/s41419-018-1173-x
61. Phatak P, Dai F, Butler M, Nandakumar MP, Gutierrez PL, Edelman MJ, Hendriks H, Burger AM (2008) KML001 cytotoxic activity is associated with its binding to telomeric sequences and telomere erosion in prostate cancer cells. Clin Cancer Res 14(14):4593. https://doi.org/10.1158/1078-0432.CCR-07-4572
62. Frydrýšková K, Mašek T, Pospíšek M (2020) Changing faces of stress: Impact of heat and

arsenite treatment on the composition of stress granules. Wiley Interdis Rev RNA 11(6):e1596. https://doi.org/10.1002/wrna.1596
63. Hu Y, Li J, Lou B, Wu R, Wang G, Lu C, Wang H, Pi J, Xu Y (2020) The role of reactive oxygen species in arsenic toxicity. Biomol Ther 10(2):240. https://doi.org/10.3390/biom10020240
64. Mohammad S, Page SJ, Sasaki T, Ayvazian N, Rakic P, Kawasawa YI, Hashimoto-Torii K, Torii M (2020) Long-term spatial tracking of cells affected by environmental insults. J Neurodev Disord 12(1):38. https://doi.org/10.1186/s11689-020-09339-w
65. Lu L, Han AP, Chen JJ (2001) Translation initiation control by heme-regulated eukaryotic initiation factor 2alpha kinase in erythroid cells under cytoplasmic stresses. Mol Cell Biol 21(23):7971–7980. https://doi.org/10.1128/MCB.21.23.7971-7980.2001
66. Guo C, Sun L, Chen X, Zhang D (2013) Oxidative stress, mitochondrial damage and neurodegenerative diseases. Neural Regen Res 8(21):2003–2014. https://doi.org/10.3969/j.issn.1673-5374.2013.21.009
67. Ježek J, Cooper KF, Strich R (2018) Reactive oxygen species and mitochondrial dynamics: the yin and yang of mitochondrial dysfunction and cancer progression. Antioxidants 7(1):13. https://doi.org/10.3390/antiox7010013
68. Leadsham Jane E, Sanders G, Giannaki S, Bastow Emma L, Hutton R, Naeimi Wesley R, Breitenbach M, Gourlay Campbell W (2013) Loss of cytochrome c oxidase promotes RAS-dependent ROS production from the ER resident NADPH oxidase, Yno1p, in yeast. Cell Metab 18(2):279–286. https://doi.org/10.1016/j.cmet.2013.07.005
69. Thompson DJ (1993) A chemical hypothesis for arsenic methylation in mammals. Chem Biol Interact 88(2):89–114. https://doi.org/10.1016/0009-2797(93)90086-E
70. Jomova K, Jenisova Z, Feszterova M, Baros S, Liska J, Hudecova D, Rhodes CJ, Valko M (2011) Arsenic: toxicity, oxidative stress and human disease. J Appl Toxicol 31(2):95–107. https://doi.org/10.1002/jat.1649
71. Shen S, Li X-F, Cullen WR, Weinfeld M, Le XC (2013) Arsenic binding to proteins. Chem Rev 113(10):7769–7792. https://doi.org/10.1021/cr300015c
72. Zhang K, Daigle JG, Cunningham KM, Coyne AN, Ruan K, Grima JC, Bowen KE, Wadhwa H, Yang P, Rigo F, Taylor JP, Gitler AD, Rothstein JD, Lloyd TE (2018) Stress granule assembly disrupts nucleocytoplasmic transport. Cell 173(4):958–971.e917. https://doi.org/10.1016/j.cell.2018.03.025
73. Lee KH, Zhang P, Kim HJ, Mitrea DM, Sarkar M, Freibaum BD, Cika J, Coughlin M, Messing J, Molliex A, Maxwell BA, Kim NC, Temirov J, Moore J, Kolaitis RM, Shaw TI, Bai B, Peng J, Kriwacki RW, Taylor JP (2016) C9orf72 dipeptide repeats impair the assembly, dynamics, and function of membrane-less organelles. Cell 167(3):774–788.e717. https://doi.org/10.1016/j.cell.2016.10.002
74. McGurk L, Gomes E, Guo L, Mojsilovic-Petrovic J, Tran V, Kalb RG, Shorter J, Bonini NM (2018) Poly(ADP-Ribose) prevents pathological phase separation of TDP-43 by promoting liquid demixing and stress granule localization. Mol Cell 71(5):703–717.e709. https://doi.org/10.1016/j.molcel.2018.07.002
75. Daigle JG, Lanson NA Jr, Smith RB, Casci I, Maltare A, Monaghan J, Nichols CD, Kryndushkin D, Shewmaker F, Pandey UB (2013) RNA-binding ability of FUS regulates neurodegeneration, cytoplasmic mislocalization and incorporation into stress granules associated with FUS carrying ALS-linked mutations. Hum Mol Genet 22(6):1193–1205
76. Bosco DA, Lemay N, Ko HK, Zhou H, Burke C, Kwiatkowski TJ Jr, Sapp P, McKenna-Yasek D, Brown RH Jr, Hayward LJ (2010) Mutant FUS proteins that cause amyotrophic lateral sclerosis incorporate into stress granules. Hum Mol Genet 19(21):4160–4175
77. Kedersha N, Anderson P (2002) Stress granules: sites of mRNA triage that regulate mRNA stability and translatability. Biochem Soc Trans 30(Pt 6):963–969
78. Kedersha NL, Gupta M, Li W, Miller I, Anderson P (1999) RNA-binding proteins TIA-1 and TIAR link the phosphorylation of eIF-2 alpha to the assembly of mammalian stress granules. J Cell Biol 147(7):1431–1442
79. Sukarieh R, Sonenberg N, Pelletier J (2009) The eIF4E-binding proteins are modifiers of cytoplasmic eIF4E relocalization during the heat shock response. Am J Physiol Cell Physiol 296(5):C1207–C1217. https://doi.org/10.1152/ajpcell.00511.2008
80. Baler R, Welch WJ, Voellmy R (1992) Heat shock gene regulation by nascent polypeptides and denatured proteins: hsp70 as a potential autoregulatory factor. J Cell Biol 117(6):1151–1159. https://doi.org/10.1083/jcb.117.6.1151

81. Wallace Edward WJ, Kear-Scott Jamie L, Pilipenko Evgeny V, Schwartz Michael H, Laskowski Pawel R, Rojek Alexandra E, Katanski Christopher D, Riback Joshua A, Dion Michael F, Franks Alexander M, Airoldi Edoardo M, Pan T, Budnik Bogdan A, Drummond DA (2015) Reversible, specific, active aggregates of endogenous proteins assemble upon heat stress. Cell 162(6):1286–1298. https://doi.org/10.1016/j.cell.2015.08.041
82. Liu Y, Sakamoto H, Adachi M, Zhao S, Ukai W, Hashimoto E, Hareyama M, Ishida T, Imai K, Shinomura Y (2012) Heat stress activates ER stress signals which suppress the heat shock response, an effect occurring preferentially in the cortex in rats. Mol Biol Rep 39(4):3987–3993. https://doi.org/10.1007/s11033-011-1179-2
83. Xu X, Gupta S, Hu W, McGrath BC, Cavener DR (2011) Hyperthermia induces the ER stress pathway. PLoS One 6(8):e23740. https://doi.org/10.1371/journal.pone.0023740
84. Cherkasov V, Hofmann S, Druffel-Augustin S, Mogk A, Tyedmers J, Stoecklin G, Bukau B (2013) Coordination of translational control and protein homeostasis during severe heat stress. Curr Biol 23(24):2452–2462. https://doi.org/10.1016/j.cub.2013.09.058
85. Bentmann E, Neumann M, Tahirovic S, Rodde R, Dormann D, Haass C (2012) Requirements for stress granule recruitment of fused in sarcoma (FUS) and TAR DNA-binding protein of 43 kDa (TDP-43). J Biol Chem 287(27):23079–23094. https://doi.org/10.1074/jbc.M111.328757
86. Didiot MC, Subramanian M, Flatter E, Mandel JL, Moine H (2009) Cells lacking the fragile X mental retardation protein (FMRP) have normal RISC activity but exhibit altered stress granule assembly. Mol Biol Cell 20(1):428–437
87. Kedersha N, Stoecklin G, Ayodele M, Yacono P, Lykke-Andersen J, Fritzler MJ, Scheuner D, Kaufman RJ, Golan DE, Anderson P (2005) Stress granules and processing bodies are dynamically linked sites of mRNP remodeling. J Cell Biol 169(6):871–884
88. Finan JD, Guilak F (2010) The effects of osmotic stress on the structure and function of the cell nucleus. J Cell Biochem 109(3):460–467. https://doi.org/10.1002/jcb.22437
89. Kühn T, Ihalainen TO, Hyväluoma J, Dross N, Willman SF, Langowski J, Vihinen-Ranta M, Timonen J (2011) Protein diffusion in mammalian cell cytoplasm. PLoS One 6(8):e22962. https://doi.org/10.1371/journal.pone.0022962
90. Fulton AB (1982) How crowded is the cytoplasm? Cell 30(2):345–347. https://doi.org/10.1016/0092-8674(82)90231-8
91. Bourque CW (2008) Central mechanisms of osmosensation and systemic osmoregulation. Nat Rev Neurosci 9(7):519–531. https://doi.org/10.1038/nrn2400
92. Feldherr CM, Akin D (1993) Regulation of nuclear transport in proliferating and quiescent cells. Exp Cell Res 205(1):179–186. https://doi.org/10.1006/excr.1993.1073
93. Kim YB, Yu J, Lee SY, Lee MS, Ko SG, Ye SK, Jong HS, Kim TY, Bang YJ, Lee JW (2005) Cell adhesion status-dependent histone acetylation is regulated through intracellular contractility-related signaling activities. J Biol Chem 280(31):28357–28364. https://doi.org/10.1074/jbc.M412608200
94. Tóth KF, Knoch TA, Wachsmuth M, Frank-Stöhr M, Stöhr M, Bacher CP, Müller G, Rippe K (2004) Trichostatin A-induced histone acetylation causes decondensation of interphase chromatin. J Cell Sci 117(Pt 18):4277–4287. https://doi.org/10.1242/jcs.01293
95. Fürst J, Gschwentner M, Ritter M, Bottà G, Jakab M, Mayer M, Garavaglia L, Bazzini C, Rodighiero S, Meyer G, Eichmüller S, Wöll E, Paulmichl M (2002) Molecular and functional aspects of anionic channels activated during regulatory volume decrease in mammalian cells. Pflugers Arch 444(1–2):1–25. https://doi.org/10.1007/s00424-002-0805-1
96. Owsianik G, D'Hoedt D, Voets T, Nilius B (2006) Structure-function relationship of the TRP channel superfamily. Rev Physiol Biochem Pharmacol 156:61–90
97. Dmitrieva NI, Michea LF, Rocha GM, Burg MB (2001) Cell cycle delay and apoptosis in response to osmotic stress. Comp Biochem Physiol A Mol Integr Physiol 130(3):411–420. https://doi.org/10.1016/s1095-6433(01)00439-1
98. Copp J, Wiley S, Ward MW, van der Geer P (2005) Hypertonic shock inhibits growth factor receptor signaling, induces caspase-3 activation, and causes reversible fragmentation of the mitochondrial network. Am J Physiol Cell Physiol 288(2):C403–C415. https://doi.org/10.1152/ajpcell.00095.2004
99. Desai BN, Myers BR, Schreiber SL (2002) FKBP12-rapamycin-associated protein associates with mitochondria and senses osmotic

stress via mitochondrial dysfunction. Proc Natl Acad Sci U S A 99(7):4319–4324. https://doi.org/10.1073/pnas.261702698

100. Hans F, Glasebach H, Kahle PJ (2020) Multiple distinct pathways lead to hyperubiquitylated insoluble TDP-43 protein independent of its translocation into stress granules. J Biol Chem 295(3):673–689. https://doi.org/10.1074/jbc.RA119.010617

101. Cai Q, Brooks HL (2011) Phosphorylation of eIF2α via the general control kinase, GCN2, modulates the ability of renal medullary cells to survive high urea stress. Am J Physiol Renal Physiol 301(6):F1202–F1207. https://doi.org/10.1152/ajprenal.00272.2011

102. Farabaugh KT, Majumder M, Guan B-J, Jobava R, Wu J, Krokowski D, Gao X-H, Schuster A, Longworth M, Chan ED, Bianchi M, Dey M, Koromilas AE, Ramakrishnan P, Hatzoglou M (2017) Protein kinase R mediates the inflammatory response induced by hyperosmotic stress. Mol Cell Biol 37(4):e00521–e00516. https://doi.org/10.1128/MCB.00521-16

103. Bevilacqua E, Wang X, Majumder M, Gaccioli F, Yuan CL, Wang C, Zhu X, Jordan LE, Scheuner D, Kaufman RJ, Koromilas AE, Snider MD, Holcik M, Hatzoglou M (2010) eIF2alpha phosphorylation tips the balance to apoptosis during osmotic stress. J Biol Chem 285(22):17098–17111. https://doi.org/10.1074/jbc.M110.109439

104. Schwarz DS, Blower MD (2016) The endoplasmic reticulum: structure, function and response to cellular signaling. Cell Mol Life Sci 73(1):79–94. https://doi.org/10.1007/s00018-015-2052-6

105. Svitkin YV, Pause A, Haghighat A, Pyronnet S, Witherell G, Belsham GJ, Sonenberg N (2001) The requirement for eukaryotic initiation factor 4A (elF4A) in translation is in direct proportion to the degree of mRNA 5' secondary structure. RNA 7(3):382–394. https://doi.org/10.1017/s135583820100108x

106. Naineni SK, Itoua Maïga R, Cencic R, Putnam AA, Amador LA, Rodriguez AD, Jankowsky E, Pelletier J (2020) A comparative study of small molecules targeting eIF4A. RNA 26(5):541–549. https://doi.org/10.1261/rna.072884.119

107. Sun Y, Atas E, Lindqvist LM, Sonenberg N, Pelletier J, Meller A (2014) Single-molecule kinetics of the eukaryotic initiation factor 4AI upon RNA unwinding. Structure 22(7):941–948. https://doi.org/10.1016/j.str.2014.04.014

108. Harms U, Andreou AZ, Gubaev A, Klostermeier D (2014) eIF4B, eIF4G and RNA regulate eIF4A activity in translation initiation by modulating the eIF4A conformational cycle. Nucleic Acids Res 42(12):7911–7922. https://doi.org/10.1093/nar/gku440

109. Markmiller S, Soltanieh S, Server KL, Mak R, Jin W, Fang MY, Luo EC, Krach F, Yang D, Sen A, Fulzele A, Wozniak JM, Gonzalez DJ, Kankel MW, Gao FB, Bennett EJ, Lécuyer E, Yeo GW (2018) Context-dependent and disease-specific diversity in protein interactions within stress granules. Cell 172(3):590–604.e513. https://doi.org/10.1016/j.cell.2017.12.032

110. Padrón A, Iwasaki S, Ingolia NT (2019) Proximity RNA labeling by APEX-seq reveals the organization of translation initiation complexes and repressive RNA granules. Mol Cell 75(4):875–887.e875. https://doi.org/10.1016/j.molcel.2019.07.030

111. Vu L, Ghosh A, Tran C, Tebung WA, Sidibé H, Garcia-Mansfield K, David-Dirgo V, Sharma R, Pirrotte P, Bowser R, Velde CV (2021) Defining the Caprin-1 interactome in unstressed and stressed conditions. bioRxiv:2021.2001.2006.425453. https://doi.org/10.1101/2021.01.06.425453

112. Namkoong S, Ho A, Woo YM, Kwak H, Lee JH (2018) Systematic characterization of stress-induced RNA granulation. Mol Cell 70(1):175–187.e178. https://doi.org/10.1016/j.molcel.2018.02.025

113. Marmor-Kollet H, Siany A, Kedersha N, Knafo N, Rivkin N, Danino YM, Moens TG, Olender T, Sheban D, Cohen N, Dadosh T, Addadi Y, Ravid R, Eitan C, Toth Cohen B, Hofmann S, Riggs CL, Advani VM, Higginbottom A, Cooper-Knock J, Hanna JH, Merbl Y, Van Den Bosch L, Anderson P, Ivanov P, Geiger T, Hornstein E (2020) Spatiotemporal proteomic analysis of stress granule disassembly using APEX reveals regulation by SUMOylation and links to ALS pathogenesis. Mol Cell 80(5):876–891.e876. https://doi.org/10.1016/j.molcel.2020.10.032

114. Kershaw CJ, Nelson MG, Lui J, Bates CP, Jennings MD, Hubbard SJ, Ashe MP, Grant CM (2020) Integrated multi-omics reveals common properties underlying stress granule and P-body formation. bioRxiv:2020.2005.2018.102517. https://doi.org/10.1101/2020.05.18.102517

115. Hubstenberger A, Courel M, Bénard M, Souquere S, Ernoult-Lange M, Chouaib R, Yi Z, Morlot J-B, Munier A, Fradet M, Daunesse M, Bertrand E, Pierron G,

Mozziconacci J, Kress M, Weil D (2017) P-body purification reveals the condensation of repressed mRNA regulons. Mol Cell 68(1): 144–157.e145. https://doi.org/10.1016/j.molcel.2017.09.003

116. Khong A, Kerr CH, Yeung CH, Keatings K, Nayak A, Allan DW, Jan E (2017) Disruption of stress granule formation by the multifunctional cricket paralysis virus 1A. Protein 91 (5):e01779-16. https://doi.org/10.1128/jvi.01779-16
117. Berchtold D, Battich N, Pelkmans L (2018) A systems-level study reveals regulators of membrane-less organelles in human cells. Mol Cell 72(6):1035–1049.e1035. https://doi.org/10.1016/j.molcel.2018.10.036
118. Ohn T, Kedersha N, Hickman T, Tisdale S, Anderson P (2008) A functional RNAi screen links O-GlcNAc modification of ribosomal proteins to stress granule and processing body assembly. Nat Cell Biol 10(10): 1224–1231
119. Buchan JR, Parker R (2009) Eukaryotic stress granules: the ins and outs of translation. Mol Cell 36(6):932–941
120. Roux KJ, Kim DI, Burke B, May DG (2018) BioID: a screen for protein-protein interactions. Curr Protoc Protein Sci 91: 19.23.11–19.23.15. https://doi.org/10.1002/cpps.51
121. Han S, Li J, Ting AY (2018) Proximity labeling: spatially resolved proteomic mapping for neurobiology. Curr Opin Neurobiol 50: 17–23. https://doi.org/10.1016/j.conb.2017.10.015
122. Kim DI, Birendra KC, Zhu W, Motamedchaboki K, Doye V, Roux KJ (2014) Probing nuclear pore complex architecture with proximity-dependent biotinylation. Proc Natl Acad Sci U S A 111(24): E2453–E2461. https://doi.org/10.1073/pnas.1406459111
123. Hung V, Zou P, Rhee HW, Udeshi ND, Cracan V, Svinkina T, Carr SA, Mootha VK, Ting AY (2014) Proteomic mapping of the human mitochondrial intermembrane space in live cells via ratiometric APEX tagging. Mol Cell 55(2):332–341. https://doi.org/10.1016/j.molcel.2014.06.003
124. Hung V, Udeshi ND, Lam SS, Loh KH, Cox KJ, Pedram K, Carr SA, Ting AY (2016) Spatially resolved proteomic mapping in living cells with the engineered peroxidase APEX2. Nat Protoc 11(3):456–475. https://doi.org/10.1038/nprot.2016.018
125. Smith JA, Curry EG, Blue RE, Roden C, Dundon SER, Rodríguez-Vargas A, Jordan DC, Chen X, Lyons SM, Crutchley J, Anderson P, Horb ME, Gladfelter AS, Giudice J (2019) Regulation of FXR1 by alternative splicing is required for muscle development and controls liquid-like condensates in muscle cells. bioRxiv:818476. https://doi.org/10.1101/818476
126. Smith BN, Topp SD, Fallini C, Shibata H, Chen HJ, Troakes C, King A, Ticozzi N, Kenna KP, Soragia-Gkazi A, Miller JW, Sato A, Dias DM, Jeon M, Vance C, Wong CH, de Majo M, Kattuah W, Mitchell JC, Scotter EL, Parkin NW, Sapp PC, Nolan M, Nestor PJ, Simpson M, Weale M, Lek M, Baas F, Vianney de Jong JM, Ten Asbroek A, Redondo AG, Esteban-Pérez J, Tiloca C, Verde F, Duga S, Leigh N, Pall H, Morrison KE, Al-Chalabi A, Shaw PJ, Kirby J, Turner MR, Talbot K, Hardiman O, Glass JD, De Belleroche J, Maki M, Moss SE, Miller C, Gellera C, Ratti A, Al-Sarraj S, Brown RH Jr, Silani V, Landers JE, Shaw CE (2017) Mutations in the vesicular trafficking protein annexin A11 are associated with amyotrophic lateral sclerosis. Sci Transl Med 9(388). https://doi.org/10.1126/scitranslmed.aad9157
127. Smith BN, Ticozzi N, Fallini C, Gkazi AS, Topp S, Kenna KP, Scotter EL, Kost J, Keagle P, Miller JW, Calini D, Vance C, Danielson EW, Troakes C, Tiloca C, Al-Sarraj S, Lewis EA, King A, Colombrita C, Pensato V, Castellotti B, de Belleroche J, Baas F, Ten Asbroek AL, Sapp PC, McKenna-Yasek D, McLaughlin RL, Polak M, Asress S, Esteban-Perez J, Munoz-Blanco JL, Simpson M, van Rheenen W, Diekstra FP, Lauria G, Duga S, Corti S, Cereda C, Corrado L, Soraru G, Morrison KE, Williams KL, Nicholson GA, Blair IP, Dion PA, Leblond CS, Rouleau GA, Hardiman O, Veldink JH, van den Berg LH, Al-Chalabi A, Pall H, Shaw PJ, Turner MR, Talbot K, Taroni F, Garcia-Redondo A, Wu Z, Glass JD, Gellera C, Ratti A, Brown RH Jr, Silani V, Shaw CE, Landers JE (2014) Exome-wide rare variant analysis identifies TUBA4A mutations associated with familial ALS. Neuron 84(2):324–331. https://doi.org/10.1016/j.neuron.2014.09.027
128. Blasco H, Mavel S, Corcia P, Gordon PH (2014) The glutamate hypothesis in ALS: pathophysiology and drug development. Curr Med Chem 21(31):3551–3575
129. Khong A, Parker R (2018) mRNP architecture in translating and stress conditions reveals an ordered pathway of mRNP compaction. J Cell Biol 217(12):jcb.201806183

130. Mateju D, Eichenberger B, Eglinger J, Roth G, Chao JA (2020) Single-molecule imaging reveals translation of mRNAs localized to stress granules. bioRxiv:2020.2003.2031.018093. https://doi.org/10.1101/2020.03.31.018093
131. Unsworth H, Raguz S, Edwards HJ, Higgins CF, Yagüe E (2010) mRNA escape from stress granule sequestration is dictated by localization to the endoplasmic reticulum. FASEB J 24(9):3370–3380. https://doi.org/10.1096/fj.09-151142
132. Tsang CK, Zheng XF (2007) TOR-in(g) the nucleus. Cell Cycle 6(1):25–29
133. Anders M, Chelysheva I, Goebel I, Trenkner T, Zhou J, Mao Y, Verzini S, Qian SB, Ignatova Z (2018) Dynamic m(6)A methylation facilitates mRNA triaging to stress granules. Life Sci Alliance 1(4): e201800113. https://doi.org/10.26508/lsa.201800113
134. Ries RJ, Zaccara S, Klein P, Olarerin-George-A, Namkoong S, Pickering BF, Patil DP, Kwak H, Lee JH, Jaffrey SR (2019) m6A enhances the phase separation potential of mRNA. Nature 571(7765):424–428. https://doi.org/10.1038/s41586-019-1374-1
135. Aulas A, Stabile S, Vande Velde C (2012) Endogenous TDP-43, but not FUS, contributes to stress granule assembly via G3BP. Mol Neurodegener 7:54
136. Loschi M, Leishman CC, Berardone N, Boccaccio GL (2009) Dynein and kinesin regulate stress-granule and P-body dynamics. J Cell Sci 122(21):3973. https://doi.org/10.1242/jcs.051383
137. Franchini D-M, Lanvin O, Tosolini M, Patras de Campaigno E, Cammas A, Péricart S, Scarlata C-M, Lebras M, Rossi C, Ligat L, Pont F, Arimondo PB, Laurent C, Ayyoub M, Despas F, Lapeyre-Mestre M, Millevoi S, Fournié J-J (2019) Microtubule-driven stress granule dynamics regulate inhibitory immune checkpoint expression in T cells. Cell Rep 26(1):94–107.e107. https://doi.org/10.1016/j.celrep.2018.12.014

Chapter 14

Detecting Stress Granules in Drosophila Neurons

Fabienne De Graeve, Nadia Formicola, Kavya Vinayan Pushpalatha, Akira Nakamura, Eric Debreuve, Xavier Descombes, and Florence Besse

Abstract

Stress granules (SGs) are cytoplasmic ribonucleoprotein condensates that dynamically and reversibly assemble in response to stress. They are thought to contribute to the adaptive stress response by storing translationally inactive mRNAs as well as signaling molecules. Recent work has shown that SG composition and properties depend on both stress and cell types, and that neurons exhibit a complex SG proteome and a strong vulnerability to mutations in SG proteins. *Drosophila* has emerged as a powerful genetically tractable organism where to study the physiological regulation and functions of SGs in normal and pathological contexts. In this chapter, we describe a protocol enabling quantitative analysis of SG properties in both larval and adult *Drosophila* CNS samples. In this protocol, fluorescently tagged SGs are induced upon acute ex vivo stress or chronic in vivo stress, imaged at high-resolution via confocal microscopy and detected automatically, using a dedicated software.

Key words Central nervous system, Confocal imaging, Fluorescent stress granule proteins, Automated detection, *Drosophila melanogaster*

1 Introduction

Cellular stress induces a translational shutdown within minutes, characterized by inhibition of translation initiation and polysome disassembly. Cytoplasmic release of translationally inactive mRNAs in turn triggers the assembly of hundreds of nanometer-sized membraneless compartments enriched in stalled housekeeping transcripts and associated proteins, and referred to as stress granules (SGs) [1, 2]. These higher order ribonucleoprotein (RNP) assemblies behave as dynamic condensates: they form through the self-association of their constituents into dense networks of transient RNA-RNA, RNA-protein and protein-protein interactions and get actively disassembled upon stress release [3–5]. The rapid and

Fabienne De Graeve, Nadia Formicola and Kavya Vinayan Pushpalatha contributed equally.

Daniel Matějů and Jeffrey A. Chao (eds.), *The Integrated Stress Response: Methods and Protocols*,
Methods in Molecular Biology, vol. 2428, https://doi.org/10.1007/978-1-0716-1975-9_14,

reversible mode of SG assembly is thought to play important roles in the adaptive stress response, first by promoting translational reprogramming through transient sequestration of unnecessary RNAs, and second by rewiring cellular pathways through recruitment of signaling molecules [6, 7]. Consistent with the functional importance of SG dynamics, extensive links have recently been established between alterations of SG material properties and neurodegenerative diseases [8–10]. Abnormally stable inclusions enriched in SG components, for example, have been observed in pathological contexts and defined as a characteristic signature of amyotrophic lateral sclerosis (ALS) or frontotemporal dementia (FTD) patient samples [10, 11]. Furthermore, mutations in an increasing number of SG components, including the RNA binding proteins TDP-43, FUS, or TIA1, have been causally linked to disease progression and shown to promote the transition of RNP assemblies into irreversible solid-like condensates [9, 10, 12–15]. As revealed by a recent systematic study, the pathological entities formed upon expression of ALS mutant proteins also have a composition distinct from their dynamic and reversible counterparts [16], highlighting their capacity to recruit, and potentially titrate molecules involved in RNA homeostasis. More work is now required to decipher if and how pathological SGs induce toxicity in neuronal cells, which, as long-lived nondividing cells, appear to be particularly vulnerable to the chronic stress induced by mutant SG proteins [9]. Importantly, proteomic studies have uncovered that variations in the composition of SGs are also observed in normal contexts in function of cell types and nature of the stress [16, 17]. While a core set of obligatory components, including factors essential for SG nucleation (e.g., G3BP1, TIA-1), has been found in the different cell types analyzed, a significant fraction of the SG proteome was shown to be recruited exclusively in certain cell types, particularly in neurons [16, 17]. Together, these studies have uncovered an unexpected diversity of SG composition and highlighted the limits of working with standard immortalized cell lines. They have raised the need to develop alternative biological models in which SG regulation and function can be studied under physiological conditions, in differentiated tissues.

Drosophila represents an excellent model organism in which advanced genetics can be combined with high-resolution imaging to unravel the mechanisms underlying SG assembly, as well as SG function in adaptation to environmental stress or disease-associated chronic stress. Fly orthologs of mammalian SG components, indeed, were shown to accumulate within cytoplasmic condensates in response to different acute stresses including oxidative stress, Endoplasmic Reticulum (ER) stress or hypoxia [18–23]. Furthermore, various *Drosophila* ALS models have been developed, in which SG proteins with disease-causing mutations are chronically expressed in the nervous system [24–27]. These models were shown to recapitulate many aspects of the disease, among which cytoplasmic accumulation of pathological SG-like assemblies [25, 27–29]. Here, we describe a protocol that enables induction of SGs in the nervous system of *Drosophila*, either chronically in response to in vivo expression of pathological SG proteins, or acutely upon treatment of explants with stress inducers (e.g., arsenite). This protocol includes the procedure to perform high-resolution confocal imaging of fluorescently tagged SG markers and to accurately and automatically detect SGs using the *Obj.MPP* software [28]. The described method is compatible with analysis of both larval and adult central nervous system (CNS) samples, and is particularly adapted to the quantitative analysis of SG properties in complex tissues.

2 Materials

2.1 Fly Lines for Expression of Fluorescent SG Proteins

1. Gal4 and UAS transgenic flies for conditional ectopic expression of fluorescent pathological SG proteins in the nervous system (e.g., OK371-Gal4 and UAS-TDP-43 fly lines; *see* Table 1).
2. Knockin lines expressing fluorescent SG proteins from the endogenous locus (e.g., GFP-Rasputin (Rin; the fly ortholog of G3BP); *see* Table 1).

2.2 Arsenite Treatment

1. Chambered slide, four wells (*see* **Note 1**).
2. 40 mM stock solution of sodium (meta)arsenite dissolved in freshly prepared HL3 (*see* Subheading 2.3.2); store at room temperature (*see* **Note 2**). Alternatively, purchase commercially available aqueous solution.

2.3 Dissection and Fixation of Drosophila CNS Samples

2.3.1 Dissection and Fixation of Larval CNS

1. A pair of dissection forceps.
2. 60 mm dissection petri dishes.
3. 1× Phosphate-Buffered Saline (PBS) (*see* **Note 3**).
4. Fixing solution: 4% formaldehyde in 1× PBS (*see* **Note 3**).

2.3.2 Dissection and Fixation of Adult Brains

1. A pair of dissection forceps.
2. Minutien pins.
3. 60 mm dissection petri dishes covered with 2% agarose.
4. HL3 buffer: 70 mM NaCl, 5 mM KCl, 4 mM $MgCl_2$, 5 mM trehalose, 115 mM sucrose, 5 mM HEPES, 10 mM $NaHCO_3$, pH 7.20–7.25 (*see* **Note 4**).

Table 1
Useful *Drosophila* lines for detection of wild-type or pathological fluorescent SG proteins

Genotype	Description	Functionality	Generation	Fly line source
w;EGFP::Rin	EGFP inserted in the endogenous *rasputin (rin)* locus, right after the ATG	Homozygous viable	CRISPR/Cas9 editing, as described in [30]	Akira Nakamura
UAS-Venus::TDP-43 wild type	Venus fused N-terminally to human TDP-43; construct cloned downstream of UAS sequences	rescues the lethality of *tbph* null mutant flies	Transgenesis, random insertion	Paul Taylor [29]
UAS-Venus::TDP-43M337V	Venus fused N-terminally to an ALS-causing form of the human TDP-43; cloned downstream of UAS sequences	does not rescue the lethality of *tbph* null mutant flies	Transgenesis, random insertion	Paul Taylor [29]
OK371-Gal4	Expresses Gal4 in approx. 40 glutamatergic neurons (5 dorsal neurons identified as aCC and RP1–4 and 35 lateral and ventral neurons) and 6 glutamatergic interneurons per hemisegment.	NA	Enhancer-trap screen	Serge Birman [31]

5. Fixing solution: 4% formaldehyde in HL3.
6. Wash buffer: PBS; 0.5% Triton X.

2.4 Mounting of Drosophila CNS Samples

1. Antifade mounting medium with DAPI (Vectashield).
2. 10-well glass slides (black Teflon coating).
3. 1.5, 24 × 60 mm coverslips.

2.5 Image Acquisition

1. Scanning confocal microscope with highly sensitive detectors.
2. 63× 1.4 NA oil objective.
3. Immersion oil.

2.6 Image Analysis

1. ImageJ/Fiji (https://imagej.net/Fiji).
2. *Obj.MPP* software [28] (https://gitlab.inria.fr/edebreuv/Obj.MPP).

3 Methods

In this protocol, stress can either be applied endogenously (Subheading 3.1.1) or exogenously (Subheading 3.1.2) (Fig. 1). Note that dissection of *Drosophila* nervous system (Subheading 3.2) is performed after stress induction in case of endogenous stress and before stress in case of exogenous stress (Fig. 1).

3.1 Induction of Stress

3.1.1 Ectopic In Vivo Expression of Pathological Proteins

1. Cross transgenic flies expressing a fluorescently tagged pathological SG protein under UAS control (e.g., UAS-Venus::TDP-43 M337V; *see* Table 1) with flies expressing a neuronal Gal4 driver (e.g., motoneuron OK371-Gal4; *see* Table 1) (Fig. 1, *upper left*).
2. Maintain the flies at 25 °C and transfer them in a new vial every 3–4 days (*see* **Note 5**).

3.1.2 Ex Vivo Treatment with Arsenite

1. Freshly prepare the working solution of 0.4 mM sodium arsenite by diluting the stock solution in HL3 (*see* **Note 2**).
2. Transfer dissected samples in a multiwell chambered slide (Fig. 1, *upper right*). At least 15 samples should be treated per condition.
3. Incubate in 500 μL of HL3 or arsenite solution for 1 h at 25 °C, covered from light.

3.2 Dissection of Drosophila CNS Samples

3.2.1 Dissection of Larval CNS

1. Collect wandering third instar larvae expressing normal or pathological fluorescent SG proteins.
2. Transfer them into a 60 mm petri dish filled with PBS (*see* **Note 3**).
3. Tear the larvae in two using a pair of forceps.
4. Turn the anterior half of larvae inside out, remove the fat tissue while keeping the CNS attached to the cuticle. Collect samples with forceps in microtubes containing PBS (*see* **Note 3**).

3.2.2 Dissection of Adult Brains

1. Collect 7–10 day-old flies expressing normal or pathological SG proteins and anesthetize them with CO_2.
2. Dissect brains in HL3 buffer, as described in [32, 33]. Briefly, immobilize the flies ventral side up by pinning them in a dissecting dish filled with HL3. Pull the proboscis upward with one forceps and insert the tips of the other forceps underneath, in a closed position. Slowly open the forceps so as to tear apart the head cuticle. Carefully remove the cuticle and the retina, without damaging the underlying optic lobes and central brain.
3. Complete the dissection by thoroughly removing the air sacs (*see* **Note 6**).
4. Separate the brains from the rest of the body. Collect the dissected brains using a glass pipette or a filter tip (*see* **Note 7**).

3.3 Fixation of Drosophila CNS Samples

This step comes right after dissection in case of endogenous stress induction or after treatment of brain explants in case of ex vivo arsenite treatment.

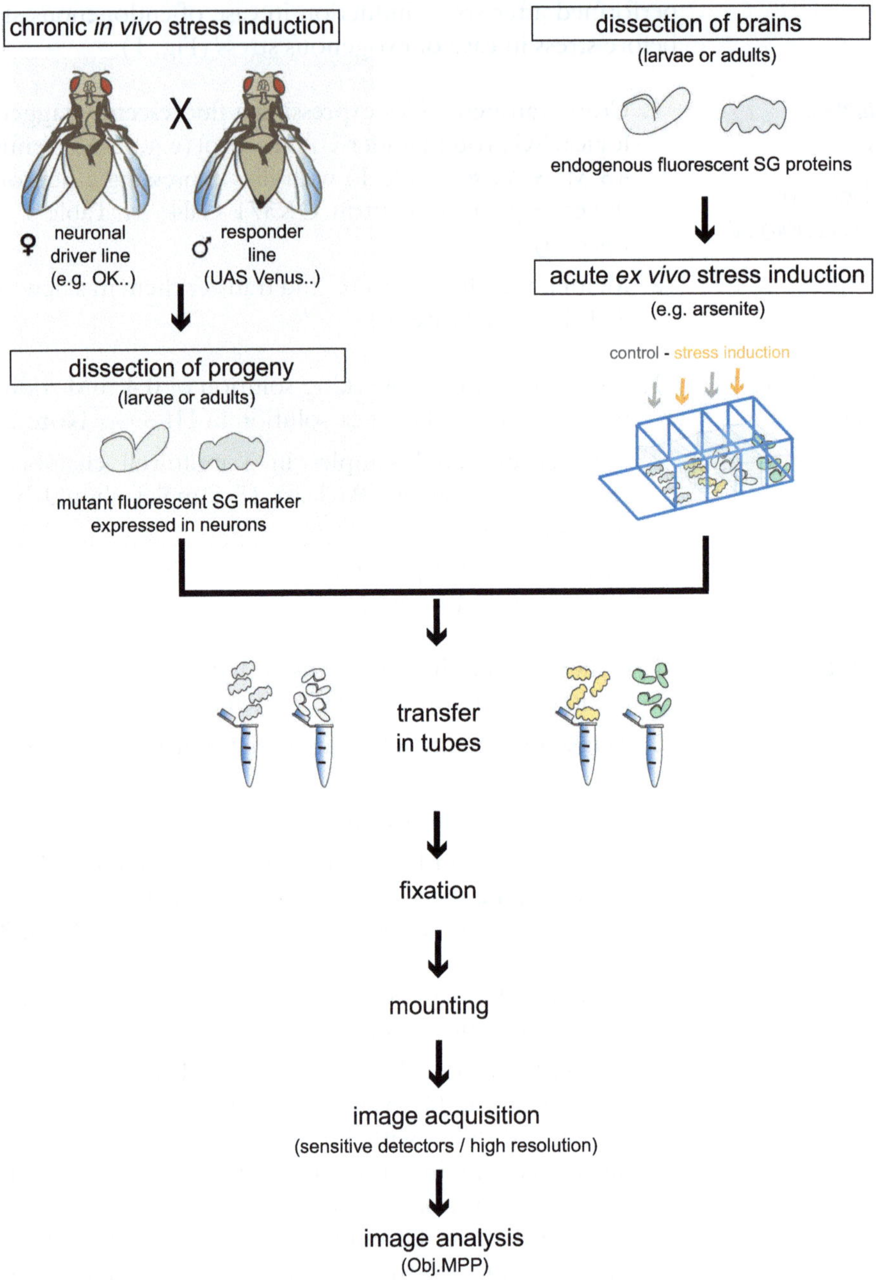

Fig. 1 Method workflow. Induction of stress in *Drosophila* nervous system was performed either endogenously (left) or exogenously (right). For the endogenous strategy, expression of fluorescent pathological SG proteins is induced chronically in vivo using the Gal4/UAS system. Larval or adult progenies expressing the mutant fluorescent SG markers in neurons are dissected and their CNS/brain collected. Ex vivo stress induction is achieved through acute arsenite treatment of larval CNS/brain explants previously dissected from larvae or adults expressing endogenous fluorescent SG proteins. In both procedures, stress induction and dissection are followed by sample fixation, mounting and confocal imaging (lower panel). Automated detection of SGs is performed via the *Obj.MPP* software

3.3.1 Fixation of Larval CNS

1. Remove 1× PBS.
2. Add 500 μL of fixing solution and gently rock the samples for 20 min at room temperature (RT).
3. Replace the fixing solution with 1 mL of 1× PBS and gently rock the samples for 30 min at RT.
4. Repeat **step 3** twice.
5. Remove 1× PBS and add a drop of antifade mounting medium supplemented with DAPI.
6. Keep at 4 °C for a minimum of 2 h (preferentially overnight).

3.3.2 Fixation of Adult Brains

1. Remove HL3.
2. Add 300 μL of fixing solution and gently rock the samples for 25 min at room temperature (RT).
3. Remove the fixing solution, replace with 800 μL of wash buffer and gently rock the samples for 30 min at RT.
4. Repeat **step 3** twice.
5. Remove the wash buffer and add a drop of antifade mounting medium supplemented with DAPI.
6. Keep at 4 °C for a minimum of 2 h (preferentially overnight).

3.4 Mounting of Drosophila CNS Samples

3.4.1 Mounting of Larval CNS

1. Transfer the samples onto a dissection dish using a 1 mL end-cut tip and dissect the samples further by detaching the CNS from the cuticle and removing eye-antenna imaginal discs. Recover the brain lobes and ventral nerve cord.
2. Transfer the clean CNS to a multiwell slide (~5 CNS per well) (*see* **Notes 8** and **9**).
3. Orient the larval CNS with forceps, such that the dorsal side of the ventral cord is up.
4. Carefully place a 24 × 60 mm coverslip on top of the slide and seal the coverslip with clear nail varnish.

3.4.2 Mounting of Adult Brains

1. Transfer the brains to a multiwell slide using a 1 mL end-cut tip (~ 5 CNS per well) (*see* **Notes 8** and **9**).
2. Orient the brains with forceps, such that their dorsal side is up.
3. Carefully place a 24 × 60 mm coverslip on top of the slide and seal the coverslip with clear nail varnish.

3.5 Imaging of Drosophila CNS Samples

1. Acquire images from larval CNS or adult brains with a confocal microscope equipped with high-sensitivity detectors, and appropriate laser lines (*see* **Note 10**).
2. Image with optimal resolution (*see* **Note 11**), using a 63× 1.4 NA oil objective.

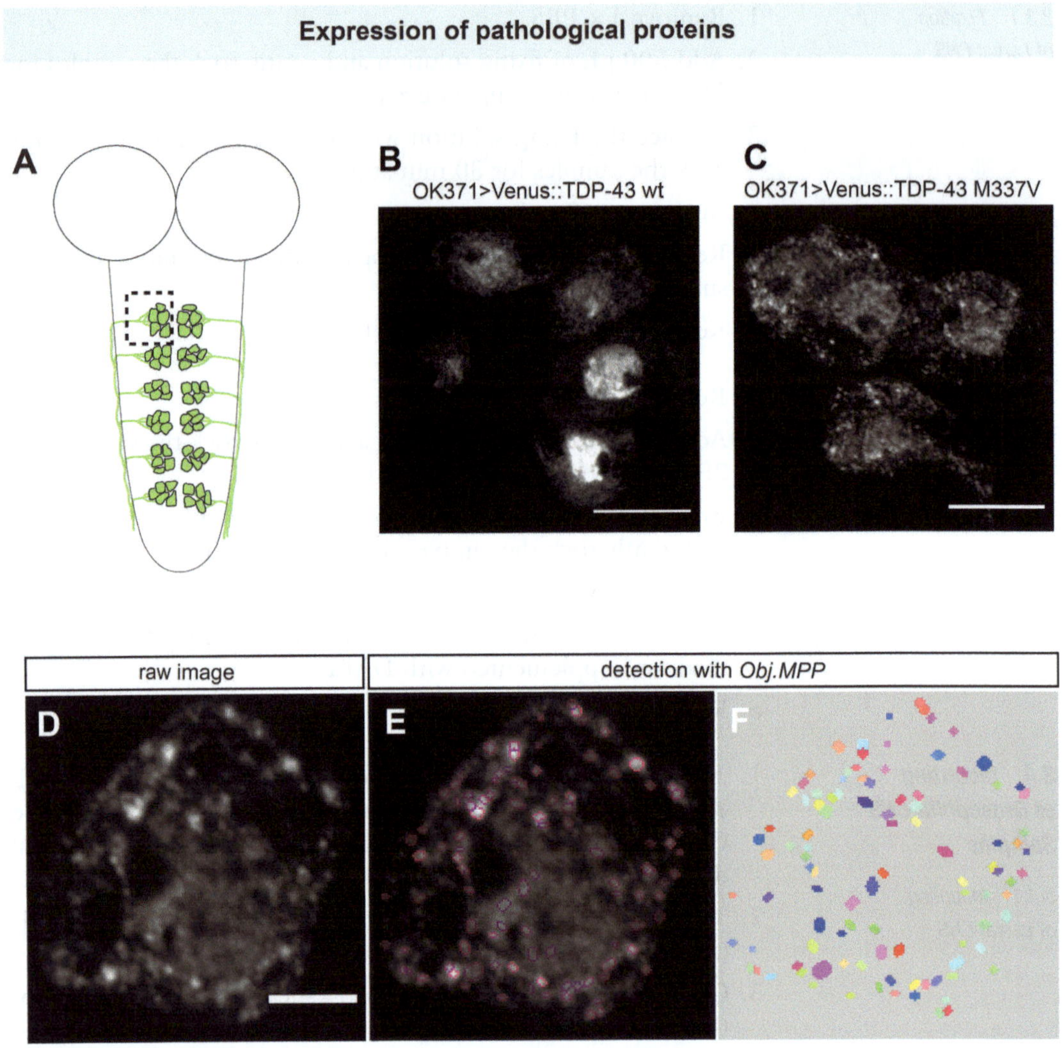

Fig. 2 Imaging and automated detection of pathological SGs in larval CNS. (**a**) Schematic representation of a third instar larva ventral nerve cord with the soma of a subset of OK371-Gal4-expressing motoneuron highlighted in green. The box delimits the region imaged in **b**, **c**. (**b–d**) Single confocal section of motoneuron cell bodies chronically expressing wild-type Venus::TDP-43 (**b**) or Venus::TDP-43 M337V (**c**, **d**) in motoneurons (OK371-Gal4/+; UAS-Venus::TDP43 M337V/+). Scale bar: 10 μm in **b**, **c** and 3 μm in **d**. Note the presence of pathological aggregates in motoneuron cytoplasm in **c**, **d**. (**e**) Overlay of the raw confocal image and the *Obj.MPP* detections. (**f**) Mask of the detected objects

3. SGs appear as discrete, bright cytoplasmic foci with a typical diameter of hundreds of nanometers (Figs. 2 and 3).

3.6 Image Analysis: Detection of Stress Granules

1. Using ImageJ/Fiji, select single optical sections and crop to generate stereotypic regions of interest. Save images in .tif format, in a single dedicated folder.
2. Launch the *Obj.MPP* software (*see* **Note 12**).

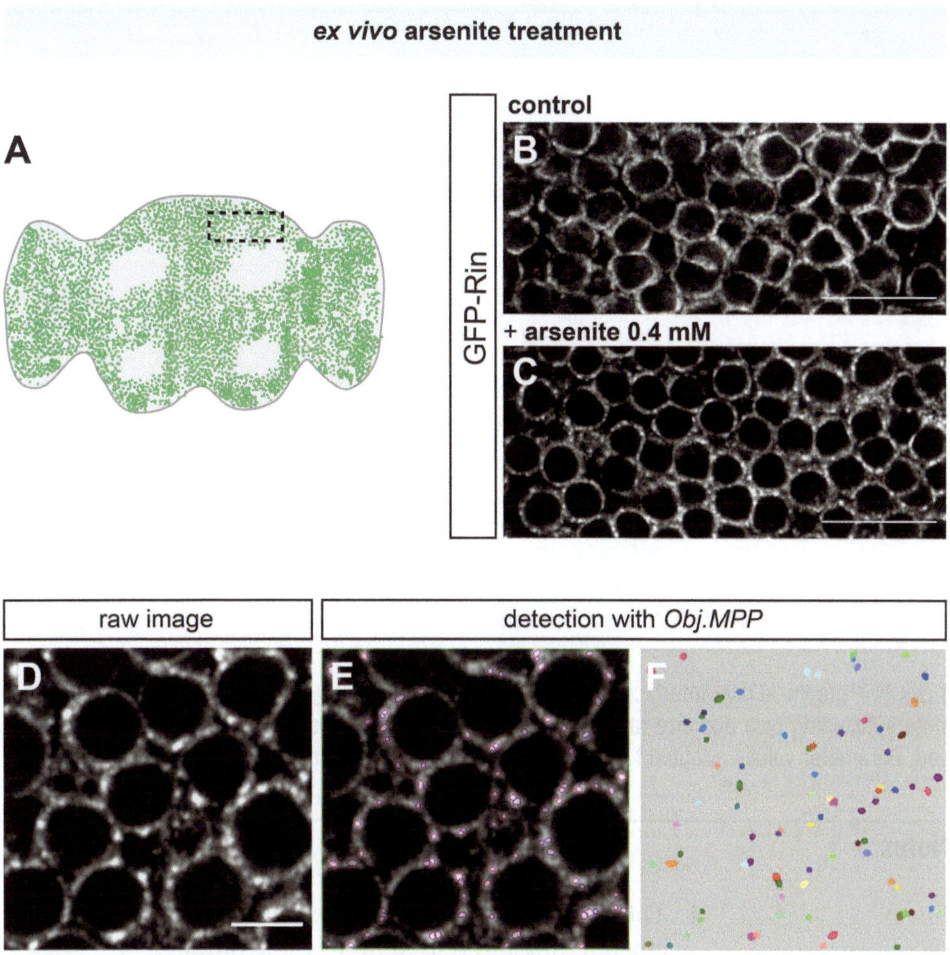

Fig. 3 Imaging and automated detection of arsenite-induced SGs in adult *Drosophila* brain. (**a**) Schematic representation of an adult brain expressing GFP-Rasputin (Rin) proteins from the endogenous locus (green). The box delimits the region imaged in **b**, **c**. (**b–d**) Single confocal section of Mushroom Body neurons expressing GFP-Rin, treated (**c**, **d**) or not (**b**) with arsenite. Scale bar: 10 μm in **b**, **c** and 3 μm in **d**. While GFP-Rin exhibits diffuse cytoplasmic distribution in the absence of stress, it localizes to SGs upon arsenite treatment. (**e**) Overlay of the raw confocal image and the *Obj.MPP* detections. (**f**) Mask of the detected objects

3. Select images to be analyzed in the first tab of the GUI.
4. Select the detection parameters in the second tab of the GUI (Fig. 4). These parameters include object type and expected size range (*see* **Note 13**), as well as object radiometric properties (defined by the quality function; *see* **Note 14**).
5. Set the number of iterations in the third tab of the GUI (*see* **Note 15**).
6. Select output files (*see* **Note 16**) and output path in the fourth tab of the GUI.

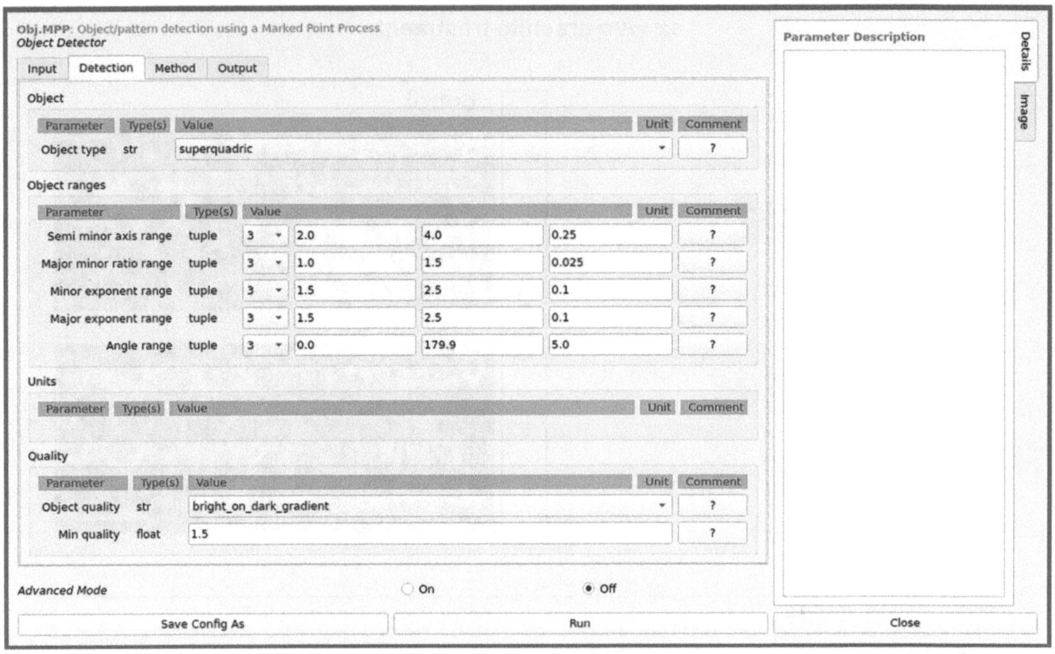

Fig. 4 *Obj.MPP* graphical user interface (GUI). The second tab of the *Obj.MPP* GUI is shown, in which detection parameters including type and size ranges of objects, as well as threshold for the quality function, must be selected. Parameter values adapted to the detection of SGs in larval motoneurons are displayed

4 Notes

1. The multiwell chambered slides can be rigorously washed with ethanol 80% and reused up to three times.
2. Sodium arsenite is a hazardous substance classified as carcinogen, mutagen and teratogen; it should be handled safely, under a chemical hood. When solubilized, sodium arsenite should be stored as sealed aliquots covered from light to avoid oxidation.
3. Prefer HL3 in case long incubations are required (if applying ex vivo stress).
4. HL3 buffer contains sugars (sucrose and trehalose) and can easily get contaminated. Store at 4 °C in aliquots sealed with parafilm. Opened aliquots should not be kept for more than 2 months.
5. Temperature should be adapted so as to permit high expression level while preventing toxicity.
6. If air sacs are not removed, brains will float, making it difficult to not pipet them away.
7. Prewetting the pipette tip or the glass pipette with HL3 prevents the brains from sticking to the plastic/glass wall.

8. Do not place samples in wells close to the edge of the slide; they will not be accessible on regular microscope stages.
9. Transfer samples in a drop of mounting medium only, as excess medium can make brains float over the edge of the wells.
10. We used a confocal microscope equipped with ultrasensitive detectors (Zeiss LSM 880 with gallium arsenide phosphide (GaAsP) detectors).
11. Imaging with a *xy* pixel size of less than 80 nm is recommended. We imaged larval CNS with a *xy* pixel size of 74 nm (regular confocal microscopy), and adult brains with a *xy* pixel size of 45 nm (Airy scan confocal microscopy).
12. *Obj.MPP* can be used either through the graphical user interface (GUI) or through a terminal console (Command-Line Interface (CLI)). More parameters can be adjusted when using the latter mode (*see* https://edebreuv.gitlabpages.inria.fr/Obj.MPP/).
13. Object types and their corresponding parameters (notably size and orientation) are described under: https://edebreuv.gitlabpages.inria.fr/Obj.MPP/contents/users/object-types.html. Superquadrics are typically recommended for detection of objects with potentially complex shapes such as SGs. We used the following parameter ranges for detection of SGs from larval CNS: semi_minor_axis_range (2, 4, 0.25), major_minor_ratio_range (1, 1.5, 0.025), major_ and minor_exponent_range (1.5, 2.5, 0.1), angle_degree_range (0.0, 179.9, 5.0) and the following parameters for detection of SGs in adult brains: semi_minor_axis_range (3, 4, 0.25), major_minor_ratio_range (1, 2, 0.025), major_ and minor_exponent_range (1, 2, 0.1), angle_degree_range (0, 179.9, 5.0) (*see* **Note 14** about mpp_quality_chooser.py).
14. Available quality functions and associated signal transformations are described under: https://edebreuv.gitlabpages.inria.fr/Obj.MPP/contents/users/quality-measures.html. Note that mpp_quality_chooser.py (https://edebreuv.gitlabpages.inria.fr/Obj.MPP/contents/users/mpp-quality-chooser.html) can be used to identify the best quality function and parameter ranges to detect objects of interest. We used the bright-on-dark gradient quality function with a min_quality of 1.5 for larval CNS and 2.5 for adult brains.
15. The number of iterations and the number of births per iteration should be set so that best objects are all reproducibly retained at the end of the process. We used 1.500 iterations with 50 births per iteration for larval CNS and 1.000 iterations with 550 births per iteration for adult brains.

16. Different outputs can be selected in the last tab of the GUI, including: CSV files containing the characteristics of the detected granules (geometrical parameters, intensity), raw images with granule contours highlighted, or masks of the detected granules, each having its own label (Figs. 2d–f and 3d–f).

Acknowledgments

Development of this protocol was supported by the CNRS, as well as grants from the ANR (ANR-15-CE12-0016 and ANR-20-CE16-0010-01) and the Fondation pour la Recherche Médicale (Equipe FRM; grant #DEQ20180339161) to F.B. Part of this work was also supported by the Joint Usage/Research Center for Developmental Medicine, IMEG, Kumamoto University. N.F. and K.P. were supported by fellowships from the LABEX SIGNALIFE program (#ANR – 11 – LABX –0028 – 01). K.P. was in addition supported by a 1 year- La Ligue contre le cancer fellowship. We thank the iBV PRISM Imaging facility for use of their microscopes and support (especially B. Monterroso), and L. Palin for excellent technical assistance.

References

1. Protter DSW, Parker R (2016) Principles and properties of stress granules. Trends Cell Biol 26(9):668–679. https://doi.org/10.1016/j.tcb.2016.05.004
2. Riggs CL, Kedersha N, Ivanov P, Anderson P (2020) Mammalian stress granules and P bodies at a glance. J Cell Sci 133(16). https://doi.org/10.1242/jcs.242487
3. Mittag T, Parker R (2018) Multiple modes of protein-protein interactions promote rnp granule assembly. J Mol Biol 430(23):4636–4649. https://doi.org/10.1016/j.jmb.2018.08.005
4. Hofmann S, Kedersha N, Anderson P, Ivanov P (2020) Molecular mechanisms of stress granule assembly and disassembly. Biochim Biophys Acta Mol Cell Res 1868(1):118876. https://doi.org/10.1016/j.bbamcr.2020.118876
5. Protter DSW, Rao BS, Van Treeck B, Lin Y, Mizoue L, Rosen MK, Parker R (2018) Intrinsically disordered regions can contribute promiscuous interactions to RNP granule assembly. Cell Rep 22(6):1401–1412. https://doi.org/10.1016/j.celrep.2018.01.036
6. Buchan JR, Parker R (2009) Eukaryotic stress granules: the ins and outs of translation. Mol Cell 36(6):932–941. https://doi.org/10.1016/j.molcel.2009.11.020
7. Kedersha N, Ivanov P, Anderson P (2013) Stress granules and cell signaling: more than just a passing phase? Trends Biochem Sci 38(10):494–506. https://doi.org/10.1016/j.tibs.2013.07.004
8. Formicola N, Vijayakumar J, Besse F (2019) Neuronal ribonucleoprotein granules: dynamic sensors of localized signals. Traffic 20(9):639–649. https://doi.org/10.1111/tra.12672
9. Li YR, King OD, Shorter J, Gitler AD (2013) Stress granules as crucibles of ALS pathogenesis. J Cell Biol 201(3):361–372. https://doi.org/10.1083/jcb.201302044
10. Wolozin B, Ivanov P (2019) Stress granules and neurodegeneration. Nat Rev Neurosci 20(11):649–666. https://doi.org/10.1038/s41583-019-0222-5
11. Neumann M, Sampathu DM, Kwong LK, Truax AC, Micsenyi MC, Chou TT, Bruce J, Schuck T, Grossman M, Clark CM, McCluskey LF, Miller BL, Masliah E, Mackenzie IR, Feldman H, Feiden W, Kretzschmar HA, Trojanowski JQ, Lee VM (2006) Ubiquitinated TDP-43 in frontotemporal lobar degeneration

and amyotrophic lateral sclerosis. Science 314(5796):130–133. https://doi.org/10.1126/science.1134108

12. Vance C, Rogelj B, Hortobagyi T, De Vos KJ, Nishimura AL, Sreedharan J, Hu X, Smith B, Ruddy D, Wright P, Ganesalingam J, Williams KL, Tripathi V, Al-Saraj S, Al-Chalabi A, Leigh PN, Blair IP, Nicholson G, de Belleroche J, Gallo JM, Miller CC, Shaw CE (2009) Mutations in FUS, an RNA processing protein, cause familial amyotrophic lateral sclerosis type 6. Science 323(5918):1208–1211. https://doi.org/10.1126/science.1165942
13. Mackenzie IR, Nicholson AM, Sarkar M, Messing J, Purice MD, Pottier C, Annu K, Baker M, Perkerson RB, Kurti A, Matchett BJ, Mittag T, Temirov J, Hsiung GR, Krieger C, Murray ME, Kato M, Fryer JD, Petrucelli L, Zinman L, Weintraub S, Mesulam M, Keith J, Zivkovic SA, Hirsch-Reinshagen V, Roos RP, Züchner S, Graff-Radford NR, Petersen RC, Caselli RJ, Wszolek ZK, Finger E, Lippa C, Lacomis D, Stewart H, Dickson DW, Kim HJ, Rogaeva E, Bigio E, Boylan KB, Taylor JP, Rademakers R (2017) TIA1 mutations in amyotrophic lateral sclerosis and frontotemporal dementia promote phase separation and alter stress granule dynamics. Neuron 95(4):808–816.e809. https://doi.org/10.1016/j.neuron.2017.07.025
14. Sreedharan J, Blair IP, Tripathi VB, Hu X, Vance C, Rogelj B, Ackerley S, Durnall JC, Williams KL, Buratti E, Baralle F, de Belleroche J, Mitchell JD, Leigh PN, Al-Chalabi A, Miller CC, Nicholson G, Shaw CE (2008) TDP-43 mutations in familial and sporadic amyotrophic lateral sclerosis. Science 319(5870):1668–1672. https://doi.org/10.1126/science.1154584
15. Patel A, Lee HO, Jawerth L, Maharana S, Jahnel M, Hein MY, Stoynov S, Mahamid J, Saha S, Franzmann TM, Pozniakovski A, Poser I, Maghelli N, Royer LA, Weigert M, Myers EW, Grill S, Drechsel D, Hyman AA, Alberti S (2015) A liquid-to-solid phase transition of the ALS protein FUS accelerated by disease mutation. Cell 162(5):1066–1077. https://doi.org/10.1016/j.cell.2015.07.047
16. Markmiller S, Soltanieh S, Server KL, Mak R, Jin W, Fang MY, Luo EC, Krach F, Yang D, Sen A, Fulzele A, Wozniak JM, Gonzalez DJ, Kankel MW, Gao FB, Bennett EJ, Lécuyer E, Yeo GW (2018) Context-dependent and disease-specific diversity in protein interactions within stress granules. Cell 172(3):590–604.e513. https://doi.org/10.1016/j.cell.2017.12.032
17. Advani VM, Ivanov P (2020) Stress granule subtypes: an emerging link to neurodegeneration. Cell Mol Life Sci 77(23):4827–4845. https://doi.org/10.1007/s00018-020-03565-0
18. Jevtov I, Zacharogianni M, van Oorschot MM, van Zadelhoff G, Aguilera-Gomez A, Vuillez I, Braakman I, Hafen E, Stocker H, Rabouille C (2015) TORC2 mediates the heat stress response in Drosophila by promoting the formation of stress granules. J Cell Sci 128(14): 2497–2508. https://doi.org/10.1242/jcs.168724
19. Gareau C, Houssin E, Martel D, Coudert L, Mellaoui S, Huot ME, Laprise P, Mazroui R (2013) Characterization of fragile X mental retardation protein recruitment and dynamics in Drosophila stress granules. PLoS One 8(2): e55342. https://doi.org/10.1371/journal.pone.0055342
20. Farny NG, Kedersha NL, Silver PA (2009) Metazoan stress granule assembly is mediated by P-eIF2alpha-dependent and -independent mechanisms. RNA 15(10):1814–1821. https://doi.org/10.1261/rna.1684009
21. Buddika K, Ariyapala IS, Hazuga MA, Riffert D, Sokol NS (2020) Canonical nucleators are dispensable for stress granule assembly in Drosophila intestinal progenitors. J Cell Sci 133(10). https://doi.org/10.1242/jcs.243451
22. van der Laan AM, van Gemert AM, Dirks RW, Noordermeer JN, Fradkin LG, Tanke HJ, Jost CR (2012) mRNA cycles through hypoxia-induced stress granules in live Drosophila embryonic muscles. Int J Dev Biol 56(9): 701–709. https://doi.org/10.1387/ijdb.103172al
23. Bakthavachalu B, Huelsmeier J, Sudhakaran IP, Hillebrand J, Singh A, Petrauskas A, Thiagarajan D, Sankaranarayanan M, Mizoue L, Anderson EN, Pandey UB, Ross E, VijayRaghavan K, Parker R, Ramaswami M (2018) RNP-granule assembly via Ataxin-2 disordered domains is required for long-term memory and neurodegeneration. Neuron 98(4):754–766.e754. https://doi.org/10.1016/j.neuron.2018.04.032
24. McGurk L, Berson A, Bonini NM (2015) Drosophila as an in vivo model for human neurodegenerative disease. Genetics 201(2): 377–402. https://doi.org/10.1534/genetics.115.179457
25. Estes PS, Boehringer A, Zwick R, Tang JE, Grigsby B, Zarnescu DC (2011) Wild-type and A315T mutant TDP-43 exert differential neurotoxicity in a Drosophila model of ALS.

Hum Mol Genet 20(12):2308–2321. https://doi.org/10.1093/hmg/ddr124
26. Lanson NA Jr, Maltare A, King H, Smith R, Kim JH, Taylor JP, Lloyd TE, Pandey UB (2011) A Drosophila model of FUS-related neurodegeneration reveals genetic interaction between FUS and TDP-43. Hum Mol Genet 20(13):2510–2523. https://doi.org/10.1093/hmg/ddr150
27. Chen Y, Yang M, Deng J, Chen X, Ye Y, Zhu L, Liu J, Ye H, Shen Y, Li Y, Rao EJ, Fushimi K, Zhou X, Bigio EH, Mesulam M, Xu Q, Wu JY (2011) Expression of human FUS protein in Drosophila leads to progressive neurodegeneration. Protein Cell 2(6):477–486. https://doi.org/10.1007/s13238-011-1065-7
28. De Graeve F, Debreuve E, Rahmoun S, Ecsedi S, Bahri A, Hubstenberger A, Descombes X, Besse F (2019) Detecting and quantifying stress granules in tissues of multicellular organisms with the Obj.MPP analysis tool. Traffic 20(9):697–711. https://doi.org/10.1111/tra.12678
29. Alami NH, Smith RB, Carrasco MA, Williams LA, Winborn CS, Han SSW, Kiskinis E, Winborn B, Freibaum BD, Kanagaraj A, Clare AJ, Badders NM, Bilican B, Chaum E, Chandran S, Shaw CE, Eggan KC, Maniatis T, Taylor JP (2014) Axonal transport of TDP-43 mRNA granules is impaired by ALS-causing mutations. Neuron 81(3):536–543. https://doi.org/10.1016/j.neuron.2013.12.018
30. Kina H, Yoshitani T, Hanyu-Nakamura K, Nakamura A (2019) Rapid and efficient generation of GFP-knocked-in Drosophila by the CRISPR-Cas9-mediated genome editing. Dev Growth Differentiat 61:265–275
31. Mahr A, Aberle H (2006) The expression pattern of the Drosophila vesicular glutamate transporter: a marker protein for motoneurons and glutamatergic centers in the brain. Gene Expr Patterns 6(3):299–309. https://doi.org/10.1016/j.modgep.2005.07.006
32. Tito AJ, Cheema S, Jiang M, Zhang S (2016) A simple one-step dissection protocol for whole-mount preparation of adult Drosophila brains. J Vis Exp (118). https://doi.org/10.3791/55128
33. Williamson WR, Hiesinger PR (2010) Preparation of developing and adult Drosophila brains and retinae for live imaging. J Vis Exp (37). https://doi.org/10.3791/1936

Chapter 15

Monitoring and Quantification of the Dynamics of Stress Granule Components in Living Cells by Fluorescence Decay After Photoactivation

Anna-Carina Söhnel, Nataliya I. Trushina, and Roland Brandt

Abstract

Stress granules (SGs) are cytosolic, nonmembranous RNA-protein (RNP) complexes that form in the cytosol of many cells under various stress conditions and can integrate responses to various stressors. Although physiological SG formation appears to be an adaptive and survival-promoting mechanism, inappropriate formation or chronic persistence of SGs has been linked to aging and various neurodegenerative diseases. The quantitative monitoring of the dynamics of SG components in living nerve cells can therefore be an important tool for identifying conditions that disrupt SG function and lead to disease-related attacks in the cells. Here, we describe a method for the quantitative determination of the distribution and shuttling dynamics of components of SGs in living model neurons by fluorescence decay after photoactivation (FDAP) measurements using a standard confocal laser scanning microscope. The method includes lipofection of photoactivatable green fluorescent protein (paGFP) fused to an SG protein of interest in a neural cell line, differentiation of the cells into a neuronal phenotype, focal activation using a blue diode (405 nm), and recording of decay curves with a 488 nm laser line. By modeling the decay measurements with FDAP functions, the approach enables estimating the residence time of the SG protein of interest, determining the proportion of the respective component in SGs, and the detection of possible changes after experimental manipulation.

Key words Stress granules, Liquid-liquid phase separation, Fluorescence decay after photoactivation, RNA-binding proteins, G3BP1, IMP1, Model neurons

1 Introduction

Stress granules (SGs) are a class of RNA-protein (RNP) complexes that form in the cytosol of many cells in response to environmental stressors such as heat or oxidative stress [1]. SG formation is

---

Anna-Carina Söhnel and Nataliya I. Trushina contributed equally to the work.

Supplementary Information The online version of this chapter (https://doi.org/10.1007/978-1-0716-1975-9_15) contains supplementary material, which is available to authorized users.

Daniel Matějů and Jeffrey A. Chao (eds.), *The Integrated Stress Response: Methods and Protocols*,
Methods in Molecular Biology, vol. 2428, https://doi.org/10.1007/978-1-0716-1975-9_15,

considered to be a survival-promoting mechanism for adapting the translatome of a cell to adverse environmental conditions in a fast, adjustable, and reversible manner [2].

The formation of SGs is thought to be driven by a process called liquid-liquid phase separation (LLPS), which creates subcellular microcompartments in which RNAs and proteins are concentrated in droplet-like structures [3]. A special feature of these droplets, which distinguishes them from membrane-surrounded microcompartments, is that they do not have lipid membranes and that macromolecules are only held together by weak intermolecular interactions. As a result, LLPS favors the interactions between the SG components in a dynamic and adjustable manner and can integrate responses to various stressors.

The RNA-binding protein (RBP) Ras-GTPase-Activating Protein SH3-Domain-Binding Protein 1 (G3BP1) is regarded as a paradigmatic RBP of SGs and is involved in the regulation of SG assembly and function [4]. G3BP1 contains multiple protein-protein interaction domains and a single RNA-binding domain and dynamically shuttles in and out of SGs. Insulin-Like Growth Factor 2 mRNA-Binding Protein 1 (IMP1) is an RBP that may be present in SGs but differs from G3BP1 in that it contains multiple RNA-binding domains and only a single protein-protein interaction module [5].

Understanding SG formation in neuronal cells is of particular interest as impaired, or aberrant SG composition and dynamics have been linked to several neurodegenerative diseases such as amyotrophic lateral sclerosis, frontotemporal dementia, and Alzheimer's disease. The determination of disease-related changes in SG formation and dynamics of individual SG proteins could therefore provide important information about the conditions that cause the change from physiological to pathological SG behavior [6].

The following chapter describes a method for the quantitative determination of the distribution and shuttling dynamics of components of SGs in living neuronal cells by fluorescence decay after photoactivation measurements (FDAP). Rat PC12 cells have been used as a well-established and robust neural model [7]. As a cell line, they are susceptible to efficient gene transfer by lipofection and can then be differentiated into cells with a neuron-like phenotype by treatment with nerve growth factor (NGF) under reduced serum conditions. Differentiated PC12 cells express typical neuronal marker proteins and react to chemically induced stress (treatment with arsenite) with the formation of SGs [8].

We describe the approach for exogenously expressed human G3BP1 genetically labeled with photoactivatable GFP (paGFP) as a paradigmatic component of SGs. The workflow of the experimental approach is shown in Fig. 1. Photoactivation was performed using a standard confocal laser scanning microscope equipped with a blue laser diode (405 nm). The fluorescence of locally activated paGFP

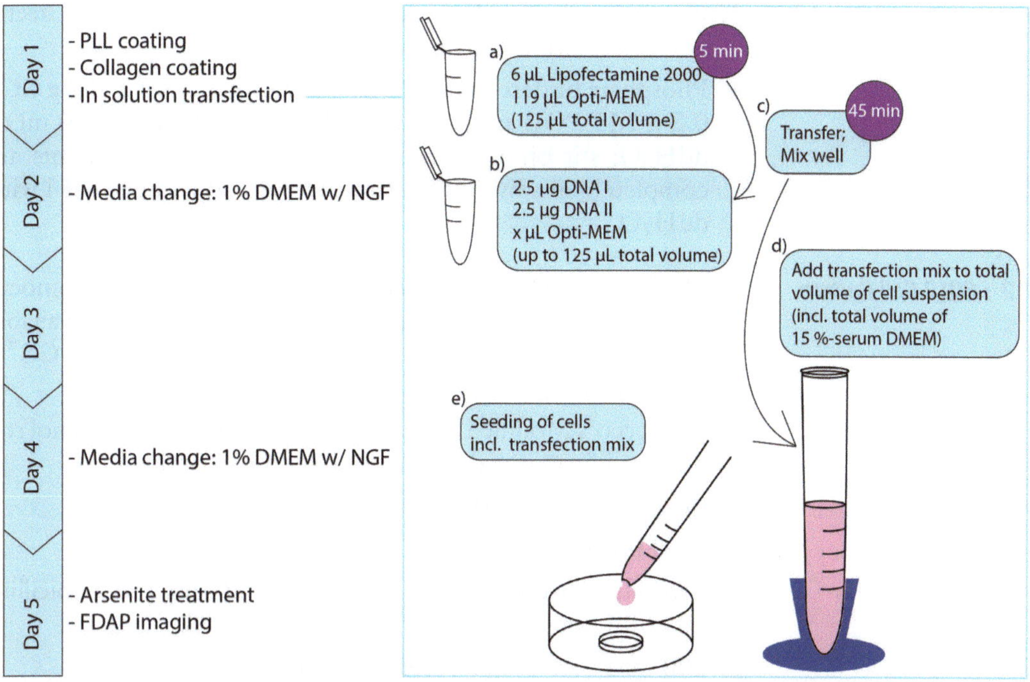

Fig. 1 Workflow of the experimental approach and the transfection process. Left: Timeline with typical sample preparation steps for FDAP experiments. Right: Schematic representation of the in-solution transfection. The lipofection reagent is mixed with the respective medium and incubated for 5 min (**a**). During the incubation, the plasmid DNA is diluted in the respective medium in a second cup (**b**). Lipofection reagent and DNA are mixed well and incubated for 45 min (**c**). After a short incubation (**d**), the transfection mix is added to the cell suspension, and the complete mixture is transferred to the reservoir of the glass-bottom dish (**e**)

was recorded using a 488 nm laser line. Decay measurements from FDAP experiments were modeled with functions for two populations, representing both a free and a granule-associated population with a dynamic exchange. The approach enables the residence time of the protein of interest to be estimated, the proportion of the respective component in the granules to be determined, and possible changes after experimental manipulation to be evaluated. The approach should be easily adjustable for other SG components and the use of other cell types.

2 Materials

2.1 Coating of Glass-Bottom Culture Dishes

1. Glass-bottom culture dishes, for example, MatTek Corporation, 35 mm petri dish, 14 mm microwell, No. 1 cover glass (*see* **Note 1**).
2. 100 µg/ml poly-L-lysine (PLL, Sigma-Aldrich) in borate buffer, pH 8.5, store at 4 °C (*see* **Note 2**).

3. Collagen solution: Prepared from rat tails with 20 mM acetic acid diluted to 50 μg/ml, sterile filtered, store at 4 °C.
4. Phosphate buffered saline (PBS): Dissolve 8 g NaCl, 0.2 g KCl, 0.2 g KH_2PO_4, and 1.15 g $Na_2HPO_4 \times 2H_2O$ in 800 ml of ddH_2O; stir on a magnetic stirrer until the components are completely dissolved; adjust to pH 7.4 and add up to 1 l with ddH_2O.

2.2 PC12 Cell Culture

1. PC12 cells (cell line derived from an adrenal pheochromocytoma of *Rattus norvegicus*, first described by [7]); available from American Type Culture Collection (ATCC®), CRL-1721™, Manassas, USA.
2. Dulbecco's Modified Eagle Medium (DMEM) with phenol red prepared from powder, containing 4.5 g/l D-glucose.
3. Phenol red-free DMEM.
4. Fetal bovine serum (FBS).
5. Penicillin/Streptomycin (Pen/Strep): 10 mg/ml penicillin, 10,000 U/ml streptomycin, store at −20 °C.
6. 200 mM L-glutamine, store at −20 °C.
7. 15% serum-DMEM: DMEM with 10% (v/v) FBS, 5% (v/v) horse serum (HS), 1% (v/v) 200 mM L-glutamine, and 1% (v/v) Pen/Strep, store at 4 °C.
8. 1% serum-DMEM: DMEM with 0.67% (v/v) FBS, 0.33% (v/v) HS, 1% (v/v) 200 mM L-glutamine and 1% (v/v) Pen/-Strep, store at 4 °C (*see* **Note 3**).
9. Nerve growth factor (NGF): 10 μg/ml 7S mouse NGF (Alomone Laboratories) in phenol red-free 1% serum-DMEM, store at −80 °C (*see* **Note 4**).
10. Opti-MEM™ I Reduced Serum Medium (Life Technologies), store at 4 °C.
11. Transfection reagent: Lipofectamine® 2000 (Invitrogen), store at 4 °C (*see* **Note 5**).
12. 0.4% Trypan blue solution (Sigma-Aldrich) for detecting dead cells during cell counting.
13. Eukaryotic expression plasmids (pSems-paGFP-G3BP1 and pSems-mCherry-IMP1) for human G3BP1 and IMP1 with aminoterminally fused paGFP or mCherry-tags (*see* **Note 6**); the vectors were constructed in pRc/cytomegalovirus (CMV)-based expression vectors (Invitrogen) containing a CMV promoter and kanamycin and neomycin resistance genes [8] (*see* **Note 7**).

2.3 Stress Induction

0.05 mol/l sodium arsenite ($NaAsO_2$) solution, store at room temperature (RT), protect from light (*see* **Note 8**).

2.4 Equipment for Image Acquisition

A schematic representation of FDAP measurements of SGs is shown in Fig. 2. Live cell imaging was performed using a confocal laser scanning microscope (Eclipse Ti2, inverted, Nikon) equipped with a 60× oil immersion objective with NA 1.40 (PlanApo VC, Nikon). The system was enclosed in an incubation chamber, maintaining a temperature of 37 °C (fine mechanics, University of Osnabrück) and humidified with 10% CO_2 (Solent Scientific). A fluorescent lamp (Lumencor) was used to identify cells coexpressing paGFP-G3BP1 and mCherry-IMP1. For FDAP measurements, a 405 nm laser diode (LD), and a 488 nm laser (LD), which are incorporated in the Nikon Laser Unit LU-N4/LU-N4S, were used as an excitation source. Photoactivation of paGFP was carried out using the 405 nm laser. The following automated image acquisition was carried out as previously described [9] using suitable image acquisition software (NIS Elements, Nikon). For each experiment, 300 images were collected at a rate of 1 frame per second (1 fps) using the 488 nm laser. The images were detected camera-based (C2, Nikon). A resolution of 256 × 256 pixels and a pixel dwell time of 3.8 μs/pixel were used. The pinhole was open completely (150 μm, *see* Fig. 3a).

2.5 Software and Hardware

1. Operating system: Any system that R can run on (Microsoft Windows 7, Microsoft Windows 10, any Mac OSX, or any Linux distribution).
2. Software requirements
   (a) NotePad++ (or other text editor application).
   (b) R (and RStudio for convenience).
   (c) Software for statistical data analysis (Graph Pad Prism, Origin, or SciDaVis).

3 Methods

3.1 Glass Coverslip Coating

The following steps should be performed under a sterile workbench at RT. The incubator used works at 37 °C, 10% CO_2, and 100% humidity.

1. Put 0.5 ml PLL in the reservoir of the glass-bottom dish on the glass coverslip. Incubate for 30 min at RT (*see* **Note 9**).
2. Replace the PLL with 2 ml of sterile ddH_2O, gently rock the dish and incubate for 60 min at RT. Repeat this procedure a second time (*see* **Note 10**).
3. Aspirate the ddH_2O and add 0.5 ml collagen into the glass-bottom dish's reservoir on the glass coverslip. Incubate for 45 min in the incubator at 37 °C.

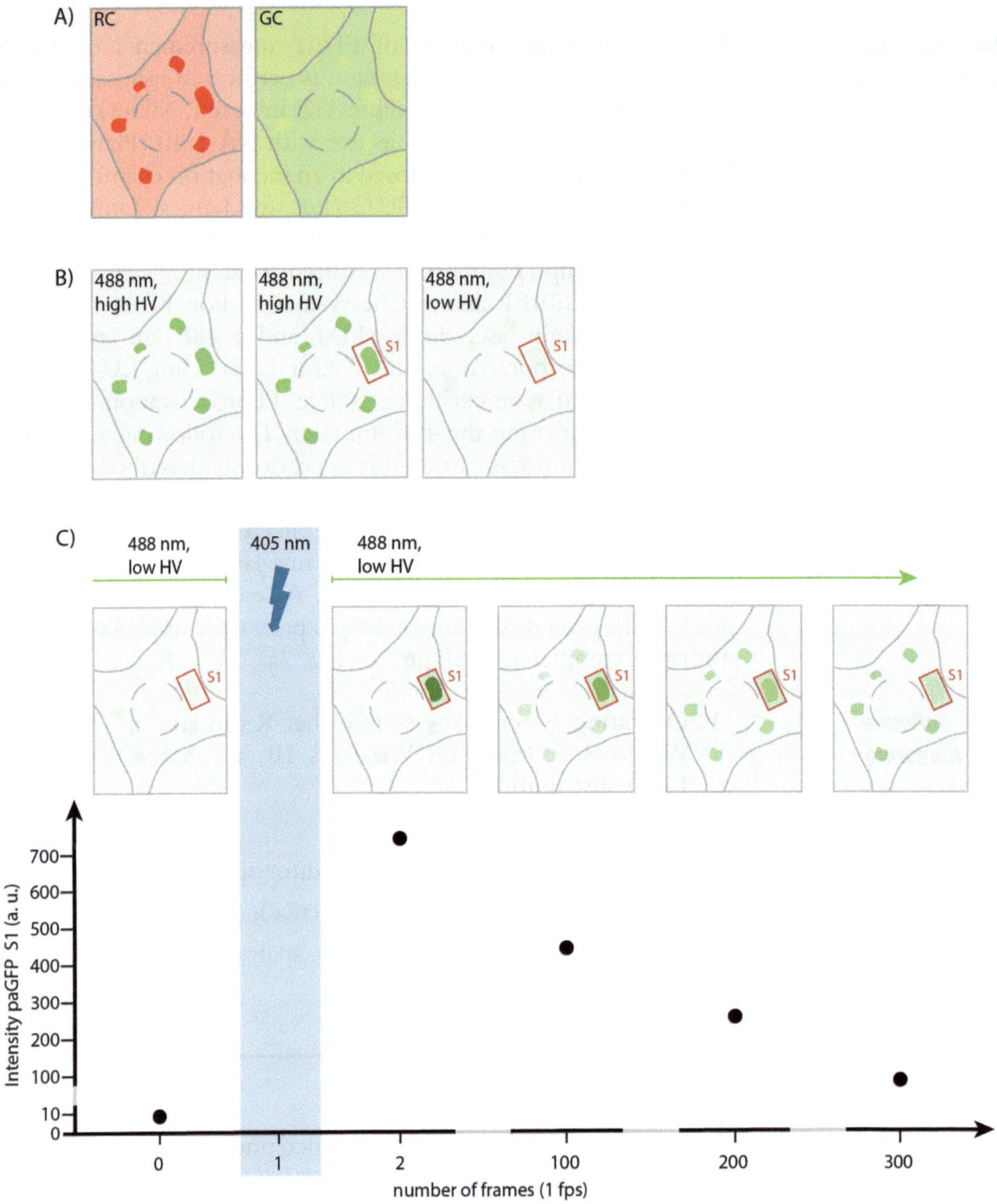

Fig. 2 Schematic representation of FDAP measurements of stress granules. (**a**) Identification of cells expressing both mCherry using the red fluorescence channel (RC) and paGFP using the green fluorescence channel (GC). (**b**) Focusing a granule expressing paGFP by adjusting the high voltage (HV) of the camera, centering the granule in a created stimulation ROI (S1, red box, size: $3.0 \times 5.0\ \mu m$), followed by a reduction of the HV, whereby the signal of paGFP in S1 just disappears. (**c**) Generating the FDAP curve starting with a background intensity measurement using a 488 nm laser before photoactivation (frame 0). Photoactivation of paGFP with 405 nm in S1 then takes place. After photoactivation, the paGFP intensity in S1 is imaged over time using a 488 nm laser and a rate of 1 frame per second (1 fps) for 300 s. The paGFP intensity is given in arbitrary units (a.u.)

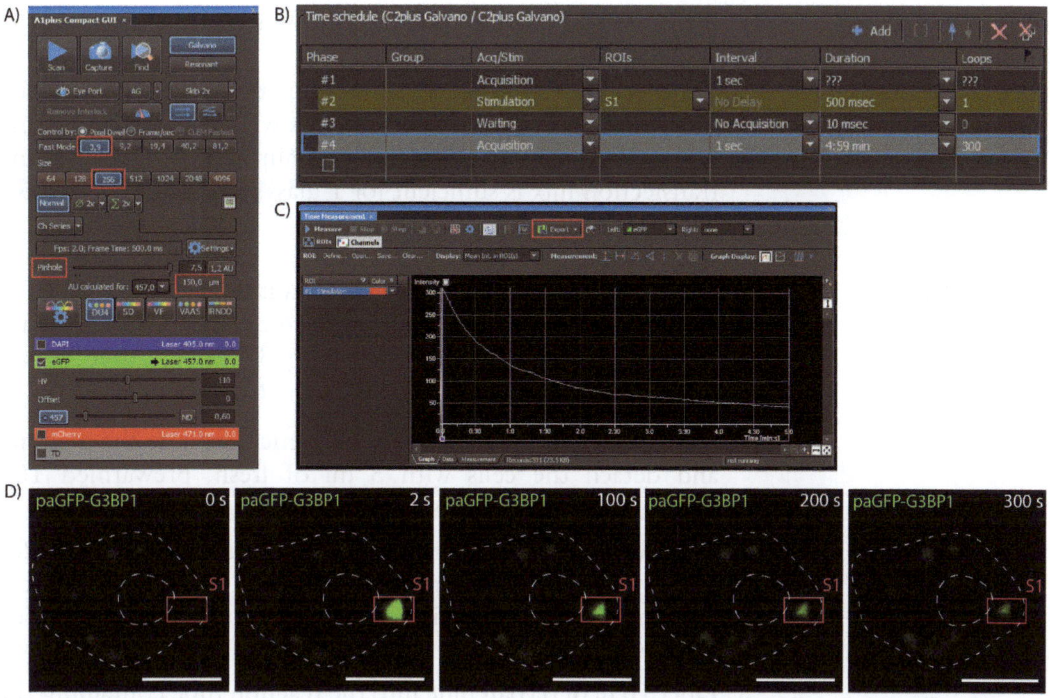

Fig. 3 Imaging parameters for FDAP measurements and representative fluorescence images. (**a**) Screenshot of the NIS Elements Software with settings for resolution, pixel dwell time, and pinhole size (highlighted in red boxes). (**b**) Acquisition sequence including background intensity measurement (#1), photoactivation (#2), and image acquisition over time (#4). (**c**) Single FDAP curve of a stress granule expressing paGFP-G3BP1. The export button for exporting intensity values with the corresponding number of frames is highlighted in a red box. (**d**) Raw intensity images of a single living PC12 cell coexpressing mCherry-IMP1 (intensity channel not shown) and paGFP-G3BP1 after 20 min of treatment with sodium arsenite. First, the background intensity measurement in the stimulation ROI (S1, red box) is displayed at *t*0. Then paGFP is activated in S1 and the paGFP intensity in S1 is tracked over time. Pictures are shown for representative times (2 s, 100 s, 200 s, 300 s). The outline of the cell body and the cell nucleus is shown in dashed lines. Scale bar: 10 μm

4. Wash the coverslip twice with 2 ml PBS. The PBS should remain in the culture dish until the cells are plated, up to a maximum of 24 h.

3.2 PC12 Cell Culture

The following steps should take place under a sterile workbench at RT. The incubator used works at 37 °C, 10% CO_2, and 100% humidity. The workflow of the experimental approach and the transfection procedure is described in Fig. 1.

3.2.1 Transfection

1. Pipet 120 μl Opti-MEM™ into a conical 15 ml polystyrene tube and add 6 μl Lipofectamine® 2000. Incubate for 5 min at RT.
2. Pipet 120 μl Opti-MEM™ into a 1.5 ml reaction tube and add 2.5 μg plasmid I and 2.5 μg plasmid II for a cotransfection, here

pSems-paGFP-G3BP1 (plasmid I) and pSems-mCherry-IMP1 (plasmid II).

3. Transfer the Opti-MEM™–DNA mixture from the reaction tube to the polystyrene tube and mix well by pipetting up and down 10 times. Incubate for 45 min at RT. The resulting transfection mix is sufficient for 1 glass-bottom dish (*see* **Note 11**).

3.2.2 Cell Plating

PC12 cells are grown in a 10 cm tissue culture dish with 10 ml of 15% serum-DMEM. Splitting is performed after the cells have reached 80–90% confluence (usually every 3–4 days). After 25 passages, a new batch of cells is grown from a frozen stock.

1. When the cells reach 80–90% confluence, aspirate the medium and detach the cells with 5 ml of fresh, prewarmed 15% serum-DMEM.
2. Transfer 10 μl of the suspension into a 1.5 ml reaction tube (for cell counting) and the remainder into a 15 ml conical tube. Store the conical tube in the incubator with the cap closed loosely.
3. Pipet 10 μl of trypan blue into the reaction tube containing the cell suspension and mix by pipetting up and down. Load 10 μl of the suspension into a hemocytometer and determine the cell concentration.
4. Prepare 1.5 ml suspension at 1.2×10^5 cells/ml in a reaction tube using prewarmed 15% serum-DMEM and the appropriate amount of cell suspension from the conical tube (*see* **Note 12**).
5. Transfer the appropriate concentration of the cell solution to the polystyrene tube containing the transfection mix prepared in the previous section and incubate for 5 min in the incubator.
6. Aspirate the PBS completely from the glass-bottom dish and carefully transfer the cell–transfection mix to the glass coverslip of the dish. Incubate in the incubator for at least 5 h.

3.2.3 Neuronal Differentiation

The differentiation of PC12 cells is induced by treatment with NGF, which leads to the development of axon-like processes.

1. The next day after plating, aspirate the medium and replace it with 1.5 ml of prewarmed 1% serum-DMEM.
2. Add 15 μl mouse NGF on top of the glass-bottom dish reservoir and gently swirl the culture dish. Incubate for 2 days in the incubator.
3. Replace the medium containing NGF with fresh 1.5 ml 1% serum-DMEM and add fresh NGF as described before. Incubate for another day in the incubator.

3.2.4 Stress Induction

Sodium arsenite solution is used to induce the formation of SGs. Add 15 μl of 0.05 mol/l sodium arsenite solution directly into the reservoir of the glass-bottom dish. Incubate for 20 min in the incubator. Sodium arsenite is not washed off before imaging (*see* **Note 13**).

3.3 Image Acquisition

The microscope's heating system should be turned on at least 2 to 3 h before imaging to ensure a stable and even temperature distribution throughout the setup. A steady temperature is not only critical to cell viability, but it also guarantees proper image acquisition.

1. Focus on the cells using bright-field illumination.
2. Use the fluorescent lamp to screen for cotransfected cells. Begin screening for a cell showing mCherry-IMP1-positive granules with the red channel (RC). Then use the green fluorescent channel (GC) to check whether the same cell also has paGFP-G3BP1 positive granules (*see* Fig. 2a, **Note 14**).
3. Use the 488 nm laser to focus on the granules of the cell and increase the camera voltage to detect the paGFP autofluorescence signal (*see* Fig. 2b, left).
4. Create a Region of Interest (ROI) for laser stimulation and imaging with a size of 3.0 × 5.0 μm.
5. Align the stimulation ROI 1 (termed "S1") with a granule of interest and refocus the granule to its maximum extent (Fig. 2b, middle, **Note 15**).
6. Reduce the voltage of the camera until you can only observe a weak signal of the paGFP autofluorescence (Fig. 2b, right).
7. Run the time FDAP measurement as described in Fig. 3b starting with the background intensity measurement in S1 using 488 nm. A schematic representation of the complete FDAP measurement is shown in Fig. 2c.
8. Photoactivate paGFP within S1 by using the 405 nm laser (*see* **Note 16**).
9. After activation, continue with automated image acquisition (488 nm). Collect 300 images at a rate of 1 frame per second (fps) until the fluorescence decay reaches a plateau in intensity. The imaging parameters used are shown in Fig. 3a (*see* **Note 17**). A single FDAP decay and corresponding, exemplary fluorescence intensity images are shown in Fig. 3c, d.
10. Save the measurement in the original format as nd2-file. Also, save it as a txt-file using an editor application. Press the "export" button in the NIS Elements software (*see* Fig. 3c, red box). A list in which each measuring point is correlated with the respective paGFP intensity is copied to the clipboard. Open the editor application. Paste the list into the editor and

save the document under the name of the nd2-file (*see* **Note 18**).

11. Create one folder with only the txt-files for a particular series of measurements for further analysis in R, as well as a separate folder containing the nd2-files of the respective measurements.

3.4 Image Analysis

3.4.1 Intensity Measurement in Fiji

As described in Subheading 3.3, paGFP-intensity values in the activation area S1 are directly measured in NIS elements, and the corresponding txt-files can be used for subsequent analysis in R. Here, the intensity values perfectly correspond to the activation ROI S1, making the analysis highly precise. However, due to the slight movement of SGs within the cell as well as SG movement provoked by cellular movement upon sodium arsenite treatment, sometimes a broader ROI is required to measure the intensity of the chosen granule over time correctly. The Fiji macro "FDAP_batch_image_processing.ijm" enables the setting of a new ROI per image series, as well as automatic analysis of intensity values, which are saved as csv-file (*see* **Note 19**). Additionally, the Fiji macro usage enables visual quality control of raw data appropriately and helps exclude those affected by a focal drift.

The evaluation of image stacks containing FDAP data to extract the respective fluorescence intensity values is done by the following steps:

1. Start the Fiji program, open the Fiji macro "FDAP_batch_image_processing.ijm" via drag and drop and click "Run" (*see* **Note 20**).
2. Select a folder containing only the image series for analysis as the input directory using "Browse" (Fig. 4a). Ensure that only files with the selected extension (nd2 or tiff) are located in the selected folder and its subfolders.
3. Create a second folder and select it as the output directory to save analysis output in corresponding intensity files.
4. Before the automatic opening of the respective image series, the "Bio-Formats Input options" are displayed. Always select "Split Channels" and press "OK."
5. The second frame is selected, and the brightness and contrast for the whole image series are automatically adjusted.
6. Follow the instructions on the new window "Action Required."
7. Use the selection tool "Rectangle" and draw an ROI around the activated SG. Check the previous frame to control the fluorescence gain. Review the selected ROI in all frames (Fig. 4b). If necessary, adjust the ROI so that it *only* contains the activated SG during the entire acquisition time. Use the selection tool "Polygon" if this helps to meet the requirements

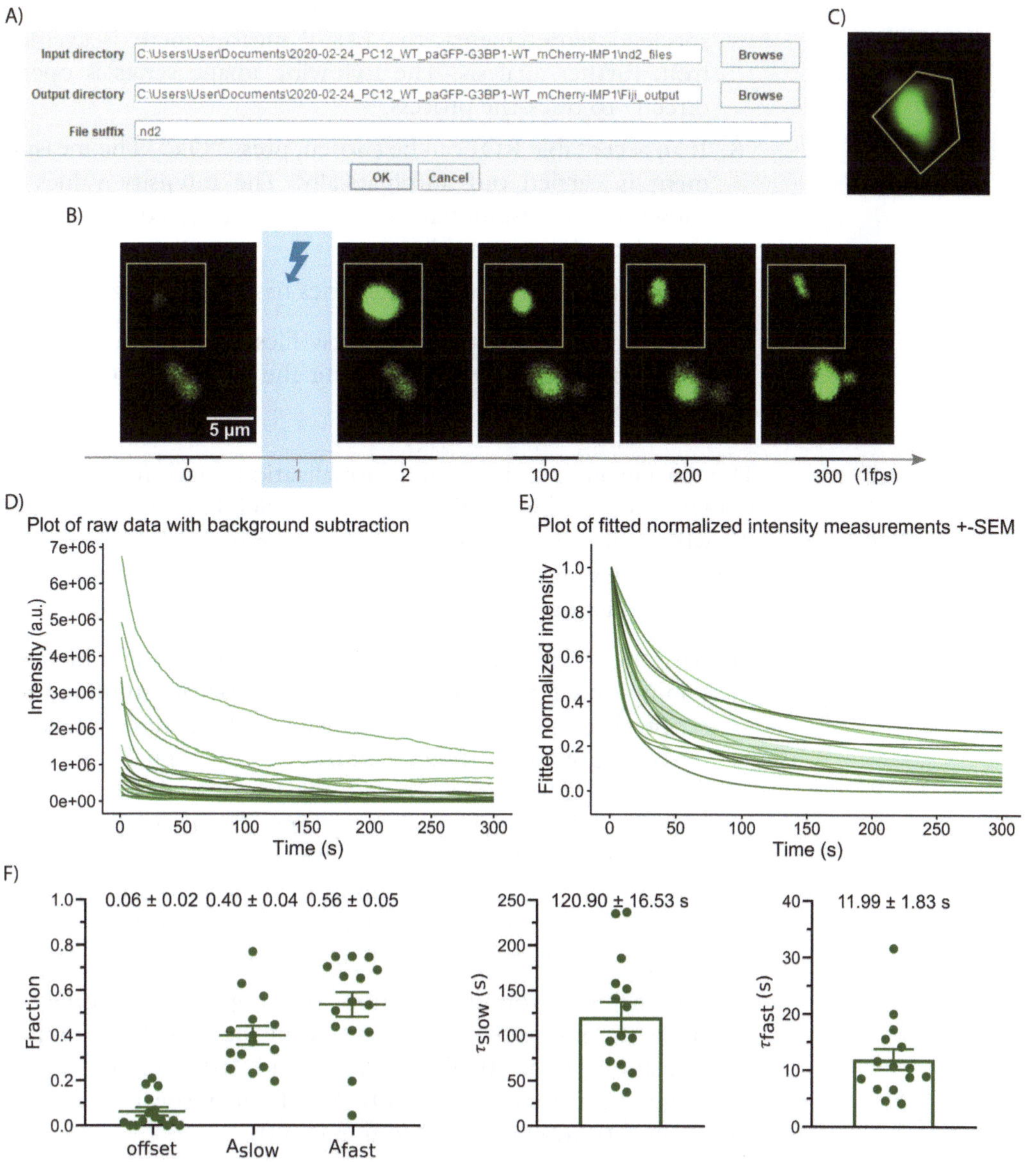

Fig. 4 Intensity measurement, background extraction, normalization, and fitting of FDAP curves. (**a**) Screenshot with an example of the selected input and output folders. (**b**) Selection and verification of the ROI around the activated stress granule. (**c**) An example of the "Polygon" selection tool used to include only the activated SG during the entire acquisition. (**d**) Example of measured raw intensities before extrapolation and normalization. (**e**) Extrapolated and normalized intensities from the exemplary FDAP experiments shown in **d**. (**f**) The graphic representations of the residence time (τ_{slow} and τ_{fast}) and the proportion inside and outside of the granules (A_{slow} and A_{fast}) are shown as mean $\pm$ SEM. Typical mean values for τ_{slow}, τ_{fast}, A_{slow}, and A_{fast} are shown and illustrate the highly dynamic properties of paGFP-G3BP1

(Fig. 4c); close the image series if a proper ROI cannot be set; for example, if the chosen SG moves out of any drawn ROI, a surrounding SG moves into the ROI, or if a focal drift is observable. By closing an image series that does not meet the specified requirements, the FDAP measurement is excluded from further analysis. The following image series is opened directly to track the process.

8. If an acceptable ROI can be chosen, press "Ok". The measurement is carried out automatically. The intensity values are shown in the results table and are directly saved as csv-file in the selected output folder.
9. Continue until the last image series has been analyzed.

The raw intensity data stored in csv-files in the output directory folder can be further processed using the R script "SG_FDAP_for_RStudio.R".

3.4.2 Background Extraction, Normalization, and Fitting of FDAP Curves

The background extraction, normalization, and fitting of the FDAP curves is done with the R script "SG_FDAP_for_RStudio.R" written for the respective purpose (*see* Fig. 4d, e, **Note 19**). Fig. 4d shows an example of measured raw intensities after background extraction.

1. Save the script "SG_FDAP_for_RStudio.R" in the folder containing only txt- or csv-files for analysis. Before running the R script, three configuration settings must be entered: construct or condition name (e.g., "G3BP1wt", *see* line 31 in the R script), number of rows that needs to be analyzed (e.g., "300" as the frame number, *see* line 32 in the script, **Note 21**), and the analysis type (e.g., "NIS" for NIS-elements output or "Fiji" if the output files of the described Fiji macro will be used for R analysis, *see* line 33 in the R script). Save all changes before running the R script (*see* **Note 22**).
2. For the background extraction, the script automatically subtracts the intensity value measured in the ROI before photoactivation (frame 0, Figs. 2c, 3d, and 4b) from all intensity values measured in the ROI S1 of the respective SG after photoactivation (frame 2 to frame 300).
3. The data are then normalized using the maximal value at $t = 0$ s (here the time of photoactivation). To calculate the maximum intensity value at $t = 0$ s, the script extrapolates the measured intensities over the entire time with a biexponential fit:

$$I(t) = I_0 + a_1 * e^{-\frac{t}{b_1}} + a_2 * e^{-\frac{t}{b_2}}$$

At $t = 0$ s (time of photoactivation):

$$I(0) = I_0 + a_1 * e^{-\frac{0}{b_1}} + a_2 * e^{-\frac{0}{b_2}} \\ = I_0 + a_1 + a_2$$

4. The FDAP data from each measurement are divided by the respective extrapolated value for $t = 0$ s. The extrapolated and normalized intensities from the exemplary FDAP experiments are shown in Fig. 4e.
5. The normalized data are modeled using a biphasic exponential decay function for two populations that represents the presence of a free and a granule-associated population with dynamic exchange (*see* **Note 23**).

$$I(t) = \text{offset} + A_{\text{fast}} * e^{-\frac{t}{\tau_{\text{fast}}}} + A_{\text{slow}} * e^{-\frac{t}{\tau_{\text{slow}}}}$$

The decay time of the slow component (τ_{slow}) reflects the retention time of the protein of interest in the SG, and the decay time of the fast component (τ_{fast}) reflects the mobility of the respective protein in the cytosol. A_{slow} and A_{fast} denote the relative fractions ranging from 0 to 1 of the slow and fast fractions, respectively. The offset depicts both the dynamics of supposed immobile proteins as well as the steady-state of proteins shuttling between the SGs continuously.

6. Additionally, modeling with a monophasic exponential decay function for one population is done to confirm that the decay data follows a model for two populations better.

$$I(t) = \text{offset} + A * e^{-\frac{t}{\tau}}$$

This equation is a simplification of the two-phase exponential decay model in which the fraction of one of the two populations is set to 0.

7. After running the script, the individual curve plots can be found in folders "jpeg_fits_extrapolation_check" and "jpeg_-fits" and can be filtered by visual analysis. All measurements could be modeled with biphasic exponential decay and are collected in respective figures and tables. The calculated coefficients and statistical parameters can be found in the output table "all_fit_coefs.csv".
8. A χ^2 statistical test is used to compare the results of the modeling for the monophasic and biphasic exponential decay fits. Based on the expected output of two populations—a highly dynamic and a slower protein fraction, the FDAP measurements can be modeled by a two-phase exponential decay suitably (*see* **Note 24**).
9. Average the resulting model coefficients for each curve to obtain a mean and the standard error of the mean (SEM) for each construct and experimental condition. Typical values for the residence time of G3BP1 and its proportion in SGs are shown in Fig. 4f.

4 Notes

1. In general, any type of glass-bottom dish for microscopy or any glass coverslip for microscopy that can be built into a corresponding microscopy chamber can be used. Only the media change to 1% serum-DMEM with NGF, and the sodium arsenite treatment must be guaranteed.
2. Instead of PLL and collagen, it is also possible to coat the surface with an RGD-functionalized poly-L-lysine graft (poly-ethylene glycol)-copolymer [10]. However, PC12 cells appear to have an increased tolerance to the detachment under stressful conditions when plated on PLL and collagen-coated surfaces.
3. Prepared 1% serum-DMEM can also be stored at −20 °C for several months.
4. The preparation of 500 μl NGF stocks in 1.5 ml reaction cups is recommended. Working solutions can be stored at −20 °C for short term use.
5. In our laboratory, transfection of PC12 cells by means of lipofection has proven effective, while calcium phosphate transfection has a toxic effect on PC12 cells. In principle, however, any other transfection reagent could be used after the protocol has been optimized with respect to the amount of transfection reagent, cell density, and incubation time.
6. For further information regarding technology and application of paGFP to monitor protein dynamics in living cells, check the application note "Photoactivatable green fluorescent protein (PA-GFP): a valuable tool for selective photolabelling of proteins and cells" on the web page of the Department of Neurobiology, University of Osnabrück (https://www.neurobiologie.uni-osnabrueck.de/index.php?cat=Research&page=Ressources%20and%20Materials).
7. Here the pSems-vector is used for transfection. Virtually any vector plasmid should, however, be useful as long as the cell line used tolerates the method of gene transfer and the expression rate of the exogenous protein.
8. Sodium arsenite is characterized as toxic by ingestion, inhalation, or absorption through the skin. It is therefore recommended that small aliquots be made for direct use under the hood. In general, all chemicals should be used according to the CLP ("Classification, Labeling, and Packaging") regulation, and the hazard (H-) and precaution (P-) measures should be checked before use.
9. When using glass coverslips, these should be cleaned by passing them through the flame of a Bunsen burner or by using a

plasma cleaner (e.g., Diener electronics FEMTO) for 15 min and three subsequent washing steps with ddH_2O before the PLL coating.

10. Instead of using the PLL-coated glass-bottom dish after the washing steps directly for the collagen coating, it can be stored at 4 °C for up to a month. Aspirate the ddH_2O and leave the dish to dry on the bench at RT until the remaining ddH_20 has completely evaporated. Seal the glass-bottom dish with parafilm and store at 4 °C.
11. If more than one glass-bottom dish needs to be transfected, it is possible to increase the volume of reagents used and to prepare the transfection mixture in a single polystyrene tube. The transfection protocol described is specific for the use of Lipofectamine 2000 and can vary considerably in case of the cell line or plasmid used.
12. If more than one coverslip has to be made, the amount of cell suspension can be increased (*see* also **Note 11**).
13. PC12 cells tend to detach easily after treatment with sodium arsenite. The vibrations from carrying the glass-bottom dish to the microscope can lead to significant cell loss. To avoid cell loss, the treatment and incubation with sodium arsenite should be carried out directly on the microscope. Imaging can be performed for approximately 1–1.5 h after the sodium arsenite treatment before the cells completely detach.
14. While the fluorescence signal for mCherry is unmistakable, the emission of paGFP hardly differs from the autofluorescence background signal. You can be sure not before successful photoactivation of paGFP with 405 nm irradiation.
15. Select a granule between the nucleus and the plasma membrane to reduce SG movement during the measurement. The SG should also be separated from other SGs to avoid fusion events during FDAP measurements.
16. Photoactivation of paGFP successfully reaches an activation ratio (paGFP) ≥ 10. After activation, the paGFP-signal should not reach saturation. Before and after photoactivation, the paGFP intensity values can be followed in the time measurements graph in NIS Elements (*see* Fig. 3c). In order to recognize a possible paGFP saturation after 405 nm stimulation, activate the "saturation indicator"-button in NIS Elements. Here, pixels that reach saturation are shown in pink. FDAP measurements that do not meet the specified requirements are aborted and therefore directly excluded from further analysis. If the activation rate is low, the entire cell is excluded from additional measurements. No second activation step is carried out for the same granule. If the activation is successful, only one granule per cell is measured.

17. The FDAP measurement is complete when the decay reaches an intensity plateau. If imaging is stopped before this property, molecules with slow decay will be underrepresented, shifting the results to lower values and creating a measurement artifact. After reaching the intensity plateau, the measurement time should not be artificially increased since molecules with long decay are overrepresented, which in this case leads to a shift of the results to lower values. In addition, the cell is exposed to phototoxicity during long-term imaging, which can cause cell detachment and granule movement. Since the paGFP intensity is immensely reduced after long-term imaging, the photobleach content increases over time, which can also affect the results obtained.
18. When using a different program than NIS Elements, save the measurement in the respective original file and export image series as tiff, additionally. Further analysis should be done as described in Subheading 3.4.1 using the Fiji macro "FDAP_batch_image_processing.ijm".
19. The Fiji macro "FDAP_batch_image_processing.ijm", as well as the R script "SG_FDAP_for_RStudio.R" (for Linux users: "Rscript SG_FDAP_for_linux.R") described in Subheading 3.4.2 in detail, are available on request from the authors and can be downloaded at the web page of the Department of Neurobiology, University of Osnabrück (https://www.neurobiologie.uni-osnabrueck.de/index.php?cat=Research&page=Ressources%20and%20Materials), or as Electronic Supplementary Material included with this chapter. For a thorough understanding of the scripts, please check the respective headers and comments given in the Fiji macro and the R script.
20. For the shown data set in Fig. 4, the main channel for paGFP is "C = 0" (*see* line 33 in the Fiji macro "FDAP_batch_image_processing.ijm"). The main channel may need to be changed in case of using other image formats, other microscope systems, or other photoactivatable (pa) proteins like pa-mCherry, for example, to "C = 1" or "C = 2" before running the macro.
21. To ensure proper execution of the script, the entered frame number should not be higher than the shortest measurement acquisition.
22. In case of using the script "SG_FDAP_for_Linux.R", run the script from the terminal using the command "Rscript SG_FDAP_for_linux.R G3BP1wt 300 NIS" for exemplary configuration settings.
23. Initially, the coefficients should be adapted manually close to their expected values (*see* lines 180–186 in the "SG_FDAP_for_RStudio.R" script). The closer you set the coefficients to

their final values, the better the result of the modeling algorithm.

24. FDAP curves of components that do not localize into granules, such as a 3×paGFP control construct, are better modeled with a monophasic exponential decay function. Note that both models do not contain any information about the geometry of the activation region or the molecular kinetics underlying the decay.

Acknowledgments

We thank Mina Bakharzi for the contribution of image data and Rainer Kurre for his help with data analysis, particularly curve modeling. This work has been supported by the Deutsche Forschungsgemeinschaft (DFG Grant SFB 944, Project P1 (to R.B.)) and a fellowship of the graduate college "EvoCell" of the University of Osnabrück (to N.T.).

References

1. Riggs CL, Kedersha N, Ivanov P, Anderson P (2020) Mammalian stress granules and P bodies at a glance. J Cell Sci 133(16):jcs242487. https://doi.org/10.1242/jcs.242487
2. Rabouille C, Alberti S (2017) Cell adaptation upon stress: the emerging role of membraneless compartments. Curr Opin Cell Biol 47: 34–42. https://doi.org/10.1016/j.ceb.2017.02.006
3. Banani SF, Lee HO, Hyman AA, Rosen MK (2017) Biomolecular condensates: organizers of cellular biochemistry. Nat Rev Mol Cell Biol 18(5):285–298. https://doi.org/10.1038/nrm.2017.7
4. Tourriere H, Chebli K, Zekri L, Courselaud B, Blanchard JM, Bertrand E, Tazi J (2003) The RasGAP-associated endoribonuclease G3BP assembles stress granules. J Cell Biol 160(6): 823–831. https://doi.org/10.1083/jcb.200212128
5. Niewidok B, Igaev M, Pereira da Graca A, Strassner A, Lenzen C, Richter CP, Piehler J, Kurre R, Brandt R (2018) Single-molecule imaging reveals dynamic biphasic partition of RNA-binding proteins in stress granules. J Cell Biol 217(4):1303–1318. https://doi.org/10.1083/jcb.201709007
6. Wolozin B, Ivanov P (2019) Stress granules and neurodegeneration. Nat Rev Neurosci 20(11):649–666. https://doi.org/10.1038/s41583-019-0222-5
7. Greene LA, Tischler AS (1976) Establishment of a noradrenergic clonal line of rat adrenal pheochromocytoma cells which respond to nerve growth factor. Proc Natl Acad Sci U S A 73(7):2424–2428. https://doi.org/10.1073/pnas.73.7.2424
8. Moschner K, Sundermann F, Meyer H, da Graca AP, Appel N, Paululat A, Bakota L, Brandt R (2014) RNA protein granules modulate tau isoform expression and induce neuronal sprouting. J Biol Chem 289(24): 16814–16825. https://doi.org/10.1074/jbc.M113.541425
9. Weissmann C, Reyher HJ, Gauthier A, Steinhoff HJ, Junge W, Brandt R (2009) Microtubule binding and trapping at the tip of neurites regulate tau motion in living neurons. Traffic 10(11):1655–1668. https://doi.org/10.1111/j.1600-0854.2009.00977.x
10. Wedeking T, Lochte S, Birkholz O, Wallenstein A, Trahe J, Klingauf J, Piehler J, You C (2015) Spatiotemporally controlled reorganization of signaling complexes in the plasma membrane of living cells. Small 11(44):5912–5918. https://doi.org/10.1002/smll.201502132

their final values, the better the result of the modeling algorithm.

3. FDAP curves of compartments that do not localize to granules, such as a 3 × [illegible] constant, [illegible] with a monophasic exponential decay function. Note that both [illegible]

Acknowledgments

[illegible]

Chapter 16

Probing Protein Solubility Patterns with Proteomics for Insight into Network Dynamics

Xiaojing Sui, Mona Radwan, Dezerae Cox, and Danny M. Hatters

Abstract

Proteome solubility contains latent information on the nature of protein interaction networks in cells and changes in solubility can provide information on rewiring of networks. Here, we report a simple one-step ultracentrifugation method to separate the soluble and insoluble fraction of the proteome. The method involves quantitative proteomics and a bioinformatics strategy to analyze the changes that arise. Because protein solubility changes are also associated with protein misfolding and aggregation in neurodegenerative disease, we also include a protocol for isolating disease-associated protein aggregates with pulse shape analysis (PulSA) by flow cytometry as a complementary approach that can be used alongside the more general measure of solubility or as a stand-alone approach.

Key words Solubility, Ultracentrifugation, Pulse shape analysis (PulSA), Protein aggregate, Proteomics, Bioinformatics

1 Introduction

Proteins form different sized assembly states depending on ligand interactions and functional state. These states can include very large structures such as biological condensates (e.g., nucleoli, stress granules, and germ granules), polyribosome-mRNA complexes [1] and pathological protein aggregates in the case of neurodegenerative diseases such as Alzheimer's, Parkinson's, and Huntington's disease [2]. Upon stimuli, protein assemblies can be remodeled as part of their function. For example, heat shock induces the aminoacyl-tRNA synthetase complex to form in a transient manner through the recovery process [3]. T-cell receptor complexes also can involve a massive coalescence of signaling proteins upon induction of the trigger for signaling [4].

We and others have shown that insight to the function and remodeling of large complexes can be derived by analysis of the proteins that change solubility with proteomics [3, 5]. While we

Daniel Matějů and Jeffrey A. Chao (eds.), *The Integrated Stress Response: Methods and Protocols*, Methods in Molecular Biology, vol. 2428, https://doi.org/10.1007/978-1-0716-1975-9_16,

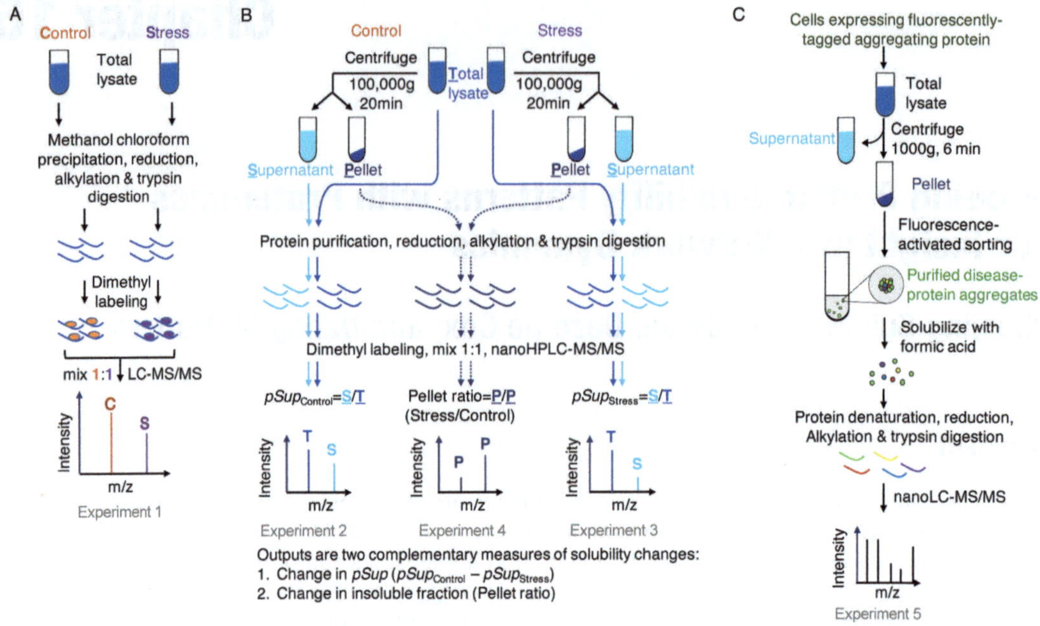

Fig. 1 Dimethyl labeling-based proteomics workflow for measurement of proteome solubility changes under stimulus of interest. (**a**) Strategy to quantify total proteome abundances changes due to stimulus (Experiment 1). (**b**) Strategies to measure proteome solubility changes by supernatant and pellet experiments (Experiments 2–4). (**c**) Strategy to quantify disease-associated protein aggregates (Experiment 5)

have previously explored the effect of stress on solubility of the proteome [5], in principle the methods would be applicable to any type of stimulus or change in cellular state including cytokine responses, signaling cascades or comparing different stages of differentiation state. The method involves first separating the soluble and insoluble subproteomes by one-step ultracentrifugation. Using proteomics, the method then quantifies the differential partitioning of proteins into the soluble or insoluble fractions between control and treatment (e.g., stimulus of choice) with two complementary measurements. The first, *Δ pSup*, defines the changes in proportions of a protein in supernatants between control and treatment. The second is the pellet ratio, which describes the ratio of the masses of a protein in the insoluble fraction between control and treatment (Fig. 1a, b). Defining both *Δ pSup* and pellet ratios together enables a complementary perspective to the changes in solubility of proteomes and covers proteins of different baseline levels of solubility and abundances. Bioinformatic analysis of proteins that change their solubility is then used to garner information on the biological processes and protein complexes associated with the changes in solubility. This method has been developed by the authors with mammalian cell models. In principle the method should be adaptable to lysates from yeast and *Caenorhabditis elegans*, although this has not been tested by the authors directly.

The inappropriate formation of protein aggregates is a hallmark of neurodegenerative diseases and may arise from misfolding of disease proteins and aberrant interactions with surrounding proteins [6–9]. Here, we also describe a protocol for directly isolating disease-associated protein aggregates (Fig. 1c). The essence of the approach is to express a fluorescently labeled protein known to inappropriately aggregate in the cell line of interest and then separate the aggregates that arise from the cell lysates using a flow cytometry assay called pulse shape analysis (PulSA) [10]. Here we describe how to identify the proteins enriched in PulSA-purified aggregates with quantitative proteomics, as a sister approach to the solubility assays described in this article.

2 Materials

1. Biological samples ready for lysis. For example, mammalian cells.
2. Ultracentrifuge (Beckman Coulter Optima Max), Beckman MLA-130 ultracentrifuge rotor for 1.5 mL ultracentrifuge tubes and microcentrifuge.
3. Low binding protein microcentrifuge tubes and filter pipette tips.
4. 21 gauge (G) and 31 G needles.
5. Lysis buffer 1: 50 mM Tris–HCl (pH 7.5), 150 mM NaCl, 1% (v/v) IGEPAL CA-630, 10 units/mL DNase I, EDTA-free protease inhibitor cocktail tablet.
6. Lysis buffer 2: Lysis buffer 1 with 4% SDS and 4 mM DTT (1,4-dithiothreitol).
7. Lysis buffer 3: 20 mM Tris (pH 8.0), 2 mM $MgCl_2$, 150 mM NaCl, 1% (w/v) Triton X-100, 20 units/mL Benzonase (Novagen), EDTA-free protease inhibitor cocktail tablet.
8. Buffer A: 1 × phosphate buffered saline (PBS) + EDTA-free protease inhibitor tablet (Roche).
9. 0.5 M EDTA.
10. Bicinchoninic acid (BCA) protein assay kit.
11. 25, 50, 100 mM and 1 M triethylammonium bicarbonate (TEAB).
12. 10 mM DTT.
13. 55 mM iodoacetamide (IAA).
14. 8 M urea, 100 mM TEAB.
15. 3 M Tris.

16. Formic acid (HPLC grade).
17. Glass slides and widefield fluorescence microscope.
18. 5 mL round-bottom polystyrene test tube (Falcon).
19. Flow cytometer (e.g., BD FACS Aria III).
20. Pierce trypsin protease, sequencing grade (Thermo Scientific).
21. Solid-phase extraction cartridge (Oasis HLB 1 cc Vac Cartridge, 10 mg sorbent, Waters Corp., USA).
22. 1% (vol/vol) trifluoroacetic acid (TFA) (HPLC grade).
23. 80% acetonitrile (ACN) containing 0.1% TFA (HPLC grade).
24. Freeze dryer and freeze dryer flask.
25. Isobaric or isotopic labeling reagents (*see* **Note 1**). For example, tandem mass tag (TMT), isobaric tags for relative and absolute quantitation (iTRAQ) or dimethyl labeling reagents.
26. Liquid chromatography–nano electrospray ionization tandem mass spectrometry (LC-nESI-MS/MS): Orbitrap Fusion Lumos mass spectrometer or Orbitrap Q Exactive Plus mass spectrometer (Thermo Fisher Scientific).
27. Nano-LC system, Ultimate 3000 RSLC (Thermo Fisher Scientific) was equipped with an Acclaim Pepmap nano-trap enrichment column (C18, 100 Å, 75 μm × 2 cm, Thermo Fisher Scientific) and an Acclaim Pepmap RSLC analytical column (C18, 100 Å, 75 μm × 50 cm, Thermo Fisher Scientific).
28. Software: Proteome Discoverer 2.4 or MaxQuant (https://www.maxquant.org/), Perceus (https://maxquant.net/perseus/), Cytoscape v3.8.2 (https://cytoscape.org/), vector graphics editing software.

3 Methods

3.1 Isolation of Soluble and Insoluble Proteins by Ultracentrifugation

For precise quantitation and to enable multiplexing, peptides are isobarically or isotopically labeled prior to mass spectrometry. The choice of specific labeling methods is in large part dictated by the starting material and experimental design, which are discussed further in **Note 1**. Here, we will focus on material derived from mammalian cell culture of immortalized cell lines which are prepared using dimethyl isotopic labeling. The volumes of reagents described below are optimized for 9×10^6 (seeding amount) Neuro2a cells and may need optimization depending on particular systems.

1. Grow cells under control and treatment conditions.
2. Harvest cell pellets and resuspend them in 500 μL ice-cold lysis buffer 1.
3. On ice, extrude the suspended cells through a 27 G needle 25 times, followed by a 31 G needle 10 times.
4. Add 0.5 M EDTA to a final concentration of 2 mM (*see* **Note 2**).
5. Transfer 300 μL of the cell lysate solution to a chilled 1.5 mL ultracentrifuge tube and centrifuge at 100,000 × *g* with the Beckman MLA-130 ultracentrifuge rotor for 20 min at 4 °C with acceleration and deceleration both set to 9. Store the remaining cell lysate at 4 °C as the representative total (T) sample.
6. To minimize the risk of disrupting the pellet following centrifugation, only collect 250 μL of the resultant supernatant into a fresh Eppendorf tube. This is reserved as the supernatant (S) sample.
7. Add 450 μL lysis buffer 1 to the pellet without resuspension, followed by ultracentrifugation at 100,000 × *g* for 20 min at 4 °C.
8. Carefully remove 450 μL of the supernatant without disturbing the pellet (*see* **Note 3**).
9. Repeat **steps** 7 and **8** twice.
10. The resultant pellet is designated as the pellet (P) fraction.
11. Dilute the T and S samples with the same volume of lysis buffer 2 as their initial volume (measured for T with a pipette, 250 μL for S), followed by incubation at 95 °C for 20 min (*see* **Note 4**).
12. Add 50 μL lysis buffer 2 to the P fraction and transfer it to a fresh microcentrifuge tube. Heat the pellet resuspension at 95 °C for 20 min.
13. Cool all samples at room temperature for 10 min.
14. Determine the concentration of proteins from T, S, and P fractions via BCA assay according to the manufacturer's instructions. We suggest a dilution factor of 20 as a starting point to obtain concentrations within the dynamic range of the standard curve (*see* **Note 5**).
15. Take 100 μg of protein from each sample and adjust the concentration to 1 μg/μL with 25 mM TEAB.
16. Reduce the cysteine residues with 10 mM DTT for 1 h at 37 °C.
17. Alkylate the reduced cysteine residues with 55 mM IAA for 1 h at 37 °C.

18. Purify proteins by chloroform/methanol precipitation [11]: Add 300 μL methanol to each sample and vortex, followed by the addition of 100 μL chloroform and vortexing.
19. Add 300 μL H_2O and vigorous vortexing, followed by centrifugation at 21,000 × *g* for 1 min. Carefully remove the upper aqueous layer without disturbing the precipitated protein at the interface.
20. Add 300 μL methanol and vortex, followed by centrifugation for 1 min at 21,000 × *g*. Remove the supernatant down to a drop so as not to disturb the pellet. Air dry the pellet (*see* **Note 6**).
21. Dissolve the precipitated protein in 100 μL of 8 M urea, 100 mM TEAB with vigorous vortexing.
22. Adjust the resuspension to a final concentration of 1 M urea, 100 mM TEAB with addition of 700 μL 100 mM TEAB buffer. Proceed to Subheading 3.3 for sample preparation for mass spectrometry analysis.

3.2 Purification of Disease-Associated Protein Aggregates

1. Grow and harvest cells with fluorescently labeled protein of interest (*see* **Note 7**).
2. Resuspend cell pellets in 500 μL ice-cold lysis buffer 3.
3. On ice, extrude the suspended cells through a 27 G needle 25 times, followed by a 31 G needle 10 times.
4. For efficient lysis, incubate cell lysates on ice for 30 min and vortex samples at regular intervals six times for 45 s each.
5. Dilute cell lysates with ice-cold 500 μL buffer A, followed by spinning the crude lysates at 1000 × *g* for 6 min at 4 °C and discard the supernatant (*see* **Note 8**).
6. Wash pellets twice by resuspension in 1 mL buffer A and spin at 1000 × *g* for 6 min at 4 °C.
7. Resuspend the pellets in 1 mL buffer A and pass the solution through a cell strainer snap cap round-bottom polystyrene 5 mL test tube. This step is important to maintain a uniform single-aggregate suspension and prevent clogging of flow cytometer. Store the resuspended pellets on ice until sorting.
8. Vortex pellet suspensions for 45 s immediately prior to fluorescence-activated cell sorting (FACS).
9. Sort fluorescent particles using the appropriate excitation and emission filter set with PulSA by flow cytometry [12]. If using a BD FACS Aria III equipped with a 100 μm diameter nozzle at 20 psi, set sort rates to 500 events per second and collect one million aggregates per condition. These parameters will need to be optimized for other instruments. For detailed PulSA settings, please refer to [12].

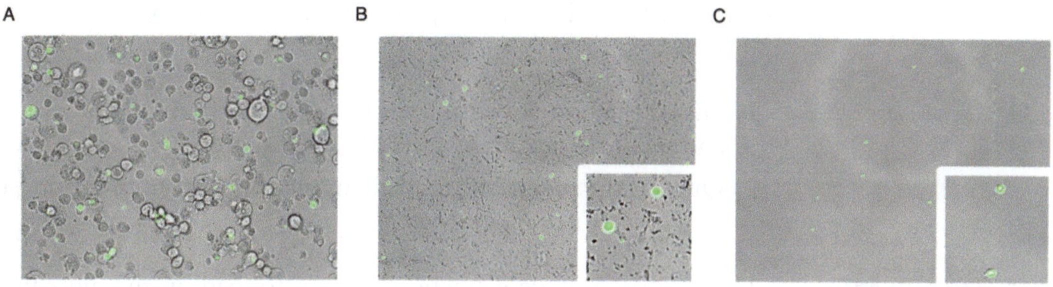

Fig. 2 Images of purification of EGFP-polyGA aggregates with a cell sorter. Microscopic observation of Neuro2a cells expressing EGFP-101GA (**a**), their pelleted lysates (**b**) and purified aggregates after sorting (**c**) presented by a merged image of phase contrast and the corresponding fluorescence images. Zoom-ins at the right corner of **b** and **c** images depict the efficient elimination of all contaminants spun down with polyGA aggregates

10. Lay 10 μL of the purified aggregate suspension on a glass slide and observe the fluorescent aggregates by fluorescence microscopy. *See* Fig. 2 for example images of purified poly(glycine–alanine) (polyGA) aggregates.
11. Transfer the remaining purified aggregate suspensions into microcentrifuge tubes, then pellet aggregates at 21,000 × *g* for 6 min at 4 °C. Discard the supernatant (*see* **Note 5**).
12. Dissolve the aggregate pellets in 20 μL neat formic acid.
13. Incubate the mixture in the Eppendorf thermal mixer (Cat. # 022670051) for 30 min at 37 °C with shaking at 1000 rpm.
14. Vortex the mixture three times for 20 s then sonicate for 1 min.
15. Repeat **steps 13** and **14** twice, for three rounds total of vortexing and sonication.
16. Adjust the pH of dissolved aggregates by adding 210 μL of 3 M Tris, and verify the solution has been neutralized to pH 7.0 using pH strips.
17. Determine the protein concentration using BCA assay according to the manufacturer's protocol.
18. Aliquot 100 μg of protein sample into a fresh tube, then add 1 M TEAB to a final concentration of 50 mM TEAB, pH 8 (*see* **Note 5**).
19. Add urea to a final concentration of 4 M urea and mix well by vortexing. Reduction and alkylation as described in **steps 16** and **17** at Subheading 3.1. Dilute the samples 1:3 (vol/vol) with 50 mM TEAB (pH 8.0) to reduce the urea concentration to 1 M. Proceed to Subheading 3.3 for preparation for mass spectrometry.

3.3 Peptide Preparation for Mass Spectrometry

This protocol is performed over the course of 3 days. All procedures are performed at either room temperature (20 °C) or higher (where indicated) to avoid precipitation of detergent or urea from solutions. For this purpose, preequilibrate the microcentrifuge to 20 °C.

3.3.1 Trypsin Digestion

1. Digest the proteins with trypsin at 1:40 (trypsin: protein ratio) and incubate overnight at 37 °C at 600 rpm.
2. Quench the digestion by acidifying to a final concentration of 1% (vol/vol) TFA (*see* **Note 9**).

3.3.2 Peptide Desalting

1. Use a solid-phase extraction column to load, wash, and elute the peptide mixtures, as recommended by the manufacturer, using 0.1% TFA for washing and 80% ACN containing 0.1% TFA for elution. The recommended eluting volume is 800 μL.
2. Lyophilize peptides by freeze drying overnight.
3. Resuspend freeze dried peptides in distilled water.
4. Quantify the peptide concentration by bicinchoninic acid assay according to the manufacturer's instructions. Adjust the peptide concentration to 0.25 μg/μL per sample.

3.3.3 Stable Isotope Dimethyl Labeling of Peptides

There are multiple options for isobaric/isotopic labeling. Here we describe the strategy to compare various samples based on dimethyl labeling. For larger sample comparisons, this workflow is also compatible with multiplexing using TMT or iTRAQ labeling systems (*see* **Note 1**).

1. For solubility experiments, prepare four separate sets of samples for dimethyl labeling: (a) Peptides from T fraction (control & treatment) for quantifying protein total abundance changes induced by treatment (Experiment 1 in Fig. 1a); (b) Peptides from P fraction (control & treatment) for quantifying protein abundance changes in the pellet fraction induced by treatment (Experiment 2 in Fig. 1b); (c) Peptides from S and T fractions from the control for quantifying the proportion of protein in supernatant over total (*pSup*) in the proteome (Experiment 3 in Fig. 1b); (d) Peptides from S and T fractions from the treatment for quantifying the proportion of protein in supernatant over total (*pSup*) in the proteome (Experiment 4 in Fig. 1b). For disease-associated aggregates, prepare peptides from control and disease conditions for dimethyl labeling (Experiment 5 in Fig. 1c).
2. Label peptides for each paired comparison with light and medium dimethyl labeling reagents respectively, following the manufacturer's instructions.
3. Mix equal amounts of labeled peptides for each comparison.
4. Perform bottom-up proteomic measurement as described previously [5]. Typically, 4 μL of dimethyl labeled peptide mixtures at a concentration of 0.25 μg/μL is injected, followed by

peptide separation via nano-HPLC and peptide identification and quantitation by MS/MS.

3.4 Proteomic Data Analysis

Two common software packages to analyze proteomic datasets are Proteome Discoverer and MaxQuant. Here, we describe in detail a workflow for the analysis of a dimethyl-labeled dataset using Proteome Discoverer 2.4 software. A detailed workflow for MaxQuant can be found elsewhere [13].

1. Download an organism-specific sequence database in FASTA format. We recommend using reference proteomes available from UniProtKB/Swiss-Prot, a manually annotated and non-redundant protein sequence database that can be accessed online (http://www.uniprot.org/downloads). For analyzing disease-associated aggregate experiments, append the database with the fluorescent tag sequence, which can be obtained from UniProt using the specific ID (e.g., P42212 for GFP).
2. Create a new study in Proteome Discoverer, load. RAW files and select "Dimethylation 2plex (C2H6, C2H2D4)" or "Dimethylation 3plex (C2H6, C2H2D4, 13C2D6)" as the quantitation method to be used in the study.
3. Create the appropriate processing and consensus workflows using the Workflow Editor. Figure 3 shows the recommended workflow we routinely use for identification and quantification of dimethyl-labeled peptides. We use the Mascot search engine for the database search and the precursor ion quantifier node for relative peptide quantitation.
4. Define the parameters for each node and start the search. In most cases, we use the default parameters for each node except for the setting shown in Fig. 2. The specific parameters include organism-specific UniProtKB/Swiss-Prot (e.g., *Mus musculus*), 20 ppm MS tolerance, 0.2 Da MS/MS tolerance, 2 missed cleavages, carbamidomethyl (C) as a fixed modification, oxidation (M) and acetylation (Protein N-term) as variable modifications. The false discovery rate (FDR) maximum is set to 1% at the peptide identification level and 1% at the protein identification level. Specify the numerator/denominator for ratio calculation.
5. In the analysis tab, check "As Batch." Launch the search by clicking "Run."
6. Upon completion of the search and quantification, combine ". pdResult" files for replicates belonging to same experiment into single ".pdResult" file to obtain quantitation ratios (i.e., L/H or M/L) for each identified protein. To do so, select the different result files in the "Analysis Results" tab you want to combine, then from "Reprocess" drop down list select "Use

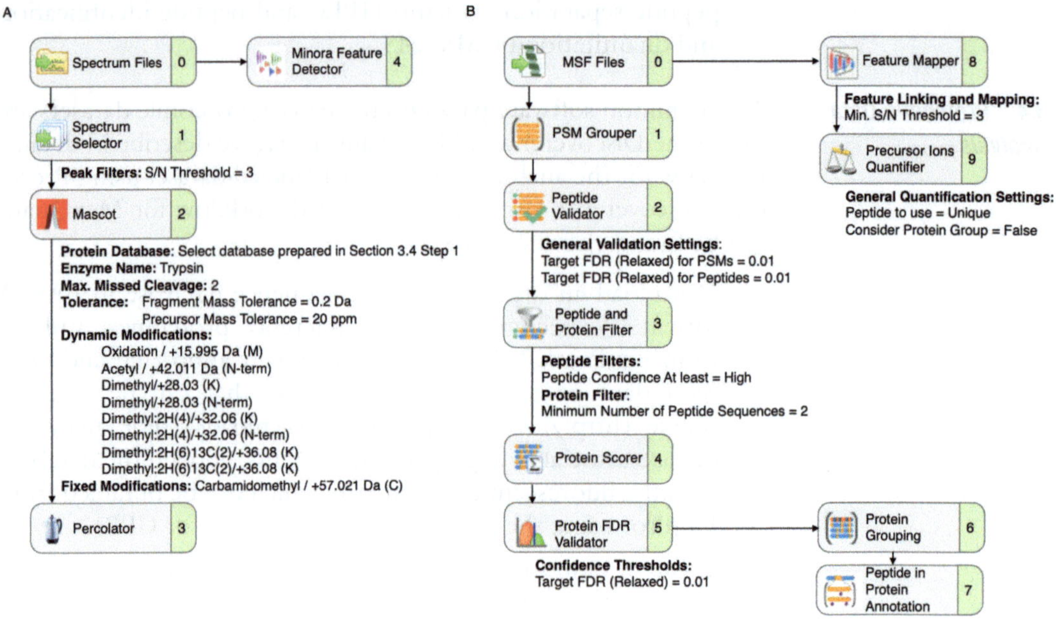

Fig. 3 Schematic representation of Proteome Discoverer 2.4 workflows for identification and quantitation of dimethyl-labeled peptides and the recommended parameters for each node. (**a**) Processing workflow. (**b**) Consensus workflow

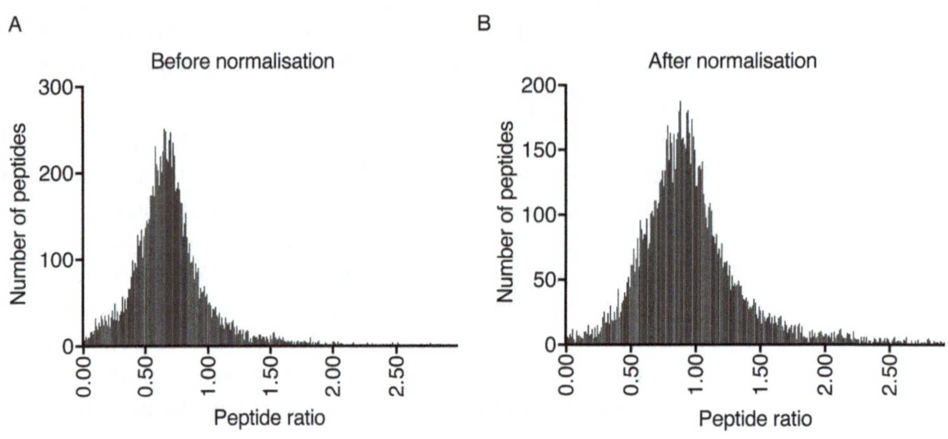

Fig. 4 Plots of peptide ratio distribution before (**a**) and after (**b**) normalization

Result to Create New (Multi) Consensus" and run the search again.

7. Open the final annotated result file and assign filters using the filter menu. Filter proteins for those containing at least two unique peptides in all biological replicates. Exclude the common contaminant "Keratin" and export the resultant protein and peptide groups to Microsoft Excel (*see* **Note 10**).

8. The generated datasets will contain peptide abundance ratios for each comparison according to the light/medium labels. Evaluate each comparison by plotting a histogram of the abundance ratios, which should approximate a Gaussian distribution (Fig. 4). Calculate the median value of peptide ratios for each comparison, which will then serve as the first normalization factor for the following analysis.
9. Calculate the second normalization factor by dividing the median value of peptide ratios by the ratio of total protein in the supernatant and total lysate samples as determined by the BCA assay in **step 14** at Subheading 3.1.
10. Apply the normalization factors according to each comparison type. (a) For the total proteome analysis, normalize the protein ratios between control and treatment by the normalization factor calculated in **step 8**. The median of normalized protein ratios should now be 1. (b) For the pellet proteome analysis, normalize protein ratios (treatment/control) with the corresponding normalization factor calculated in **step 8**. The median of normalized protein ratios should now be 1. (c) For the *pSup* analysis, apply the second normalization factor calculated in **step 9**.
11. Determine statistically significant changes using a t-test. In the case of total or pellet proteome analyses, perform a one-sample t-test for replicate measurements of each protein against the expected mean of 1. In the case of *pSup* comparisons, perform a two-sample t-test comparing replicate measures of individual proteins between the control and treatment samples. We recommend using the Perseus software package, which is capable of handling these tests with large datasets and for which

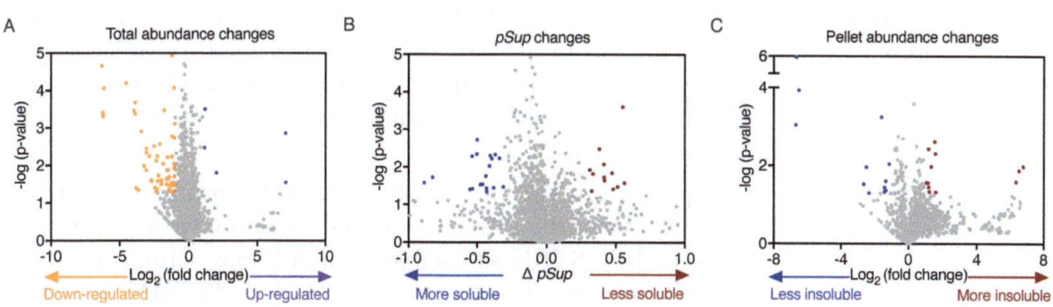

Fig. 5 Volcano plots of proteome abundance and solubility changes between control and treatment. (**a**) Total proteome abundance changes. Purple dots are proteins significantly twofold upregulated in the treatment group, while orange indicates downregulated. (**b**) Protein solubility changes measured by *pSup* changes. Red dots are those less soluble in the treatment group (Δ *pSup* > 0.3, *P*-value <0.05) whereas those more soluble are colored blue. (**c**) Protein solubility changes measured by pellet ratios. Red dots show proteins more insoluble in the treatment group (pellet ratio > 2, *P*-value <0.05) whereas blue indicates proteins more soluble. Gray-colored points indicate proteins below the threshold of significance

detailed instructions are available (http://coxdocs.org/doku.php?id=perseus:user:tutorials).

12. Calculate the logarithm transformed average total ratio and pellet ratio, and *pSup* for each protein from the biological replicates.
13. Results can then be visualized using a volcano plot, which shows the fold change and significance of changes in protein solubility or total protein abundance (Fig. 5). Biologically relevant protein solubility/abundance changes may be defined using suitable cutoffs. We recommend using a $\log_2$ (total/ratio) of 1/−1 (abundance increase/decrease), $\log_2$ (pellet ratio) of 1/−1 (more/less insoluble), Δ *pSup* ($pSup_{\text{Control}}$ − $pSup_{\text{Stress}}$) of 0.3/−0.3 (less/more soluble) and p-value <0.05.
14. Collect proteins which pass the filtering cutoff in **step 13** for further bioinformatic analysis.

3.5 Bioinformatics Analysis and Visualization Using Cytoscape v3.8.2

3.5.1 Gene Ontology (GO) Enrichment Analysis

1. To determine if any GO terms are enriched in the proteins of interest, perform a PANTHER overrepresentation test using the online tool [14].
2. Perform a Fisher's exact or binomial test with FDR correction for multiple testing against the PANTHER GO-Complete Biological Process, Cellular Component and Molecular Function datasets.
3. Filter GO terms to select those containing at least four proteins, fold enrichment >2 and $p < 0.05$. By default, terms are grouped hierarchically into families; we recommend selecting the least-redundant (left-most) terms wherever possible as the most specific and therefore functionally informative.

3.5.2 Protein–Protein Interaction Network Analysis Using Cytoscape

1. Prepare three Excel files containing Uniprot ID as the first column and $\log_2$(total ratio), $\log_2$(pellet ratio), or Δ *pSup* as the second column, respectively. We recommend the table only includes biologically relevant proteins according to the cutoffs described above.
2. Import a protein interaction network for proteins of interest derived from the total ratio analysis within Cytoscape [15] via the built-in STRING database search ("File > Import > Network from public database," then select "STRING: protein query"). Set the active interaction source parameters as Experiments, Databases, Coexpression neighborhood, Gene Fusion, and Cooccurrence. Set the minimum required interaction score (we recommend 0.4; medium confidence). Set the maximum additional interactors to 0. Optionally, toggle the option to retrieve gene ontology enrichment as an additional parameter table.

3. Proceed with the network import, then repeat for the remaining pellet ratio and *pSup* comparisons.
4. Add the prepared quantitative data for each comparison to Cytoscape using "File > Import > Table from file." During the import process, use the query name parameter as identifiers to map quantitative data to the imported interaction network.
5. The imported $\log_2$(total ratio), $\log_2$(pellet ratio), or Δ *pSup* values can then be used as node color attributes to visualize the changes across protein interacting partners.
6. Nodes can then be arranged either manually, based on GO terms of interest, or using the built-in tools available under the "Layout" menu. Optionally, various clustering applications are available for Cytoscape which group protein nodes based on various parameters, which are discussed in more detail elsewhere (https://www.ebi.ac.uk/sites/ebi.ac.uk/files/content.ebi.ac.uk/materials/2013/131104_Berlin/bio lnetworksanalysis_tutorial.pdf).
7. The finalized protein interaction network can then be exported into vector graphics or bitmap formats.

4 Notes

1. The choice of isobaric labeling reagents depends on the number of samples to be processed and requires weighing the cost of reagents against mass spectrometry machine time. For example, TMT reagents are expensive but significantly reduce mass spectrometry machine time by facilitating simultaneous quantitation of up to 16 samples. In contrast dimethyl reagents are less expensive; however, they can only label up to 3 conditions at a time. In the case of this workflow, for small numbers of comparisons such as described here between a single control and treatment group, we have found dimethyl labeling to be sufficient for robust and reproducible quantitation without requiring unreasonable instrument time.
2. EDTA interferes with DNase activity, so it needs to be added after lysis.
3. The volumes used in the above steps need to be done consistently across all treatments within experiments for data that will be compared because absolute solubility will be dependent on volumes.
4. After heating, white pellets tend to appear in T samples. Immediately transfer the supernatant to a new tube before the white pellets dissolve when it cools down. Otherwise, the sample will become very sticky.

5. At this point, samples may be stored at −80 °C after snap-freezing in liquid N_2.
6. After methanol/chloroform precipitation, the time for air drying of pellets needs to be carefully controlled and should be limited to the disappearance of the last liquid droplet. If left to dry too long, the protein pellet will be very difficult to redissolve in the urea buffer.
7. The disease-associated protein must be fluorescently labeled by fusion to a fluorescent protein. The Emerald form of GFP containing the A206K, S72A, N149K, M153T, and I167T mutations [16] is monomeric and bright, hence is well suited for sorting aggregates in mammalian cell lysates using FACS. A preliminary PulSA experiment is recommended to estimate the required number of cells to collect one million aggregates. For transient transfection of a relatively abundant protein (e.g., using a cytomegalovirus promotor to drive high yielding expression), we recommend 12×10^6 cells as a starting point for Neuro2A cells to yield one million aggregates from flow cytometry sorting. Higher levels of expression can be obtained with HEK AD293 cells.
8. This spin condition has resulted in recovery of 99% of aggregates in lysates without forming a very compact pellet that is hard to resuspend. Spinning at $500 \times g$ results in a recovery of around 95% of aggregates in lysates.
9. Using pH strips to check the pH of the solution is <3. If not, it indicates some errors in previous steps. Prepare new samples from the beginning.
10. Export the final annotated result to PepXML and ProtXML which will be needed for final submission to Proteome Xchange.

References

1. Banani SF, Lee HO, Hyman AA, Rosen MK (2017) Biomolecular condensates: organizers of cellular biochemistry. Nat Rev Mol Cell Biol 18(5):285–298. https://doi.org/10.1038/nrm.2017.7
2. Ross CA, Poirier MA (2004) Protein aggregation and neurodegenerative disease. Nat Med 10(Suppl):S10–S17. https://doi.org/10.1038/nm1066
3. Wallace EW, Kear-Scott JL, Pilipenko EV, Schwartz MH, Laskowski PR, Rojek AE, Katanski CD, Riback JA, Dion MF, Franks AM, Airoldi EM, Pan T, Budnik BA, Drummond DA (2015) Reversible, specific, active aggregates of endogenous proteins assemble upon heat stress. Cell 162(6):1286–1298. https://doi.org/10.1016/j.cell.2015.08.041
4. Huse M (2009) The t-cell-receptor signaling network. J Cell Sci 122(Pt 9):1269–1273. https://doi.org/10.1242/jcs.042762
5. Sui X, Pires DEV, Ormsby AR, Cox D, Nie S, Vecchi G, Vendruscolo M, Ascher DB, Reid GE, Hatters DM (2020) Widespread remodeling of proteome solubility in response to different protein homeostasis stresses. Proc Natl Acad Sci U S A 117(5):2422–2431. https://doi.org/10.1073/pnas.1912897117
6. Radwan M, Lilley JD, Ang CS, Reid GE, Hatters DM (2020) Immiscible inclusion bodies formed by polyglutamine and poly(glycine-alanine) are enriched with distinct proteomes

but converge in proteins that are risk factors for disease and involved in protein degradation. PLoS One 15(8):e0233247. https://doi.org/10.1371/journal.pone.0233247

7. Radwan M, Ang CS, Ormsby AR, Cox D, Daly JC, Reid GE, Hatters DM (2020) Arginine in c9orf72 dipolypeptides mediates promiscuous proteome binding and multiple modes of toxicity. Mol Cell Proteomics 19(4):640–654. https://doi.org/10.1074/mcp.RA119.001888
8. Cox D, Raeburn C, Sui X, Hatters DM (2020) Protein aggregation in cell biology: an aggregomics perspective of health and disease. Semin Cell Dev Biol 99:40–54. https://doi.org/10.1016/j.semcdb.2018.05.003
9. Kim YE, Hosp F, Frottin F, Ge H, Mann M, Hayer-Hartl M, Hartl FU (2016) Soluble oligomers of polyq-expanded huntingtin target a multiplicity of key cellular factors. Mol Cell 63(6):951–964. https://doi.org/10.1016/j.molcel.2016.07.022
10. Ramdzan YM, Polling S, Chia CP, Ng IH, Ormsby AR, Croft NP, Purcell AW, Bogoyevitch MA, Ng DC, Gleeson PA, Hatters DM (2012) Tracking protein aggregation and mislocalization in cells with flow cytometry. Nat Methods 9(5):467–470. https://doi.org/10.1038/nmeth.1930
11. Fonslow BR, Yates JR III (2012) Proteolytic digestion methods for shotgun proteomics. In: Comprehensive sampling and sample preparation. https://doi.org/10.1016/B978-0-12-381373-2.10082-1
12. Ramdzan YM, Wood R, Hatters DM (2013) Pulse shape analysis (pulsa) to track protein translocalization in cells by flow cytometry: applications for polyglutamine aggregation. Methods Mol Biol 1017:85–93. https://doi.org/10.1007/978-1-62703-438-8_6
13. Tyanova S, Temu T, Cox J (2016) The maxquant computational platform for mass spectrometry-based shotgun proteomics. Nat Protoc 11(12):2301–2319. https://doi.org/10.1038/nprot.2016.136
14. Mi H, Muruganujan A, Casagrande JT, Thomas PD (2013) Large-scale gene function analysis with the panther classification system. Nat Protoc 8(8):1551–1566. https://doi.org/10.1038/nprot.2013.092
15. Shannon P, Markiel A, Ozier O, Baliga NS, Wang JT, Ramage D, Amin N, Schwikowski B, Ideker T (2003) Cytoscape: a software environment for integrated models of biomolecular interaction networks. Genome Res 13(11):2498–2504. https://doi.org/10.1101/gr.1239303
16. Tsien RY (1998) The green fluorescent protein. Annu Rev Biochem 67:509–544. https://doi.org/10.1146/annurev.biochem.67.1.509

Chapter 17

Analyzing the Composition and Organization of Ribonucleoprotein Complexes by APEX-Seq

Alejandro Padrón and Nicholas Ingolia

Abstract

Diverse protein–RNA complexes assemble in cells, and their composition and localization regulate the fate of mRNAs. Here, we detail APEX-Seq, an experimental strategy to capture protein–RNA interactions and profile their sub-cellular organization by in vivo proximity labeling and high-throughput sequencing. APEX-Seq relies on direct proximity labeling of RNAs by the peroxidase enzyme APEX2, which can be targeted to specific sites in the cell or fused to proteins of interest. Direct RNA proximity labeling promises new insights into the dynamic behavior of RNA, addressing length scales beyond direct physical contact but too short for microscopy. APEX-Seq should be widely applicable to diverse biological questions and in many cell types, enabling comprehensive studies of the spatial transcriptome and its dynamics over time.

Key words APEX-seq, Spatial transcriptomics, Ribonucleoprotein complexes, Protein–RNA interactions, Proximity labeling, APEX2

1 Introduction

RNA localization influences almost every aspect of post-transcriptional regulation, ranging from RNA synthesis through processing, transport, and translation, and ultimately to decay. Transcripts are also accompanied by numerous RNA binding proteins that influence the fate of RNAs. In order to understand RNA fate decisions, we therefore require tools that comprehensively capture RNAs and reveal their localization dynamics relative to particular proteins or overall sub-cellular organization.

Optical microscopy provides the classic approach for monitoring sub-cellular RNA localization, although historically it has been best suited for detailed analysis of a few specific transcripts of interest [1, 2]. More recently, various techniques have been developed for direct imaging of thousands of RNAs simultaneously by relying on a barcoding strategy [3, 4]. Although powerful, these techniques require cell fixation and permeabilization, which can

Daniel Matejů and Jeffrey A. Chao (eds.), *The Integrated Stress Response: Methods and Protocols*, Methods in Molecular Biology, vol. 2428, https://doi.org/10.1007/978-1-0716-1975-9_17, affect the localization of cellular components [5, 6]. The spatial resolution of light microscopy also limits our ability to determine how RNAs localize in relation to various cellular structures. Moreover, transcriptome-wide optical imaging is challenging to implement and requires specialized instrumentation not readily available in typical molecular biology laboratories [7].

Recent years have seen the emergence of in vivo proximity labeling strategies to map out cellular organization of proteins and nucleic acids. Techniques to directly label RNAs located near a query protein, such as TRIBE (targets of RNA-binding proteins identified by editing) have proven valuable [8]. However, direct RNA labeling depends on direct enzymatic modification, which operates at short defined distances from the query protein, with steric and geometric constraints, and suffers from other technical limitations. Indirect proximity labeling strategies, exemplified by BioID and APEX, rely on an enzyme that creates a diffusible, reactive intermediate [9]. Both BioID and APEX protein labeling have provided new biological insights in a range of settings, including membraneless organelles that form by phase separation [10–13].

We and others have recently shown that APEX labeling extends to RNA, allowing us to implement an APEX-Seq strategy for studying RNA localization [7, 12]. Our approach relies on the APEX2 enzyme [14], a peroxidase enzyme that functions in the reducing environment within cells. APEX2 creates a reactive radical derivative of biotin that will modify neighboring proteins and RNAs wherever the enzyme is localized in the cell [12]. Here, we present the APEX-Seq protocol in detail as a reference for those interested in implementing it in their own laboratories.

2 Materials

2.1 RNA Proximity Labeling

1. DMEM with 10% fetal bovine serum (FBS).
2. Tetracycline.
3. Biotin-tyramide.
4. Hydrogen peroxide.
5. DPBS.
6. 2× Quenching Buffer: 10 mM Trolox, 20 mM sodium ascorbate in DPBS
7. 1× Quenching Buffer: 5 mM Trolox, 10 mM sodium ascorbate in DPBS
8. cell scraper.

2.2 Total RNA Extraction

1. TRIzol.
2. Chloroform.
3. 3 M sodium acetate, pH 5.5 (Invitrogen)
4. Isopropanol.
5. GlycoBlue (optional).
6. Ethanol.
7. DNase I (NEB).
8. DNase I Reaction Buffer (NEB).
9. Multitube vortex mixer.
10. Nuclease-free water.

2.3 Biotinylated RNA Enrichment

1. MyOne C1 Streptavidin Dynabeads (ThermoFisher).
2. 50× Denhardt's Solution (Thermo Fisher Scientific)
3. 10× PBS
4. Buffer 1: 1 mM MgCl2, 0.5% sodium deoxycholate in PBS.
5. Blocking Buffer: 5× Denhardt's Solution, 150 μg/mL Poly I: C in Buffer 1.
6. Buffer 2: 6 M urea pH 8, 0.1% SDS in PBS.
7. Buffer 3: 2% (w/v) SDS in PBS.
8. Buffer 4: 750 mM NaCl, 0.5% (w/v) sodium deoxycholate, 0.1% (w/v) SDS, in PBS.
9. Buffer 5: 150 mM NaCl, 0.5% (w/v) sodium deoxycholate, 0.1% (w/v) SDS, in PBS.
10. Optional: Direct-zol kit.

2.4 RNA Purification and Fragmentation

1. RNAClean XP Beads (Beckman Coulter).
2. RNase T1/A (Thermo Fisher Scientific).
3. 1 M Tris pH 7.5 (Invitrogen)
4. 500 mM EDTA, pH 8 (Invitrogen)
5. Agilent 2200 TapeStation.
6. High Sensitivity RNA ScreenTape (Agilent).
7. High Sensitivity RNA ScreenTape Sample Buffer (Agilent).
8. High Sensitivity RNA ScreenTape Ladder (Agilent).

2.5 RNA Sequencing

1. Optional: NEBNext rRNA depletion kit (Human/Mouse/Rat).
2. NEBNext Ultra II RNA Library Prep Kit for Illumina.
3. NEBNext® Multiplex Oligos for Illumina.
4. AMPure XP Beads.

3 Methods

3.1 Cell Growth and Preparation

1. Grow cells expressing APEX fusion construct in 15 cm dishes. For Flp-In T-Rex cells, which express single-copy integrations of APEX fusions under tetracycline inducible CMV promoters, induce APEX fusion expression for 3 days using 1 μg/mL of tetracycline (*see* **Note 1**).
2. In order to start an APEX labeling experiment, add 20 mL of fresh, prewarmed cell culture media containing biotin-tyramide (500 μM) to each 15 cm dish of cells and return to the incubator for 30 min.
3. Prepare a 2× quenching solution (10 mM Trolox and 20 mM sodium ascorbate) as well as a 1× quenching solution (5 mM Trolox and 10 mM sodium ascorbate) in DPBS (*see* **Note 2**).

3.2 Labeling Reaction

1. The following steps should be done in immediate succession (*see* **Note 3**).
2. Add hydrogen peroxide to a 15 cm dish of cells using a pipette, to a final concentration of 1 mM. Mix the contents of the 15 cm dish gently, but quickly (*see* **Note 4**).
3. Immediately quench the reaction by adding 20 mL of 2× quenching solution to the dish.
4. Aspirate the solution in the dish and perform two washes with 20 mL of 1× quenching solution.
5. Aspirate the quenching solution and add 800 μL of TRIzol to the dish.
6. Tilt the dish and use a cell scraper to harvest cells in the TRIzol solution. Collect the TRIzol lysate with a pipette and transfer it to a 1.5 mL eppendorf tube. Place the samples on ice.
7. Vortex the samples thoroughly using a VWR Signature Pulsing Vortex Mixer, making sure that any cell clumps are disrupted before proceeding.

3.3 RNA Extraction

1. Add 200 μL of chloroform to the TRIzol sample.
2. Vortex for 10–15 s and place the sample at RT for 5 min.
3. Centrifuge the solution at 20,000 × *g* for 20 min in a 4 °C chilled centrifuge and gently collect the aqueous phase of the solution, taking care to avoid disrupting the interface between the aqueous and organic phases.
4. Transfer ~500 μL of aqueous phase to a new eppendorf tube. Add 1/10 volume of 3 M sodium acetate (pH 5.5) to the aqueous phase, followed by 1 volume of isopropanol (*see* **Note 5**).
5. Place the samples at −80 °C for 1 h to overnight.

6. Warm the samples on ice and centrifuge at 20,000 × *g* for 20 min in a 4 °C chilled centrifuge. During the spin, prepare a 70% EtOH solution and chill on ice.
7. Wash in 1 mL ice-cold 70% ethanol and vortex briefly.
8. Spin the samples at 20,000 × *g* for 5 min in a 4 °C chilled centrifuge.
9. Repeat the 1 mL wash in ice-cold 70% ethanol once more.
10. Remove the final wash solution. Spin the samples briefly on a tabletop centrifuge and collect any excess ethanol wash that may be present with a P10 pipette.
11. Uncap the samples and place them on a tilted rack to let any residual EtOH evaporate, being careful not to let the pellets dry out totally (roughly 5 min).
12. Resuspend the pellets in 30 μL RNase-free water (*see* **Note 6**).

3.4 DNase I Digest

1. Prepare a DNase I (NEB) digestion reaction (*see* **Note 7**): 100 μg of RNA, 10 μL NEB DNase I Reaction Buffer (10×), 4 μL NEB DNase I (RNase-free), and bring the reaction up to 100 μL using nuclease-free water.
2. Incubate reactions for 1.5 h at 37 °C.

3.5 RNA Purification

1. Make a fresh 70% EtOH solution.
2. Shake the bottle of Agencourt RNAClean XP vigorously to resuspend any magnetic particles that may have settled.
3. Add 180 μL of RNAClean XP to 100 μL of DNase digestion reaction (*see* **Note 8**). Mix thoroughly by pipetting up and down 10 times.
4. Incubate the mixture of DNase-digested RNA samples and RNAClean XP for 5 min at room temperature.
5. Place the mixture on a magnetic stand and leave in place for 5–10 min to separate beads from solution. The separation time depends on the size of the reaction; wait for the solution to clear before proceeding.
6. Leave the microfuge tubes in place on the magnetic stand and slowly aspirate the cleared solution. Discard the solution. *Do not disrupt the separated magnetic beads.*
7. Add 500–1000 μL of 70% ethanol to the microfuge tube, pipetting onto the opposite side of the tube from the magnetic beads. Incubate for 30 s at room temperature. Aspirate and discard the ethanol wash without disturbing the beads.
8. Repeat the wash with 500–1000 μL of 70% ethanol for a total of three washes (*see* **Note 9**).

9. Let the reaction tube air-dry for 10 min on the magnetic stand with the cap open (*see* **Note 10**).
10. Remove the microfuge tube from the magnetic rack and resuspend the beads in 30 μL of nuclease-free water to elute the purified RNA.
11. Return the microfuge tube to the magnetic rack and allow the beads to separate from the eluted sample. Recover the eluted sample.

3.6 RNA Fragmentation

1. Prepare a 10× RNA digestion buffer (500 mM Tris, pH 7.5, 10 mM EDTA in RNase-free water) (*see* **Note 11**).
2. Set up reaction in PCR strip tubes, on ice.
3. Dilute RNase T1/A to 10^{-5} U/μL.
4. Determine the RNA concentration in the samples using a Nanodrop and prepare a reaction with a final RNA concentration of 1 μg/μL of RNA. For 100 μg of RNA, prepare the following reaction: 100 μg of sample, 10 μL of 10X RNA digestion buffer, 10 μL diluted RNase T1/A, and RNase-free water up to 100 μL. Add RNase T1/A to reaction last (*see* **Note 12**).
5. Briefly flick PCR strip tubes to mix the reactions and spin down again to collect the liquid at the bottom of the tubes.
6. Incubate in thermocycler for 10 min at 37 °C.
7. Immediately place the samples back on ice and quickly transfer reaction into 400 μL TRIzol (*see* **Note 13**).
8. Perform a Direct-Zol clean up following the manufacturer's instructions. Use IIIC columns and elute in 60 μL of RNase-free water (*see* **Note 14**).
9. After the RNA purification, assess the quality of the RNA by analyzing a small sample on a High Sensitivity RNA ScreenTape with an Agilent TapeStation. The RNA integrity number (RIN) reported by the TapeStation analysis software will be used to determine the fragmentation time needed in later steps of the protocol.
10. Set aside 20 ng of RNA for use in generating an "input" library.

3.7 Biotinylated RNA Enrichment

1. Resuspend MyOne C1 Streptavidin Dynabeads by shaking or vortexing the bottle. Transfer 10 μL of bead slurry to a nonstick microfuge tube.
2. Place the tube of beads on a magnetic stand and allow the beads to collect on the side of the tube for 1 min. Remove the storage solution.
3. Remove the tube of beads from the magnetic stand and resuspend the beads in 1.0 mL Buffer 1.

4. Return the beads to the magnetic rack and allow the beads to collect on the side of the tube again for 1 min. Remove the wash buffer.
5. Repeat for a total of three washes.
6. Remove the tube of beads from the magnetic strand and resuspend the beads in 1.0 mL Blocking Buffer. Incubate the beads with Blocking Buffer for 30 min at room temperature on a nutator.
7. Place the tube of beads on a magnetic stand and allow the beads to collect on the side of the tube for 1 min. Remove the Blocking Buffer.
8. Resuspend the beads with 940 μL of fresh Blocking Buffer, and add 60 μL of RNA sample.
9. Incubate samples at room temperature for 1 h on a nutator.
10. Place the tube of beads on a magnetic stand and allow the beads to collect on the side of the tube for 1 min. Remove the unbound RNA in Blocking Buffer.
11. Remove the tube from the magnetic stand, resuspend the beads in 1.0 mL Buffer 2, return to the magnetic stand, allow beads to collect on the side for 1 min, and remove the wash.
12. Repeat a second wash of Buffer 2.
13. Wash once as described above, with Buffer 3.
14. Wash once as described above, with Buffer 4.
15. Wash once as described above, with Buffer 5.
16. Elute RNA from streptavidin beads by resuspending the beads in 300 μL of TRIzol (*see* **Note 15**).
17. Perform a Direct-zol clean up using IC columns by adding 300 μL of 100% EtOH, and following manufacturer recommendation.
18. Resuspend purified RNA in 6 μL of RNase-free water.

3.8 Fragmentation and Priming Reaction

1. The steps below describe the use of an NEBNext Ultra II Directional RNA Library Prep Kit according to the manufacturer's instructions.
2. Prepare one fragmentation reaction for each biotinylated RNA enrichment sample, along with one reaction for each matched input sample, by combining the following in a PCR tube on ice: 5 μL of RNA sample, 4 μL of NEBNext First Strand Synthesis Buffer, and 1 μL of Random Primers for a total of 10 μL.
3. Mix thoroughly by pipetting up and down.
4. Incubate the sample in a thermocycler at 94 °C for 7 min. *Note these fragmentation conditions are suitable for partially*

degraded RNA with RIN values between 2 and 7). Consult the NEB protocol for RIN values outside this range. Highly degraded samples (RIN = 1 to 2) do not require additional fragmentation.

5. Transfer the PCR tube strip to ice immediately after the fragmentation finishes.

3.9 First and Second Strand cDNA Synthesis

1. Prepare the cDNA synthesis reaction by adding the following to each fragmentation reaction, on ice: 10 μL of fragmented and primed RNA sample, 8 μL of Strand Specificity Reagent, and 2 μL of First Strand Synthesis Enzyme mix for a total of 20 μL.
2. Mix thoroughly by pipetting up and down.
3. Preheat a thermal cycler with a heated lid at ≥80 °C.
4. Transfer the PCR tubes to the preheated thermocycler and incubate 10 min at 25 °C, 15 min at 42 °C, and 15 min at 70 °C. Hold at 4 °C after the reaction finishes.
5. Remove the PCR tubes from the thermocycler and place on ice.
6. Add to the PCR tubes on ice: 20 μL of First Strand Synthesis reaction, 8 μL of 10× NEBNext Second Strand Synthesis Reaction Buffer with dUTP mix, 4 μL of Second Strand Synthesis Enzyme Mix, and 48 μL of nuclease-free water for a total of 80 μL.
7. With the samples on ice, mix thoroughly by pipetting the reaction up and down.
8. Chill a thermocycler to 16 °C with the heated lid off.
9. Transfer the PCR tubes to the chilled thermocycler and incubate 1 h at 16 °C.

3.10 Purification of Double-Stranded cDNA

1. Resuspend AMpure XP beads by vortexing or vigorously shaking.
2. Add 144 μL of AMpure beads to 80 μL of cDNA synthesis reaction in PCR tubes and mix thoroughly by pipetting up and down at least ten times.
3. Incubate PCR tubes 5 min at room temperature. Prepare a fresh 80% ethanol solution during this incubation.
4. Spin down the PCR tubes to gather the samples at the bottom of the tubes, place the PCR tubes on a magnetic stand, and allow the beads to collect on the side of the tube.
5. Leave the microfuge tubes in place on the magnetic stand while slowly aspirating the cleared solution. Discard the solution. *Do not disrupt the separated magnetic beads.*
6. Add 200 μL of 80% ethanol to the microfuge tube, pipetting onto the opposite side of the tube from the magnetic beads.

Incubate for 30 s at room temperature. Aspirate and discard the ethanol wash without disturbing the beads.

7. Repeat the wash with 200 μL of 80% ethanol for a total of two washes.
8. Let the reaction tube air-dry for 5 min on the magnetic stand with the cap open (*see* **Note 16**).
9. Remove the PCR tubes from the magnetic rack and resuspend the beads in 53 μL of 0.1× TE to elute the purified DNA. Incubate 2 min at room temperature.
10. Return the PCR tube to the magnetic rack and allow the beads to separate from the eluted sample.
11. Remove 50 μL of the supernatant and transfer to a clean nuclease-free PCR tube.
12. If you need to stop at this step in the protocol, samples can be stored in −20 °C.

3.11 End Processing of cDNA Library and Adaptor Ligation

1. Assemble the end repair reaction on ice by adding the following components to the second strand synthesis product: 50 μL of Second Strand Synthesis Product, 7 μL of NEBNext Ultra II End Prep Reaction Buffer, and 3 μL of NEBNext Ultra II End Prep Enzyme Mix for a total of 60 μL.
2. Pipette the entire volume up and down at least 10 times to mix thoroughly.
3. Incubate the reaction in a thermocycler with a heated lid: 30 min at 20 °C, 30 min at 65 °C, hold at 4 °C.
4. During the incubation, dilute NEBNext Adapter in Adapter Dilution buffer and keep on ice. For an RNA input of 100–10 ng, such as a typical "input" sample, dilute 25-fold in Adaptor Dilution Buffer. For a total RNA input of 5 ng, such as a typical purified APEX-Seq sample, dilute Adaptor 200-fold in Adaptor Dilution Buffer.
5. Remove the PCR tubes from the thermocycler and place on ice. Then add the following: 2.5 μL of diluted adapter, 1 μL of NEBNext Ligation Enhancer, and 30 μL of NEBNext Ultra II Ligation Master Mix for a total of 93.5 μL.
6. Mix well by pipetting at least 10 times (*see* **Note 17**).
7. Return PCR tubes to a thermocycler and incubate 15 min at 20 °C.
8. Add 3 μL of USER Enzyme and mix well.
9. Return PCR tubes to a thermocycler and incubate 15 min at 37 °C with a heated lid.

3.12 Ligation Product Purification

1. Resuspend AMpure XP beads by vortexing or vigorously shaking.

2. Add 87 μL of resuspended AMpure XP beads to the ligation reaction in PCR tubes and mix thoroughly by pipetting up and down at least 10 times.
3. Incubate PCR tubes 10 min at room temperature.
4. Briefly spin down the PCR tubes to gather the samples at the bottom of the tubes, place the PCR tubes on a magnetic stand, and allow the beads to collect on the side of the tube (~5 min).
5. During the incubation prepare a fresh 80% ethanol solution.
6. Leave the microfuge tubes in place on the magnetic stand while slowly aspirating the cleared solution. Discard the solution. *Do not disrupt the separated magnetic beads.*
7. Add 200 μL of 80% ethanol to the microfuge tube, pipetting onto the opposite side of the tube from the magnetic beads. Incubate for 30 s at room temperature. Aspirate and discard the ethanol wash without disturbing the beads.
8. Repeat the wash with 200 μL of 80% ethanol for a total of two washes.
9. Completely remove the residual ethanol, and air-dry beads for 5 min on the magnetic stand with the cap open (*see* **Note 16**).
10. Remove the PCR tubes from the magnetic rack and resuspend the beads in 17 μL of 0.1× TE to elute the purified DNA. Mix well by pipetting up and down.
11. Briefly spin the tube and incubate for 2 min at room temperature.
12. Return the PCR tube to the magnetic rack and allow the beads to separate from the eluted sample. Recover 15 μL of eluted sample into a new PCR tube.
13. If you need to stop at this point in the protocol samples can be stored at −20 °C.

3.13 PCR Amplification

1. Choose distinct NEBNext i7 and NEBNext i5 primers from the NEBNext Multiplex Oligos for each sample and record which primers are chosen for which samples.
2. Add to each purified ligation reaction the following (*see* **Note 18**): 25 μL of NEBNext Ultra II Q5 Master Mix, 5 μL of NEBNext i5 Primer (10 μM), and 5 μL of NEBNext i7 Primer (10 μM) for a total of 50 μL.
3. Mix by pipetting and spin briefly to collect the reaction at the bottom of the PCR tube.
4. Place the PCR tubes in a thermocycler with a heated lid and carry out the PCR amplification using the following conditions (*see* **Note 19**): 30 s at 98 °C, followed by X cycles of 10 s at 98 °C followed by 75 s at 65 °C, and a final extension of 5 min at 65 °C. Hold at 4 °C.

5. Purify PCR product as described for the ligation product, *except* add only 45 μL of AMpure XP beads to 50 μL of PCR, and elute with 23 μL of 0.1× TE to recover 20 μL of PCR product.
6. Assess the library using a High Sensitivity DNA ScreenTape on an Agilent TapeStation.

4 Notes

1. It is important to assess expression levels of the APEX fusion and maintain consistent fusion protein levels.
2. We recommend aliquoting 20 mL of quenching solution into 50 mL falcon tubes. That way, the user can pour the 20 mL quenching solution at the end of the 1 min incubation onto the plate in a rapid manner.
3. Importantly, we recommend doing no more than a few plates at a time in order to maintain proper timing of these rapid, time-sensitive steps.
4. The ascorbate peroxidase-mediated labeling reaction is complete within 15–30 s. However, typical in vivo proximity labeling reactions are 1 min reactions.
5. Option to add 1.5 μL of GlycoBlue in order to more easily visualize the RNA pellet.
6. The pellet may require some time to dissolve in RNase-free water. If you notice that the pellet does not dissolve within a few minutes, incubate the samples in 4 °C until the pellet dissolves.
7. More than one DNase I digest reaction per sample can be set up in order to attain 100 μg of RNA going into the RNA fragmentation reaction. The DNase I digest reactions should be pooled during the subsequent RNA clean up step.
8. To determine the volume of Agencourt RNAClean XP for a given reaction can be derived from the following equation: (Volume of Agencourt RNAClean XP per reaction) = 1.8 × (Reaction Volume).
9. The volume of EtOH required for washing will depend on the size of the original reaction. The wash solution must completely cover the entire bead mass on the side of the tube.
10. Air-dry the tubes until the last visible traces of EtOH evaporate. Over drying the sample will result in lower recovery.
11. Buffer can be stored in −20 °C and used in subsequent reactions.

12. We prefer to add RNase T1/A to the caps of the PCR strips and then spin briefly to ensure all tubes receive RNase T1/A at the same time.
13. We prefer to add 100 μL of TRIzol to each sample using a multichannel pipette. This way, the reaction is quenched faster. The 200 μL solution is then collected and placed into a 1.5 mL eppendorf tube containing 300 μL of TRIzol for a total of 500 μL.
14. Alternatively, recover RNA by phase separation. Add an additional 500 μL of TRIzol to the 500 μL solution (100 μL sample and 400 μL TRIzol), followed by 200 μL chloroform and purify the sample as described previously. Resuspend in 60 μL of RNase-free water. *Option to add 1.5 μL of GlycoBlue in order to clearly distinguish the pellet.*
15. Purify the sample using either phase separation and precipitation, as described above, or Direct-zol kit, following the manufacturer's instructions.
16. Do not over-dry the beads. This may result in lower recovery of DNA. Elute the samples when the beads are still dark brown and glossy looking. Do not let the beads turn lighter brown and start to crack.
17. The NEBNext Ultra II Ligation Master Mix is viscous, and it is important to mix the ligation reaction well in order to ensure efficient ligation.
18. It is important to select a number of PCR cycles that will produce enough library for sequencing, but avoids biases that can arise when PCR approaches saturation. A reaction with half (7.5 μL) of the Adaptor Ligated DNA volume can be performed if the samples are precious. Make up the remaining reaction volume with 7.5 μL of nuclease-free water.
19. Determine the number of cycles to perform based on the amount of RNA going into the library prep stage of the protocol to avoid overamplification: 5 ng, 15–16 cycles; 10 ng, 14–15 cycles; 100 ng, 11–12 cycles; 1000 ng, 7–8 cycles.

References

1. Bertrand E, Chartrand P, Schaefer M et al (1998) Localization of ASH1 mRNA particles in living yeast. Mol Cell 2:437–445
2. Femino AM, Fay FS, Fogarty K, Singer RH (1998) Visualization of single RNA transcripts in situ. Science 280:585–590
3. Chen KH, Boettiger AN, Moffitt JR et al (2015) RNA imaging. Spatially resolved, highly multiplexed RNA profiling in single cells. Science 348:aaa6090
4. Shah S, Lubeck E, Zhou W, Cai L (2016) In situ transcription profiling of single cells reveals spatial organization of cells in the mouse hippocampus. Neuron 92:342–357
5. Fox CH, Johnson FB, Whiting J, Roller PP (1985) Formaldehyde fixation. J Histochem Cytochem 33:845–853
6. Schnell U, Dijk F, Sjollema KA, Giepmans BNG (2012) Immunolabeling artifacts and

the need for live-cell imaging. Nat Methods 9: 152–158
7. Fazal FM, Han S, Parker KR et al (2019) Atlas of subcellular RNA localization revealed by APEX-Seq. Cell 178:473–490.e26
8. McMahon AC, Rahman R, Jin H et al (2016) TRIBE: hijacking an RNA-editing enzyme to identify cell-specific targets of RNA-binding proteins. Cell 165:742–753
9. Qin W, Cho KF, Cavanagh PE, Ting AY (2021) Deciphering molecular interactions by proximity labeling. Nat Methods 18(2): 133–143. https://doi.org/10.1038/s41592-020-01010-5
10. Marmor-Kollet H, Siany A, Kedersha N et al (2020) Spatiotemporal proteomic analysis of stress granule disassembly using APEX reveals regulation by SUMOylation and links to ALS pathogenesis. Mol Cell 80:876–891.e6
11. Markmiller S, Soltanieh S, Server KL et al (2018) Context-dependent and disease-specific diversity in protein interactions within stress granules. Cell 172:590–604.e13
12. Padrón A, Iwasaki S, Ingolia NT (2019) Proximity RNA labeling by APEX-Seq reveals the organization of translation initiation complexes and repressive RNA granules. Mol Cell 75: .875–887.e5
13. Youn J-Y, Dunham WH, Hong SJ et al (2018) High-density proximity mapping reveals the subcellular organization of mRNA-associated granules and bodies. Mol Cell 69:517–532.e11
14. Rhee H-W, Zou P, Udeshi ND et al (2013) Proteomic mapping of mitochondria in living cells via spatially restricted enzymatic tagging. Science 339:1328–1331

Chapter 18

Differential Analysis of the Nuclear and the Cytoplasmic RNA Interactomes in Living Cells

Michael Backlund and Andreas E. Kulozik

Abstract

RNA-binding proteins are key mediators of many of the RNA-regulatory functions throughout the RNA life cycle in the nucleus and in the cytoplasm. The invention and the recent refinement of the RNA-interactome capture technology has now enabled the analysis of the global RNA-interactome in living cells in the nucleus and in the cytoplasm separately. This technology thus allows an unprecedented differential view on the function of RNA-binding proteins in these compartments. Here we describe a method combining nucleo-cytoplasmic fractionation and enhanced RNA-interactome capture (eRIC) for studying RBPs binding to polyadenylated RNAs separately in the cytoplasmic and in the nuclear compartments.

Key words RNA, RNA-binding protein, RNA Interactome capture, Nucleo-cytoplasmic fractionation, Locked nucleic acid

1 Introduction

RNA-binding proteins (RBP) play a central role in processing, regulation, and function of various RNAs. Within the pursue to identify the full repertoire of eukaryotic RBPs and their target RNAs, that is, the RNA interactome, several methods have been developed to capture either the total cellular RNA or a specific RNA species such as mRNA. RNA-interactome studies have substantially increased our understanding on RBPs and their functions and have identified previously unknown RNA-binding proteins, many without typical RNA-binding domains. Furthermore, they have unveiled surprising second functions of RNA-binding proteins such as those also acting as metabolic enzymes [1]. The RNA-interactome capture (RIC) technology enables the identification of the entire population of polyadenylated (p(A)) RNA-binding proteins in living cells. The critical step of the procedure is the in vivo UV irradiation of cells to cross-link endogenous RBP–RNA

Daniel Matějů and Jeffrey A. Chao (eds.), *The Integrated Stress Response: Methods and Protocols*,
Methods in Molecular Biology, vol. 2428, https://doi.org/10.1007/978-1-0716-1975-9_18,

complexes followed by a specific capture of p(A) RNA (mostly mRNA) and the interacting proteins, that is, the mRNA interactome, using antisense probes against the p(A) tail [2–5]. Recently, the Hentze laboratory at EMBL developed a refinement of the RIC methodology. This method, referred to as enhanced RIC (eRIC), employs locked nucleic acid (LNA) and DNA mixmer probes instead of DNA-based oligo d(T) probes [6]. Compared to methods using fully DNA-based probes, the incorporation of LNAs results in a substantially increased melting temperature (Tm) of the hybrid. This modification enables the use of much more stringent washing conditions thus profoundly increasing the capture specificity, reducing the signal-to-noise ratio, and allowing for a more sensitive identification of p(A) RNA interactions.

With this highly sensitive and specific technology, RNA–protein interactions can now be studied even when the input material is scarce and the potential of unspecific interactions substantial. Such conditions are found in the nucleus, where the abundance of proteins is low and the potential of unspecific interactions may be driven by the abundance of ribosomal RNA, genomic DNA, and their associated proteins. We have recently developed the methodology of using eRIC for the differential analysis of the nuclear and the cytoplasmic RNA interactomes and were thus able to analyze the compartmentalized response to arsenite-induced oxidative stress [7]. Many RBPs are multifunctional proteins that can shuttle between different cellular compartments [8, 9]. This can be exemplified by a canonical nucleo-cytoplasmic shuttling RBP, HuR (also known as ELAVL1) [10]. In the cytoplasm HuR functions in regulating mRNA stability and translation, while in the nucleus it is associated with regulation of alternative pre-mRNA splicing [11–13]. Thus, in order to fully understand the regulatory functions of RBPs in either the nucleus or the cytoplasm it is crucial to study the RNA interactomes in these compartments separately.

Here, we describe a protocol combining nucleo-cytoplasmic fractionation and eRIC to study both the cytoplasmic and nuclear p(A) RNA interactomes from the same cells (Fig. 1).

2 Materials

2.1 Equipment and Disposables

1. Thermomixer.
2. Rotator mixer.
3. Microcentrifuge.
4. Swinging bucket centrifuge with conical tube inserts.
5. UV crosslinker with 254 nm UV light bulbs.
6. Incubator at +37 °C.
7. Water bath.

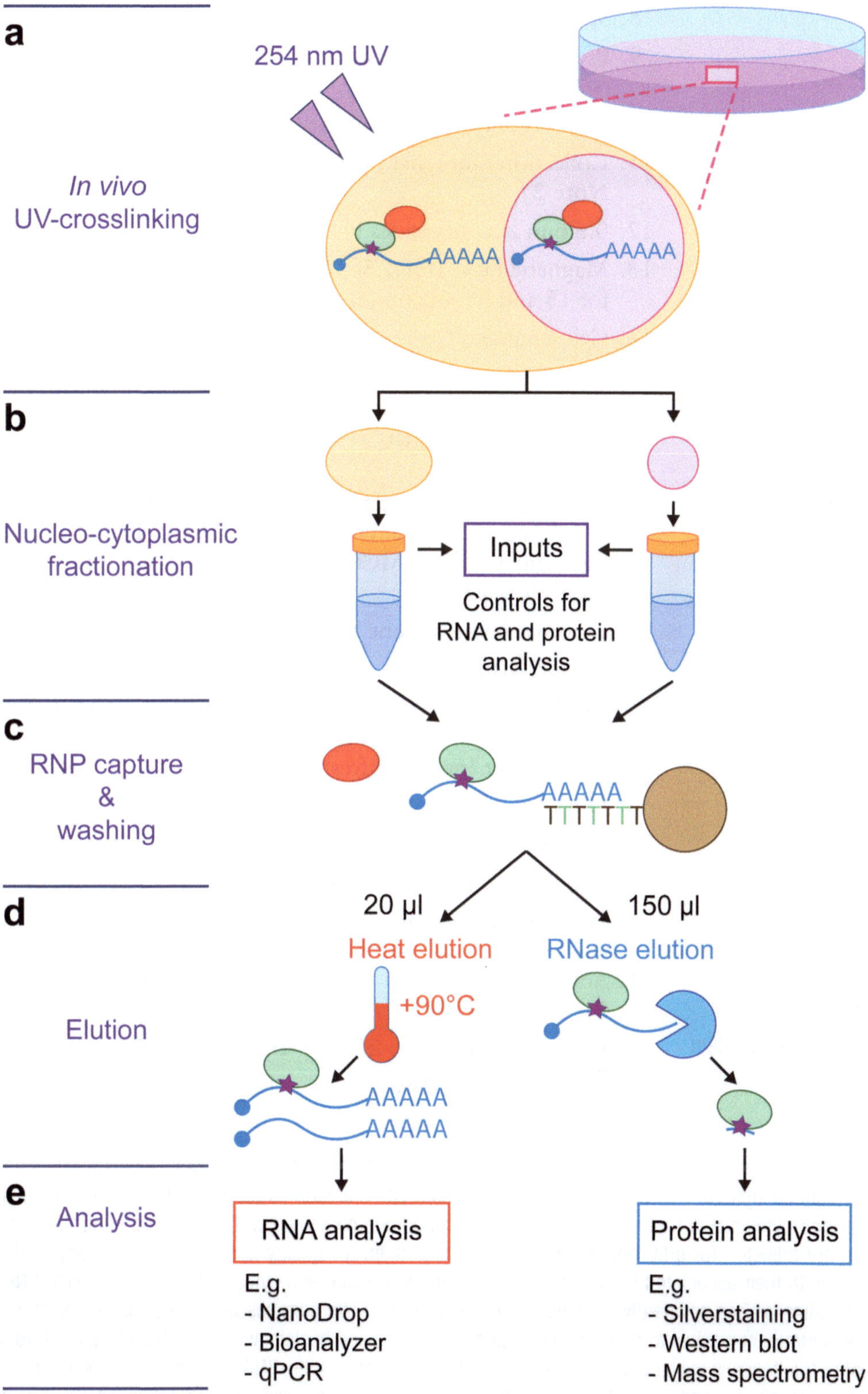

Fig. 1 A methodological overview of nucleo-cytoplasmic fractionation coupled RNA-interactome capture by eRIC. (**a**) Subconfluent monolayer cultures of adherent cells are irradiated with 150 mJ/cm^2 of 254 nm UV light

8. SpeedVac concentrator.
9. Precellys tissue homogenizer (Bertin Instruments) at +4–6 °C (*see* **Note 1**).
10. 7 ml Precellys sample tubes
11. Collection tubes that can fit the 7 ml Precellys sample tubes (*see* **Note 2**).
12. 0.5 mm zirconia/silica beads
13. Magnetic racks (for 1.5–2 ml microcentrifuge tubes as well as for 15 and 50 ml conical tubes).
14. Cell scrapers.
15. 24 gauge needles
16. DNA low-binding microcentrifuge (1.5 and 2 ml) and conical (15 and 50 ml) tubes.

2.2 Coupling of LNA Oligos to Carboxylated Beads

1. LNA oligos consisting of a 5′ amino modifier C6 (/5AmMC6/) and 20 nucleotides altering between DNA (T) and LNA (+T) thymidines: (/5AmMC6/+TT + TT + TT + TT + TT + TT + TT + TT + TT + TT).
2. Carboxylated magnetic beads (*see* **Note 3**).
3. 50 mM MES pH 6
4. EDC solution: 40 mg/ml N-(3-dimethylaminopropyl)-N-′-ethylcarbodiimide hydrochloride (EDC) in 50 mM MES pH 6 buffer (prepared fresh just before use) (*see* **Note 4**).
5. Phosphate buffered saline (PBS).
6. 1 M NaCl
7. PBST: 0.1% Tween 20 in PBS.
8. 200 mM ethanolamine, pH 8.5.

2.3 Cell Culture, Fractionation, and Lysate Preparation

1. Adherent cells grown on plates with removable lids (*see* **Note 5**).
2. Fractionation buffer: 20 mM Tris–HCl pH 7.5, 10 mM KCl, 0.2% Nonidet P-40.

Fig. 1 (continued) for in vivo cross-linking of endogenous RBPs to their target RNAs. (**b**) Cells are processed to nuclear and cytoplasmic fractions using mild lysis conditions combined with low speed centrifugation. The lysates are adjusted for eRIC and aliquots (Inputs) are taken for quality controls. (**c**) Polyadenylated RNAs together with their associated RBPs are captured from the nuclear and cytoplasmic samples using LNA/DNA polythymidine probes covalently coupled to magnetic beads. The samples are further subjected to several rounds of stringent washes to remove any unspecific proteins and nucleic acids from the complexes. (**d**, **e**) An aliquot of the beads is subjected to heat elution of the captured protein-RNA complexes for downstream RNA analysis. For elution and downstream protein analysis of the captured RBPs, majority of the beads are subjected to RNase elution

3. DNase buffer: 20 mM Tris–HCl pH 7.5, 2.5 mM $MgCl_2$, 0.5 mM $CaCl_2$.
4. DNase stop buffer: 20 mM Tris–HCl pH 7.5, 25 mM EDTA.
5. Roche cOmplete protease inhibitor cocktail, EDTA-free.
6. DNase I (2 U/μl, RNase free).
7. Protein concentration measurement assay.
8. Liquid nitrogen.

2.4 eRIC

1. 5× Interactome buffer: 20 mM Tris-HCl pH 7.5, 2.5 M LiCl, 5 mM EDTA, 2.5% lithium dodecyl sulfate (LiDs), 25 mM dithiothreitol (DTT) (*see* **Note 6**)
2. Lysis buffer: 20 mM Tris–HCl pH 7.5, 500 mM LiCl, 1 mM EDTA, 5 mM DTT, 0.5% (w/v) LiDs.
3. Buffer 1: 20 mM Tris–HCl pH 7.5, 500 mM LiCl, 1 mM EDTA, 5 mM DTT, 0.1% (w/v) LiDs.
4. Buffer 2: 20 mM Tris–HCl pH 7.5, 500 mM LiCl, 1 mM EDTA, 5 mM DTT, 0.02% (v/v) Nonidet P-40.
5. Buffer 3: 20 mM Tris–HCl pH 7.5, 200 mM LiCl, 1 mM EDTA, 5 mM DTT, 0.02% (v/v) Nonidet P-40.
6. 10× RNase buffer: 100 mM Tris–HCl pH 7.5, 1.5 M NaCl
7. RNase A (~2 Kunitz units/μl).
8. RNase T_1 (~200 U/μl).
9. RNase elution buffer: 1× RNase buffer, 5 mM DTT, 0.01% Nonidet P-40, 1 μl/ml RNase T_1, and 1 μl/ml RNase A. Prepare freshly before use from the stock 10× RNase buffer.

3 Methods

This protocol describes a large scale experiment used for characterizing the complete p(A) RNA interactome from 20 mg of cytoplasmic lysate and from 3.5 mg of nuclear lysate from adherent HuH7 cells. For small-scale experiments, or for different cell types, the sample and reagent amounts should be adjusted accordingly.

3.1 Coupling of LNA Oligos to Carboxylated Magnetic Beads

The LNA-coupled beads can be prepared beforehand and stored at +4 °C. The beads can also be washed and reused several times. Throughout the procedure, use sterile, nuclease-free, DNA low-binding tubes and pipette tips when possible.

1. Resuspend the LNA probe in nuclease-free water to a final concentration of 100 μM and vortex for 30 s (*see* **Note 7**).
2. For one coupling reaction, 200 μl of 100 μM LNA probes are coupled to 200 μl of carboxlyated beads. For multiple

reactions, transfer the required total amount of beads to a clean, low-binding tube that is at least five times the volume of the bead slurry used, and place the tube on a magnet (*see* **Note 8**).

3. Remove the commercial storage buffer and wash the beads three times with 1 ml of MES buffer per 200 μl of the original bead slurry.
4. After the last wash, resuspend the beads to the original volume in MES buffer and aliquot 200 μl of beads per tube in 2 ml DNA low-binding tubes.
5. Dilute required amount of LNA probes in freshly prepared EDC solution. For coupling 200 μl of beads, add 200 μl of 100 μM LNA probes in 1 ml EDC solution making a total of 1200 μl LNA–EDC solution per sample.
6. Magnetize the 200 μl aliquots of the beads, aspirate the supernatant, and resuspend each aliquot into 1200 μl of the LNA–EDC solution.
7. Incubate the samples in a thermomixer for 5 h at +50 °C with 800 rpm mixing (*see* **Note 9**).
8. Place the samples on a magnet, aspire the supernatant, and wash the coupled beads twice with 1.5 ml PBS.
9. For inactivating any uncoupled reactive groups on the beads, resuspend the beads in 1.2 ml of 200 mM ethanolamine per tube and incubate in a thermomixer for 1 h at +37 °C with 800 rpm mixing.
10. Place the beads on a magnet, aspirate the supernatant, and wash the beads three times with 1.5 ml of 1 M NaCl per tube.
11. After the last wash, resuspend the beads in 200 μl 0.1% PBST per tube and store at +4 °C.

3.2 Cell Culture and In Vivo UV Cross-Linking

The method below is for adherent cells. For suspension cells one needs to optimize the procedure.

1. Grow adherent cells under standard cell culture conditions.
2. 48 h prior to harvest, seed the cells on 500 cm^2 square plates at a density that will lead to 80–90% confluence by the time of harvest (*see* **Note 10**)
3. At the time of harvest, wash the cell culture plates twice with ice-cold PBS and place the plates on a cold block. For non–cross-linked controls, skip the following steps and proceed directly to cell fractionation (*see* **Note 11**).
4. Remove the lid from the culture plate and place the plate on a cold block in the UV crosslinker.
5. Irradiate the cells with 150 mJ/cm^2 of 254 nm UV light.

6. Replace the lid on the culture plate and place the plate on the cold block back on ice. Proceed directly to cell fractionation.

3.3 Cell Fractionation

Perform all steps with cooled buffers and centrifuges. Keep the samples on ice at all times.

1. For each 500 cm^2 culture plate, add 4 ml of ice-cold fractionation buffer supplemented with protease inhibitors. Collect the lysates by scraping them gently and pooling the same conditions to low-binding conical tubes (*see* **Note 12**).
2. Incubate the lysates on ice for 5 min and gently swirl the tube a few times in between.
3. Centrifuge the lysates 1000 × *g* for 3 min at +4 °C in a swinging bucket centrifuge.
4. Transfer the cytoplasmic fraction to a new low-binding tube without disrupting the nuclear pellet. Take an aliquot of the cytoplasmic fraction for analysis (e.g., protein concentration, western blot) and keep it on ice, snap-freeze the rest in liquid nitrogen and store at −80 °C until eRIC.
5. Resuspend the nuclear pellet gently in 12 ml of ice-cold fractionation buffer and transfer to a clean low-binding tube.
6. Centrifuge the nuclei 300 × *g* for 2 min at +4 °C.
7. Remove the fractionation buffer, resuspend the nuclear pellet to another 12 ml of fractionation buffer and centrifuge 300 × *g* for 2 min at +4 °C.
8. Carefully aspirate the fractionation buffer without disrupting the nuclei (*see* **Note 13**).
9. Resuspend the nuclei in 3.5 ml of ice-cold DNase buffer.
10. Transfer the nuclear suspension to cold 7 ml Precellys sample tubes containing 500 μl of 0.5 mm zirconia/silica beads.
11. Place the samples in a Precellys tissue homogenizer cooled down to +4–6 °C and disrupt the nuclei with 2 × 20 s bursts with a 5 s interval (*see* **Note 14**).
12. Place the tubes inside the collection tubes and keep on ice.
13. Puncture two holes at the bottom of the 7 ml Precellys sample tubes with a hot 24 gauge needle and place the tubes back inside the collection tubes.
14. Collect the lysate by centrifugation with 450 × *g* for 5 min at +4 °C.
15. Transfer 3.5 ml of the nuclear lysate to 15 ml low-binding tubes, add 500 μl of DNase I, and incubate for 30 min at +37 °C in a rotator mixer.
16. To each nuclear lysate, add 1 ml of DNase I stop buffer, take an aliquot for analysis (e.g., protein concentration, western blot),

and snap-freeze the rest in liquid nitrogen. Store the lysates at −80 °C until proceeding to the eRIC.

17. Measure the protein concentration of the cytoplasmic and nuclear lysates using the collected aliquots and a protein concentration measurement assay. If the rest of the aliquots are not processed directly for any downstream quality controls, snap-freeze them and store at −80 °C.

3.4 Preparation of Lysates for eRIC

Prepare the lysates at the day of the capture.

1. Partly thaw the lysates in a + 37 °C water bath and transfer on ice to thaw completely.
2. Adjust the samples to equal protein concentrations (*see* **Note 15**).
3. Add required amount of 5× interactome buffer to the lysates (1 ml per every 4 ml of lysate) and mix (*see* **Note 16**).
4. Incubate the samples in a + 60 °C water bath for 10–15 min. Invert the samples several times during the incubation.
5. Place the samples on ice for a few minutes and add fresh DTT to a final concentration of 5 mM and mix.
6. Centrifuge the samples with 1700 × *g* for 10 min at +4 °C. and transfer the samples to new low-binding tubes. Take an aliquot of the lysates as "input" samples (*see* **Note 17**).

3.5 Capture of Polyadenylated RNAs by eRIC

Use sterile, nuclease-free, DNA low-binding tubes and pipette tips throughout the procedure when possible.

1. Pool the required amount of LNA beads (300 μl per sample) into a low-binding tube at least five times the final volume of the beads, place the tube on a magnet and remove the supernatant.
2. Wash the beads three times with 1 ml of lysis buffer per 300 μl of beads.
3. After the last wash, resuspend the beads in lysis buffer to the original volume of the bead slurry.
4. Add 300 μl of the washed LNA beads to the lysates and incubate for 1 h at +37 °C in a rotator mixer.
5. Quickly spin down the samples, place the tubes on a magnet and pellet the beads for 5 min at +37 °C.
6. Transfer the supernatants to new low-binding tubes and store them at +4 °C for the second capture round (*see* **Note 18**).
7. For every 300 μl of beads, add 8 ml of pre-warmed lysis buffer and resuspend the beads with a pipette and by inverting the tube. Incubate the beads in a rotator mixer for 5 min at +37 °C, place the tubes on a magnet and pellet the beads for 5 min at

+37 °C. Remove the supernatant and repeat the washing steps twice with pre-warmed buffer 1, buffer 2, and buffer 3, each. After the last wash, quickly spin down the tubes prior to placing them on a magnet (*see* **Note 19**).

8. Resuspend the washed beads in 1 ml of buffer 3, transfer them into 1.5 ml low-binding tubes, pellet the beads on a magnet, and aspirate the buffer.
9. Resuspend the beads in 170 μl of RNase-free H_2O and incubate in a thermomixer for 5 min at +40 °C and 800 rpm.
10. Transfer 20 μl aliquots of the beads to new 1.5 ml tubes for heat elution of the captured RNAs while the remaining 150 μl is used for RNase elution of the captured RBPs. Place both of the tubes immediately on a magnet and aspirate the H_2O (*see* **Note 20**).

3.6 Heat-Elution of the Captured RNA

1. For heat elution of the captured RNAs, resuspend the 20 μl aliquots of the beads to another 20 μl of RNase-free H_2O and incubate in a thermomixer at +90 °C for 10 min and 800 rpm.
2. Quickly spin down the samples, pellet the beads on a magnet, and recover the eluates to clean low-binding tubes. If not used directly for further analysis (e.g., NanoDrop, bioanalyzer, qPCR), snap-freeze the samples in liquid nitrogen and store at −80 °C.
3. Resuspend the beads in RNase-free water for pooling and washing with the rest of the beads later.

3.7 RNase Elution of Captured RBPs

1. For RNase elution of the captured RBPs, resuspend the 150 μl aliquots of the beads in 150 μl of freshly prepared RNase elution buffer.
2. Incubate the samples in a thermomixer at +37 °C for 45 min and 800 rpm.
3. Quickly spin down the samples, pellet the beads with a magnet, and transfer the supernatant into a new tube. To remove any carryover beads, place the tubes again on a magnet and transfer the supernatant again to a new tube. If not processed directly further, snap-freeze the samples in liquid nitrogen and store at −80 °C.
4. Resuspend the beads in 1 ml of RNase-free H_2O.

3.8 Second Capture Round

1. Combine the beads from the heat and RNase elutions, place them on a magnet, aspirate the H_2O, and resuspend in 300 μl of fresh RNase-free H_2O. Do not combine beads from different conditions.
2. Incubate the beads in a thermomixer at +90 °C for 10 min and 800 rpm.

3. Quickly spin down the tubes to remove evaporation from the lids, pellet the beads with a magnet, and aspirate the H_2O.
4. Wash the beads 3 times with 1 ml of lysis buffer per 300 µl of beads.
5. After the last wash, resuspend the beads in the lysates from the same conditions as in the first capture round.
6. Repeat the capture, wash, and elution steps (from **step** 4 in Subheading 3.5 to **step 4** in Subheading 3.7).
7. For downstream RNA and protein analysis, pool the heat and RNase eluates from the two capture rounds together, respectively. Add 2 µl of 10% SDS to the pooled RNase eluates and concentrate them to 100 µl using a SpeedVac (*see* **Note 21**).

3.9 Washing and Storing the Beads for Reuse

1. After the second capture round, combine and wash the beads with H_2O as described after the first capture. Put the beads on a magnet and aspirate the H_2O.
2. Wash the beads 3 times with 1 ml of H_2O per 300 µl of beads.
3. Wash the beads 3 times with 1 ml of lysis buffer per 300 µl of beads.
4. Resuspend the beads in 0.1% PBST, pool the beads together, and store at +4 °C.

4 Notes

1. The Precellys homogenizer can be cooled down by either using a Cryolys cooling unit (Bertin Instruments) or operating the homogenizer in a cold room.
2. Regular 15 ml conical tubes tend to be too narrow but the 15 ml screw cap tubes from Sarstedt (product number 60.732.001) works well.
3. Various suppliers provide different carboxylated magnetic beads. We have routinely used the carboxylated M-PVA C11 magnetic beads with 0.5–1 µm particle size from PerkinElmer chemagen Technology with good results.
4. We have successfully used 20–40 mg/ml EDC (~0.1–0.2 M) for the coupling reactions. For coupling M-PVA C11 magnetic beads, an EDC concentration of 0.1–0.5 M is recommended by the manufacturer.
5. While regular cell culture plates (e.g., 15 cm) can be used, for large scale experiment we strongly recommend using the Nunc™ 500 cm^2 Square BioAssay Dishes (Thermo Fisher Scientific) for ease of handling.

6. Autoclaved and filter sterilized stocks of the buffers can be prepared and stored at +4 °C up till few months. Do not add DTT, LiDs, or Nonidet P-40 to the buffers before autoclaving. DTT should be added freshly to the buffers before use. LiDs should not be added directly to the 5× interactome buffer as it will precipitate. Instead, it should be added directly to the lysates to a correct final concentration.
7. For a 100 μM working solution, resuspend each 100 nmol of LNA probe in 1 ml nuclease free water. Store the resuspended probes at −20 °C if not proceeding directly to coupling, or if not all of the probe is used.
8. The volume of the beads refers to the volume of the original bead slurry, not the packed volume of the beads.
9. To ensure that the beads do not pack at the bottom of the tubes during the incubations, you may vortex them gently from time to time, avoiding to get the beads sticking to the lids. As some buffer will condensate to the lids, you may quickly spin down the samples before the vortexing.
10. The seeding density and culture conditions should be individually determined for each cell line used. As a general rule, avoid growing the plates over confluent as, in addition to effects on cell metabolism, the UV cross-linking efficiency may be reduced if the cells grow in multiple layers. We have routinely seeded 5.5×10^6 HuH7 cells per 500 cm^2 square plates to have about 80–90% confluent cultures 48 h later, at the time of harvesting. For large scale experiments with HuH7 cells we have used 4 square plates per condition. This should yield enough material for a eRIC with 20 mg cytoplasmic and 3.5 mg nuclear lysates as inputs.
11. After the second wash, remove the PBS by pouring. The remaining liquid layer on top of the cells is sufficient to inhibit them from drying during the UV cross-linking step. Also, if working with several plates and/or conditions it is advisable to process a workable amount of plates at a time to avoid prolonged incubation times before or during the fractionation workflow.
12. HuH7 cells need to be scraped directly into the fractionation buffer. Scraping them into PBS tend to break the cells leading to significant sample loss. For other adherent cells, collecting them by scraping and centrifugation before lysis may be tested if preferred. To avoid breaking the nuclei during the fractionation be gentle when scraping or resuspending them. Also, try to work quickly and without delays to avoid extended incubation times in the fractionation buffer. For analysis of nuclear integrity, mix an aliquot of the resuspended nuclei before the last centrifugation step with trypan blue and observe the

stained nuclei under a microscope. If you observe damaged nuclei, or enrichment of nuclear proteins in the cytoplasmic fractions of your quality controls (e.g., western blot), the processing has been too harsh. If you on the other hand observe high degree of cytoplasmic proteins in the nuclear fraction, consider adding an extra washing step or using longer incubation times with the fractionation buffer.

13. If not proceeding directly to nuclear lysis, snap-freeze the nuclear pellet in liquid nitrogen and store short term at −80 °C. Prior to proceeding, thaw the frozen pellet on ice.
14. Other means of disrupting the nuclei and fragmenting the genomic DNA (e.g., sonication or syringe) can be used but the conditions need to be optimized accordingly.
15. Adjust the cytoplasmic and nuclear fractions to equal concentrations, respectively. For adjusting cytoplasmic fractions use fractionation buffer and for adjusting nuclear fractions use mixture consisting of: 7 parts DNase buffer, 2 parts DNase stop buffer, and 1 part H_2O. For large scale experiments with HuH7 cells we have usually worked with final volumes between 10 and 20 ml.
16. Add 5× interactome buffer without LiDs to the lysates and finally add the required amount of LiDs. For example, to 4 ml of lysate add 0.75 ml 5× interactome buffer (with fresh DTT) and 0.25 ml of 10% LiDs.
17. If there is no pellet after the centrifugation, it is not essential to transfer the lysate to a new tube. A pellet after the centrifugation may indicate traces of intact genomic DNA in the samples. No pellet should be observed in the cytoplasmic samples after a successful fractionation. Also, no pellet is usually observed in the nuclear fractions when using the Precellys for disrupting the nuclei and fragmenting the genomic DNA, followed by the DNase I treatment. However, this might not be the case if using other means to disrupt the nuclei (e.g., syringe). In this case it would be advised to try to optimize the nuclear lysis conditions to further minimize the presence of intact genomic DNA in the samples. Part of the input aliquots can be used for protein analysis (e.g., silver staining and western blot) and the rest subjected to RNA extraction followed by various RNA analyses (e.g., NanoDrop, Bioanalyzer, qPCR), as controls for the eRIC eluates.
18. Do not store the lysates on ice for prolonged time as they will cool down too much and the LiDs in the buffers may precipitate. Instead, store the lysates at +4 °C (e.g., in a refrigerator).
19. Only prewarm the required amount of buffers shortly before use (e.g., 5 min in a + 37 °C water bath). Do not store the entire amount of the washing buffers at +37 °C throughout the

protocol as DTT may not be stable for extended periods at these temperatures.

20. To avoid rebinding of the unspecific RNAs and proteins removed by the water wash, try to aspirate the H_2O without delays.

21. The eluates from the two capture rounds can also be analyzed separately before pooling them. An initial separate analysis of the heat eluates from the successive capture rounds can be helpful for interpreting the RNA capture efficiency and quality. Typically, a large scale experiment from HuH7 cells yields a total of ~30 μg of captured RNA from both the cytoplasmic and the nuclear fractions when extrapolated from the concentrations of the heat eluates. Suggested analysis and quality controls for the RNA capture include measuring the RNA concentration (e.g., NanoDrop), analysis of RNA quality (e.g., Bioanalyzer), and qPCR analysis of selected targets. If required, a proteinase K digestion can be performed with the heat eluates of the UV cross-linked samples prior to any RNA analysis. Suggested quality controls for the RNase eluates include western blot against known RBPs and silver staining. RBPs should be primarily detected from the inputs and from the eluates of the UV cross-linked samples, while the lanes with the non–cross-linked eluates should be mostly clear. Similarly, silver staining should be observed mainly in the lanes of the inputs and the eluates of the UV cross-linked samples. Depending on the scale of the experiment and the ultimate aim of the project, the RNase eluates can be further subjected to, for example, mass spectrometry identification of the total RNA interactome, or to a more in-depth analysis of a protein of interest by other experimental approaches. Throughout any protein analysis, it is recommended to use the non–cross-linked samples as controls for unspecific background. The details about the various suggested quality controls and downstream analysis mentioned above have been extensively discussed and described in-depth in a recent publication [14].

Acknowledgments

The authors would like to thank the past and present members of both the Hentze and the Kulozik labs for all their support. This work was supported by the Deutsche Forschungsgemeinschaft (SFB1036 TP5 to A.E.K.).

References

1. Hentze MW, Castello A, Schwarzl T, Preiss T (2018) A brave new world of RNA-binding proteins. Nat Rev Mol Cell Biol 19(5):327–341. https://doi.org/10.1038/nrm.2017.130
2. Castello A, Fischer B, Eichelbaum K, Horos R, Beckmann BM, Strein C, Davey NE, Humphreys DT, Preiss T, Steinmetz LM, Krijgsveld J, Hentze MW (2012) Insights into RNA biology from an atlas of mammalian mRNA-binding proteins. Cell 149(6):1393–1406. https://doi.org/10.1016/j.cell.2012.04.031
3. Castello A, Horos R, Strein C, Fischer B, Eichelbaum K, Steinmetz LM, Krijgsveld J, Hentze MW (2013) System-wide identification of RNA-binding proteins by interactome capture. Nat Protoc 8(3):491–500. https://doi.org/10.1038/nprot.2013.020
4. Castello A, Fischer B, Frese CK, Horos R, Alleaume AM, Foehr S, Curk T, Krijgsveld J, Hentze MW (2016) Comprehensive identification of RNA-binding domains in human cells. Mol Cell 63(4):696–710. https://doi.org/10.1016/j.molcel.2016.06.029
5. Baltz AG, Munschauer M, Schwanhäusser B, Vasile A, Murakawa Y, Schueler M, Youngs N, Penfold-Brown D, Drew K, Milek M, Wyler E, Bonneau R, Selbach M, Dieterich C, Landthaler M (2012) The mRNA-bound proteome and its global occupancy profile on protein-coding transcripts. Mol Cell 46(5):674–690. https://doi.org/10.1016/j.molcel.2012.05.021
6. Perez-Perri JI, Rogell B, Schwarzl T, Stein F, Zhou Y, Rettel M, Brosig A, Hentze MW (2018) Discovery of RNA-binding proteins and characterization of their dynamic responses by enhanced RNA interactome capture. Nat Commun 9(1):4408. https://doi.org/10.1038/s41467-018-06557-8
7. Backlund M, Stein F, Rettel M, Schwarzl T, Perez-Perri JI, Brosig A, Zhou Y, Neu-Yilik G, Hentze MW, Kulozik AE (2020) Plasticity of nuclear and cytoplasmic stress responses of RNA-binding proteins. Nucleic Acids Res 48(9):4725–4740. https://doi.org/10.1093/nar/gkaa256
8. Dreyfuss G, Kim VN, Kataoka N (2002) Messenger-RNA-binding proteins and the messages they carry. Nat Rev Mol Cell Biol 3(3):195–205. https://doi.org/10.1038/nrm760
9. Lunde BM, Moore C, Varani G (2007) RNA-binding proteins: modular design for efficient function. Nat Rev Mol Cell Biol 8(6):479–490. https://doi.org/10.1038/nrm2178
10. Grammatikakis I, Abdelmohsen K, Gorospe M (2017) Posttranslational control of HuR function. Wiley Interdiscip Rev RNA 8(1). https://doi.org/10.1002/wrna.1372
11. Brennan CM, Steitz JA (2001) HuR and mRNA stability. Cell Mol Life Sci 58(2):266–277. https://doi.org/10.1007/pl00000854
12. Izquierdo JM (2008) Hu antigen R (HuR) functions as an alternative pre-mRNA splicing regulator of Fas apoptosis-promoting receptor on exon definition. J Biol Chem 283(27):19077–19084. https://doi.org/10.1074/jbc.M800017200
13. Kuwano Y, Kim HH, Abdelmohsen K, Pullmann R Jr, Martindale JL, Yang X, Gorospe M (2008) MKP-1 mRNA stabilization and translational control by RNA-binding proteins HuR and NF90. Mol Cell Biol 28(14):4562–4575. https://doi.org/10.1128/mcb.00165-08
14. Perez-Perri JI, Noerenberg M, Kamel W, Lenz CE, Mohammed S, Hentze MW, Castello A (2020) Global analysis of RNA-binding protein dynamics by comparative and enhanced RNA interactome capture. Nat Protoc 16(1):27–60. https://doi.org/10.1038/s41596-020-00404-1

Chapter 19

Identification of RNA–RBP Interactions in Subcellular Compartments by CLIP-Seq

Sonu Sahadevan, Manuela Pérez-Berlanga, and Magdalini Polymenidou

Abstract

Cross-linking immunoprecipitation and high-throughput sequencing (CLIP-seq) allows the identification of RNA targets bound by a specific RNA-binding protein (RBP) in in vivo and ex vivo experimental models with high specificity. Due to the little RNA yield obtained after cross-linking, immunoprecipitation, polyacrylamide gel electrophoresis, membrane transfer, and RNA extraction, CLIP-seq is usually performed from relatively large amounts of starting material, like cell lysates or tissue homogenates. However, RBP binding of its specific RNA targets depends on its subcellular localization, and a different set of RNAs may be bound by the same RBP within distinct subcellular sites. To uncover these RNA subsets, preparation of CLIP-seq libraries from specific subcellular compartments and comparison to CLIP-seq datasets from total lysates is necessary, yet there are currently no available protocols for this. Here we describe the adaptation of CLIP-seq to identify the specific RNA targets of an RBP (FUS) at a small subcompartment, that is, neuronal synapses, including subcompartment isolation, RBP–RNA complex enrichment, and upscaling steps.

Key words CLIP-seq, Cross-linking immunoprecipitation, Immunoprecipitation, Protein–RNA interactions, RNA, RNA-binding proteins, High throughput sequencing

1 Introduction

Cross-linking immunoprecipitation and high-throughput sequencing (CLIP-seq) enables genome-wide identification of the RNA targets of an RNA-binding protein (RBP) [1]. In order to identify the RNA targets and specific sites directly bound by an RBP of interest, biological samples are cross-linked using ultraviolet radiation with a wavelength of 254 nm, creating covalent links between the nucleotide bases in close contact with the interacting amino acids of the RBP [2]. Immunoprecipitation with a specific antibody against the RBP of interest, followed by mild enzymatic treatment

Sonu Sahadevan and Manuela Pérez-Berlanga contributed equally to this work.

Daniel Matějů and Jeffrey A. Chao (eds.), *The Integrated Stress Response: Methods and Protocols*,
Methods in Molecular Biology, vol. 2428, https://doi.org/10.1007/978-1-0716-1975-9_19,

to shorten the RNA fragments, is performed to purify the RNA targets associated with it. While conventional RNA immunoprecipitation (RIP) does not allow stringent immunoprecipitation washing conditions in order to prevent RNA–RBP dissociation, initial cross-linking in CLIP enables harsh washes with a buffer containing high salt concentration, resulting in specific identification of the precise RNA-binding sites of the RBP of interest [3].

The subcellular distribution of RNAs is necessary for local translation linked to their function in a cell type-specific manner. Therefore, the RNA targets bound by a specific RBP can change depending on where within the cell the given RBP is localized. Due to its stringency, CLIP-seq can be used to identify not only the general RNA targets of an RBP of interest, but also its specific RNA targets within cellular subcompartments. Here, we present an optimized protocol, which allows the identification of RNA targets of an RBP in a defined subcellular compartment, following its enrichment prior to performing CLIP-seq. In particular, we identified the specific synaptic RNA targets of the amyotrophic lateral sclerosis and frontotemporal dementia-linked protein Fused in Sarcoma (FUS). Involved in RNA metabolism, FUS is primarily nuclear in localization, where it binds RNA targets that have been previously identified by several independent CLIP-seq experiments [4–9]. Interestingly, apart from its abundant presence in the nucleus, a small proportion of FUS is present at the neuronal synapses with a function yet to be determined [10, 11]. In order to better understand the synaptic roles of FUS, we sought to identify the specific RNAs bound by FUS at this compartment, which may be different to that of nuclear FUS and could have been otherwise missed in CLIP-seq datasets from total brain homogenates due to the small RNA and FUS protein levels at the synapses. Below, we describe the CLIP-seq protocol that we employed to map RNA targets of FUS protein from synaptic fractions [12].

After isolation of synaptoneurosomes (SNS) from mouse brain with an optimized protocol compatible with ultraviolet (UV) cross-linking of protein-RNA interactions and lysis, RNA–FUS complexes were captured with magnetic anti-FUS-coated beads. A mild nuclease treatment to shorten RNA fragments was performed before radiolabeling them for visualization according to their size after polyacrylamide gel electrophoresis (PAGE) and transfer onto a nitrocellulose membrane. A smear of a wide-size range of RNA–FUS complexes was observed and these RNAs were purified directly from the membrane, followed by cDNA library generation, sequencing, and computational analysis to determine the entire set of FUS-associated RNA targets at the synapses (Fig. 1). To our knowledge, this is the first time that CLIP-seq was successfully adapted to a small specialized subcompartment such as the synapse, and this work shows the feasibility of this approach that could be expanded to other biologically relevant subcellular sites.

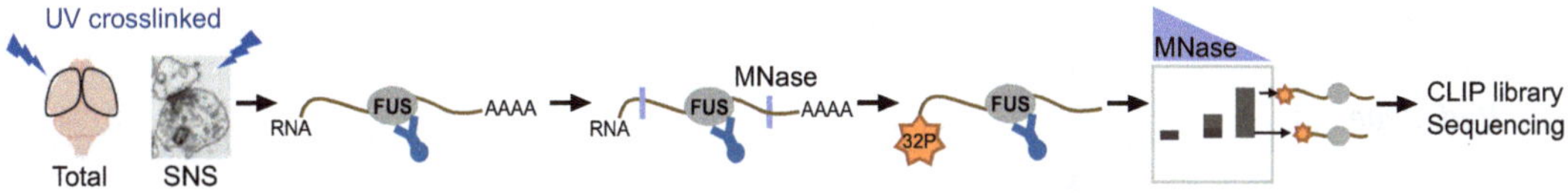

Fig. 1 Schematic of the CLIP-seq protocol allowing stringent purification of RNAs bound to FUS protein. UV irradiation led to the RNP cross-linking ex vivo. Lysate was treated with mild nuclease conditions before immunoprecipitation of RNA–protein complexes with FUS antibody. Autoradiograph of SDS-PAGE depicting RNA–RBP complexes as a smear, which progressively decreases as the amount of nuclease used in the previous step increases, until full collapse to the size of the full-length protein. RNA of the appropriate size is isolated directly from the membrane, and used for the generation of the cDNA libraries that are finally sequenced

2 Materials

All solutions are prepared using nuclease-free-certified distilled water and stored at room temperature, unless indicated otherwise.

2.1 SNS Preparation

1. SNS buffer: pH 7.4, 10 mM 4-(2 hydroxyethyl)-1-piperazineethanesulfonic acid (HEPES), 0.35 M sucrose, 1 mM ethylenediaminetetraacetic acid (EDTA), 0.25 mM dithiothreitol (DTT), 30 U/ml RNaseOUT™ Recombinant Ribonuclease Inhibitor (Thermo Fisher Scientific), EDTA-free protease inhibitors, PhosSTOP™ (Roche).
2. SNS buffer without sucrose: pH 7.4, 10 mM HEPES, 1 mM EDTA, 0.25 mM DTT, 30 U/ml RNase inhibitor, EDTA-free protease inhibitors, PhosSTOP™.
3. 100 μm nylon net filter, 25 mm diameter (Millipore).
4. 11 μm nylon net filter, 25 mm diameter (Millipore).
5. Swinnex filter holder, 25 mm diameter.
6. 15 ml glass Dounce homogenizer.
7. 10 ml syringes.
8. 10 cm petri dish.
9. 1-month-old C57/Bl6 mice.

2.2 Cross-Linking

1. 254 nm UV cross-linker.
2. Ice-cold phosphate-buffered saline (PBS): 136.9 mM NaCl, 2.7 KCl, 10.14 mM Na_2HPO_4, 1.8 mM KH_2PO_4, pH 7.4 (*see* **Note 1**).
3. 100 μm cell strainer.
4. 10 cm petri dish.
5. P1000 pipette.
6. Liquid nitrogen.

2.3 Immunoprecipitation and Radiolabeling

1. Antibody conjugation buffer: PBS, 0.02% Tween 20.
2. RQ1 RNase-free DNase I (Promega).
3. RNaseOUT™ Recombinant Ribonuclease Inhibitor (Thermo Fisher Scientific).
4. PXL buffer: PBS, 0.1% sodium dodecyl sulfate (SDS), 0.5% sodium deoxycholate, 0.5% NP-40.
5. PXL+ buffer: PXL buffer with added protease inhibitors. Prepare freshly on the day of the experiment and store at 4 °C.
6. 5× PXL buffer: 684.5 mM NaCl, 13.4 KCl, 52.8 mM Na_2HPO_4, 8.8 mM KH_2PO_4, 0.1% SDS, 0.5% sodium deoxycholate, 0.5% NP-40 (*see* **Note 2**).
7. 5× PXL+ buffer: 5× PXL buffer with added protease inhibitors. Prepare freshly on the day of the experiment and store at 4 °C.
8. PNK buffer: 50 mM Tris, 10 mM $MgCl_2$, 0.5% NP-40, pH 7.4. Store at 4 °C.
9. PNK/EGTA buffer: 50 mM Tris, 20 mM egtazic acid (EGTA), 0.5% NP-40, pH 7.4. Store at 4 °C.
10. 1 mM adenosine triphosphate (ATP). Store at −20 °C.
11. 10 mCi/ml, 3000 Ci/mmol ^{33}P-γ-ATP solution.
12. Thermomixer.
13. MNase buffer: 50 mM Tris–HCl, 5 mM $CaCl_2$, pH 7.9.
14. Rotating wheel.
15. Protein G-conjugated magnetic beads. Stored at 4 °C.
16. 15 U/μl S7 micrococcal nuclease (MNase) (Roche) stock concentration. Aliquoted and stored at −20 °C.
17. T4 polynucleotide kinase (PNK) enzyme.
18. 10× PNK buffer (variable composition depending on the enzyme brand).
19. Nuclease-free 1.5 ml microfuge tubes.
20. Nuclease-free 2 ml microfuge tubes.
21. Antibody against the RBP of interest.

2.4 PAGE and Immunoblotting

1. 4–12% gradient gel (Novex NuPAGE 4–12% Bis-Tris Gel).
2. XCell SureLock Mini System.
3. Nitrocellulose membrane (0.45 μM pore size).
4. iBlot™ Gel Transfer Device (Life Technologies).
5. Phosphor screen.
6. NuPAGE™ 20× MOPS SDS Running buffer.

7. 4× NuPAGE™ LDS sample buffer (Life Technologies) (*see* **Note 3**).
8. 10× Bolt™ Sample Reducing Agent (Life Technologies) (*see* **Note 3**).
9. Transfer buffer: 48 mM Tris, 39 mM glycine, 20% methanol pH 9.2 (*see* **Note 4**).
10. Prestained protein ladder for SDS-PAGE.
11. Typhoon FLA 9500 apparatus.
12. PBS-T: PBS, 0.05% Tween 20.
13. Blocking solution: PBS-T, 5% (w/v) skimmed milk.
14. Antibody diluent: PBS-T, 1% (w/v) skimmed milk (*see* **Note 5**).
15. Primary antibody against the RBP of interest (*see* **Note 6**).
16. Horseradish peroxidase (HRP)-conjugated secondary antibodies (*see* **Note** 7).
17. Chemiluminescence-based digital imager (*see* **Note** 7).
18. Luminol-based developing solution for immunoblots (*see* **Note** 7).
19. Autoradiograph exposure cassette.

2.5 RNA Isolation

1. Clean scalpel.
2. Nuclease-free 1.5 ml microfuge tubes.
3. 20 mg/ml proteinase K (PK) stock solution. Store at −20 °C.
4. PK buffer: 100 mM Tris-Cl, 50 mM NaCl, 10 mM EDTA, pH 7.5.
5. PK/Urea buffer: 100 mM Tris–HCl, 10 mM EDTA, 50 mM NaCl, 7 M Urea, pH 7.5. Prepare freshly on the day of the experiment and store at room temperature.
6. Acid phenol–chloroform–isoamyl alcohol solution (25:24:1), pH 4.5.
7. 3 M sodium acetate, pH 5.2.
8. 20 mg/ml glycogen stock solution. Stored at −20 °C.
9. Ethanol–isopropanol solution (1:1).

2.6 DNA Library Preparation

1. SMARTer Stranded Total RNA-Seq Kit—Pico Input Mammalian (Takara Bio).
2. TapeStation (Agilent).
3. cDNA library buffer: 10 mM Tris-Cl, 0.1% Tween 20, pH 8.5.

3 Methods

3.1 SNS Isolation and RNA–RBP Cross-Linking

1. Sacrifice an adult (1-month-old or older) C57/BL6 mouse and dissect the brain out on a 10 cm petri dish covered with ice-cold PBS. If handled by a single person, process five brains at a time, so as not to compromise the quality of the tissue. The amount of starting sample required for the CLIP-seq experiment depends on the abundance of the protein of interest at the desired subcellular location and the nature of the biological sample (*see* **Note 8**).
2. Excise the cortex and place it on a petri dish containing ice-cold SNS buffer enough to cover the tissue.
3. Homogenize the pool of cortices from five brains in SNS buffer (10% w/v) using a prechilled Dounce homogenizer. Move the piston a total of twelve times in two rounds of six, waiting for 1 min in between rounds, and taking care of not pulling the piston out completely to avoid bubbles.
4. Remove cell debris and nuclei by centrifugation at 1000 × *g* for 10 min at 4 °C. Discard the nuclear pellet (*see* **Note 9**).
5. Strain the resulting supernatant sequentially through three 100 μm nylon net filters and one 5 μm filter.
6. Dilute the final filtrate in three volumes of SNS buffer without sucrose and centrifuge at 2000 × *g* for 15 min at 4 °C to collect the SNS in the pellet.
7. Resuspend each SNS pellet in 3 ml of PBS and transfer onto a 10 cm petri dish without a lid placed on an ice tray.

   Steps 1–7 can be replaced with other RNA-compatible subcellular compartment isolation methods (*see* **Note 10**).
8. Place the tray with the petri dish in a 254 nm UV cross-linker and irradiate at 100 mJ/cm^2 for two cycles (*see* **Note 11** for changes in the cross-linking conditions). Gently shake the dish between the cycles to ensure a homogenous lysate distribution and prevent sedimentation.
9. Transfer the SNS suspension into a 15 ml centrifuge tube with a p1000 pipette. Rinse the plates with 1 ml per dish and add it to the UV cross-linked lysate.
10. Pellet UV- and nonirradiated SNS at 840 × *g* for 3 min at 4 °C.
11. Remove supernatant and resuspend the pelleted SNS in 1 ml of ice-cold PBS.
12. Transfer 1 ml of SNS suspension into 1.5 ml microfuge tubes.
13. Spin at 17000 × *g* for 5 min at 4 °C, remove supernatant and snap-freeze pellets in liquid nitrogen. SNS pellets can be stored at −80 °C until further use.

3.2 Total Homogenate Preparation and RNA–RBP Cross-Linking

In order to compare the RNAs bound by the RBP of interest in general and specifically at a given subcellular compartment, CLIP from the total lysate should be carried throughout the experiment as a control.

1. Sacrifice two adult (age-matched) C57/BL6 mice, dissect the brains out on a 10 cm petri dish with ice-cold PBS and excise the cortices.
2. Place a cell 100 μm strainer on a 10 cm petri dish and add 1 ml of ice-cold PBS to wet the strainer.
3. Place the cortices on the wet cell strainer and, using a syringe, plunge to mash the tissue and force the cells through the strainer.
4. Add 1 ml PBS to the strainer to force any remaining sample out.
5. Transfer the suspension to a fresh 15 ml centrifuge tube and rinse the petri dish with 1 ml PBS to collect any remaining cells.
6. Transfer the suspension onto a 10 cm petri dish without a lid placed on an ice tray.
7. Irradiate at 100 mJ/cm^2 (keep aside the non–UV cross-linked sample – 3 ml) on ice for 2 times with short breaks in between to move the plates.
8. Collect suspension with a p1000 pipette and rinse the plates with 1 ml PBS per plate.
9. Collect the cells at 840 × *g* for 3 min at 4 °C.
10. Remove supernatant and resuspend pellet in 3 ml ice-cold PBS and transfer it into a nuclease-free 1.5 ml microfuge tube.
11. Spin at 4 °C for 17,000 × *g* for 5 min, remove supernatant and snap-freeze pellets in liquid nitrogen. Pellets can be stored at −80 °C until further use.

3.3 Immunoprecipitation: Part 1 — Antibody Coupling to Magnetic Beads

1. Wash 400 μl of protein G-conjugated magnetic beads per 1.5 ml nuclease-free tubes with 1 ml of antibody conjugation buffer for a total of three times with the help of a magnetic rack.
2. Resuspend the beads in 500 μl of antibody conjugation buffer containing the appropriate amount of primary antibody against the RBP of interest (*see* **Notes 12** and **13**).
3. Incubate in a rotating wheel for at least 45 min at 4 °C.

3.4 Immunoprecipitation: Part 2 — Lysate Preparation

1. Thaw the pellets prepared in 3.1.15 and 3.2.10 on ice.
2. Resuspend and pool the SNS pellets corresponding to 200 mice in 5 ml PXL+ buffer and distribute the lysate into three fresh nuclease-free 2 ml microfuge tubes. Resuspend total homogenate pellet corresponding to two mice in 1 ml PXL+

buffer in the same nuclease-free microfuge tube it was frozen in.

3. Incubate the lysates on ice for 10 min.
4. 4.Add 100 U/tube and 36 U/tube of RNase-free DNase to each tube of SNS and total homogenate lysates, respectively. Incubate for 5 min at 37 °C in a ThermoMixer C (Eppendorf) shaking at 1100 rpm (*see* **Note 14**).
5. Remove debris from the lysates by centrifugation at full speed (~17,200 × *g*) for 20 min at 4 °C in a tabletop microcentrifuge.
6. Transfer the supernatant into fresh 2 ml nuclease-free tubes. A small aliquot of lysate can be removed at this point to control for the immunoprecipitation efficiency (*see* **Note 15**).
7. Proceed with the remainder of the lysate to immunoprecipitation.

3.5 Immuno-precipitation: Part 3—Pull-Down

1. Wash the now antibody-conjugated beads thrice in 1 ml of PXL + buffer on a magnetic rack.
2. Add the cleared supernatant to the beads and incubate the mixture on a rotating wheel at 4 °C for 3 h. After incubation, a small aliquot of supernatant can be removed to control for the immunoprecipitation efficiency (*see* **Note 15**).
3. Sequentially wash the beads twice each with 1 ml ice-cold PXL +, 5× PXL+, PNK buffer. Leave the tubes on ice in the final wash buffer while preparing the MNase solution.

3.6 Nuclease Treatment

1. Dilute the S7 MNase to a total amount of 100 U and 0.02 U in 500 μl of MNase buffer per tube.
2. Remove the final wash buffer from the beads and add 500 μl of diluted MNase to washed beads.
3. Incubate on the ThermoMixer C (Eppendorf) at 37 °C and 1400 rpm for 1 min.
4. Let the tubes rest on ice for 3 min.
5. Incubate on the thermomixer at 37 °C and 1200 rpm for 15 s.
6. Repeat **steps 4** and **5** twice more.
7. Sequentially wash the beads twice each with 1 ml ice-cold PNK/EGTA buffer (which will quench the MNase activity) and 1 ml wash buffer, followed by three washes with 1 ml PNK buffer.

3.7 RNA Radiolabeling (See Note 16)

1. Add 400 μl of PNK buffer per tube. Mix thoroughly, but carefully, with a p1000 pipette and reserve 40 μl in a fresh nuclease-free 1.5 ml tube in order to visualize the immunoprecipitated protein by immunoblot (Subheading 3.9). Remove the PNK buffer of the aliquot with the help of a magnetic rack

and resuspend the beads in 40 μl 1× 4× NuPAGE™ LDS sample buffer in PNK/EGTA buffer with reducing agent if required (*see* **Notes 17** and **18**).

2. Remove the remaining buffer in the tube with the help of a magnetic rack and resuspend the beads with 40 μl of kinase reaction mixture per tube (4 μl 10× PNK buffer, 2 μl T4 PNK, 32 μl nuclease-free water and 2 μl ^{33}P-γ-ATP solution).
3. Incubate at 37 °C and 1100 rpm for 20 min on a tabletop thermomixer.
4. Place the tubes back on ice and add 5 μl of 1 mM ATP.
5. Return the tube to the thermomixer and incubate at 37 °C and 1100 rpm for further 5 min.
6. Place the tube on the magnet rack and discard the solution.
7. Wash the beads three times with 1 ml each of cold PNK/EGTA buffer.
8. Elute the radiolabeled RNA–RBP complexes from the bead-conjugated antibodies with 40 μl of 1× 4× NuPAGE™ LDS sample buffer in PNK/EGTA buffer with reducing agent if required (*see* **Note 18**) by mixing at 70 °C and 1000 rpm for 10 min in a tabletop ThermoMixer C (Eppendorf).
9. Shortly spin the samples and recover the supernatant into a new tube with the help of the magnetic rack.

3.8 SDS-PAGE and Transfer onto a Nitrocellulose Membrane

1. Load each 40 μl sample into two lanes (20 μl each) on a Novex NuPAGE 4–12% Bis-Tris Gel in parallel with 3 μl of a pre-stained protein ladder of choice (*see* **Note 19**).
2. Run the PAGE at 100 V in 1× NuPAGE™ MOPS SDS Running buffer until the loading buffer front reaches the bottom of the gel.
3. Transfer the gel to a nitrocellulose membrane (0.45 μm pore size) using a semidry transfer apparatus at 20 V for 1.5 h.
4. After the transfer, rinse the nitrocellulose membrane with PBS and gently tap it on a tissue paper to remove the excess buffer.
5. Place the membrane and wrap it in a plastic film (*see* **Note 20**) before exposure under a phosphor screen for 2 h at room temperature in an autoradiograph exposure cassette.
6. Develop the phosphor image with a Typhoon FLA 9500 apparatus (*see* **Note 21**). The autoradiograph should show a sharp RNA band at the expected molecular weight of the single molecule of the RBP of interest (in this case, FUS is 70 kDa) in the sample treated with 100 U MNase, and an RNA–RBP smear of lower mobility complexes in the sample treated with 0.02 U MNase (Fig. 2a). Nitrocellulose membrane can be stored at −80 °C until isolation of the RNA (Subheading 3.10) (*see* **Note 22**).

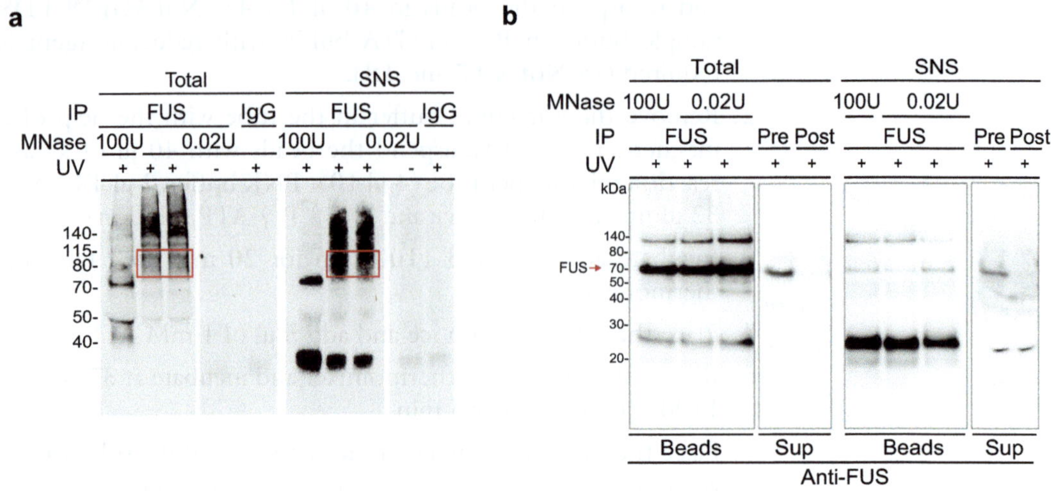

Fig. 2 Representative autoradiograph and corresponding immunoblot of FUS-CLIP in total cortex and cortical synaptoneurosomes. (**a**) Autoradiograph of FUS–RNA complexes immunoprecipitated from total cortex and cortical synaptoneurosomes (SNS) trimmed by different concentrations of MNase (100 U and 0.02 U). The red box highlights the region of the RNA smear corresponding to the RNA–RBP complexes from a single FUS molecule (70 kDa). (**b**) Immunoblot from the same samples as in a, showing the efficiency of FUS immunoprecipitation from both total cortex and SNS, resulting in protein depletion in the lysate after immunoprecipitation (post), compared to input (pre)

3.9 SDS-PAGE and Immunoblot

1. Perform an SDS-PAGE as described in steps1 and 2, Subheading 3.8 with the immunoprecipitation aliquot samples from step 1, Subheading 3.7. Samples saved for visualizing the efficiency of immunoprecipitation could be run similarly in parallel (*see* **Note 15**).
2. Transfer the gel onto a nitrocellulose membrane (0.2 μm pore size) following a standard semidry transfer protocol (*see* **Note 23**).
3. Saturate the membrane with blocking solution for at least 15 min at room temperature.
4. Incubate the primary antibody against the RBP of choice in the antibody diluent overnight at 4 °C (*see* **Note 5**).
5. Wash the membrane with abundant PBS-T thrice for 5 min at room temperature.
6. Incubate the appropriate HRP-conjugated secondary antibody in the antibody diluent for 1 h at room temperature.
7. Wash the membrane with abundant PBS-T thrice for 5 min at room temperature.
8. Develop the immunoblot with a luminol-based solution using a digital imager to confirm the presence and exact position of the free RBP protein in the immunoprecipitated sample (*see*

Note 7). Only the free protein, and not the RNA–RBP complex smear, is detected by immunoblot (Fig. 2b).

3.10 RNA Isolation

1. Using a clean scalpel blade, cut out the fragment of the nitrocellulose membrane just above the size of the free protein to purify the small gel-shifted RNA (*see* **Note 24**). Cut each band into smaller pieces and place into separate microtubes.
2. Fragment the nitrocellulose membrane in tiny pieces with a clean scalpel blade and place them in a nuclease-free 1.5 ml microfuge tube.
3. Freshly prepare a 4 mg/ml proteinase K solution in PK buffer. Preincubate it at 37 °C for 20 min to kill any RNases and add 200 μl of this solution to each tube of isolated nitrocellulose fragments.
4. Incubate at 37 °C and 1000 rpm for 20 min.
5. Add 200 μl of PK/Urea buffer and further incubate for 20 min.
6. Add 400 μl acid phenol–chloroform–isoamyl alcohol solution and continue the incubation for an additional 20 min.
7. Spin tubes at full speed (~17,200 × *g*) for 5 min at room temperature in a tabletop microcentrifuge.
8. Collect the aqueous (upper) phase and transfer to a clean microfuge tube.
9. Precipitate the RNA by adding 1 ml of ethanol–isopropanol 1: 1 solution, 40 μl 3 M NaOAc, pH 5.2 and 0.8 μl of glycogen.
10. Place the tube overnight at −20 °C.
11. Centrifuge for 10 min at 12000 × *g* and 4 °C.
12. Carefully discard the supernatant and wash the pellet with 1 ml of 70% ethanol.
13. Air-dry the pellet and resuspend in 11 μl water (*see* **Note 25**).
14. Incubate the tube at 65 °C for 5 min.

3.11 cDNA Library Preparation and Sequencing

1. Generate the cDNA libraries suitable for high throughput sequencing with the SMARTer Stranded Total RNA-Seq Kit—Pico Input Mammalian following the manufacturer's instructions as the amount of RNA obtained from SNS is small (*see* **Note 26**).
2. Validate the quality and quantity of the enriched libraries using a TapeStation.
3. Normalize libraries in cDNA library buffer.
4. Strand-specific paired-end CLIP libraries were sequenced on HiSeq 2500 for 15 cycles to proceed with data analysis. Each read was 76 bp long. In general, paired-end reads are not needed for CLIP-seq and the shortest, widely mappable reads are ideal to define the RBP binding site.

4 Notes

1. Slight deviations in the composition of the PBS are well tolerated in this protocol. For convenience, one can also use commercial cell-culture grade PBS.
2. The basis of the 5× PXL buffer is 5× PBS. Slight deviations in the composition of the 5× PBS are well tolerated in this protocol. For convenience, one can also use commercial cell-culture grade 10× PBS diluted by half to prepare the solution.
3. Classical 4× Laemmli loading buffer with DTT as reducing agent (200 mM Tris, 8% w/v SDS, 40% v/v glycerol, 0.2% bromophenol blue, 400 mM DTT, pH 6.8) should also work well for this technique.
4. Do not add acid or base to adjust pH. The buffer will range from pH 9.0 to 9.4, depending on the quality of the Tris, glycine, nuclease free water and methanol used. Methanol should be analytical reagent grade.
5. Blocking solution can be adapted according to the RBP and its primary antibody of choice. For example, antibodies against a phospho-RBP give better specificity when incubated in 1% BSA in PBS-T after thoroughly washing the blocking solution containing skimmed milk [13]. Secondary antibodies can nonetheless be incubated in PBS-T 1% (w/v) skimmed milk.
6. If available, the primary antibody raised against the RBP of interest used for the immunoblot should be of a different species to that used for the immunoprecipitation in order to prevent unwanted detection of the heavy and light chains of the bead-conjugated antibody in the immunoblot. If the antibody is covalently cross-linked to the beads during the pull down, then the same antibody could be used for both the immunoprecipitation and the immunoblot (*see* **Note 17**).
7. Enzymatic-conjugated secondary antibodies are recommended over fluorophore-conjugated ones due to their higher sensitivity for protein detection thanks to signal amplification. However, fluorophore-conjugated antibodies could be used if the amount of immunoprecipitated protein is high enough to allow detection. Immunoblot developing must therefore be based on the initial sample amount (*see* **Note 8**) [14].
8. The amount of isolated RNA obtained after the CLIP-seq procedure is relatively low, which represents one of the major drawbacks of this technique. It is important to obtain a sufficient amount of good-quality RNA in order to generate a successful cDNA library for sequencing. Therefore, the amount of starting material has to be optimized depending on the abundance of the protein in the subcellular location of

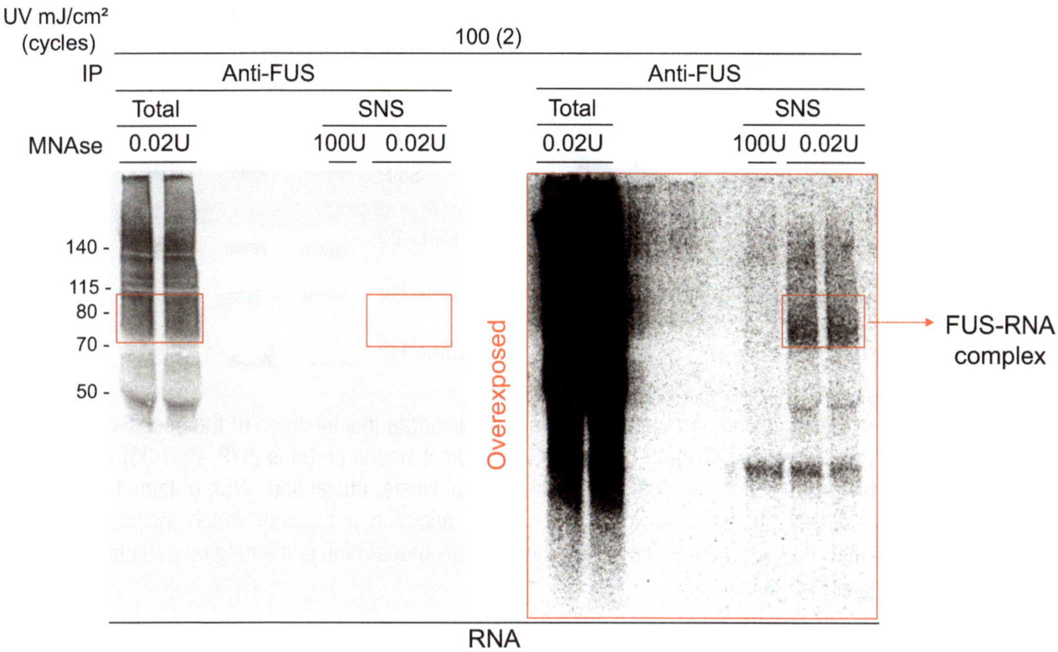

Fig. 3 Observation of RNA–RBP complexes obtained by CLIP depends on the amount and quality of the starting material. Autoradiograph of FUS–RNA complexes immunoprecipitated from total cortex (from 2 mice) and cortical SNS (from five mice) and trimmed by different concentrations of MNase as indicated, only showed the smear of RNA–FUS complexes for the total cortex sample (left). However, longer exposure of the same autoradiograph (red outline, right) revealed the presence of the RNA–FUS complexes also in the SNS-CLIP sample

interest and the nature of the sample used for CLIP-seq (tissue, cultured cells). In the particular case of FUS, only a small proportion is present at the synapses, which caused the isolation of synaptic FUS RNA targets to be challenging. As observed by autoradiograph, the amount of RNA obtained from total cortex homogenates is so noticeably higher than that from SNS, due to the differences in protein amounts (Fig. 4b), that RNA–RBP complexes could not be visualized at the same exposure (Fig. 3). As FUS is minimally present at the synapses, we scaled up the starting material in order to obtain enough RNA sufficient for cDNA library preparation. SNS isolated from cortices of 200 mice were used for CLIP-seq to identify synaptic FUS RNA targets.

9. In order to obtain the accurate RNA targets of the protein of interest at a given cellular location, it is of the utmost importance to ensure a clean enrichment of such fraction, avoiding contamination with protein from other compartments that may be binding more abundant RNA targets. Being a majorly nuclear protein, FUS is minimally present in the synapses, which required the enrichment protocol to render SNS

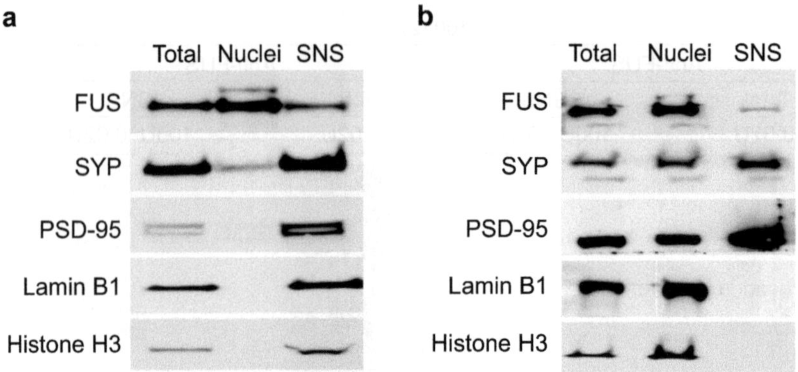

Fig. 4 Importance of clean subcompartment enrichment for accurate identification of the specific subcellular targets of the RBP of interest. (**a**) Western blot of FUS, specific synaptic proteins (SYP, PSD-95) and nuclear proteins (Lamin B1, Histone H3) in the different fractions (total lysate, nuclei and SNS), obtained during SNS isolation with a previously published protocol [15]. (**b**) By adapting a protocol which included an extra centrifugation step prior to the filtration step [16], a cleaner SNS preparation containing no detectable nuclear proteins by immunoblot was obtained

fractions devoid of nuclear fractions. By optimizing previously published protocols [15, 16], nuclei depletion in the SNS fraction was ensured by immunoblot of total, nuclear, and SNS fractions from sections 1.3, 1.4, and 1.10, respectively (Fig. 4a, b).

10. Different cellular subcompartments can be isolated and/or enriched as a preliminary step before performing CLIP-seq, such as nuclear or cytoplasmic fractions via simple fractionation [17, 18] or different organelles via immunomagnetic isolation [19, 20]. In any case, the isolation should be as clean as possible and CLIP-seq from total lysate should be carried as control in order to compare the RNAs bound in the given compartment as a fraction of the total RNA targets bound by the RBP of interest.

11. UV irradiation to covalently cross-link RNA–protein interactions can greatly affect protein stability itself, depending on both the UV energy intensity and the number of irradiation cycles. Since protein degradation will result in a low RNA yield after CLIP, it is essential to optimize the UV cross-linking conditions to ensure a balance between protein stability and effective RNA–RBP cross-linking. Cross-linking at a fixed UV intensity for a different number of cycles (1, 2, or 3), followed by immunoprecipitation and western blot, results in an increased degradation of the RBP directly proportional to the number of irradiation cycles (Fig. 5a). Additionally, optimization of the UV intensity used for cross-linking allowed us to observe a much intense RNA–RBP complex smear with smaller energy amounts (Fig. 5b). Of important note, a non–UV

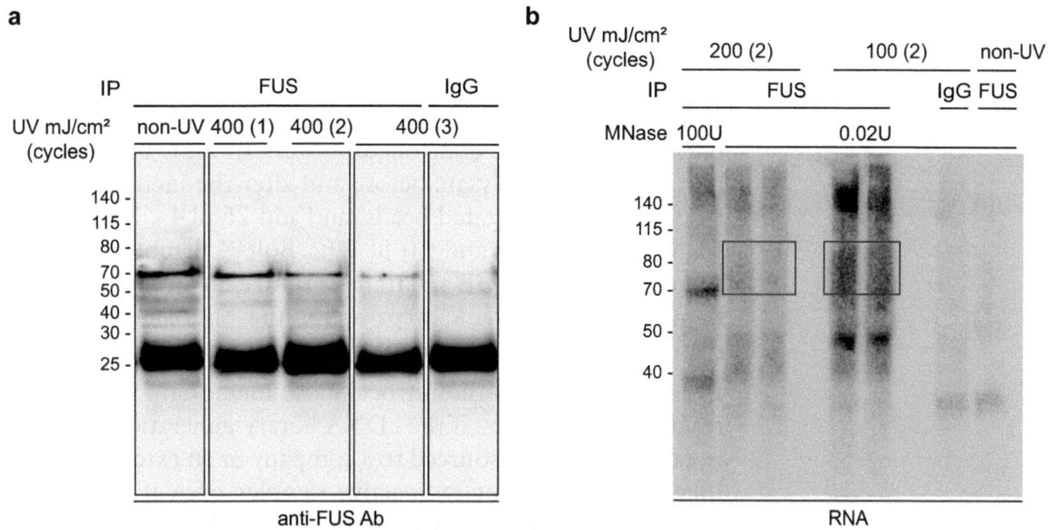

Fig. 5 Finding a balance between efficient protein–RNA cross-linking and protein stability. (**a**) Immunoblot of immunoprecipitated FUS from SNS cross-linked at 400 mJ/cm^2 for a different number of cycles (1, 2, or 3) showed degradation of FUS protein with increased number of UV cycles. (**b**) Autoradiograph of FUS–RNA complexes immunoprecipitated from cortical SNS treated with different UV conditions (200 mJ/cm^2 and 100 mJ/cm^2). The red box marks the FUS–RNA complexes at the expected molecular weight of a single FUS molecule (70 kDa)

cross-linked control sample is indispensable for this optimization and for further autoradiograph visualization and analysis. Cross-linking energy and the number of cycles can greatly vary among the experimental models. For example, for monolayer cultured cells, one cross-linking cycle at 150 mJ/cm^2 is enough to achieve a good compromise between cross-linking efficiency and protein degradation [21].

12. Since immunoprecipitation beads may nonspecifically bind both proteins and nucleic acids, it is highly recommended that the antibody-to-beads ratio is optimized beforehand. To ensure saturation of the beads with the antibody, and to minimize the amount of free protein G that might nonspecifically bind other factors, it is recommended to determine the antibody concentration that leaves detectable antibody levels in the supernatant after incubation with the beads.
13. In order to ensure the specific pull down of the RBP of interest with its corresponding RNA targets, it is of utmost importance to perform a control experiment (western blot and autoradiograph) with beads in the presence of an IgG isotype control to check for immunoprecipitation specificity by western blot (Fig. 2a, b).
14. DNase concentration and incubation times may need optimization depending on the DNA concentration in the subcellular

compartment of choice. It is critical that all DNA has been thoroughly removed from the sample before continuing with immunoprecipitation.

15. Depletion of the RBP of interest in the lysate after immunoprecipitation can be controlled by western blot. For this purpose, save 65 μl of lysate before and after the incubation with the antibody-conjugated beads, and add 25 μl 4× NuPAGE™ LDS sample buffer and 10 μl 10× Bolt™ Sample Reducing Agent. For SDS-PAGE and transfer onto a nitrocellulose membrane, follow the same steps as for the radiolabeled sample (from Subheading 3.6, **step 9** until Subheading 3.7, **step 3**).
16. Isolated RNA from the nitrocellulose membrane will contain residual radioactivity. If the cDNA library generation from the isolated RNA is outsourced to a company or an external facility, they may not accept radioactive samples even if the Geiger counter shows no counts. In that case, the sample treated with 0.02 U MNase should not be radiolabeled, and the radiolabeled 100 U sample will serve as a guide of the molecular size for the free unbound protein. Isolation of the RNA from the nitrocellulose membrane will thus be done blindly. It is therefore recommended to perform all optimization steps with radiolabeled RNA in order to determine the steps that yield the best RNA–RBP complex smear.
17. Pulled down aliquots in 1× loading buffer can be kept at −80 °C until use. For SDS-PAGE and transfer onto a nitrocellulose membrane, the same steps as for the radiolabeled sample should be followed (from Subheading 3.6, **step 9** until Subheading 3.7, **step 3**).
18. Since the primary antibody used for immunoprecipitation is coupled to the magnetic beads through interaction to protein G, addition of a reducing agent (DTT, beta-mercaptoethanol or a commercial solution) to the loading buffer will result in the dissociation of the light (~25 kDa) and heavy (~50 kDa) immunoglobulin chains, which can affect the migration of the RBP of interest. Including a reducing agent in the loading buffer will thus depend on the size of the RBP of interest and, consequently, the size of the RNA–RBP complexes. For proteins smaller than 70 kDa, it would be recommended to skip the reducing agent. If pull down is otherwise performed with an antibody covalently linked to the beads, a nanobody or through a high-affinity interaction (e.g., biotin–streptavidin), no antibody has to be taken into account and addition of a reducing buffer will generally result in neater immunoblots and autoradiographs.
19. The main sample treated with 0.02 U of MNase is loaded onto two lanes in order not to saturate the signal and to observe a

clear separation of the RNA–RBP complexes from potentially other RNA–protein complexes. For the 100 U sample and other controls, the whole 40 μl can be loaded on one lane. It is also recommended to leave empty lanes between different conditions to prevent cross-contamination of the different samples.

20. Plastic wrapping allows to confine the radioactivity not bound to the membrane and prevents the contamination of the cassette and, particularly, the apparatus, which can be used for nonradioactive experiments.
21. An alternative to using phosphor screens for exposing the radioactive nitrocellulose membrane would be to use X-ray films. However, the advantage of using phosphor imager screen is that this is more sensitive compared to X-ray films and thus requires shorter developing time.
22. Immunoprecipitation, RNA labeling, SDS-PAGE, and transfer onto a nitrocellulose membrane are performed on the same day. However, the radioactive membrane can be either exposed on the same day and then stored at −80 °C until RNA isolation takes place or stored at −80 °C and then exposed on the following day, when RNA isolation should also be performed.
23. Gel transfer onto a nitrocellulose membrane for immunoblot can be done following a standard semidry transfer, not necessarily following the same conditions as for the transfer of the radiolabeled complexes. If desired, the process can be sped up using commercially available stacks for a fast semidry transfer (*see* Subheading 2.4).
24. In order to obtain the accurate RBP binding sites on its targets, excised RNA–RBP complexes from the nitrocellulose membrane should correspond to the RBP bound to ssRNA molecules of 50–75 bases in size, which roughly corresponds to a shift of 16–24 kDa in their molecular weight.
25. Caution must be taken not to overdry the RNA pellet as excessive dehydration may result in poor resolubilization of RNA, which may affect the downstream steps.
26. Selection of the kit for cDNA library generation is based on the amount of RNA obtained.

Acknowledgments

We gratefully acknowledge the support from the animal facility of the University Hospital of Zurich (USZ) and the Short Read Sequencing group from the Genomics/Transcriptomics team at the Functional Genomics Center Zurich (FGCZ) from the

University of Zurich (UZH). We thank Julien Weber for experimental support and the National Center of Competence in Research (NCCR), RNA & Disease funded by the Swiss National Science Foundation (SNF) for financial support.

References

1. Ule J, Jensen KB, Ruggiu M, Mele A, Ule A, Darnell RB (2003) CLIP identifies nova-regulated RNA networks in the brain. Science 302:1212–1215. https://doi.org/10.1126/science.1090095
2. König J, Zarnack K, Luscombe NM, Ule J (2012) Protein-RNA interactions: new genomic technologies and perspectives. Nat Rev Genet 13(2):77–83. https://doi.org/10.1038/nrg3141
3. Murigneux V, Saulière JÔ, Roest Crollius H, Le Hir H (2013) Transcriptome-wide identification of RNA binding sites by CLIP-seq. Methods 63(1):32–40. https://doi.org/10.1016/j.ymeth.2013.03.022
4. Ishigaki S, Masuda A, Fujioka Y, Iguchi Y, Katsuno M, Shibata A, Urano F, Sobue G, Ohno K (2012) Position-dependent FUS-RNA interactions regulate alternative splicing events and transcriptions. Sci Rep 2:1–8. https://doi.org/10.1038/srep00529
5. Lagier-Tourenne C, Polymenidou M, Hutt KR et al (2012) Divergent roles of ALS-linked proteins FUS/TLS and TDP-43 intersect in processing long pre-mRNAs. Nat Neurosci 15:1488–1497. https://doi.org/10.1038/nn.3230
6. Masuda A, Takeda JI, Okuno T, Okamoto T, Ohkawara B, Ito M, Ishigaki S, Sobue G, Ohno K (2015) Position-specific binding of FUS to nascent RNA regulates mRNA length. Genes Dev 29:1045–1057. https://doi.org/10.1101/gad.255737.114
7. Rogelj B, Easton LE, Bogu GK et al (2012) Widespread binding of FUS along nascent RNA regulates alternative splicing in the brain. Sci Rep 2:1–10. https://doi.org/10.1038/srep00603
8. Zhou Y, Liu S, Liu G, Öztürk A, Hicks GG (2013) ALS-associated FUS mutations result in compromised FUS alternative splicing and autoregulation. PLoS Genet 9:e1003895
9. Hoell JI, Larsson E, Runge S, Nusbaum JD, Duggimpudi S, Farazi TA, Hafner M, Borkhardt A, Sander C, Tuschl T (2011) RNA targets of wild-type and mutant FET family proteins. Nat Struct Mol Biol 18(12):1428–1431. https://doi.org/10.1038/nsmb.2163
10. Schoen M, Reichel JM, Demestre M et al (2016) Super-resolution microscopy reveals presynaptic localization of the ALS/FTD related protein FUS in hippocampal neurons. Front Cell Neurosci 9:1–16. https://doi.org/10.3389/fncel.2015.00496
11. Deshpande D, Higelin J, Schoen M, Vomhof T, Boeckers TM, Demestre M, Michaelis J (2019) Synaptic FUS localization during motoneuron development and its accumulation in human ALS synapses. Front Cell Neurosci 13:1–17. https://doi.org/10.3389/fncel.2019.00256
12. Sahadevan S, Hembach KM, Tantardini E et al (2020) Synaptic accumulation of FUS triggers age-dependent misregulation of inhibitory synapses in ALS-FUS mice. bioRxiv 2020.06.10.136010. https://doi.org/10.1101/2020.06.10.136010
13. Laferriere F, Maniecka Z, Perez-Berlanga M et al (2019) TDP-43 extracted from frontotemporal lobar degeneration subject brains displays distinct aggregate assemblies and neurotoxic effects reflecting disease progression rates. Nat Neurosci 22:65–77. https://doi.org/10.1038/s41593-018-0294-y
14. Bass JJ, Wilkinson DJ, Rankin D, Phillips BE, Szewczyk NJ, Smith K, Atherton PJ (2017) An overview of technical considerations for Western blotting applications to physiological research. Scand J Med Sci Sports 27:4–25. https://doi.org/10.1111/sms.12702
15. Muddashetty RS, Kelić S, Gross C, Xu M, Bassell GJ (2007) Dysregulated metabotropic glutamate receptor-dependent translation of AMPA receptor and postsynaptic density-95 mRNAs at synapses in a mouse model of fragile X syndrome. J Neurosci 27(20):5338–5348. https://doi.org/10.1523/JNEUROSCI.0937-07.2007
16. Williams C, Shai RM, Wu Y, Hsu YH, Sitzer T, Spann B, McCleary C, Mo Y, Miller CA (2009) Transcriptome analysis of synaptoneurosomes identifies neuroplasticity genes overexpressed in incipient Alzheimer's disease. PLoS One 4(3):e4936. https://doi.org/10.1371/journal.pone.0004936
17. Suzuki K, Bose P, Leong-Quong RYY, Fujita DJ, Riabowol K (2010) REAP: a two minute

cell fractionation method. BMC Res Notes 3: 294. https://doi.org/10.1186/1756-0500-3-294

18. Sanford JR, Coutinho P, Hackett JA, Wang X, Ranahan W, Caceres JF (2008) Identification of nuclear and cytoplasmic mRNA targets for the shuttling protein SF2/ASF. PLoS One 3: 1–10. https://doi.org/10.1371/journal.pone.0003369
19. Kausch AP, Owen TPJ, Narayanswami S, Bruce BD (1999) Organelle isolation by magnetic immunoabsorption. BioTechniques 26: 336–343. https://doi.org/10.2144/99262rr04
20. Tchikov V, Fritsch J, Kabelitz D, Schütze S (2010) Immunomagnetic isolation of subcellular compartments. In: Kabelitz D, Kaufmann SHE (eds) Immunology of Infection. Academic Press, New York, pp 21–33. https://doi.org/10.1016/S0580-9517(10)37002-4
21. Jutzi D, Campagne S, Schmidt R et al (2020) Aberrant interaction of FUS with the U1 snRNA provides a molecular mechanism of FUS induced amyotrophic lateral sclerosis. Nat Commun 11:6341. https://doi.org/10.1038/s41467-020-20191-3

Chapter 20

Monitoring Virus-Induced Stress Granule Dynamics Using Long-Term Live-Cell Imaging

Vera Magg, Philipp Klein, and Alessia Ruggieri

Abstract

The integrated stress response is a highly regulated signaling cascade that allows cells to react to a variety of external and internal stimuli. Activation of different stress-responsive kinases leads to the phosphorylation of their common downstream target, the eukaryotic translation initiation factor 2 alpha (eIF2α), which is a critical component of functional translation preinitiation complexes. As a consequence, stalled ribonucleoprotein complexes accumulate in the cytoplasm and condense into microscopically visible cytoplasmic stress granules (SGs). Over the past years, numerous microscopy approaches have been developed to study the spatiotemporal control of SG formation in response to a variety of stressors. Here, we apply long-term live-cell microscopy to monitor the dynamic cellular stress response triggered by infection with chronic hepatitis C virus (HCV) at single-cell level and study the behavior of infected cells that repeatedly switch between a stressed and unstressed state. We describe in detail the engineering of fluorescent SG-reporter cells expressing enhanced yellow fluorescent protein (YFP)-tagged T cell internal antigen 1 (TIA-1) using lentiviral delivery, as well as the production of mCherry-tagged HCV *trans*-complemented particles, which allow live tracking of SG assembly and disassembly, SG number and size in single infected cells over time.

Key words Integrated stress response, Stress granule dynamics, Stress granule reporter cell line, Virus infection, Fluorescent reporter virus, Hepatitis C virus, Long-term live-cell imaging, Spinning disc confocal microscopy

1 Introduction

Cellular homeostasis is maintained by highly specialized stress kinases that detect imbalances of vital cellular processes and trigger a rapid cellular response referred to as the integrated stress response (ISR) [1]. The ISR ensures rapid adaptation to a wide range of stressors such as nutrient deprivation, virus infection, oxidative damage or endoplasmic reticulum (ER) stress caused by the accumulation of unfolded proteins. A central regulatory node of the ISR is the phosphorylation of the eukaryotic translation initiation factor 2α (eIF2α), which is part of the ternary complex and delivers the

Daniel Matějů and Jeffrey A. Chao (eds.), *The Integrated Stress Response: Methods and Protocols*, Methods in Molecular Biology, vol. 2428, https://doi.org/10.1007/978-1-0716-1975-9_20,

initiator tRNA methionine to the small 40S ribosomal subunit, thereby leading to stalling of translation initiation and polysome disassembly [2]. In mammals, four distinct kinases phosphorylate eIF2α in response to different types of stress [3]. Among them, Protein Kinase R (PKR) is an interferon (IFN)-induced kinase that senses viral double-stranded (ds) RNA replication intermediates in the cytoplasm of infected cells. PKR-like ER kinase (PERK) can sense the accumulation of viral proteins in the ER. Both kinases mediate translation suppression in response to the replication of many RNA viruses [4]. The repression of global protein synthesis and likely the accumulation of stalled preinitiation complexes promote the formation of stress granules (SGs), which are droplets that separate from the cytosol by phase separation. RNA binding proteins such as RasGAP-associated endoribonuclease 1 (G3BP1), Caprin1, T cell internal antigen 1 (TIA1), fragile X mental retardation protein (FMRP) and ubiquitin-associated protein 2-like (UBAP2L) engage into protein–protein and protein–RNA interactions via their intrinsically disordered regions and promote SG condensation [5].

The highly dynamic nature of SGs and SG resident proteins and mRNAs has been addressed by several studies. Notably, fluorescence recovery after photobleaching (FRAP) [5, 6] and fluorescence loss in photobleaching (FLIP) [6–8] experiments have revealed the transient character of SG composition, as SGs undergo constant exchanges with the cytosol and P-bodies. SG dynamics was monitored in living cells using overexpression of monomeric fluorescent proteins fused to SG components such as TIA-1, PABP-1, eIF4E, and G3BP1, or components of the translation machinery such as subunits of eIF3, and time-lapse microscopy [9–12]. SG assembly begins with the appearance of small foci at the cell periphery that become microscopically visible within 15–30 min and later grow by fusion and move on microtubules toward the perinuclear region [6, 13]. In mammalian cells, SG size varies depending on the type and duration of stress and can reach up to 5 μm^3 [12]. SG disassembly and disappearance can occur within a few hours after stress relief and translational reactivation [14]. Using long-term live-cell imaging, we discovered that in contrast to other stress types, which mainly induce an acute and strong translation repression, infection with chronic hepatitis C virus (HCV) induces highly dynamic cycles of SG assembly and disassembly (Fig. 1) concomitant with alternating phases of stalled and active host cell translation. This singular behavior is long-lasting, conserved among several RNA viruses and governed on the molecular level by the antagonistic actions of PKR and Growth Arrest and DNA-Damage-inducible 34 (GADD34) [12], the stress-induced regulatory subunit of protein phosphatase 1, which mediates eIF2α dephosphorylation to resume translation initiation [15].

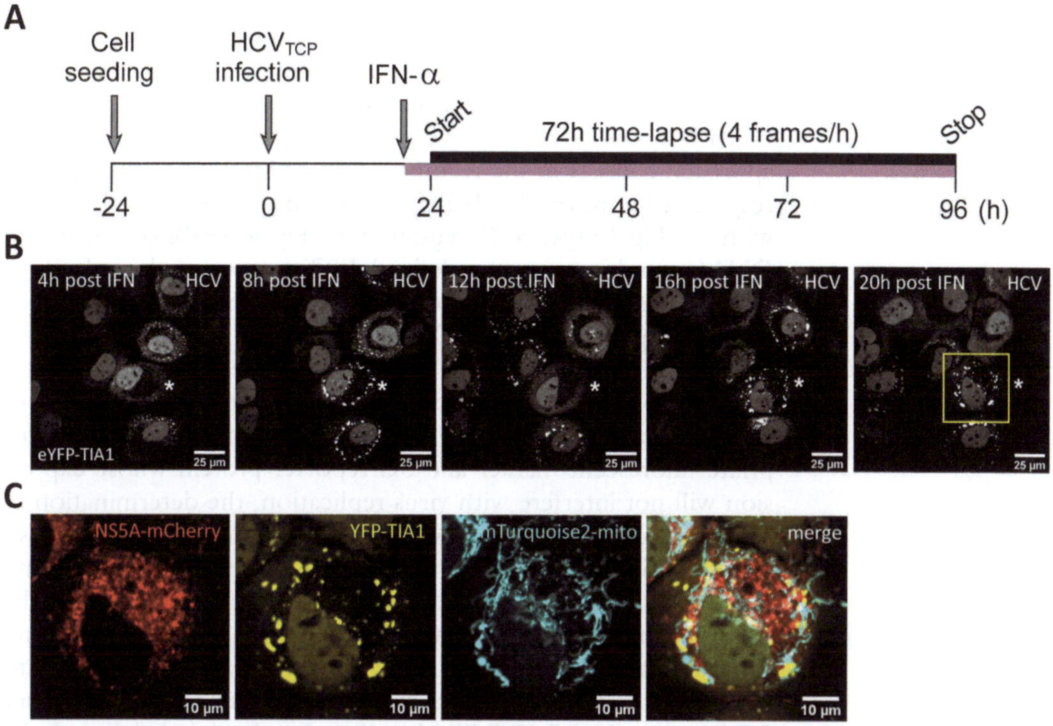

Fig. 1 Multichannel long-term live-cell imaging of HCV_{TCP} infected YFP-TIA1 reporter cells. Huh7 cells stably expressing yellow fluorescent protein fused to the SG protein TIA-1 (YFP-TIA1) and mTurquoise2-mito, a mitochondria-targeting peptide fused to the cyan fluorescent protein mTurquoise2, were infected with mCherry-tagged HCV *trans*-complemented particles (HCV_{TCP}). (**a**) Experimental set-up. Cell are seeded 24 h before infection and treated with 100 IU/ml IFN-α at 48 h post infection. Images are acquired for 72 h at 15 min interval (4 frames per hour). (**b**) Still images of a representative movie at the time points indicated at the top post IFN-α treatment. The asterisk indicates a representative infected cell in which SGs appear and disappear over time. The yellow square represents a cropped section shown as an example in (**c**). Scale bar, 25 μm. (**c**) Cropped section of the movie showing all recorded channels (mCherry-NS5A, YFP-TIA1, mTurquoise2-mito). Scale bar, 10 μm

SG composition as well as the interaction network of G3BP1 vary upon different stress conditions, including virus infection [16, 17]. Recent work by Sanders and colleagues highlighted the importance of stoichiometry-dependent interactions during SG phase separation and how they shape SG composition [18]. As a consequence, the overexpression of SG proteins might alter this interaction network and shift the balance toward promoting or impairing SG assembly. Indeed, transient or stable ectopic expression of several SG resident proteins such as G3BP1 [19], G3BP2 [20], TIA-1 [14], tristetraprolin (TTP) [21], Fas-activated serine/threonine kinase (FAST) [22], or Caprin1 [23] can induce spontaneous SG nucleation. On the other hand, overexpression of the SG proteins HSP70 and Staufen1 impairs SG formation in response to stress [24, 25]. Thus, although less time and cost efficient, alternatives to overexpression should be envisaged in the future. The insertion of a fluorescent tag at endogenous gene loci using the CRISPR-mediated knock-in technology is widely expanding [26, 27] and has already been successfully applied to visualize G3BP1 in cell lines [28] and in vivo in zebrafish [29]. Another approach consists in the reconstitution of knockout cells with the respective fluorescently labeled SG protein at physiological levels or with the Flp-In system (integration via Flp recombinase-mediated DNA recombination at a defined FRT site inserted in the host genome) as recently used to study the dynamics of the unfolded protein response sensor IRE1α [30].

The successful implementation of long-term live-cell microscopy to monitor virus-induced stress response requires a number of aspects to be considered that include the identification of an appropriate fluorescent marker and SG-reporter protein whose expression will not interfere with virus replication, the determination of the duration of the experiment and the frequency of image acquisition, and finally the selection of the most suitable equipment and settings to maintain cells in an optimal environment during the 72-h imaging period. Here, we describe the materials, methods and "tricks" that we optimized to monitor SG dynamics in response to HCV infection. This includes the generation of stable fluorescent SG-reporter Huh7 YFP-TIA1 cells using lentiviral delivery, the production and purification of mCherry-labeled HCV *trans*-complemented particles (HCV_{TCP}) and the description of our microscopy equipment and settings. While the procedure of virus production described in this chapter has specifically been optimized for HCV, the following live-cell imaging protocol can easily be adapted for infections with other full-length fluorescent reporter viruses [31–33] or for studying SG dynamics upon other stress conditions.

2 Materials

2.1 Cell Lines

Cell lines should be regularly tested for contamination with mycoplasma.

1. For lentivirus production: HEK 293 T cells (ATCC Cat. #CRL-3216).
2. For the generation of SG reporter cell line: Huh7 cells [34]: highly passaged (passage 130).
3. For live-cell imaging: Huh7 YFP-TIA1 Neo [12]: cells stably express yellow fluorescent protein (YFP) fused to the SG protein TIA-1 (*see* **Note 1**) and the neomycin resistance gene.
4. For the production of mCherry-labeled HCV_{TCP}: Huh7.5 [CE1][E2p7NS2] cells [35]: cells stably express HCV capsid

(C), envelope (E1 and E2), the viroporin p7 and nonstructural (NS) protein 2 for the *trans*-encapsidation of HCV RNA genome harboring mCherry (*see* **Note 2**).

2.2 Cell Culture Reagents

1. Cell culturing medium: Dulbecco's modified Eagle medium (DMEM) containing 4.5 g/l D-glucose, 2 mM L-glutamine is supplemented with 1x nonessential amino acids, 100 U/ml penicillin, 100 μg/ml streptomycin, and 10% fetal calf serum (FCS). This complete medium can be stored at 4 °C for up to several months.
2. Phosphate-buffered saline (PBS).
3. Trypsin–EDTA solution (1×).
4. G418 (Geneticin): 100 mg/ml stock solution prepared in H_2O and filter-sterilized through a 0.22 μm pore size membrane. Culturing medium is supplemented with G418 at a final concentration of 1 mg/ml for selection of Huh7 cells after lentiviral transduction and of 0.5 mg/ml for maintenance of cells in culture.

2.3 Production of Lentiviruses and Generation of Fluorescent SG-Reporter Cell Lines

1. Lentiviral packaging plasmids: (1) pCMVΔ8.91, vector encoding human immunodeficiency virus type 1 genes *gag*, *pol*, *tat*, and *rev* under the control of cytomegalovirus immediate early promoter [36, 37], (2) pMD2.G, plasmid encoding the envelope glycoprotein (G) of vesicular stomatitis virus, which provides a broad tropism to the pseudotyped lentiviral particles and stability upon purification [38]. All were kindly provided by Didier Trono (*see* **Note 3**).
2. Biscistronic lentiviral transfer plasmid pWPI YFP-TIA1 Neo [12], this plasmid allows for the packaging of a transgene in virus particles, transduction and stable integration of the transgene into the host genome and simultaneous expression of YFP-TIA1 and of the neomycin resistance gene (*see* **Note 3**).
3. Polyethylenimine (PEI) solution (polyethylenimine 25 kD linear, Polysciences Inc.): dissolve the PEI powder in H_2O. Adjust the pH to 7.0–7.2 with HCl by stirring until complete dissolution. The solution will become clear as the pH is adjusted. This process can take several hours. Adjust the volume to the final concentration of 1 mg/ml. Filter-sterilize the solution through a 0.22 μm pore size membrane. Aliquots can be kept at 4 °C for up to 1 month or stored at −20 °C for longer periods (*see* **Note 4**).
4. Reduced-serum medium for the transfection of lentiviral plasmids: Opti-MEM (Gibco, Thermo Fisher).
5. 0.45 μm pore size membrane filters.
6. Fluorescence-assisted cell sorting (FACS).

2.4 Production of HCV_{TCP}

1. Plasmid expressing HCV strain JFH1 subgenomic RNA: pFKi389 Neo NS3-3' JFH NS5A aa383 mCherry K1402Q [12]. This plasmid harbors a T7 polymerase promoter that allows for the in vitro synthesis of HCV self-replicative bicistronic RNA (HCV replicon) (*see* [12] for detailed protocol). The latter encodes for the neomycin resistance gene under the control of HCV internal ribosome entry site (IRES) followed by HCV replicase genes (nonstructural proteins NS3 to NS5B) under the control of the encephalomyocarditis virus (EMCV) IRES. The mutation K1402Q [39] was inserted in the NS3 sequence to enhance HCV_{TCP} production. The insertion of the fluorescent protein mCherry in NS5A allows for the visualization of HCV-infected cells.
2. Cytomix solution [40] for electroporation of in vitro transcribed HCV replicon RNA into cells: 120 mM KCl, 0.15 mM $CaCl_2$, 10 mM K_2HPO_4/KH_2PO_4, 25 mM HEPES, 2 mM EGTA, 5 mM $MgCl_2$. Adjust pH to 7.6 with KOH. Filter-sterilize the solution through a 0.22 μm pore size membrane and store at room temperature. Supplement freshly with 2 mM ATP and 5 mM glutathione before use (stock solutions adjusted to pH 7.6 with KOH, sterilized by filtration and stored in aliquots at −20 °C). This solution enhances cells survival after electroporation.
3. Electroporation device: Bio-Rad Gene Pulser II Electroporator and corresponding electroporation cuvettes (gap width: 0.4 cm).
4. Vacuum-driven 0.45 μm pore size membrane filter units for larger volumes.
5. Disposable centrifugal filter devices: Centricon Plus-70 centrifugal filter unit and Amicon Ultra-15 centrifugal filter unit (15 ml sample volume) with Ultracel-100 regenerated cellulose membrane and 100 kDa cut-off (Millipore).
6. Density gradient medium: Iodixanol, for example, OptiPrep (60% iodixanol in water (w/v)). For the two-layer cushion, prepare a 10% OptiPrep solution: 15 ml OptiPrep (19.8 g) in 150 ml PBS and a 28% OptiPrep solution: 42 ml OptiPrep (55.4 g) in 150 ml DMEM without FCS. Filter-sterilize the solutions through a 0.22 μm pore size membrane and store at 4 °C.
7. Ultracentrifuge, swinging-bucket rotor (SW32 Ti) and corresponding open-top thin-wall Ultra-Clear tubes (25 × 89 mm/6 × 38 ml).

2.5 Live-Cell Microscopy

1. 12-well glass bottom plates with black polystyrene frame (Cellvis) or 35 mm glass-bottom dishes with 20 mm glass diameter

(MatTek), both with cover glass #1.5 (thickness 0.16–0.19 mm). We highly recommend the usage of glass over polymer bottoms for live-cell microscopy when using objectives with high numerical aperture as they provide an increased image quality also at low signal intensities. The use of 12-well glass-bottom plates has the advantage of allowing the parallel acquisition of different conditions, for example, noninfected and infected cells or treated and untreated cells.

2. Microscopy medium: phenol red-free DMEM with 4.5 g/l D-glucose and 25 mM HEPES, supplemented before use with 100 U/ml penicillin, 100 μg/ml streptomycin, and 10% FCS.
3. Nikon Eclipse Ti2/Andor Revolution W1 spinning disc confocal microscope (*see* **Note 5**) equipped with a 20× air objective CFI Plan Apo lambda (NA = 0.75) (*see* **Note 6**), Nikon Perfect Focus System (PFS), Andor lasers 514 nm (40 mW) and 561 nm (50 mW), triple line dichroic beamsplitter 445/514/561 (*see* **Note 7**), emission filters for YFP (540/30) and mCherry (600/50), Nikon motorized stage with linear encoder, EM CCD Camera Andor iXon DU-888, 13 μm × 13 μm pixel size.
4. Okolab environmental control chamber enclosing the microscope and the objectives, which provides constant temperature (37 °C) and a humidified CO_2 atmosphere to maintain cells (*see* **Note 8**). Environmental control is required for long-term imaging to avoid medium evaporation and associated changes in pH that will cause stress and affect cell survival.
5. Nikon Imaging software NIS-Elements Advanced Research version 5.21.
6. ilastik: interactive machine learning for (bio)image analysis (Linux v1.3) (*https://www.ilastik.org/*) [41].

3 Methods

The following steps are executed at room temperature if not otherwise indicated. Handling of lentiviruses and HCV_{TCP} requires biosafety level 2 conditions and appropriate permissions. A graphical summary of the workflow is presented in Fig. 2.

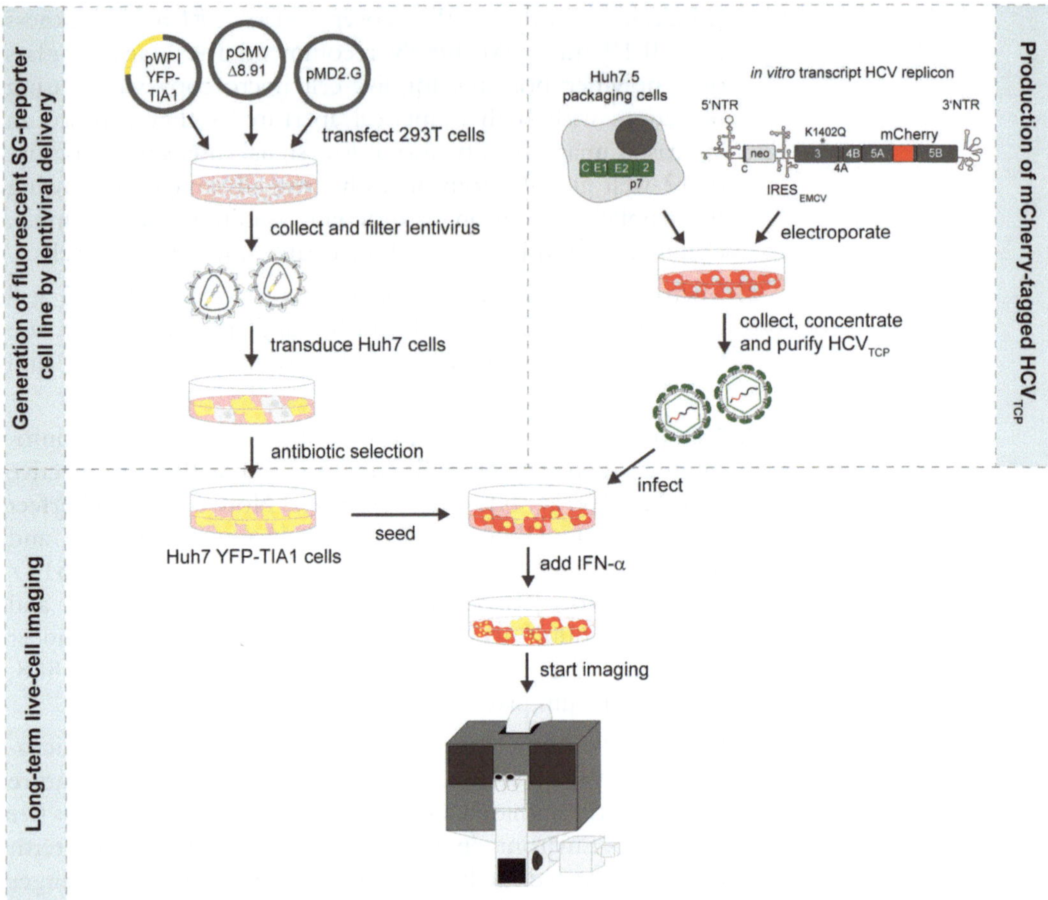

Fig. 2 Live-cell imaging workflow. Stable fluorescently labeled SG-reporter cell lines are generated by lentiviral delivery (upper left panel). For lentivirus production, 293 T cells are transfected with the packaging plasmids pCMVΔ8.91 and pMD2.G and the lentiviral transfer plasmid pWPI YFP-TIA1 Neo encoding the fluorescently labeled SG-reporter TIA-1 and a neomycin resistance cassette. Huh7 cells are transduced with filtered lentiviral supernatant and subsequently subjected to antibiotic selection. mCherry-tagged HCV *trans*-complemented particles (HCV_{TCP}) are generated by electroporation of Huh7.5 packaging cell lines encoding HCV structural proteins with in vitro transcripts of HCV subgenomic replicon encoding mCherry-tagged NS5A (upper right panel). Stable Huh7 YFP-TIA1 cells are infected with purified HCV_{TCP} and SG dynamics is analyzed using long-term live-cell imaging (bottom panel)

3.1 Generation of a Fluorescent SG-Reporter Cell Line Using Lentiviral Delivery

3.1.1 Production of VSV-G Pseudotyped Lentiviral Particles (See ***Note 9****)*

1. Seed 5×10^6 HEK 293 T cells in a 10-cm cell culture dish (to be nearly 50% confluent).
2. Sixteen to 24 h after seeding, replenish with 6 ml of fresh culturing medium per dish. This step should be performed at least 1 h before transfection to allow equilibration to the appropriate pH and temperature.
3. Prepare the plasmid mix by diluting the lentiviral packaging, lentiviral transfer and envelope plasmids to a 3:3:1 ratio (6.42 μg of pCMVΔ8.91, 6.42 μg of pWPI YFP-TIA1 Neo,

and 2.16 μg of pMD2.G, 15 μg DNA in total) in 1.5 or 2 ml tube. Add Opti-MEM to a final volume of 400 μl per transfection. Mix by gently vortexing.

4. Prepare the PEI mix by adding 45 μl PEI solution to 355 μl Opti-MEM per transfection in a new 1.5 or 2 ml tube. This corresponds to a DNA–PEI ratio of 1:3. Mix by gently vortexing.
5. Add the PEI mix to the plasmid mix and vortex immediately for about 30 s. Incubate for 20 min at room temperature.
6. Add the 800 μl of transfection mix dropwise to HEK 293 T cells to prevent their detachment and gently swirl the dish before placing it back into the incubator.
7. Five to six hours after transfection, remove the transfection medium and replace it with 10 ml of fresh culturing medium by slowly pipetting on the corner of the dish.
8. Forty-eight hours after transfection, check the transfection efficiency by assessing YFP expression levels using a fluorescence microscope. Harvest the lentivirus-containing supernatant and filter through a 0.45 μm pore size membrane to remove cell debris. Supernatants can be stored at 4 °C for up to 24 h without loss of infectivity or directly used for transduction (*see* **Note 10**).
9. Estimate the amount of lentiviral particles released in the supernatant (*see* **Note 11**).

3.1.2 Generation of Stable SG Reporter Cells for Live-Cell Imaging

1. Seed 5×10^5 Huh7 cells in a 10-cm cell culture dish (to be nearly 30% confluent). Prepare one additional dish for the antibiotic killing control (nontransduced cells) (*see* **Note 12**).
2. Sixteen to 24 h after seeding, transduce the target cells with the freshly harvested or thawed lentivirus-containing supernatant. Medium can be replenished 8 h posttransduction (*see* **Note 13**).
3. Forty-eight hours after transduction, transduced cells are selected by addition of the corresponding antibiotics. Replenish medium with 8 ml fresh culturing medium containing the respective antibiotics (Huh7 YFP-TIA1 Neo cells are selected with 1 mg/ml G418). It is important to add the antibiotic selection medium to the nontransduced cells and monitor antibiotics killing kinetics.
4. Grow the transduced cells in the antibiotic selection medium until approx. 90% of non-transduced cells have died in the killing control plate. Split cells once if necessary. Note that antibiotic selection is less efficient on confluent cells.
5. Assess YFP-TIA1 expression levels using a fluorescence microscope.

6. Expand antibiotic-resistant cells, freeze and store in liquid nitrogen.
7. FACS-sort SG-reporter cells based on their fluorescence intensity (*see* **Note 14**). Expand cells after FACS sorting, freeze and store in liquid nitrogen.
8. Characterize the sorted SG-reporter cell line including formation of SGs in response to chemical stressors such as arsenite and thapsigargin [42] (*see* **Note 15**).

3.2 Production and Purification of HCV_{TCP}

1. Detach Huh7.5[CE1][E2p7NS2] cells with trypsin–EDTA and transfer the cell suspension to a 50 ml Falcon tube (*see* **Note 16**). Keep an aliquot and count live cells with a hemocytometer chamber. Fill up with PBS and wash by spinning for 5 min at 1000 × *g*.
2. Remove the supernatant and wash the cell pellet a second time with 50 ml PBS. Spin for 5 min at 1000 × *g*.
3. Resuspend cells to a final concentration of 1.5×10^7 cells/ml in the cytomix solution freshly supplemented with 2 mM ATP and 5 mM glutathione.
4. Mix 400 μl cells (6×10^6 cells) with 10 μg of HCV replicon in vitro transcript. Transfer the mix into the electroporation cuvette and the cuvette immediately into the electroporation chamber.
5. Pulse. The following settings were determined for an optimal transfection and survival of Huh7 cells and must be adjusted for different cell types: 975 μF, 270 V with an expected time constant of ~18–20 ms.
6. Resuspend the electroporated cells in 500 μl of culturing medium and transfer them into a 10-cm cell culture dish containing 6 ml of fresh culturing medium.
7. Harvest supernatants at 48, 72 and 96 h post electroporation, filter through a vacuum-driven 0.45 μm pore size membrane unit to remove cell debris. Store transiently at 4 °C until the last harvest. Replenish producer cells with 6 ml fresh culturing medium.
8. Ninety-six hours post electroporation, pool the collected supernatants and reduce the total volume using Centricon Plus-70 centrifugal filter devices (*see* **Note 17**). Load 60 ml of filtered virus-containing supernatant on the Centricon unit and spin for 25 min at 3500 × *g* at 4 °C. Extend the time by 5–10 min if the volume has not completely passed though the filter. Recover the concentrated supernatant (approx. by tenfold) by reverting the filter cups onto the dedicated collector unit and spin for 2 min at 1000 × *g*. Repeat this step until the

volume of the virus-containing supernatant is reduced by maximum 20-fold.

9. Meanwhile, precool the ultracentrifuge, rotor and buckets at 4 °C.
10. Prepare a two-layer OptiPrep cushion in which HCV_{TCP} will sediment based on their density: add 20 ml of preconcentrated virus-containing supernatant on the bottom of SW32 ultra-clear tubes. Carefully underlay the supernatant with 3 ml of 10% OptiPrep solution (use a 5 ml serological pipette filled with 4 ml to avoid any air bubble). Carefully underlay the 10%-layer with the 28% OptiPrep solution (use a 5 ml serological pipette filled with 5 ml to avoid any air bubble). Fill up the tube from the top by adding 10 ml of virus-containing supernatant. Mark the interface between the different layers to simplify virus recovery after centrifugation. Place the tubes into the SW32 buckets, close them and tare weight allowing for a maximum of 0.01 g difference. Ultracentrifuge at 28,000 rpm (SW32 Ti, 96,281 × *g*) for 4 h at 4 °C.
11. Take out the tubes and slowly aspirate the liquid from the top, leaving approx. 6 ml in the tube (middle of the marked 10%-layer). Recover the virus-OptiPrep fraction of all tubes and pool in a 50 ml Falcon tube.
12. Reduce the OptiPrep content by purifying the concentrated virus using Amicon centrifugal devices: dilute the concentrated virus-OptiPrep mixture 1:2 with phenol-red free DMEM. Load on an Amicon filtration unit and spin 3500 × *g* for 20 min at 4 °C. Discard the flow-through and rinse the concentrated virus supernatant by loading an additional 8 ml of phenol-red free DMEM on the filtration unit and spinning at 3500 × *g* for 20 min at 4 °C. Recover the concentrated virus supernatants (approx. 200 μl), pool, and store 50-μl aliquots at −80 °C.
13. Determine virus stock infectious titer using a limited dilution assay [43].

3.3 Long Term Live-Cell Imaging

3.3.1 Cell Seeding and Infection with HCV_{TCP} (See Note 18)

1. Detach and count Huh7 YFP-TIA1 Neo cells. Make sure cells are thoroughly resuspended to avoid uneven seeding. Seed 3×10^4 cells per well in a 12-well glass-bottom plate or 1×10^5 cells in a 35 mm glass-bottom dish. Cells should be seeded in a total volume of 2 ml, as cells tend to settle mainly at the well periphery when seeded in smaller volumes.
2. Twenty-four hours after seeding, infect cells with concentrated HCV_{TCP} at a multiplicity of infection (MOI) of 3 tissue culture infectious doses ($TCID_{50}$) per cell in a maximum volume of 1 ml.

3. Twenty-four hours after infection, gently wash cells twice with sterile PBS and add 3 ml of microscopy medium (*see* **Note 19**).

3.3.2 Microscope Set-up and Image Acquisition

1. Turn on the environmental control device at least 2 h before starting image acquisition to allow for thermal equilibration of the microscope equipment at 37 °C and stabilization of the stage z-movement to reduce focus drift over time. Cells will be maintained in a humidified atmosphere with 5% CO_2 during acquisition.
2. Place the 12-well plate or the 35 mm dish on the stage holder and make sure they are perfectly immobilized.
3. Determine the optimal acquisition settings for each experiment: (1) set an appropriate laser power and illumination time for each channel to be acquired. We recommend keeping laser intensities as low as possible to avoid phototoxicity. Use EM gain for low signal intensities to maximize the signal-to-noise ratio. However, high EM gain might reduce imaging performance and therefore a balance between these different parameters should be achieved and must be optimized on a case-by-case base (*see* **Note 20**). (2) Set the time interval. In our experience, a 15 min time interval is appropriate for repeated exposure over up to 96 h without inducing phototoxicity and a good compromise to visualize early SG formation events. (3) Set the total imaging duration.
4. Switch on the laser autofocus system to determine the optimal focus plan in order to visualize SGs. On our Nikon equipment, this system called "Perfect Focus System" (PFS) compensates for focus drifts during acquisition and uneven plate bottom thickness (*see* **Note 21**).
5. Select multiple (x,y) positions. Adjust the focus plan for each individual (x,y) position using the PFS. For our experiments, the definition of "suitable" positions is based on the density that cells will reach at the end of the 72 h imaging period. Cell boundaries should be distinguishable to facilitate cell tracking. Thus, a "suitable" position will be characterized by loosely scattered groups of cells expressing the SG reporter at comparable signal intensities (Fig. 3). The number of (x,y) positions that can be recorded simultaneously depends on several parameters such as the illumination time, time required for the microscope to switch optical filters and dichroic beamsplitters, and the movement speed of the stage. With our microscope setup, we usually pick up to 60 (x,y) positions per well when imaging multiple conditions or up to 120 for single conditions (*see* **Note 22**).
6. Start image acquisition (288 frames per color for each positions).

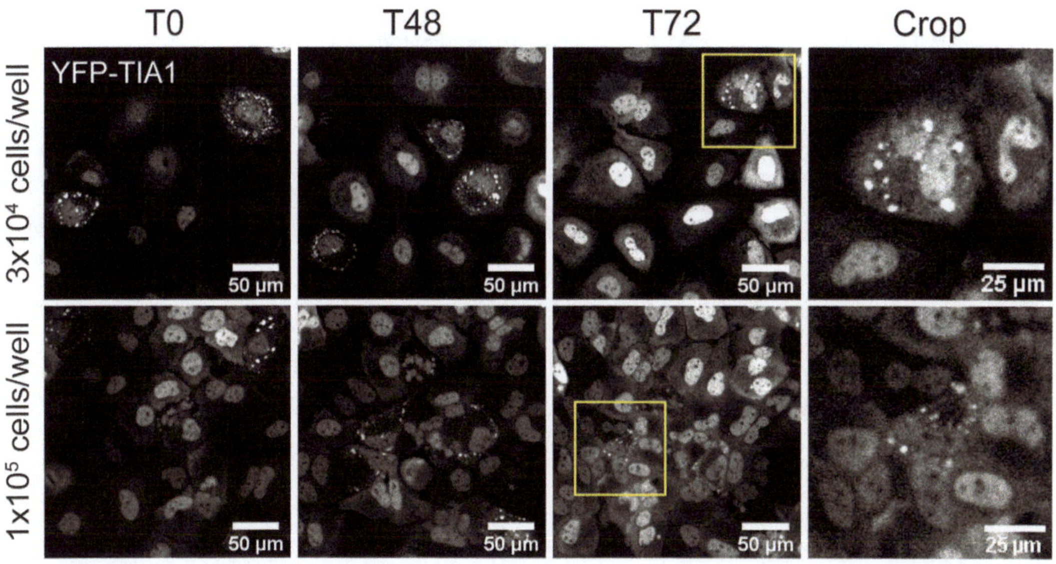

Fig. 3 Impact of cell density on image quality and SG tracking. Comparison of different cell densities from movie start (T0) to end (T72). Huh7 eYFP-TIA1 cells were seeded at the indicated densities and infected with mCherry-tagged HCV_{TCP} for 24 h before imaging start. Shown is the YFP-TIA1 channel of representative (x,y) positions. Scale bar, 50 μm. The yellow square represents a cropped section (crop, scale bar 25 μm). It will not be possible to detect cell boundaries and assign SGs to the correct cells, if cells are nearly 100% confluent at the end of the imaging period

7. At the end of the experiment, archive raw data, export and save multichannel TIF hyperstacks using the NIS Elements Advanced Research software to proceed with image analysis (*see* **Notes 23** and **24**).

3.3.3 Semiautomated Image Processing Pipeline Using Ilastik (See Note 25)

1. Convert the TIF images with ilastik Image Conversion Tool into 8-bit h5 format.
2. Pixel classification: use ilastik Pixel Classification Tool to automatically detect cell nuclei (YFP-TIA1 signal in the nucleus), SGs (YFP-TIA1 condensates in the cytosol) and infection pixels (NS5A-mCherry signal). Use random frames to manually train the software and create separate labels for pixels representing nuclei, SGs, background pixels and virus signal.
3. Create a segmentation mask for each label using ilastik Segmentation tool.
4. Generate single cell tracks with ilastik Tracking Toolkit using the nuclei segmentation masks.
5. SGs are assigned to individual cell tracks at any given point in time using a home-made script and Voronoi tessellation of each timeframe based on the nuclei segmentation center of mass. For each cell track and frame, the respective Voronoi cells are overlaid with SG and infection marker masks to generate time-

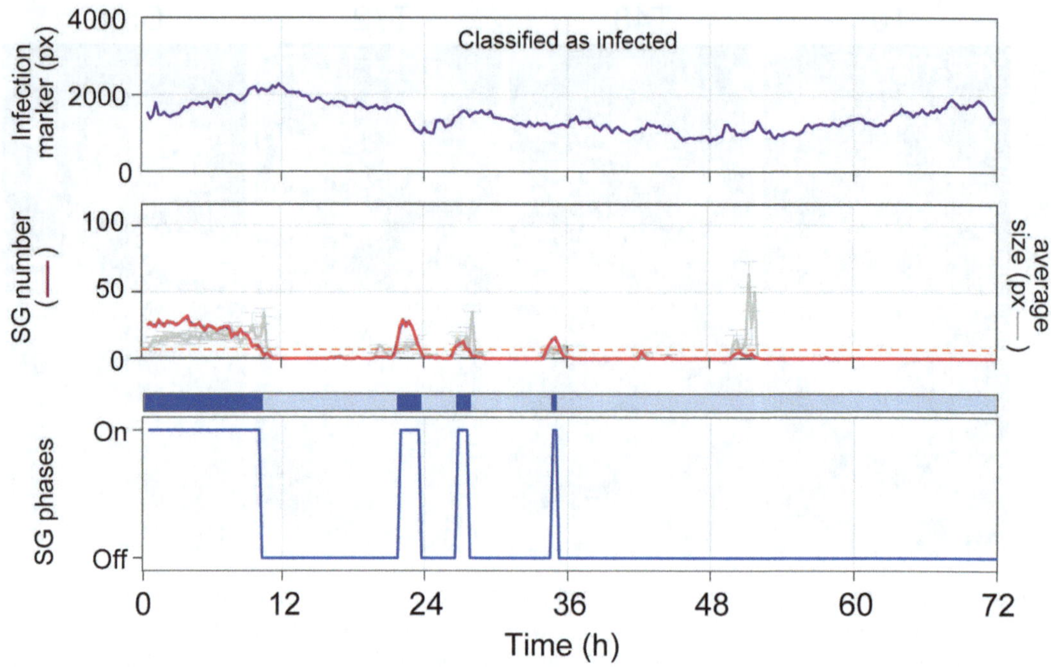

Fig. 4 Representative output from semiautomated analysis of a single-cell trajectory. Using ilastik and home-made scripts, different parameters are analyzed in single cells over time. The plots indicate (a) the levels of infection marker (HCV NS5A-mCherry, in pixel counts per cell), (b) the number of SGs per cell and average SG size (in pixel). The dotted red line represents the threshold above which the sum of pixels that are detected is considered to form a SG (15 pixel, estimated visually for each experiment series). Using these results, we create two graphical representations of single cell SG dynamics represented as plots with "on" and "off" phases (bottom panel) or as shown on the top as a line (light blue: unstressed, dark blue: stressed)

resolved single-cell trajectories of the total number of SGs, and level of infection (number of mCherry-NS5A positive pixels) (Fig. 4). Information for individual cells in each (x,y) recorded position are exported as .csv files. The analysis includes the (a) SG number per cell, (b) SG size (in pixel), and the duration and number of SG phases (Klein, Kallenberger, Ruggieri, unpublished).

6. Verify each cell track manually (we use a home-made Python/ Kivy-based graphical interface). Only tracks with a minimum length of 48 h and correctly assigned presence of SGs are further analyzed.

4 Notes

1. The palette of fluorescent proteins and tags available for live-cell microscopy is in perpetual evolution and offers many alternatives based on their brightness, maturation time, photostability or oligomerization properties [44–46] (see also the

databases fpvis and fpbase [47]). While the choice of the suitable fluorescent tag or protein is usually based on their property or obviously on the available microscope equipment, the choice of the position in which to insert it, that is, N- or C-terminus, can turn out to be complex. The insertion of fluorescent markers can abrogate the functions of proteins or modify their localization. It is therefore essential to confirm that fluorescent SG proteins maintain the same properties as endogenous proteins. Another aspect to consider when studying SG dynamics in response to virus infection is the impact of the chosen SG protein on viral replication and spread. In our previous work, we measured the impact of several SG and P-body proteins on HCV replication and identified G3BP1, TIAR, Lsm1, and Xrn1 as repressors and CNOT1 and CPEB1 as enhancers of viral replication [12]. Finally, as mentioned above, other SG resident proteins might be selected. However, we strongly recommend ascertaining that their levels of expression are not causing spontaneous SG formation when expressed ectopically.

2. The insertion of a fluorescent protein or tag sequence into virus genomes, notably of RNA viruses, often interferes with viral replication. In such a case, cell-based fluorescent reporter systems are useful and allow indirect visualization of virus infection. Several fluorescent reporters were developed based, for example, on the translocation of green fluorescent protein (GFP) from the cytosol to the nucleus upon cleavage of the reporter construct by a viral protease [48–51]. Active HIV replication can also be visualized by coinfection with an adenoviral HIV-LTR-GFP vector, which is transcriptionally activated in HIV-infected cells expressing HIV Tat transactivator protein [52]. Alternatively, infection events can be visualized by coupling the induction of the innate immune response in infected or neighboring cells to the transcriptional activation of a reporter gene [32, 53]. From a kinetics perspective, the use of cell-based fluorescent reporters might be preferable in some cases, as they allow visualization of infection events more readily after infection than when using tagged viruses. For HCV, as well as for dengue virus and SARS-CoV-2, we and others could show that cleavage by the viral protease and translocation of the fluorescent protein into the nucleus occurs 4–6 h postinfection, whereas at least 20 h are necessary before tagged viral proteins become detectable [12, 32, 48, 50, 54].

3. The plasmid pMD2.G and an alternative packaging plasmid of newer generation developed by the Trono lab, psPAX2, are available at Addgene (#12259 and #12260, respectively). The lentiviral transfer plasmid pWPI YFP-TIA1 Neo was derived from pWPI MCS GFP (kind gift from Didier Trono) also available at Addgene (#12254), a self-inactivating vector

whose 5′ long terminal repeat (LTR) was replaced with the human elongation factor 1 alpha promoter, and 3′ LTR was extensively deleted. This plasmid encodes GFP under the control of EMCV IRES.

4. We recommend preparing larger batches (100 ml) of PEI with two different pH and test transfection efficiency of HEK 293 T cells using a green fluorescent protein-expressing plasmid. Transfection efficiency should reach at least 90% to ensure sufficient lentivirus production.
5. Wide field fluorescence microscopes are also commonly used for long-term live cell imaging [32]. However, it might be a challenge, especially when signals are weak, to achieve a good signal-to-noise ratio that is required for automated image analysis, while avoiding photodamage.
6. We routinely use Nikon "lambda" air objectives, whose "Crystal Coat technology" maximizes light collection and are, in our experience, comparable to standard immersion oil objectives. In addition, they allow wide movements of the motorized stage, for example, between different wells of the same plate, without having to worry about the lack of immersion oil at the (x,y) positions. The choice of a low magnification objective (20×) represents a compromise between image resolution, the possibility to detect SGs early during their formation and the number of cells per field of view that can be tracked for at least 48 h (20 to 25 cells on average).
7. Our microscope is configured with a triple line dichroic (445/514/561), which allows for the rapid switch between three fluorescent proteins. With this particular dichroic we can record, in addition to YFP and mCherry, the signal of the fluorescent protein mTurquoise2 (cyan, excitation 445 nm), for example when simultaneously monitoring mitochondria and SG dynamics in HCV-infected cells (Fig. 1).
8. When performing live-cell experiments longer than 24 h, we strongly recommend choosing full microscope enclosures for environmental control, for example, EMBLEM incubator box, Okolab stage incubator, Pecon new lighttight incubator, as they allow thermal equilibration of parts of the microscope body. This will greatly minimize z-movements of the stage and thus reduce focus drift over time. In addition, in contrast to stage-top incubators, such enclosing chambers allow heating of the objectives to 37 °C. Of note, certain environment control enclosures, for example, EMBLEM box, are tightly sealed to the microscope body (and perfectly designed to fit the microscope model) and perfuse 5% CO_2 inside the whole box. Here, 50% humidity is provided by the presence of an additional recipient filled up with double-distilled H_2O. This

level is sufficient to image for up to 96 h without medium evaporation in the plate. Others, for example, Okolab, whose box is not sealed on the microscope body, have an additional closed top-stage incubator in which humidified 5% CO_2 is perfused, reaching humidity levels of about 90%. Such a design is to be considered when imaging cells in small volumes (less than 500 μl) over long time periods. Finally, chambers are now also available with black panels for microscopes that are not located in a dark environment.

9. Alternative to lentiviral delivery, reporter cell lines can also be engineered by plasmid transfection. However, lentiviral transduction has proven to be more efficient for some cell lines and does not necessarily require clonal selection, which eliminates naturally occurring variances in the cell population and might lead to clonal artifacts.

10. If larger volumes of lentivirus supernatant are required, for example when several target cells will be transduced in parallel, 10 ml of fresh culturing medium can be added to producer cells after the first harvest for a second 24 h production, which usually is as productive as the first one. For longer storage, we recommend making 2 ml aliquots and freeze them at −80 °C. Lentiviruses can be stored over 6 months without affecting the infectious titer.

11. Transduction efficiencies of target cells depend among others on the amount of infectious lentiviral particles released in the supernatant during production. We usually estimate the amount of lentivirus particles beforehand by performing a one-step SYBR Green based Product Enhanced Reverse Transcriptase assay (SG-PERT) [55]. This qRT-PCR-based assay measures the activity of the reverse transcriptase (RT) packaged in the lentivirus particles. The use of a recombinant RT standard dilution series allows to determine the amount of RT activity per μl of supernatant (typically 10^8 to 10^9 pU/μl). Importantly, this method does not inform about the number of "infectious" particles but rather about the amount of all physical particles (both infectious and noninfectious) released in the supernatant. Infectious particles usually represent approx. 1% of the physical particles produced. Alternatively, infectious lentivirus titration can be performed using a limited dilution assay in 96-well plates and antibiotic selection. Here, the numbers of colonies will be used to calculate the infectious titer per ml of supernatant. Typically, we obtain lentivirus titers between 10^6 and 10^7 colony forming units (CFU) per ml.

12. Antibiotic killing curves should be performed for each target cell line to determine the optimal antibiotics concentration used for the selection of transduced cells.

13. While higher rates of integration are desired for a satisfactory expression of the fluorescent SG marker, increased rates of random integration into the host genome might lead to poor target cell survival. Thus, in addition to the determination of lentiviral particles concentration, we test different volumes of lentiviral-containing supernatants in parallel. In our experiments, a single round transduction with 6 ml lentivirus-containing supernatant is usually enough. If necessary, transduction efficiency can be improved either by addition of 4 μg/μ l polybrene directly to the lentiviral supernatant during transduction or by repeating transduction rounds for up to three times before proceeding with antibiotics selection.
14. The efficacy of delivery of transgenes using lentiviruses differs between individual cells of a population and therefore results in heterogeneous expression levels. We recommend fluorescence-assisted cell sorting of the transduced cells to minimize the signal variability within the population. We usually sort and amplify medium and high expressing cells. This will greatly improve image quality due to an increased signal-to-noise ratio and thus simplify subsequent image analysis.
15. The absence of spontaneous SG formation due to SG protein overexpression (*see also* **Note 1**) should be assessed by immunofluorescence analysis. In addition, it is important to measure virus replication dynamics in the SG reporter cell lines and compare it to naïve parental cells as well as control cells transduced with lentiviruses harboring the antibiotics resistance gene only. This will rule out potential artifacts resulting from antibiotics selection or from the position of lentiviral integration, and highlight potential effects of overexpressing SG proteins on virus replication [4].
16. HCV production usually yields virus stocks with low infectious titers before concentration and purification. Infectious titers of HCV_{TCP} are usually tenfold lower [12]. Although the purification protocol described here allows to concentrate viral stocks by 100-fold, we largely upscale virus production by increasing the number of electroporations (usually 100 electroporations yield approx. 2 l of HCV-containing supernatant). The reduction of total volume is therefore necessary before proceeding with the density cushion purification of viral particles. The same procedure can be followed for the production and purification of HCV full-length fluorescent reporter viruses. However, in this case, all steps occurring after cell electroporation, including live-cell imaging, are performed in our laboratory under biosafety level 3 conditions.
17. As mentioned by the manufacturer, Centricon centrifugal filter devices are composed of three modules: a capped 70 ml-filter

cup, a collection module for the flow through, and a collector module for the concentrated supernatant. These devices are compatible with swinging-bucket high-speed centrifuges only. Before use, rinse residual glycerin traces by loading 60 ml of sterile H_2O and centrifuging for 5 min at 3500 × *g*. To avoid future contaminations during infection experiments with the concentrated virus stock, immerse the collection module used for concentrated supernatant recovery in 70% ethanol for 5 min and let it dry under the laminar flow cabinet. Passing the supernatant though the filter will also concentrate the FCS present in the supernatant. We therefore avoid concentrating the supernatant by more than 20-fold to keep its density lower than the one of the 10% OptiPrep solution.

18. Before starting with live-cell imaging experiments, viral replication rates, optimal MOI and stress response dynamics should be determined to choose the most suitable time frame and acquisition duration. We suggest performing a fixed-cell immunofluorescence-based kinetics pilot experiment to identify the most relevant condition for live-cell imaging. This will reduce the consumption of virus stocks, image data file size and the time required for subsequent image analysis. The desired cell seeding density depends on the length of the experiment, as well as the individual replication rates of the reporter viruses used and the kinetics of the induced stress response. Thus, cells should be seeded in a way that they will not overgrow toward the end of the time-lapse as SGs and cell boundaries are hardly detectable in individual cells. It is important to note that for some viruses, low cell density reduces infectivity and might require the acquisition of more (x,y) positions.

19. Huh7 cells are poorly immune competent. We discovered that the addition of IFN-α to HCV-infected cells, which mimics the natural cellular response to infection, induces a highly dynamic SG response with increased cycles of SG assembly and disassembly when compared to untreated cells [12]. We usually add 100 IU/ml IFN-α in the microscopy medium and start live-cell imaging within 4 h.

20. We avoid pixel binning during image acquisition as this will greatly impact the image quality. Signal intensity of YFP-TIA1 is rather low. We preferably use less than 50% laser power, an exposure time of 200 ms and reduced EM gain (150). Initial testing of the system and microscope settings is a prerequisite for a successful long-term imaging. We recommend (1) to test potential temperature fluctuations by inserting an additional temperature sensor in the chamber, possibly positioned in close proximity of the imaging dish or plate bottom, (2) to monitor naïve SG reporter cells, which are excellent indicators as they will form SGs if temperature rises or CO_2 lacks.

21. Alternatively, z-stack images can be acquired of all positions [12]. However, this will reduce the number of (x,y) positions that can be recorded over time, likely increase photodamage and considerably increase the size of the output data file.
22. We usually circulate between (x,y) positions on the plate in a clock-wise manner to ensure that the last selected (x,y) position is close proximity to the first one and thereby minimize stage movement between positions.
23. Unequal illumination might bias the detection of fluorescent signal toward the center of the image. This effect can be reduced by performing an illumination correction prior to image analysis. To this end, we generate an illumination bias mask by imaging an autofluorescent plastic slide (Chromas) for each channel to be acquired with the corresponding objective. This mask is further processed by applying a Gaussian blur (radius 50) using Fiji image processing package (https://fiji.sc/) [56]. For all images to be corrected, pixel intensities are divided by this mask using Fiji and subsequently multiplied with a suitable scaling factor to achieve pixel intensities well above background levels and below pixel saturation.
24. Several dynamic SG data and analysis tools were developed by several groups (see for more details [12, 57, 58]).
25. Automated single-cell tracking in such long-term time-lapses is yet challenging due to the relatively long interval between acquisition frames, cell division events, and especially the increasing cell density toward the end of the acquisition period [59, 60]. Thus, analysis often requires manual track correction. Recently, researchers have applied semiautomated correction tools and deep learning approaches to tracking software to reduce the amount of manual correction required [61–63]. We therefore implemented a semiautomated image processing pipeline primarily based on the machine-learning toolkit ilastik [41] to detect and analyze the appearance and disappearance of SGs over time in single cells (Klein, Kallenberger, and Ruggieri, unpublished).

Acknowledgments

We thank Dr. Monika Langlotz from the ZMBH Flow Cytometry & FACS Core Facility (Heidelberg University) for her support in cell sorting, Dr. Vibor Laketa from the Infectious Diseases Imaging Platform (IDIP) at the Center for Integrative Infectious Disease Research (Heidelberg), Dirk Scholz and Klaus Nettesheim from Nikon Instruments Germany for their assistance and support with live-cell imaging. We are very grateful to Karsten Roth and Prof.

Fred A. Hamprecht (HCI IWR, Heidelberg University) for developing the semiautomated image processing pipeline and to Dr. Stefan Kallenberger (BioQuant, Heidelberg University) for analyzing the resulting features. This work was funded by the Deutsche Forschungsgemeinschaft (DFG, German Research Foundation)—Projektnummer 278001972 - TRR 186 A14 and Projektnummer 240245660 - SFB 1129 TP13.

References

1. de Nadal E, Ammerer G, Posas F (2011) Controlling gene expression in response to stress. Nat Rev Genet 12(12):833–845. https://doi.org/10.1038/nrg3055
2. Pakos-Zebrucka K, Koryga I, Mnich K, Ljujic M, Samali A, Gorman AM (2016) The integrated stress response. EMBO Rep 17(10):1374–1395. https://doi.org/10.15252/embr.201642195
3. Taniuchi S, Miyake M, Tsugawa K, Oyadomari M, Oyadomari S (2016) Integrated stress response of vertebrates is regulated by four eIF2alpha kinases. Sci Rep 6:32886. https://doi.org/10.1038/srep32886
4. Eiermann N, Haneke K, Sun Z, Stoecklin G, Ruggieri A (2020) Dance with the devil: stress granules and signaling in antiviral responses. Viruses 12(9):984. https://doi.org/10.3390/v12090984
5. Hofmann S, Kedersha N, Anderson P, Ivanov P (2021) Molecular mechanisms of stress granule assembly and disassembly. Biochim Biophys Acta, Mol Cell Res 1868(1):118876. https://doi.org/10.1016/j.bbamcr.2020.118876
6. Fujimura K, Katahira J, Kano F, Yoneda Y, Murata M (2009) Microscopic dissection of the process of stress granule assembly. Biochim Biophys Acta 1793(11):1728–1737. https://doi.org/10.1016/j.bbamcr.2009.08.010
7. Fujimura K, Kano F, Murata M (2008) Dual localization of the RNA binding protein CUGBP-1 to stress granule and perinucleolar compartment. Exp Cell Res 314(3):543–553. https://doi.org/10.1016/j.yexcr.2007.10.024
8. Gao X, Fu X, Song J, Zhang Y, Cui X, Su C, Ge L, Shao J, Xin L, Saarikettu J, Mei M, Yang X, Wei M, Silvennoinen O, Yao Z, He J, Yang J (2015) Poly(A)(+) mRNA-binding protein Tudor-SN regulates stress granules aggregation dynamics. FEBS J 282(5):874–890. https://doi.org/10.1111/febs.13186
9. Kedersha N, Cho MR, Li W, Yacono PW, Chen S, Gilks N, Golan DE, Anderson P (2000) Dynamic shuttling of TIA-1 accompanies the recruitment of mRNA to mammalian stress granules. J Cell Biol 151(6):1257–1268. https://doi.org/10.1083/jcb.151.6.1257
10. Kedersha N, Anderson P (2007) Mammalian stress granules and processing bodies. Methods Enzymol 431:61–81. https://doi.org/10.1016/S0076-6879(07)31005-7
11. Kedersha N, Tisdale S, Hickman T, Anderson P (2008) Real-time and quantitative imaging of mammalian stress granules and processing bodies. Methods Enzymol 448:521–552. https://doi.org/10.1016/S0076-6879(08)02626-8
12. Ruggieri A, Dazert E, Metz P, Hofmann S, Bergeest JP, Mazur J, Bankhead P, Hiet MS, Kallis S, Alvisi G, Samuel CE, Lohmann V, Kaderali L, Rohr K, Frese M, Stoecklin G, Bartenschlager R (2012) Dynamic oscillation of translation and stress granule formation mark the cellular response to virus infection. Cell Host Microbe 12(1):71–85. https://doi.org/10.1016/j.chom.2012.05.013
13. Nadezhdina ES, Lomakin AJ, Shpilman AA, Chudinova EM, Ivanov PA (2010) Microtubules govern stress granule mobility and dynamics. Biochim Biophys Acta 1803(3):361–371. https://doi.org/10.1016/j.bbamcr.2009.12.004
14. Kedersha NL, Gupta M, Li W, Miller I, Anderson P (1999) RNA-binding proteins TIA-1 and TIAR link the phosphorylation of eIF-2 alpha to the assembly of mammalian stress granules. J Cell Biol 147(7):1431–1442. https://doi.org/10.1083/jcb.147.7.1431
15. Kojima E, Takeuchi A, Haneda M, Yagi A, Hasegawa T, Yamaki K, Takeda K, Akira S, Shimokata K, Isobe K (2003) The function of GADD34 is a recovery from a shutoff of protein synthesis induced by ER stress: elucidation by GADD34-deficient mice. FASEB J 17(11):1573–1575. https://doi.org/10.1096/fj.02-1184fje
16. Piotrowska J, Hansen SJ, Park N, Jamka K, Sarnow P, Gustin KE (2010) Stable formation of compositionally unique stress granules in

virus-infected cells. J Virol 84(7):3654–3665. https://doi.org/10.1128/JVI.01320-09

17. Brocard M, Iadevaia V, Klein P, Hall B, Lewis G, Lu J, Burke J, Willcocks MM, Parker R, Goodfellow IG, Ruggieri A, Locker N (2020) Norovirus infection results in eIF2alpha independent host translation shut-off and remodels the G3BP1 interactome evading stress granule formation. PLoS Pathog 16(1): e1008250. https://doi.org/10.1371/journal.ppat.1008250
18. Sanders DW, Kedersha N, Lee DSW, Strom AR, Drake V, Riback JA, Bracha D, Eeftens JM, Iwanicki A, Wang A, Wei MT, Whitney G, Lyons SM, Anderson P, Jacobs WM, Ivanov P, Brangwynne CP (2020) Competing protein-RNA interaction networks control multiphase intracellular organization. Cell 181(2):306–324.e28. https://doi.org/10.1016/j.cell.2020.03.050
19. Tourriere H, Chebli K, Zekri L, Courselaud B, Blanchard JM, Bertrand E, Tazi J (2003) The RasGAP-associated endoribonuclease G3BP assembles stress granules. J Cell Biol 160(6):823–831. https://doi.org/10.1083/jcb.200212128
20. Matsuki H, Takahashi M, Higuchi M, Makokha GN, Oie M, Fujii M (2013) Both G3BP1 and G3BP2 contribute to stress granule formation. Genes Cells 18(2):135–146. https://doi.org/10.1111/gtc.12023
21. Stoecklin G, Stubbs T, Kedersha N, Wax S, Rigby WF, Blackwell TK, Anderson P (2004) MK2-induced tristetraprolin:14-3-3 complexes prevent stress granule association and ARE-mRNA decay. EMBO J 23(6):1313–1324. https://doi.org/10.1038/sj.emboj.7600163
22. Kedersha N, Stoecklin G, Ayodele M, Yacono P, Lykke-Andersen J, Fritzler MJ, Scheuner D, Kaufman RJ, Golan DE, Anderson P (2005) Stress granules and processing bodies are dynamically linked sites of mRNP remodeling. J Cell Biol 169(6):871–884. https://doi.org/10.1083/jcb.200502088
23. Solomon S, Xu Y, Wang B, David MD, Schubert P, Kennedy D, Schrader JW (2007) Distinct structural features of caprin-1 mediate its interaction with G3BP-1 and its induction of phosphorylation of eukaryotic translation initiation factor 2alpha, entry to cytoplasmic stress granules, and selective interaction with a subset of mRNAs. Mol Cell Biol 27(6):2324–2342. https://doi.org/10.1128/MCB.02300-06
24. Gilks N, Kedersha N, Ayodele M, Shen L, Stoecklin G, Dember LM, Anderson P (2004) Stress granule assembly is mediated by prion-like aggregation of TIA-1. Mol Biol Cell 15(12):5383–5398. https://doi.org/10.1091/mbc.e04-08-0715
25. Thomas MG, Martinez Tosar LJ, Desbats MA, Leishman CC, Boccaccio GL (2009) Mammalian Staufen 1 is recruited to stress granules and impairs their assembly. J Cell Sci 122 (Pt 4):563–573. https://doi.org/10.1242/jcs.038208
26. Lee SH, Kim S, Hur JK (2018) CRISPR and target-specific DNA endonucleases for efficient DNA |knock-in in eukaryotic genomes. Mol Cells 41(11):943–952. https://doi.org/10.14348/molcells.2018.0408
27. Bukhari H, Muller T (2019) Endogenous fluorescence tagging by CRISPR. Trends Cell Biol 29(11):912–928. https://doi.org/10.1016/j.tcb.2019.08.004
28. Wang R, Jiang X, Bao P, Qin M, Xu J (2019) Circadian control of stress granules by oscillating EIF2alpha. Cell Death Dis 10(3):215. https://doi.org/10.1038/s41419-019-1471-y
29. Wang R, Zhang H, Du J, Xu J (2019) Heat resilience in embryonic zebrafish revealed using an in vivo stress granule reporter. J Cell Sci 132(20):jcs234807. https://doi.org/10.1242/jcs.234807
30. Belyy V, Tran NH, Walter P (2020) Quantitative microscopy reveals dynamics and fate of clustered IRE1alpha. Proc Natl Acad Sci U S A 117(3):1533–1542. https://doi.org/10.1073/pnas.1915311117
31. Schaller T, Appel N, Koutsoudakis G, Kallis S, Lohmann V, Pietschmann T, Bartenschlager R (2007) Analysis of hepatitis C virus superinfection exclusion by using novel fluorochrome gene-tagged viral genomes. J Virol 81(9):4591–4603. https://doi.org/10.1128/JVI.02144-06
32. Schmid B, Rinas M, Ruggieri A, Acosta EG, Bartenschlager M, Reuter A, Fischl W, Harder N, Bergeest JP, Flossdorf M, Rohr K, Hofer T, Bartenschlager R (2015) Live cell analysis and mathematical modeling identify determinants of attenuation of dengue virus 2'-O-methylation mutant. PLoS Pathog 11(12):e1005345. https://doi.org/10.1371/journal.ppat.1005345
33. Kanai Y, Kawagishi T, Nouda R, Onishi M, Pannacha P, Nurdin JA, Nomura K, Matsuura Y, Kobayashi T (2019) Development of stable rotavirus reporter expression systems. J Virol 93(4):e01774-18. https://doi.org/10.1128/JVI.01774-18
34. Nakabayashi H, Taketa K, Miyano K, Yamane T, Sato J (1982) Growth of human

hepatoma cells lines with differentiated functions in chemically defined medium. Cancer Res 42(9):3858–3863

35. Steinmann E, Brohm C, Kallis S, Bartenschlager R, Pietschmann T (2008) Efficient trans-encapsidation of hepatitis C virus RNAs into infectious virus-like particles. J Virol 82(14):7034–7046. https://doi.org/10.1128/JVI.00118-08
36. Naldini L, Blomer U, Gallay P, Ory D, Mulligan R, Gage FH, Verma IM, Trono D (1996) In vivo gene delivery and stable transduction of nondividing cells by a lentiviral vector. Science 272(5259):263–267. https://doi.org/10.1126/science.272.5259.263
37. Zufferey R, Donello JE, Trono D, Hope TJ (1999) Woodchuck hepatitis virus posttranscriptional regulatory element enhances expression of transgenes delivered by retroviral vectors. J Virol 73(4):2886–2892. https://doi.org/10.1128/JVI.73.4.2886-2892.1999
38. Burns JC, Friedmann T, Driever W, Burrascano M, Yee JK (1993) Vesicular stomatitis virus G glycoprotein pseudotyped retroviral vectors: concentration to very high titer and efficient gene transfer into mammalian and nonmammalian cells. Proc Natl Acad Sci U S A 90(17):8033–8037. https://doi.org/10.1073/pnas.90.17.8033
39. Kaul A, Woerz I, Meuleman P, Leroux-Roels G, Bartenschlager R (2007) Cell culture adaptation of hepatitis C virus and in vivo viability of an adapted variant. J Virol 81(23):13168–13179. https://doi.org/10.1128/JVI.01362-07
40. van den Hoff MJ, Moorman AF, Lamers WH (1992) Electroporation in 'intracellular' buffer increases cell survival. Nucleic Acids Res 20(11):2902. https://doi.org/10.1093/nar/20.11.2902
41. Berg S, Kutra D, Kroeger T, Straehle CN, Kausler BX, Haubold C, Schiegg M, Ales J, Beier T, Rudy M, Eren K, Cervantes JI, Xu B, Beuttenmueller F, Wolny A, Zhang C, Koethe U, Hamprecht FA, Kreshuk A (2019) Ilastik: interactive machine learning for (bio)image analysis. Nat Methods 16(12):1226–1232. https://doi.org/10.1038/s41592-019-0582-9
42. Aulas A, Fay MM, Szaflarski W, Kedersha N, Anderson P, Ivanov P (2017) Methods to classify cytoplasmic foci as mammalian stress granules. J Vis Exp (123):55656. https://doi.org/10.3791/55656
43. Lindenbach BD, Evans MJ, Syder AJ, Wolk B, Tellinghuisen TL, Liu CC, Maruyama T, Hynes RO, Burton DR, McKeating JA, Rice CM (2005) Complete replication of hepatitis C virus in cell culture. Science 309(5734):623–626. https://doi.org/10.1126/science.1114016
44. Thorn K (2017) Genetically encoded fluorescent tags. Mol Biol Cell 28(7):848–857. https://doi.org/10.1091/mbc.E16-07-0504
45. Xue L, Karpenko IA, Hiblot J, Johnsson K (2015) Imaging and manipulating proteins in live cells through covalent labeling. Nat Chem Biol 11(12):917–923. https://doi.org/10.1038/nchembio.1959
46. Wang L, Frei MS, Salim A, Johnsson K (2019) Small-molecule fluorescent probes for live-cell super-resolution microscopy. J Am Chem Soc 141(7):2770–2781. https://doi.org/10.1021/jacs.8b11134
47. Lambert TJ (2019) FPbase: a community-editable fluorescent protein database. Nat Methods 16(4):277–278. https://doi.org/10.1038/s41592-019-0352-8
48. Jones CT, Catanese MT, Law LM, Khetani SR, Syder AJ, Ploss A, Oh TS, Schoggins JW, MacDonald MR, Bhatia SN, Rice CM (2010) Real-time imaging of hepatitis C virus infection using a fluorescent cell-based reporter system. Nat Biotechnol 28(2):167–171. https://doi.org/10.1038/nbt.1604
49. McFadden MJ, Mitchell-Dick A, Vazquez C, Roder AE, Labagnara KF, McMahon JJ, Silver DL, Horner SM (2018) A fluorescent cell-based system for imaging Zika virus infection in real-time. Viruses 10(2):95. https://doi.org/10.3390/v10020095
50. Pahmeier F, Neufeldt CJ, Cerikan B, Prasad V, Pape C, Laketa V, Ruggieri A, Bartenschlager R, Cortese M (2020) A versatile reporter system to monitor virus infected cells and its application to dengue virus and SARS-CoV-2. J Virol 95(4):e01715–e01720. https://doi.org/10.1128/JVI.01715-20
51. Arias-Arias JL, MacPherson DJ, Hill ME, Hardy JA, Mora-Rodriguez R (2020) A fluorescence-activatable reporter of flavivirus NS2B-NS3 protease activity enables live imaging of infection in single cells and viral plaques. J Biol Chem 295(8):2212–2226. https://doi.org/10.1074/jbc.RA119.011319
52. Richman L, Meylan PR, Munoz M, Pinaud S, Mirkovitch J (2002) An adenovirus-based fluorescent reporter vector to identify and isolate HIV-infected cells. J Virol Methods 99 (1–2):9–21. https://doi.org/10.1016/s0166-0934(01)00375-5
53. Swick A, Baltes A, Yin J (2014) Visualizing infection spread: dual-color fluorescent reporting of virus-host interactions. Biotechnol

Bioeng 111(6):1200–1209. https://doi.org/10.1002/bit.25170

54. Cortese M, Lee JY, Cerikan B, Neufeldt CJ, Oorschot VMJ, Kohrer S, Hennies J, Schieber NL, Ronchi P, Mizzon G, Romero-Brey I, Santarella-Mellwig R, Schorb M, Boermel M, Mocaer K, Beckwith MS, Templin RM, Gross V, Pape C, Tischer C, Frankish J, Horvat NK, Laketa V, Stanifer M, Boulant S, Ruggieri A, Chatel-Chaix L, Schwab Y, Bartenschlager R (2020) Integrative imaging reveals SARS-CoV-2-induced reshaping of subcellular morphologies. Cell Host Microbe 28(6):853–866.e5. https://doi.org/10.1016/j.chom.2020.11.003

55. Pizzato M, Erlwein O, Bonsall D, Kaye S, Muir D, McClure MO (2009) A one-step SYBR green I-based product-enhanced reverse transcriptase assay for the quantitation of retroviruses in cell culture supernatants. J Virol Methods 156(1–2):1–7. https://doi.org/10.1016/j.jviromet.2008.10.012

56. Schindelin J, Arganda-Carreras I, Frise E, Kaynig V, Longair M, Pietzsch T, Preibisch S, Rueden C, Saalfeld S, Schmid B, Tinevez JY, White DJ, Hartenstein V, Eliceiri K, Tomancak P, Cardona A (2012) Fiji: an open-source platform for biological-image analysis. Nat Methods 9(7):676–682. https://doi.org/10.1038/nmeth.2019

57. Perez-Pepe M, Slomiansky V, Loschi M, Luchelli L, Neme M, Thomas MG, Boccaccio GL (2012) BUHO: a MATLAB script for the study of stress granules and processing bodies by high-throughput image analysis. PLoS One 7(12):e51495. https://doi.org/10.1371/journal.pone.0051495

58. Ohshima D, Arimoto-Matsuzaki K, Tomida T, Takekawa M, Ichikawa K (2015) Spatio-temporal dynamics and mechanisms of stress granule assembly. PLoS Comput Biol 11(6):e1004326. https://doi.org/10.1371/journal.pcbi.1004326

59. Coutu DL, Schroeder T (2013) Probing cellular processes by long-term live imaging--historic problems and current solutions. J Cell Sci 126(Pt 17):3805–3815. https://doi.org/10.1242/jcs.118349

60. Skylaki S, Hilsenbeck O, Schroeder T (2016) Challenges in long-term imaging and quantification of single-cell dynamics. Nat Biotechnol 34(11):1137–1144. https://doi.org/10.1038/nbt.3713

61. Van Valen DA, Kudo T, Lane KM, Macklin DN, Quach NT, DeFelice MM, Maayan I, Tanouchi Y, Ashley EA, Covert MW (2016) Deep learning automates the quantitative analysis of individual cells in live-cell imaging experiments. PLoS Comput Biol 12(11):e1005177. https://doi.org/10.1371/journal.pcbi.1005177

62. Han H, Wu G, Li Y, Zi Z (2019) eDetect: a fast error detection and correction tool for live cell imaging data analysis. iScience 13:1–8. https://doi.org/10.1016/j.isci.2019.02.004

63. Mitchell C, Caroff L, Solis-Lemus JA, Reyes-Aldasoro CC, Vigilante A, Warburton F, de Chaumont F, Dufour A, Dallongeville S, Olivo-Marin JC, Knight R (2020) Cell tracking profiler - a user-driven analysis framework for evaluating 4D live-cell imaging data. J Cell Sci 133(22):jcs241422. https://doi.org/10.1242/jcs.241422

Chapter 21

Single-Molecule Imaging of mRNA Interactions with Stress Granules

Tatsuya Morisaki and Timothy J. Stasevich

Abstract

Single-molecule imaging in living cells enables the investigation of molecular dynamics and interactions underlying the physiology of a cell. We recently developed a method to visualize translation events at single-mRNA resolution in living cells. Here we describe how we apply this method to visualize mRNA interactions with stress granules in the context of translational activity during cell stress.

Key words Single-molecule imaging, Stress granules, Translation, MS2, RNA imaging

1 Introduction

When exposed to environmental stress, cells conserve resources by shutting down the translation of the majority of mRNA molecules. Stressed cells also form stress granules (SGs) and a fraction of mRNAs translocate into SGs [1, 2]. Because these processes occur at the same time, it was not clear precisely when, where, and how mRNAs stop translation and interact with SGs. To address these questions, we recently developed technology to track individual mRNA molecules and simultaneously monitor their translational status and dynamic interactions with SGs on the seconds timescale in living cells [3–5].

At the core of our technology is the MS2 stem-loop system developed by several members of the Singer lab over the past 20 years to track individual mRNA molecules live [6, 7]. This system works as follows: The operator region of the MS2 bacteriophage genome forms an RNA stem-loop that is bound by an MS2 coat protein (MCP). By repeating the stem-loop sequence (this sequence is generally referred to as MS2) in a reporter gene (typically 24 repeats placed in the 3′-UTR) and introducing this reporter gene into cells expressing fluorescently labeled MCP, single mRNA molecules can be visualized. This is possible because of the signal

Daniel Matějů and Jeffrey A. Chao (eds.), *The Integrated Stress Response: Methods and Protocols*, Methods in Molecular Biology, vol. 2428, https://doi.org/10.1007/978-1-0716-1975-9_21,

amplification that occurs when multiple MCPs bind multiple stem loops in a single reporter mRNA. MS2 stem-loops are bound by MCP as soon as they are transcribed, so the MS2 system makes it possible to track single mRNA molecules throughout their full lifecycles, from transcription to degradation [7–10]. In addition to MS2, several other similar yet orthogonal systems have been developed to tag and track single RNA molecules in multiple colors, including the PP7 and lamdaN systems [11–13].

Although MS2 makes it possible to track individual mRNA, it is not possible to visualize their translation using MS2 alone. One workaround technique, coined TRICK (translating RNA imaging by coat protein knockoff), was developed by the Chao lab [14]. Briefly, PP7 stem-loops are placed in the coding region of an MS2-tagged reporter mRNA. With this arrangement, fluorescent PP7 coat proteins bound to the PP7 stem loops are knocked off the mRNA reporter by the first ribosome that translates it. While this is a powerful way to distinguish translated from untranslated mRNA, the TRICK assay cannot be used to continuously monitor mRNA translation. To achieve this, our lab and others created a new type of protein tag that fluoresces brightly within seconds of its translation [3, 15–18]. In essence, this new type of tag is a protein-analog of the MS2 tag. Just like MS2 stem loops in the 3′ UTR of a reporter mRNA, we fuse repeats of a short peptide epitope sequence onto the N-terminus of a reporter protein. When the reporter protein is translated in cells that coexpress fluorescent antibody-based probes against the epitope (either Fab, scFv, or nanobodies, the analogs of MCP), the nascent peptide chain can be visualized as it is being translated (analogous to the visualization of transcription using MS2). Here, the signal is amplified not only by the number of repeats of the epitope tag, but also by the number of ribosomes translating the mRNA in polysomes. By combining this new protein tag with an MS2 mRNA tag, we are able to track individual mRNA molecules in living cells and continuously monitor their translational status over multiple rounds of translation [19].

In this protocol, we describe in detail how we used this new technology in cells expressing a fluorescently labeled SG marker to quantify the dynamics of translation shut off of single mRNAs and the interaction between mRNAs and SGs after stress (Fig. 1). We first describe how to set up the experiments, including a method to introduce all of the necessary components into living cells at optimal levels as well as suitable imaging conditions. We then describe the basics of image analysis to elucidate the dynamic parameters of mRNAs during cellular stress.

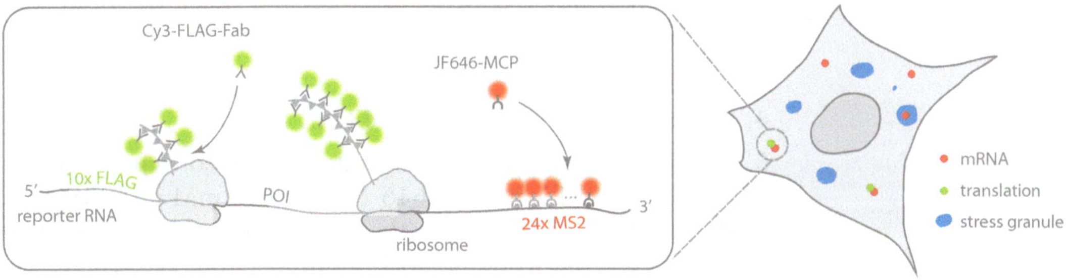

Fig. 1 Schematic illustrating a single-molecule translation reporter system in a SG marker cell

2 Materials

Prepare all solutions in sterile conditions. Sodium arsenite is a carcinogen and a teratogen. Handle with extreme caution and follow waste disposal regulations.

2.1 Sample Prep

Detailed in our recent protocol paper [20].

1. Reporter plasmid: Reporter plasmids can be acquired from Addgene (ID81082–81085). Replace promoter, 5′, 3′- UTR, and/or the protein-coding sequence of interest by subcloning (*see* **Note 1**).
2. Cy3 anti-FLAG Fab: Prepare anti-FLAG Fab from anti-FLAG antibody with the Fab preparation kit. Label anti-FLAG Fab with Cy3 NHS ester and then purify Cy3 anti-FLAG Fab with ultrafiltration. The working stock should be ~0.5 mg/mL.
3. Purified Halo-MCP: Standard protein expression and purification from E.coli should be performed. The expression vector is available upon request. Alternatively, the genetically encoded plasmid can be transfected (*see* **Note 2**).

2.2 Tissue Culture

1. U-2 OS cells stably expressing the SG marker GFP-G3BP1: Isolate GFP-G3BP1 U-2 OS cells from GFP-G3BP1/mRFP-DCP1a U-2 OS cells [21] (*see* **Note 3**).
2. Phosphate buffered saline (PBS).
3. Trypsin.
4. DMEM (+): Add 50 mL of fetal bovine serum, 5 mL Penicillin-Streptomycin (10,000 U/mL), and 5 mL of 200 mM L-Glutamine to 500 mL of Dulbecco's Modified Eagle Medium (DMEM, high glucose, no glutamine).

2.3 Imaging

1. Glass bottom dishes (*see* **Note 4**).
2. 106 μm glass beads: Sterilize glass beads in 2 M NaOH for 2 h. Wash the beads with Milli-Q water until the pH is neutral.

Then rinse the beads with 100% ethanol. Keep the beads dry (*see* **Note 5**).

3. Opti-MEM.
4. White DMEM (+): Phenol-free DMEM (+).
5. Janelia Flour 646 HaloTag Ligands [22]: Dissolve Janelia Flour 646 HaloTag Ligands (JF646-Halo) to 10 mM in DMSO for the stock solution. Dilute the stock solution to 0.2 mM in DMSO for the working stock solution. Store at −20 °C (*see* **Note 6**).
6. Puromycin: Dissolve puromycin dihydrochloride to 50 mg/mL in sterile H_2O.
7. Sodium arsenite: Dissolve sodium arsenite to 50 mM in sterile H_2O.

3 Methods

3.1 Imaging

1. Seed cells on glass-bottom dishes at 75% confluency the night before imaging (*see* **Note 7**).
2. Two hours before bead-loading, wash cells two times with PBS and exchange media to Opti-MEM supplemented with 10% FBS (v/v) (*see* **Note 8**).
3. Prepare sample for bead-loading by mixing the following: 750 ng of reporter plasmid, ~0.5 μg of Cy3 anti-FLAG Fab (~1 μL of the working stock), ~0.13 μg of purified Halo-MCP (~1 μL of the working stock), PBS to a total volume of 4 μL (*see* **Note 9**). Do not vortex.
4. Remove media from the glass-bottom dish, and set aside in a conical tube (*see* **Note 10**). Immediately after removing the media, add the 4 μL of sample mix to the glass area of the dish, then sprinkle glass beads onto the cells as a monolayer. Tap the dish ~10 times against the hood bench. Decant the media that was set aside back into the dish [23, 24].
5. 3–4 h after bead-loading, wash cells three times with PBS. Next, replace the media with white DMEM(+) supplemented with 200 nM JF646-Halo. DMSO concentration should not exceed 0.1%. Incubate the cells at 37 °C for 30 min. Wash the cells three times with PBS, and replace the media with white DMEM(+). There can be some glass beads remaining in the dish, but this should not affect the imaging provided they do not cover the area being imaged. Keep cells in an incubator until ready to image.

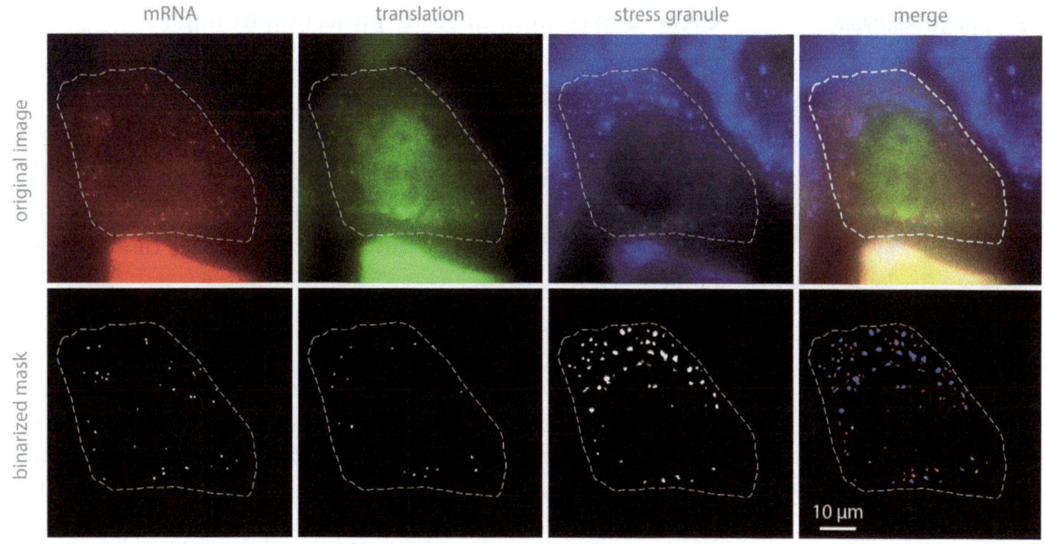

Fig. 2 Example images of cells under stress conditions in each channel and corresponding mask images

6. Place cells onto a stage top incubator at 37 °C and 5% CO_2 on a fluorescence microscope (*see* **Note 11**). Take a fraction of medium into a 1.5 mL microcentrifuge tube, and keep it at 37 °C. Place the lid of the dish loosely on such that the lid can be removed without perturbing the dish when drugs are added. A microfluidic system can be employed for this purpose.
7. Search for cells containing ~10 to ~100 mRNAs in the cytoplasm. mRNA should appear as distinct puncta excited by ~637 nm laser (red dots in Fig. 2). Some mRNAs should be colocalized with distinct puncta excited by ~561 nm laser (green dots in Fig. 2). GFP-G3BP1 should be uniformly diffused throughout the cell (blue haze) (*see* **Note 12**).
8. Acquire prestress images.
9. Dilute sodium arsenite in the medium that was set aside in **step 6**, add it back to the dish, and pipette several times to mix the medium between each volume capture. The final concentration of sodium arsenite should be 0.5 mM (*see* **Note 13**). Acquire poststress images. SGs should start forming ~10–20 min after adding sodium arsenite (blue blobs in Fig. 2).
10. To investigate translation resumption after stress relief, wash the cells ten times with white DMEM(+).
11. Image TetraSpeck Fluorescent Microspheres mounted on a slide for camera alignment if multiple cameras are used for imaging.

3.2 Image Analysis

3.2.1 Preprocessing Movies

1. Create a maximum-intensity projection through z.
2. Create a moving-average projection through time (using a few frames) if the signal-to-noise ratio is low.
3. Define a mask that contains an entire cell to analyze in order to avoid including any molecules from adjacent cells.
4. Apply a bandpass filter to individual images from movies to accentuate mRNAs, translation signals, and SGs of a given size range.
5. Binarize the bandpass-filtered images to make masks for each channel (one for mRNAs, one for translation signals, and one for SGs, Fig. 2). This is achieved by choosing an appropriate threshold value and setting all pixels above that value to one and all pixels below that value to zero. The threshold value should be chosen such that the vast majority of pixels within mRNAs (translation signals, or SGs) are above the threshold, whereas the vast majority of pixels outside mRNAs (translation signals, or SGs) are below the threshold.
6. Fit all detected mRNAs and translation signals to a 2D Gaussians of the following form to determine the precise coordinate of spots:

$$I(x,y) = I_{BG} + I\exp(-\frac{(x-x_0)^2}{2\sigma_x^2} - \frac{(y-y_0)^2}{2\sigma_y^2})$$

   where I_{BG} is the background fluorescence, I the spot intensity, (σ_x, σ_y) spreads of the spot, and (x_0, y_0) the spot location with super-resolution.
7. Apply a correction factor on either the determined centroids (for Subheadings 3.2.2, 3.2.3, and 3.2.5) or mask images (for Subheading 3.2.4) to adjust for potential camera misalignment if multiple cameras are used for imaging (*see* **Note 14**).

3.2.2 Global Translation Shut off Dynamics

1. Calculate the distance between mRNAs and their closest translation signals. Categorize mRNAs into translating mRNAs if the distance from the mRNA to the closest translation signal is less than three pixels (= 390 nm).
2. Create a mask image of mRNAs by drawing disks with a radius of 2 pixels (260 nm) centered at the super-resolved locations (determined in Subheading 3.2.1 **steps 6** or 7). Multiply this mask image with the mask image of SGs to find mRNAs in SGs.
3. Count the number of nontranslating mRNAs in the cytosol or SGs, and the number of translating mRNAs in the cytosol or SGs. Divide those numbers by the total number of mRNAs at each time point to determine the fraction of each population. Plot the fraction of each population as a function of time for each cell.

4. Normalize the number of SGs to the fraction of the maximum number of SGs detected in the cell to plot the SG formation curves for each cell. Fit the SG formation curve to the following equation such that all the curves plotted above can be aligned to correct for the cell-to-cell variability in the timing of the beginning of SG formation:

$$N_{SG} = A \times \tanh(B \times t - C)$$

where N_{SG} is the number of SGs, A, B, and C are normalization factors, and t is the time. Shift all the single cells curves such that their values at 10% of their maxima aligns from the fitted results.

3.2.3 Single mRNA Molecule Translation Shut off Dynamics

1. Track mRNA molecules with the nearest-neighbor tracking algorithm. Any single-molecule tracking package can be used, such as TrackMate [25]. Hand-curate tracks to merge tracks that get split accidentally because of dimmer signals for a few frames and/or to omit tracks that are indistinguishable from other mRNAs due to dense environments.
2. Measure the intensity of each translation signal and background signal by using a disk and a ring mask, respectively (Fig. 3). While 2D Gaussian fitting can serve this purpose, we find it sometimes fails to detect near-zero intensities of translation signals. The disk-ring mask method ensures that the intensity of the translation signal will be sensible, even when low.
3. Plot the location of mRNA with the intensity of the translation signal along with the location of SGs.

3.2.4 mRNA-SG Interaction Dynamics

1. Use the *Mathematica* command "ComponentMeasurements" (or an equivalent command in Python or Matlab) to measure the centroid position, area, total intensity, and convex vertices of each masked object determined in Subheading 3.2.1 **step 5** or 7 (i.e., each region of connected ones in the binarized images).

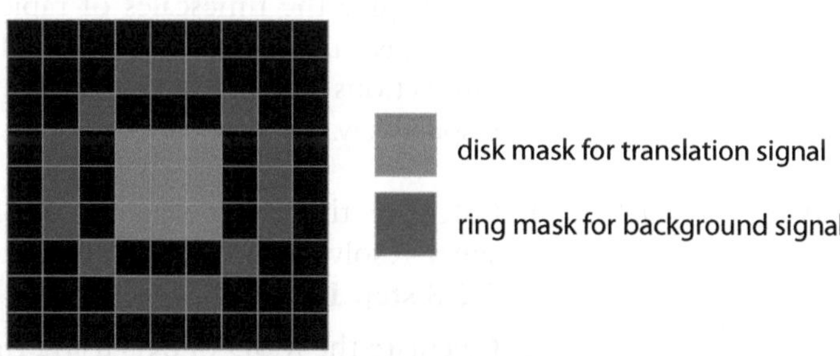

Fig. 3 Disk and ring masks to estimate translation signal and background signal, respectively, for mRNAs whose translation activity drops down to near-zero

2. Track the masked SGs (i.e., perform particle tracking, as described above in Subheading 3.2.3 **step 1**, but now on the masked images of SGs).
3. Create a sequence of 2-colored binarized sub-images from each tracked SG. This is achieved by extracting pixels within the smallest box that contains the convex vertices of each SG at each time point, as determined in **step 1** above. In each sub-image, one channel should correspond to the mRNA and the other to the SG.
4. For each tracked SG, analyze the corresponding sequence of binarized sub-images as follows: Starting from the beginning of the image sequence, check if an mRNA is overlapping with the SG. In this case, there will be pixels with ones in both channels in the binarized sub-image. If so, count how many consecutive time points this is true. Set this length of time to be the mRNA-SG interaction time. Repeat until the end of the SG track is reached (*see* **Note 15**).
5. Repeat for all tracked SG to produce a distribution of mRNA-SG interaction times. If an individual interaction persists until the final frame, ignore the interaction as its precise length of time cannot be determined (*see* **Note 16**). If an individual interaction lasts for less than two consecutive frames, ignore the interaction as it may simply reflect spurious imaging noise.
6. From the remaining SG-mRNA interaction times, calculate the interaction time survival probability curve as follows: (1) count the fraction of interactions that last longer than N seconds, with N ranging from the shortest interactions of interest (in our case 6 s) to the longest interactions of interest (in our case, 300 s); (2) Fit the fraction of tracks lasting longer than N seconds versus time to a double exponential curve of the form $A \exp(-t/t_1) + B \exp(-t/t_2)$; (3) Renormalize the fraction of interactions lasting longer than N seconds according to the fit so that the fraction is 100% at $t = 0$ s (that is $A + B = 100$).
7. From the fitted interaction time survival probability curve, t_1 and t_2 give the timescales of rapid and slow binding, respectively (assuming $t_1 < t_2$ in fit). The fraction of fast and slow interactions can be calculated as $A/(A + B)$ and $B/(A + B)$, respectively.

3.2.5 Diffusion of mRNA

1. Calculate the mean squared displacement (MSD) from the super-resolved locations of tracked mRNA in Subheading 3.2.3 **step 1**.
2. Calculate the MSD of SGs using the intensity-centroid of SGs.
3. To calculate the MSD of mRNA inside an SG, the diffusive movement of the SG must be corrected. One way to easily do

this is to use the position of another mRNA within the SG as a reference position. Thus, for each SG with more than one mRNA in it, do the following: (1) for each pair of mRNA in the SG choose one to be the reference mRNA; (2) normalize the coordinates of the other mRNA in the pair to the reference mRNA (i.e., the position of the reference mRNA defines the (0,0) position); (3) calculate the MSD from the normalized coordinates; (4) divide the obtained normalized MSD by a factor of two to remove the impact of the movement of the reference mRNA.

4 Notes

1. The epitope tag sequences are recommended to be placed at the N-terminal end of the protein to maximize the fluorescence intensity at the translation site. If placed at the C-terminal end, the ribosome will leave soon after translating the epitope tag sequence, resulting in an immediate loss of fluorescence. On the other hand, if the epitope tags are placed at the N-terminus of the protein, a ribosome remains for a while after translating the epitope tags, resulting in a sustained fluorescence signal. This also allows the researcher to estimate the translation elongation rate. Complementary epitope tags can be introduced to study noncanonical translation or dissect the translation dynamics of the relative location of two distinct translation sites, including stalling and termination rates [26, 27].
2. The mammalian expression vector for MCP is available from Addgene (ID64540). We prefer to bead-load the purified MCP because we can control the concentration of MCP in the cells in a more precise manner. The proper concentration of MCP (or anti-FLAG Fab) is critical to get the best signal-to-noise ratio due to the nature of the system relying on amplification.
3. U-2 OS cells are suitable to visualize translation from single RNA molecules because of their relatively thinner cytoplasm.
4. We use MatTek glass bottom dishes with a 14 mm glass microwell. This microwell serves to retain 4 μL of sample during bead-loading.
5. 106 μm glass beads can be harmful to the lungs, so do not inhale them.
6. Make aliquots of 10 μL to avoid freeze–thaw cycles.
7. Patches of cells can be detached while bead-loading if the confluency is close to or over 100%. Some cells (for example HEK293 cells) are easy to detach while bead-loading even though the confluency is not high.

8. This step is optional. We have empirically found that cells in Opti-MEM with FBS attach to the glass surface better and stay healthier after bead-loading.
9. This amount is sufficient for a 14 mm glass microwell. The amount of components and total volume need to be scaled accordingly.
10. Make sure to remove all media especially from the glass area so that the remaining media does not dilute the sample to bead load. The bead loading procedure should be performed promptly so the cells do not dry out.
11. Because of the nature of the system which relies on signal amplification, it is important to have thin sectioning when imaging. We therefore recommend either a HiLO microscope [28], a light-sheet microscope, or a confocal microscope.
12. In order to make sure translation signals (green dots) are bona fide translation sites, the translation elongation inhibitor puromycin can be added. All translation signals should disappear within a minute of adding 50–100 μg/mL of puromycin, while mRNA signals (red dots) should persist.
13. Avoid perturbing the dish as this could change your imaging field.
14. Binarize an image of TetraSpeck Fluorescent Microspheres to detect all spots for each channel. Fit all detected spots to 2D Gaussians. Use the *Mathematica* command "FindGeometricTransform" (or an equivalent command in Python or Matlab) to calculate a geometric transformation function.
15. Using the strategy described above, if more than one mRNA mask overlaps with an SG mask, it is still counted as a single mRNA-SG interaction. While this slightly underestimates the number of long mRNA-SG interactions, it is recommended because it is difficult to precisely determine how many mRNA are in an individual SG. This difficulty arises because mRNAs within SGs often cluster and can be partially decayed, both of which make them difficult to precisely count.
16. An individual interaction that already exists at the beginning of the observation can be included.

Acknowledgments

We thank Nancy Kedersha for providing GFP-G3BP1/mRFP-DCP1a U-2 OS cells and Esther Braselmann and Theresa Nahreini for isolating GFP-G3BP1 U-2 OS cells at the BioFrontiers Institute Flow Cytometry Core facility. We thank Dr. Luke Lavis for kindly providing JF646-labeled HaloTag ligand. We thank all members of the Stasevich lab for their support and helpful discussions. TJS and TM were supported by the NIH (grant no. R35GM119728). TJS was also supported by the NSF (MCB-1845761).

References

1. Buchan JR, Parker R (2009) Eukaryotic stress granules: the ins and outs of translation. Mol Cell 36:932–941
2. Protter DSW, Parker R (2016) Principles and properties of stress granules. Trends Cell Biol 26:668–679
3. Morisaki T et al (2016) Real-time quantification of single RNA translation dynamics in living cells. Science 352:1425–1429
4. Moon SL et al (2019) Multicolour single-molecule tracking of mRNA interactions with RNP granules. Nat Cell Biol 21(2):162–168. https://doi.org/10.1038/s41556-018-0263-4
5. Moon SL, Morisaki T, Stasevich TJ, Parker R (2020) Coupling of translation quality control and mRNA targeting to stress granules. J Cell Biol 219:e202004120
6. Bertrand E et al (1998) Localization of ASH1 mRNA Particles in Living Yeast. Mol Cell 2: 437–445
7. Tutucci E et al (2018) An improved MS2 system for accurate reporting of the mRNA life cycle. Nat Methods 15:81–89
8. Larson DR, Zenklusen D, Wu B, Chao JA, Singer RH (2011) Real-time observation of transcription initiation and elongation on an endogenous yeast gene. Science 332:475–478
9. Coulon A et al (2014) Kinetic competition during the transcription cycle results in stochastic RNA processing. eLife 3:e03939
10. Horvathova I et al (2017) The dynamics of mRNA turnover revealed by single-molecule imaging in single cells. Mol Cell 68: 615–625.e9
11. Hocine S, Raymond P, Zenklusen D, Chao JA, Singer RH (2013) Single-molecule analysis of gene expression using two-color RNA labeling in live yeast. Nat Methods 10:119–121
12. Daigle N, Ellenberg J (2007) LambdaN-GFP: an RNA reporter system for live-cell imaging. Nat Methods 4:633–636
13. Cawte AD, Unrau PJ, Rueda DS (2020) Live cell imaging of single RNA molecules with fluorogenic Mango II arrays. Nat Commun 11:1283
14. Halstead JM et al (2015) An RNA biosensor for imaging the first round of translation from single cells to living animals. Science 347: 1367–1671
15. Wu B, Eliscovich C, Yoon YJ, Singer RH (2016) Translation dynamics of single mRNAs in live cells and neurons. Science 352: 1430–1435
16. Yan X, Hoek TA, Vale RD, Tanenbaum ME (2016) Dynamics of translation of single mRNA molecules in vivo. Cell 165:976–989
17. Wang C, Han B, Zhou R, Zhuang X (2016) Real-time imaging of translation on single mRNA transcripts in live cells. Cell 165: 990–1001
18. Pichon X et al (2016) Visualization of single endogenous polysomes reveals the dynamics of translation in live human cells. J Cell Biol 214: 769–781
19. Morisaki T, Stasevich TJ (2018) Quantifying single mRNA translation kinetics in living cells. Cold Spring Harb Perspect Biol 10:a032078
20. Koch AL, Morisaki T, Stasevich TJ (2021) A multi-color bicistronic biosensor to compare the translation dynamics of different open reading frames at single-molecule resolution in Live Cells. Bio-protocol 11(14):e4096
21. Kedersha N, Tisdale S, Hickman T, Anderson P (2008) Real-time and quantitative imaging of mammalian stress granules and processing bodies. Methods Enzymol 448:521–552
22. Grimm JB et al (2015) A general method to improve fluorophores for live-cell and single-molecule microscopy. Nat Methods 12(3):244–250. https://doi.org/10.1038/nmeth.3256
23. Cialek CA, Galindo G, Koch AL, Saxton MN, Stasevich TJ (2021) Bead loading proteins and nucleic acids into adherent human cells. J Vis Exp (172):e62559. https://doi.org/10.3791/62559
24. McNeil PL, Warder E (1987) Glass beads load macromolecules into living cells. J Cell Sci 88 (Pt 5):669–678
25. Tinevez J-Y et al (2017) TrackMate: an open and extensible platform for single particle tracking. Methods 115:80–90

26. Lyon K, Aguilera LU, Morisaki T, Munsky B, Stasevich TJ (2019) Live-cell single RNA imaging reveals bursts of translational frameshifting. Mol Cell 75:172–183.e9
27. Koch A, Aguilera L, Morisaki T, Munsky B, Stasevich TJ (2020) Quantifying the dynamics of IRES and cap translation with single-molecule resolution in live cells. Nat Struct Mol Biol 27:1095–1104
28. Tokunaga M, Imamoto N, Sakata-Sogawa K (2008) Highly inclined thin illumination enables clear single-molecule imaging in cells. Nat Methods 5:159–161

Chapter 22

Image-Based Screening for Stress Granule Regulators

Katharina Hoerth, Nina Eiermann, Jürgen Beneke, Holger Erfle, and Georg Stoecklin

Abstract

Stress granule (SG)-based RNA interference (RNAi) screening is a powerful method to discover factors that control protein synthesis and aggregation, as well as regulators of SG assembly and disassembly. Here, we describe how to set up and optimize a large-scale siRNA screen, and give a detailed outline for the automated quantification of SGs as a visual readout. Hit evaluation via calculated *Z* scores provides a list of candidates for further in-depth studies.

Key words Stress granule, siRNA, Microscopic screen, Automated image analysis, *Z* score

1 Introduction

As a central part of the integrated stress response (ISR), cells inhibit global protein synthesis followed by the disassembly of polysomes and the formation of stress granules (SGs) through a liquid–liquid phase separation process in the cytosol [1]. SGs play a role in protecting cells from stress-induced damage, coordinating an effective stress response [2–4] and adjusting protein synthesis to protein folding homeostasis during recovery from stress conditions [5].

SG assembly and disassembly screens have proven to be powerful tools for the identification of proteins with essential functions in stress biology including regulators of protein translation and aggregation [6–11]. Furthermore, they revealed cross talks between SGs and a number of different cellular processes including autophagy, splicing, nucleocytoplasmic transport, and kinase-regulated signaling pathways [12–15]. Notably, a range of enzymes that catalyze or remove posttranslational modifications such as methylation, acetylation, sumoylation, ADP ribosylation, NEDDylation, glycosylation, and phosphorylation were found to control SG dynamics as

Katharina Hoerth and Nina Eiermann contributed equally to this work.

Daniel Matějů and Jeffrey A. Chao (eds.), *The Integrated Stress Response: Methods and Protocols*, Methods in Molecular Biology, vol. 2428, https://doi.org/10.1007/978-1-0716-1975-9_22,

a whole, or the recruitment of individual proteins into SGs [16]. Phosphorylation is also known to affect many translation factors, though the responsible kinases and the functional relevance of the corresponding phosphorylation events are frequently not well understood [17, 18]. A wide-scale screening approach has the advantage that the function of a large set of proteins, including proteins that are so far poorly characterized, can be analyzed in an unbiased manner, and thereby lead to the discovery of novel signaling pathways or links between pathways that were previously considered to be independent. We recently used a SG-based RNAi screen as a starting point for the identification of kinases and phosphatases that function in translation control and SG assembly, and thereby discovered that the major cell cycle kinase CDK1 serves as an activator of global protein synthesis outside of its role in mitosis [9].

Since SGs are an early sign of cellular damage, viral infections, and activation of the ISR, SGs are discussed as potential biomarkers and therapeutic targets [2, 19]. In this context, automated SG screening approaches could help to identify pharmacological compounds and antiviral drugs that elicit, or interfere with, the cellular stress response. Our RNAi-based screening approach can be applied to cells under regular growth conditions or upon exposure to various types of stress, making it a versatile tool for physiological and pathological set-ups. Since our protocol includes an automated image analysis pipeline, it can also be applied to other high-throughput approaches that use SGs as a visual read-out.

2 Materials

2.1 Reagents

1. Cell line of interest (e.g., HeLa cells, *see* **Note 1**).
2. Plasmid encoding a fluorescently tagged SG marker protein (e.g., pEYFP-C1 TIA-1, *see* **Note 2**).
3. Eukaryotic selection marker for which resistance is encoded on the plasmid (e.g., Geneticin).
4. Transfection reagent for plasmids (e.g., TurboFectin 8.0 Transfection reagent from Origene).
5. RNAi library (e.g., Dharmacon siGenome siRNA libraries GU-003505 Human.

   Protein Kinase and GU-003705 Human Phosphatase from Thermo Fisher Scientific)
6. Transfection reagent for siRNAs (e.g., Lipofectamine RNAi-MAX Transfection reagent from Invitrogen).
7. RNase-free H_2O.
8. 0.2% gelatin/1% fibronectin in solution

9. OPTIMEM transfection medium supplemented with 0.4 M sucrose.
10. Cell culture and imaging plates (e.g., 96-well MGB096-1-2-LGL matriplates from Brooks) and covering foil (gas permeable and sterile for 96-well plates, 4ti-0516/96 from Brooks).
11. Dulbecco's Modified Eagle Medium (DMEM) containing 10% FCS and 2 mM L-glutamine.
12. Phosphate-buffered saline (PBS).
13. 0.05% Trypsin–0.02% EDTA in PBS
14. 4% paraformaldehyde (PFA) in PBS
15. DRAQ5 for nuclear and cytoplasmic staining (alternatively Hoechst or DAPI in combination with an additional cytoplasmic staining, *see* **Note 11**).

2.2 Equipment

1. FACS sorter (e.g., BD FACSAria IIIu cell sorter from BD Biosciences).
2. Separate cell incubator with sufficient space for all plates (alternatively, microplate racks for stacking plates inside the incubator can also be used).
3. Automated liquid handling robot (e.g., a MICROLAB STAR from Hamilton).
4. Optional: Lyophilization or drying device for long-term storage of coated plates.
5. Microplate dispenser (e.g., a Thermo Fisher Scientific Multidrop Combi), otherwise at least a multichannel pipet.
6. Widefield fluorescence microscope with an automated stage and an automated focusing unit (e.g., Nikon TI-Eclipse microscope with Perfect Focus System (PFS) and JOBS software), a CCD camera, a 20× or 40× long working distance dry objective and an appropriate filter set (e.g., for GFP and RFP signals).
7. Fiji is just ImageJ software.
8. R Software for statistical computing and graphs.
9. High computing power (*see* **Note 17**).

3 Methods

Transfect your cell line of interest (*see* **Note 1**) with a plasmid encoding a fluorescently tagged SG marker protein (*see* **Note 2**). Here we use EYFP-TIA-1 as a marker to study SG assembly in HeLa cells.

3.1 Stable Expression of a Fluorescently Tagged SG Marker Protein in your Cell Line of Interest

1. Transfect 1.2×10^6 HeLa cells with 2 μg of plasmid DNA per well (in a 6-well format) using TurboFectin 8.0 Transfection reagent from Origene. Dilute 2 μg of pEYFP-C1-TIA-1 (p2033) in 250 μl of OPTIMEM. Add 6 μl of TurboFectin 8.0 and vortex briefly. Incubate for 20 min at room temperature (RT). Add the transfection mix dropwise to the cells.
2. 24 h after transfection, cells are subjected to antibiotic selection by supplementing the medium with 400 μg/ml geneticin. Antibiotic selection is to be continued for the next 10 days. Afterward, cells are FACS sorted in order to obtain a cell population with a homogeneous expression level of the fluorescently tagged SG marker protein. During FACS sorting, select a population of cells with medium SG marker expression levels. After FACS sorting, expand cells without selection pressure and freeze a sufficient number of aliquots. During optimization and the actual screen, make sure to always work with low passage numbers.
3. Test the expression level of your fluorescently tagged SG marker protein by Western blot analysis and fluorescence microscopy, and make sure that expression of the fluorescently tagged SG marker protein is homogeneous between cells and does not cause spontaneous formation of SGs under regular growth conditions. In case the fluorescence intensity level is too high (some cells form SGs), too low or too inhomogeneous, adapt the FACS sorting procedure. Optionally, measure and compare the global rate of translation of your stable cell line to the one of the parental cell line by performing a puromycin incorporation assay or a polysome profile analysis in order to exclude negative impacts of SG marker overexpression on global translation.

3.2 Optimization of Knockdown (kd) Efficiency

Before optimizing the kd procedure, choose a suitable RNAi library (*see* **Note 3**).

1. Your RNAi library should contain several nontargeting control sequences and at least one positive control sequence in order to assess effects of the RNAi procedure on cell viability, and estimate the transfection and silencing efficiencies. These controls might need to be added manually to the library (*see* **Notes 4** and **5**).
2. Together with your RNAi tool of choice, test several transfection reagents or alternative nucleic acid delivery methods (*see* **Note 6**) in order to determine which RNAi method is most efficient and least toxic to your cell system. This step can be tested in any format (e.g., 6-well plate) with two nontargeting and at least two additional siRNA sequences. For HeLa cells, reverse transfection (seeding cells directly onto the transfection

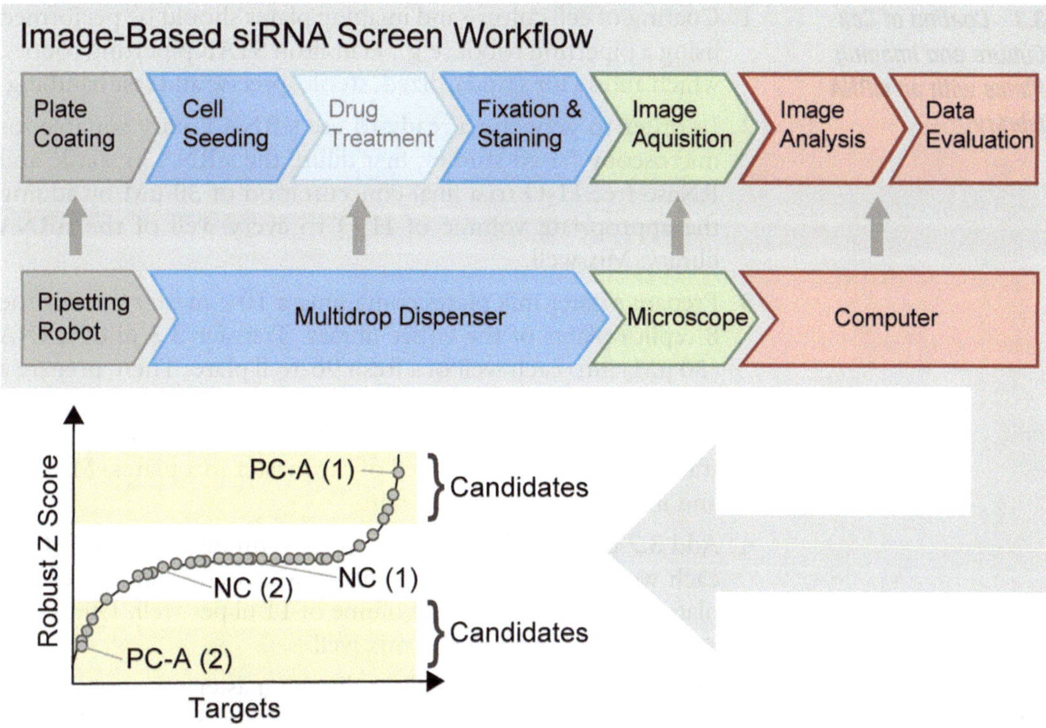

Fig. 1 Overview of the screening pipeline. Shown are the sequential steps described in Subheadings 3.3–3.8., the equipment that is needed for the different steps as well as an exemplary visualization of the screening results when using the robust *Z* score for hit identification (PC-A (1), PC-A (2): positive controls of the assay 1 and 2; NC (1), NC (2): nontargeting controls 1 and 2)

mix) of 50 nM siRNA in combination with RNAiMAX for 72 h proved to be efficient and, at the same time, gentile to the cells.

3. Perform a pilot screen in order to test all settings and operations on a small scale screening format (e.g., one 96-well plate). The pilot screen (like the actual screen) should be performed as described in Subheadings 3.3–3.8 (visualized in Fig. 1), using a small number of positive and negative control siRNAs to test different siRNA concentrations, duration times for kd, imaging plates (*see* **Note 7**), cell densities (*see* **Note 8**) and microscopic set-ups. At the same time, the settings determined in **step 2** can be verified in the actual screening format. An additional advantage is that the pilot screen will help to get familiar with the procedure and ensure fast and standardized handling later in the full screen. Finally, images generated in the pilot screen can be used as a training set for the image analysis procedure. This will allow to determine the dynamic range between positive and negative controls (*see* Subheading 3.8) and make sure that hits (positive control siRNAs) can be clearly separated from negative controls and noise.

3.3 Coating of Cell Culture and Imaging Plates with an siRNA Library

1. Coating of cell culture and imaging plates should be performed using a pipetting robot (e.g., Hamilton STAR pipetting robot), which allows for standardized, sterile, precise and fast handling.
2. In order to generate a ready-to use siRNA library suitable for microscopy-based studies, first dilute the siRNA in sterile and RNase-free H_2O to a final concentration of 30 μM by adding the appropriate volume of H_2O to every well of the siRNA library. Mix well.
3. Prepare master mix plates containing a 10× master mix for the 8 replica plates of the entire library. Transfer 3.3 μl of siRNA (30 μM) into each well of a fresh 96-well plate. Then, prepare a transfection stock containing 1.6 μl OPTIMEM supplemented with 0.4 M sucrose and 2.4 μl RNAiMAX per well. Add this transfection mix to each well of the master mix plates. Mix well and incubate at RT for 20 min.
4. Add 3.7 μl of 0.2% gelatin/1% fibronectin dissolved in H_2O to each well of the master mix plates. Mix well. The master mix plates now contain a total volume of 11 μl per well. Dilute this mix with 239 μl H_2O and mix well.
5. To obtain the replica plates, split the master mix prepared in **step 4** by pipetting 25 μl into each well of 8 separate 96-well MGB096–1-2-LGL matriplates (*see* **Note 9**). Only one of these plates is used for the primary screen. The remaining seven replicates can be used for validation screens.
6. Lyophilize (if you have the possibility), freeze or dry the transfection mix for long-term storage of the replica plates.

3.4 High-Throughput RNAi

1. Trypsinize and resuspend cells in antibiotic-free medium (DMEM supplemented with 10% FCS and 2 mM L-glutamine, *see* **Note 10**). Count your cell suspension and dilute it to the appropriate cell concentration. For HeLa cells, prepare 2000 cells per well of a 96-well MGB096–1-2-LGL matriplate. Use a total volume of 200 μl per well to reach a final siRNA concentration of approximately 50 nM.
2. Reverse transfection: Seed cells into every well of the replica plates by pipetting 200 μl of your cell suspension into each well of the 96-well matriplate. Use a microplate dispenser (Thermo Fisher Scientific Multidrop Combi) for fast and standardized seeding (if an automated dispenser is not available, you can also use a manual multichannel pipet). Seal the plates with foil in order to prevent evaporation of medium, this is especially important for wells at the edges of the plate.
3. Perform kd for 72 h by incubating the matriplates at 37 °C and 5% CO_2. Ideally, the incubator should not be opened during this period to ensure constant culturing conditions and to

minimize evaporation of medium in case plates are not sealed with foil.

4. Optional: Perform drug treatment. Premix the drug in DMEM containing 10% FCS and 2 mM L-glutamine. Shake out the medium from the wells and quickly add back the premixed medium containing the drug by using the microplate dispenser. Medium should be collected and discarded appropriately.

3.5 Cell Fixation and Staining

1. Premix 4% PFA in PBS containing DRAQ5 (1:1000 diluted, *see* **Note 11**). Shake out the medium from the wells, and quickly add 200 μl of 4% PFA in PBS with DRAQ5 to each well. Incubate for 20 min.
2. For the washing steps, shake out the fixation and staining solution from the wells, and quickly add 300 μl of PBS to each well. Incubate for 5 min. Repeat this step and wash two more times with PBS.
3. After washing, leave cells in PBS and directly proceed with image acquisition. Alternatively, you can store the plates in the dark at 4 °C until examination under the microscope. Take into account that fluorescence intensity will decline over time.

3.6 Image Acquisition

1. Randomly collect at least 16 images per well for each 96-well plate through automated acquisition. To this end, use an inverted widefield fluorescence microscope with an automated stage, an automated focusing unit and automation software (e.g., Nikon TI-Eclipse microscope with a perfect focus system (PFS) and JOBS software). Ideally, choose a 20× or a 40× long working distance dry objective for image acquisition from the bottom of the plates. For every position, acquire an image in channel one (DRAQ5 signal) first, then in channel two (YFP--TIA-1 signal).
2. The images are saved as a hyperstack (covering the two channels and the different positions), for example, in the nd2-format.

3.7 Image Analysis

Download "Fiji Is Just ImageJ" (Download at: https://imagej.net/Fiji) for the analysis of your image-based screen-dataset. Use the macro provided under the following link (https://doi.org/10.11588/data/KRPJ9A; *see* **Note 12**). Alternatively, generate your own macro via Fiji with help of the "recorder function," or use other tools and strategies for image analysis (*see* **Note 13**). The crucial steps of image analysis are depicted in Fig. 2:

1. Preprocessing

   Prepare your raw data (one stack of images including all positions and channels of one well) for image analysis by

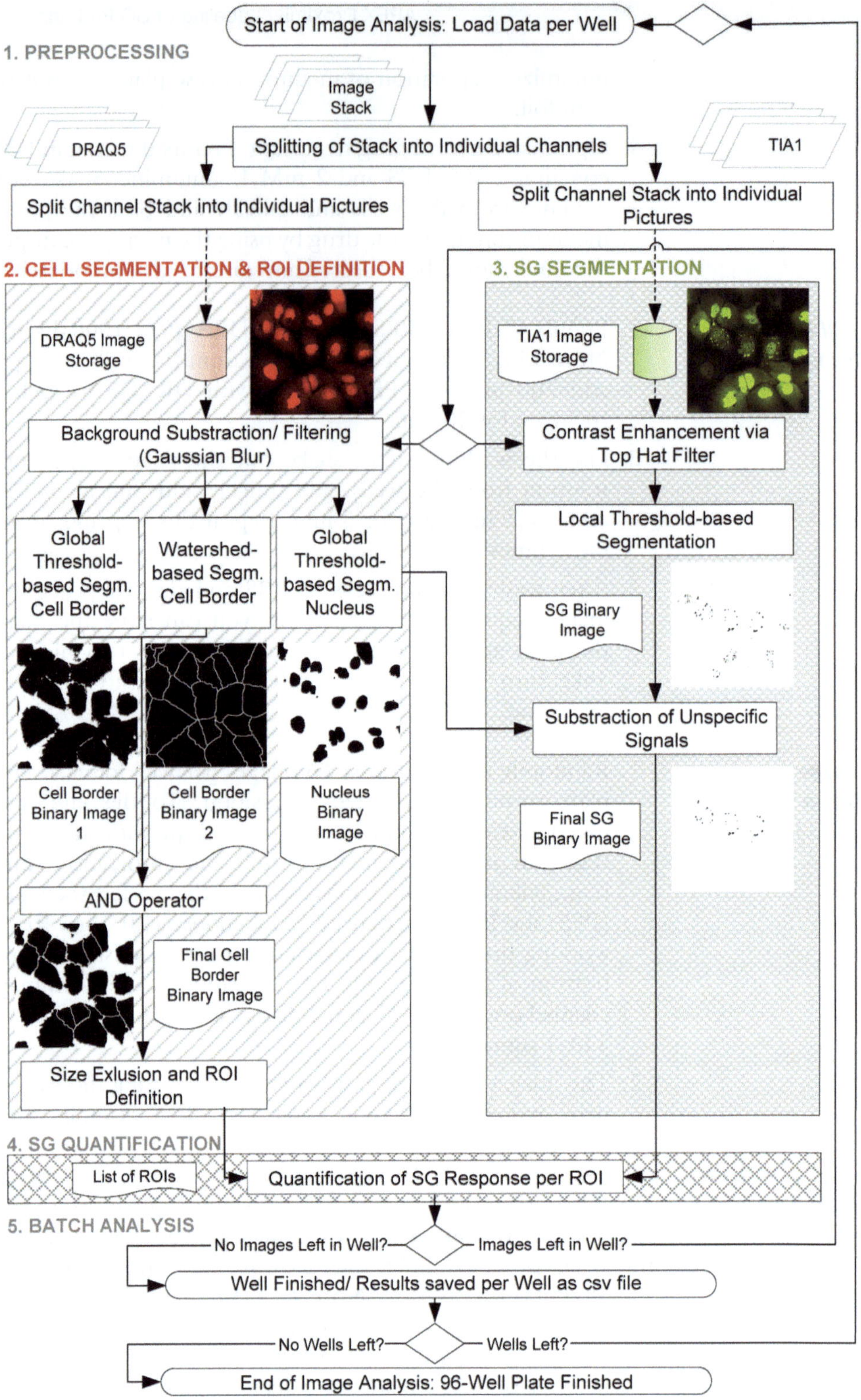

Fig. 2 Sequence diagram, illustrating the different steps of a simplified image analysis pipeline for the detection and quantification of the SG response. Image data (consisting of two channels and "x" image positions) is initially loaded for each well. Then, the pipeline executes the following steps: (1) preprocessing of image data, (2) cell segmentation and ROI definition, (3) SG segmentation (4) SG quantification within the ROIs, and (5) batch analysis over all positions and wells per 96-well plate

splitting and sorting the multistacks according to the two channels and within the channel stacks by splitting into the individual pictures.

2. Cell segmentation for Regions of Interest (ROI) definition

   Use the DRAQ5 channel (consisting of a weak cytoplasmic and strong nuclear signal) for the segmentation of the individual cells. To perform threshold-based segmentation, first subtract the background in order to improve the accuracy of segmentation. Define the optimal threshold for detecting cell areas, either in an unbiased way via autothresholding (preferred if applicable) or manually for each plate. Using this threshold, create binary image number one. To be able to differentiate between individual cells in regions with a high cell density, perform watershed-based segmentation in parallel to threshold-based segmentation. For this purpose first detect local maxima, referring to the nuclei signals in the raw DRAQ5 pictures. This might need a preliminary filtering step like the Gaussian blur filter and setting of the noise tolerance for higher accuracy. Applying the watershed-algorithm, the maxima are divided from each other by a separating line, generating binary image number two. Combine the information from both binary images using the logical "AND" operation to create an intersection of both pictures. The resulting final binary image reflects more precisely the areas covered by individual cells, independently of cell densities. Afterward, analyze the segmented regions according to cell-specific criteria (size, shape) to exclude segmented structures related to unspecific signals rather than cells. Finally add the remaining regions, referred to as ROIs, to the ROI Manager (whereby each cell will be assigned an individual number) for the subsequent analysis of the SG response within individual cells (*see* **Note 14**).

3. SG segmentation

   Use the SG marker channel for the threshold-based segmentation of SGs. Start with contrast enhancement via the top hat filter (this requires conversion into 8-bit images and downloading of the MorphoLibJ plugin) to improve the differentiation of SG signals (sequestered SG marker) from the overall cytoplasmic signal (unsequestered SG marker). To account for signal intensity variations between individual cells, due to cell-intrinsic differences in the expression level of the introduced SG marker, use local thresholding instead of global thresholding. For this purpose, select the auto threshold algorithm that performs best, define the size of the region over which the threshold should be computed and generate the SG binary image. Finally, subtract SG-unrelated segmented structures, such as the nucleus or staining artifacts, from the SG binary image such that only "real" SGs are detected. Those SG-unrelated structures can be detected and segmented using the DRAQ5 channel, similar to the ROI definition procedure (**step 2**) but using a different threshold (*see* **Note 15**).

4. SGs quantification within ROIs

   For the quantification of the detected SGs, first define the criteria of interest that should be measured (average SG size, number per cell etc.). In the following, you can analyze the results of SG segmentation (SG binary image) for the individual ROIs (*see* **Note 16**). If SG intensities should also be considered, the measurement needs to be redirected to the original SG channel image using the SG binary image as a mask.

5. Batch analysis

   To accomplish the analysis of the complete set of images for an entire plate, perform batch analysis. Loop statements will repeatedly run **steps 1–4** for all wells and all positions within a well for a complete 96-well plate (*see* **Note 17**). As an output you should receive one single result table per well which includes all measurements of the individual cells/ROIs over the different positions. The results are saved in a csv format (96 tables per plate).

3.8 Statistics and Data Validation

Fiji-derived result tables, containing information about SG measurements, are further processed via R (*see* **Note 18**).

1. Optional: Do filtering steps, for example, filter out SG marker-negative cells.
2. Calculate the parameters of interest per well (e.g., percentage of cells with SGs, total number of cells, mean SG size etc.) out of the raw data tables (csv files derived from Fiji).
3. To account for plate-to-plate variability due to random fluctuations in the performance of the screen, normalize the parameters of interest via the robust *Z* score. The robust *Z* score uses median and MAD (median absolute deviation) instead of mean and std. (standard deviation), and is therefore less sensitive to outliers compared to just the *Z* score. Assuming that most kds do not affect SG formation, the parameters of each individual well (S_i) can be rescaled by setting them in relation to all parameters on the plate (median (S_{all}), MAD (S_{all})) (*see* **Note 19**):

$$\text{Robust } Z \text{ score} = \frac{S_i - \text{median}\,(S_{all})}{\text{MAD}\,(S_{all})}$$

4. Combine the robust ***Z*** scores for the entire data set, list the *Z* score values according to plate and well number and assign the corresponding gene names.

5. Perform hit selection using the robust Z score. Therefore, select a threshold (number of MADs beyond the median), according to the strength of your effect and the variability of the data. By using stringent conditions (high threshold), one can reduce the number of false positive hits, yet the risk of losing true candidates (false negatives) increases.
6. If not pooled but rather single siRNAs are used, calculate a hit score out of the 4 individual robust Z scores that belong to the same gene but different siRNAs. A hit score of 0 means that the gene is not identified as a hit for any of the 4 siRNAs; a hit score of 1 meaning that the gene is identified as a hit for 1 of the 4 siRNAs, etc.
7. Use the hit score to sort the list of candidates and identify the most promising ones where more than one siRNA gives a change in the observed phenotype. Additionally, within one hit score category, sort the candidates according to their mean robust Z score.
8. To assess the quality and reliability of the screening results, calculate the Z factor. The Z factor takes into account variations (defined by the standard deviation, std) of sample (S_{all}) and positive control (C_{pos}) measurements, and the dynamic range between both groups (mean(C_{pos}) − mean(S_{all})). Again assuming that most kd conditions within one plate don't affect the SG phenotype, the Z factor therefore serves as an indicator of how well hits can be identified on individual plates (*see* **Note 20**):

$$Z \text{ factor} = 1 - \frac{3 \times \text{std}(C_{\text{pos}}) + 3 \times \text{std}(S_{\text{all}})}{\left|\text{mean}(C_{\text{pos}}) - \text{mean}(S_{\text{all}})\right|}$$

9. In addition, monitor positional effects per plate (edge effect or drift) by generating a heat map or a scatter plot of all the Z score values across all wells within one plate.
10. Repeat the screen to reduce the number of false positive hits. However, due to cost and time reasons, the setup of a secondary screen with only a subset of candidates, which can be more easily repeated, might be preferred. In addition, the secondary screen can be extended to exclude that the observed granules are SG-unrelated structures, to define the SG-subtype more precisely, or to monitor effects on protein synthesis in parallel (*see* **Note 21**).
11. Candidates whose effect on the SG response was confirmed in the secondary screen need to be further validated with respect to kd specificity and the observed phenotypes (*see* **Note 22**). Follow-up experiments will be needed to explore at which level (*see* **Note 23**), and via which interaction partners and pathways,

the identified candidate affects the SG response. Finally, the identified hit list can be compared to other already available datasets (*see* **Note 24**).

4 Notes

1. Since SG formation typically occurs upon an acute and strong reduction of translation, our impression is that effects can be observed best in cells with a high translation rate.
2. Table 1 provides an overview of suitable and commonly used SG marker proteins and stainings, which are used to study the different subtypes of canonical and noncanonical SGs. The use of two different fluorescently tagged SG marker proteins, or the combination of a fluorescently tagged SG marker protein and PB marker protein [6], may also be considered.
3. Several RNAi libraries are commercially available. They either contain a subset of genomic targets (e.g., druggable genome libraries, predefined gene families, and pathway targeting libraries) or target the whole genome. Custom design libraries can be purchased as well. Furthermore, different types of gene silencing libraries are available including siRNA libraries or

Table 1
List of suitable SG marker proteins and costainings. Columns represent specificity for different SG subtypes according to [3, 16, 20–23]: Type I, mediated by eIF2α phosphorylation; Type II, mediated by eIF4A-inhibition; Type III, mediated by displacing eIF4E from eIF4G (e.g., by UV, sodium selenite, nitric oxide, glucose starvation or different xenobiotic agents); Type IV, mediated by macromolecular crowding (e.g., by osmotic stress or increased LLPS, also associated with altered expression of certain RBPs, very heterogeneous and not well characterized yet, depicted markers were observed in SGs formed upon osmotic stress)

	Canonical SGs		Noncanonical SGs	
SG marker protein	**Type I**	**Type II**	**Type III**	**Type IV**
G3BP1	Yes	Yes	Yes	Yes (but not needed)
TIA-1/TIAR	Yes	Yes	Yes	Yes
eIF3b/e/g	Yes	Yes	No	Yes
Costainings				
Oligo(dT)	Yes	Yes (but less)	Variable	Yes
eIF2α	No	Yes	?	?
eIF5	No	Yes	?	?
RACK1	Yes	Yes	No	?
eIF5A	Yes	?	No (or less)	?

shRNA libraries. Moreover, different kd strategies can be chosen including pooled siRNA libraries versus single sequence RNAi libraries where several siRNAs target one gene separately. siRNAs pools possibly reduce off-target effects (due to lower concentrations of single siRNAs) and potentially increase kd efficiency. Single sequences on the other hand allow to directly validate hits with different siRNA sequences, though this approach increases the size of the screen. Consider the total library size and sampling size with respect to total costs and your sample handling as well as computing power before making your choice. Also, make sure that your cell system is susceptible to the RNAi method you choose. CRISPR-based gene silencing screens (CRISPRi [transcriptional silencing] or CRISPR KO [knockout]) can further be considered as an alternative to RNAi.

4. Make sure to spot your nontargeting and your positive controls several times on the plate, and ideally try to place them randomly across the 96-well format. Cells growing at the edges of the plate (first and last row, first and last lane) sometimes show "edge-effects," hence being more stressed than cells growing in the middle of the plate. In order to prevent edge effects, it is generally advised to use a foil to cover the 96-well plate, or to fill wells on the edges only with cell culture medium.
5. Arsenite-treated si-control and si-HRI-transfected cells (PC-A: Positive control for assay (connected to phenotype)) could serve as a negative and a positive control, respectively, for SG assembly unless you have more specific controls for your conditions. Also, kds that effect cell attachment and cell survival (PC-S: Positive control for silencing) such as kd of KIF11 or PLK1 are often used as positive controls in order to estimate kd efficiency.
6. Together with the gene silencing method of choice (siRNA, shRNA), test several transfection reagents (e.g., RNAiMAX, Lipofectamine 2000/3000, Dharmafect) or alternative nucleic acid delivery methods (viral transduction systems, electroporation) in order to determine which RNAi method is most efficient and least toxic to your cell system.
7. The final screening plates should be optimal for cell culture as well as microscopy. Importantly, they need to be sterile and permit adherent cell growth. The growth area can either be rectangular or circular. We had best results in seeding homogeneity using a rectangular growth area. Glass bottom plates are especially well suited for microscopy but are often poor substrates for cell growth. Therefore, special plate coating, for example, with gelatin and fibronectin, is advisable. Also, the plates should not emit background fluorescence, and the

variation in bottom thickness should be minimal to allow for automated focusing. For most objectives, it is also important that the plates have a flat bottom and that the rim of the plates does not protrude.

8. Special focus should be given to finding the optimal cell density within the final 96-well screening plates since this strongly impacts kd efficiency.
9. How to spot and dry the replica plates depends on the available spotting and drying devises. A detailed description for spotting with the ChipWriter "Compact" and "Pro" (Bio-Rad Laboratories) contact printer can be found in [24]. Spotting can also be performed with a multichannel pipette; this is straightforward but more labor-intensive.
10. Since performing a large-scale RNAi screen sometimes requires semisterile conditions (e.g., working outside of the laminar flow hood during coating of plates), it might be advisable to use antibiotic-containing medium. However, consider that transfection procedures often work better in the absence of antibiotics.
11. Alternatively, use Hoechst (1:10000 diluted) or DAPI (1: 10000 diluted) to stain nuclei, and use a second cytoplasmic stain to detect cell borders. If the second stain requires an antibody, make sure to include a permeabilization and blocking step before adding the antibody to the plates.
12. The supplied macro needs to be adapted with respect to the individual image features (see comment sections in macro source), validated for its functionality (by visually looking at the intermediate results) and potentially also complemented with additional steps (e.g., analysis of a specific subset of cells, additional filtering steps).
13. Alternatively, other tools can be used to generate a pipeline for image analysis such as CellProfiler [25]. In addition, there are a number of already published pipelines which can be tested [11, 14, 15, 26–28]. Depending on the number of pictures and the size of the library, another option is semiautomatic image analysis, using the following macro to facilitate the visual inspection of all images one by one.

```
Macro open and close
macro "Multiple Measurement " {
G_Sdir = getDirectory ("Choose the Directory where the file
is");
list = getFileList (G_Sdir);
for (i = 0;i<list.length;i++) {
CountSG(list[i]);
}
```

```
}
function CountSG(img_filename){
fullpath_image = G_Sdir + img_filename;
open ( fullpath_image );
sourceID = getImageID ( );
wait (1000);
close (sourceID);
run ("Close All") ;
}
```

14. Cells that are depicted only partially at the edge of the image can be excluded from the ROI list at this step.
15. Depending on the staining, SG segmentation may include nuclear signals (as for TIA-1), staining artefacts and additional SG-unrelated granular structures in the cytoplasm, all of which would be falsely detected as SGs. To exclude such structures from the analysis, they need to be subtracted from the SG binary image before quantification. Since nuclei and staining artefacts (which usually show a very intense signal in all channels) are also visible in the DRAQ5 channel, threshold-based segmentation of the nucleus in the DRAQ5 channel (which also covers staining artefacts) is used to detect and subtract most SG-unrelated structures from the SG binary image. In addition, via defining a SG-threshold (only cells containing ≥ x SGs are called SG-positive), most remaining structures unrelated to SGs can be excluded later in analysis via "R."
16. Due to variations in the expression level of the SG marker (despite FACS sorting), it might be necessary to exclude ROIs with a low SG marker signal since these cells may be falsely assigned as SG negative. This can be achieved by measurement of the SG marker intensity within the entire ROI or only the nucleus, followed by the definition of an intensity cut-off (minimal signal intensity necessary for subsequent analysis of the SG response) and the exclusion of ROIs/cells below this threshold from the analysis (as part of the final analysis via "R").
17. Note that an important limitation is the time it takes for the macro to analyze one plate. This time depends on the number of pictures and channels, but also on the computational power. For example, during one of our screens it took 24–30 h to analyze one 96-well plate from which we recorded 64 images per well, in three channels, using the following computer setup: Intel(R) Xeon(R) processor CPU E5-1650 v3; 32 GB RAM; 64-bit system. In order to improve time management, parallel execution of the macro should be considered.

18. Alternatively, one can use ready-to-use R extension packages such as highSCREEN, CellHTS, or other available tools [29–31].
19. Parameters can also be normalized to the values from the negative control wells. However, this might be problematic due to their limited number and potential edge position. As an alternative to the *Z* score/robust *Z* score, the B score, which also takes positional effects into account, can be used.
20. A *Z* factor in the range of 0.5 to 1 is excellent, while a *Z* factor in the range of 0 to 0.5 is marginal and below 0 means that positive and negative signals overlap and cannot be distinguished in the assay/for that particular plate. The *Z'* factor, a variant of the *Z* factor, compares positive and negative controls, and can already be used during assay optimization to assess the quality of the assay per se. Alternatively, the SSMD (strictly standardized mean difference), the robust SSMD or robust *Z'*-factor can be used for the quality control of an assay.
21. As an option one can combine the primary screen with a repetition screen in which the cells are cotreated with emetine (20 μg/ml) or CHX (50 μg/ml) for the last 30 min prior to fixation. Thereby, one can test if the observed granules are really SGs since emetine and CHX will hinder polysome disassembly and thus prevent SG assembly. Further, a SG-based RNAi screen using, for example, EYFP-TIA-1 as a marker protein may optionally be combined with an additional SG marker staining in order to specifically study different SG subtypes. A combination with cell survival measurements might further be interesting since canonical and noncanonical SGs were reported to have opposing effects on cell survival. In order to study the translation rate in parallel, the repetition screen could involve the cotreatment of the cells with puromycin (10 μg/ml) for the last 5 min before fixation, and staining with anti-puromycin antibody as a read-out for the overall protein synthesis rate.
22. Identification of targeted loss of function (LOF) phenotypes by RNAi can sometimes be problematic since the transfection procedure may cause sequence-dependent and -independent off-target effects by activating innate immune responses and saturating the endogenous RNAi machinery. Therefore, follow-up studies including rescue experiments (e.g., expressing an siRNA-resistant cDNA mutant) are necessary to validate initial candidate hits for a function in the regulation of translation. Additionally, the LOF phenotypes of certain proteins, especially those with long half-lives or high activities, might be obscured by residual protein activity. On the other hand, RNAi has an advantage in studying proteins with

essential functions, since residual activity might allow for cell survival. In addition to simple LOF screens, modifier screens, which enhance or suppress a specific phenotype (e.g., a LOF screen in combination with a drug treatment), are often very informative [32].

23. With respect to SG-based RNAi screens, one should consider that candidate hits can impact SG formation at different levels. Hence, the kd of a phosphotransferase might interfere with (1) stress sensing, (2) expression of translation factors or the supply of other substrates needed for protein synthesis, (3) classical signaling pathways of translation control, (4) novel signaling pathways so far not linked to translation control, or (5) the process of SG assembly and disassembly downstream of translation inhibition.

24. Results from RNAi experiments are collected and compared in databases such as http://www.genomernai.org or—more specifically for SG-based RNAi screens—http://rnagranuledb.lunenfeld.ca/. The meta-analysis of different screens may identify common principles as well as candidate genes that are specific to particular conditions.

Acknowledgments

We like to thank Maranatha Ayodele and Paul Anderson (Brigham and Women's Hospital, Boston, MA, USA) for generating and providing the EYFP-C1-TIA-1 plasmid, Amin Zehtabian and Helge Ewers (Freie Universität Berlin, Berlin, Germany) for general advice on image analysis, and Alessia Ruggieri (Heidelberg University, Heidelberg, Germany) for discussions on screening pipeline optimization. This work was supported by grants from the Deutsche Forschungsgemeinschaft, project numbers 201348542-SFB 1036 and 278001972-TRR 186 to GS. The Advanced Biological Screening Facility is supported by the Deutsche Forschungsgemeinschaft through the CellNetworks Cluster of Excellence (EXC81).

References

1. Kedersha N, Ivanov P, Anderson P (2013) Stress granules and cell signaling: more than just a passing phase? Trends Biochem Sci 38(10):494–506. https://doi.org/10.1016/j.tibs.2013.07.004
2. Mahboubi H, Stochaj U (2017) Cytoplasmic stress granules: dynamic modulators of cell signaling and disease. Biochim Biophys Acta Mol basis Dis 1863(4):884–895. https://doi.org/10.1016/j.bbadis.2016.12.022
3. Reineke LC, Neilson JR (2019) Differences between acute and chronic stress granules, and how these differences may impact function in human disease. Biochem Pharmacol 162: 123–131. https://doi.org/10.1016/j.bcp.2018.10.009
4. Eiermann N, Haneke K, Sun Z, Stoecklin G, Ruggieri A (2020) Dance with the devil: stress granules and signaling in antiviral responses.

Viruses 12(9):984. https://doi.org/10.3390/v12090984

5. Cherkasov V, Hofmann S, Druffel-Augustin S, Mogk A, Tyedmers J, Stoecklin G, Bukau B (2013) Coordination of translational control and protein homeostasis during severe heat stress. Curr Biol 23(24):2452–2462. https://doi.org/10.1016/j.cub.2013.09.058
6. Ohn T, Kedersha N, Hickman T, Tisdale S, Anderson P (2008) A functional RNAi screen links O-GlcNAc modification of ribosomal proteins to stress granule and processing body assembly. Nat Cell Biol 10(10):1224–1231. https://doi.org/10.1038/ncb1783
7. Boyd JD, Lee P, Feiler MS, Zauur N, Liu M, Concannon J, Ebata A, Wolozin B, Glicksman MA (2014) A high-content screen identifies novel compounds that inhibit stress-induced TDP-43 cellular aggregation and associated cytotoxicity. J Biomol Screen 19(1):44–56. https://doi.org/10.1177/1087057113501553
8. Fang MY, Markmiller S, Vu AQ, Javaherian A, Dowdle WE, Jolivet P, Bushway PJ, Castello NA, Baral A, Chan MY, Linsley JW, Linsley D, Mercola M, Finkbeiner S, Lecuyer E, Lewcock JW, Yeo GW (2019) Small-molecule modulation of TDP-43 recruitment to stress granules prevents persistent TDP-43 accumulation in ALS/FTD. Neuron 103(5):802–819.e11. https://doi.org/10.1016/j.neuron.2019.05.048
9. Haneke K, Schott J, Lindner D, Hollensen AK, Damgaard CK, Mongis C, Knop M, Palm W, Ruggieri A, Stoecklin G (2020) CDK1 couples proliferation with protein synthesis. J Cell Biol 219(3):e201906147. https://doi.org/10.1083/jcb.201906147
10. Yang P, Mathieu C, Kolaitis RM, Zhang P, Messing J, Yurtsever U, Yang Z, Wu J, Li Y, Pan Q, Yu J, Martin EW, Mittag T, Kim HJ, Taylor JP (2020) G3BP1 is a tunable switch that triggers phase separation to assemble stress granules. Cell 181(2):325–345.e28. https://doi.org/10.1016/j.cell.2020.03.046
11. Wheeler EC, Vu AQ, Einstein JM, DiSalvo M, Ahmed N, Van Nostrand EL, Shishkin AA, Jin W, Allbritton NL, Yeo GW (2020) Pooled CRISPR screens with imaging on microraft arrays reveals stress granule-regulatory factors. Nat Methods 17(6):636–642. https://doi.org/10.1038/s41592-020-0826-8
12. Buchan JR, Kolaitis RM, Taylor JP, Parker R (2013) Eukaryotic stress granules are cleared by autophagy and Cdc48/VCP function. Cell 153(7):1461–1474. https://doi.org/10.1016/j.cell.2013.05.037
13. Wippich F, Bodenmiller B, Trajkovska MG, Wanka S, Aebersold R, Pelkmans L (2013) Dual specificity kinase DYRK3 couples stress granule condensation/dissolution to mTORC1 signaling. Cell 152(4):791–805. https://doi.org/10.1016/j.cell.2013.01.033
14. Yang X, Shen Y, Garre E, Hao X, Krumlinde D, Cvijovic M, Arens C, Nystrom T, Liu B, Sunnerhagen P (2014) Stress granule-defective mutants deregulate stress responsive transcripts. PLoS Genet 10(11):e1004763. https://doi.org/10.1371/journal.pgen.1004763
15. Berchtold D, Battich N, Pelkmans L (2018) A systems-level study reveals regulators of membrane-less organelles in human cells. Mol Cell 72(6):1035–1049.e5. https://doi.org/10.1016/j.molcel.2018.10.036
16. Hofmann S, Kedersha N, Anderson P, Ivanov P (2021) Molecular mechanisms of stress granule assembly and disassembly. Biochim Biophys Acta Mol Cell Res 1868(1):118876. https://doi.org/10.1016/j.bbamcr.2020.118876
17. Schmidt C, Beilsten-Edmands V, Robinson CV (2016) Insights into eukaryotic translation initiation from mass spectrometry of macromolecular protein assemblies. J Mol Biol 428(2 Pt A):344–356. https://doi.org/10.1016/j.jmb.2015.10.011
18. Roux PP, Topisirovic I (2018) Signaling pathways involved in the regulation of mRNA translation. Mol Cell Biol 38(12):e00070-18. https://doi.org/10.1128/MCB.00070-18
19. Wang F, Li J, Fan S, Jin Z, Huang C (2020) Targeting stress granules: a novel therapeutic strategy for human diseases. Pharmacol Res 161:105143. https://doi.org/10.1016/j.phrs.2020.105143
20. Aulas A, Fay MM, Lyons SM, Achorn CA, Kedersha N, Anderson P, Ivanov P (2017) Stress-specific differences in assembly and composition of stress granules and related foci. J Cell Sci 130(5):927–937. https://doi.org/10.1242/jcs.199240
21. Advani VM, Ivanov P (2020) Stress granule subtypes: an emerging link to neurodegeneration. Cell Mol Life Sci 77(23):4827–4845. https://doi.org/10.1007/s00018-020-03565-0
22. Fujimura K, Sasaki AT, Anderson P (2012) Selenite targets eIF4E-binding protein-1 to inhibit translation initiation and induce the assembly of non-canonical stress granules. Nucleic Acids Res 40(16):8099–8110. https://doi.org/10.1093/nar/gks566
23. Kedersha N, Panas MD, Achorn CA, Lyons S, Tisdale S, Hickman T, Thomas M,

Lieberman J, McInerney GM, Ivanov P, Anderson P (2016) G3BP-Caprin1-USP10 complexes mediate stress granule condensation and associate with 40S subunits. J Cell Biol 212(7):845–860. https://doi.org/10.1083/jcb.201508028

24. Erfle H, Neumann B, Liebel U, Rogers P, Held M, Walter T, Ellenberg J, Pepperkok R (2007) Reverse transfection on cell arrays for high content screening microscopy. Nat Protoc 2(2):392–399. https://doi.org/10.1038/nprot.2006.483

25. Carpenter AE, Jones TR, Lamprecht MR, Clarke C, Kang IH, Friman O, Guertin DA, Chang JH, Lindquist RA, Moffat J, Golland P, Sabatini DM (2006) CellProfiler: image analysis software for identifying and quantifying cell phenotypes. Genome Biol 7(10):R100. https://doi.org/10.1186/gb-2006-7-10-r100

26. Perez-Pepe M, Slomiansky V, Loschi M, Luchelli L, Neme M, Thomas MG, Boccaccio GL (2012) BUHO: a MATLAB script for the study of stress granules and processing bodies by high-throughput image analysis. PLoS One 7(12):e51495. https://doi.org/10.1371/journal.pone.0051495

27. Mahboubi H, Kodiha M, Stochaj U (2013) Automated detection and quantification of granular cell compartments. Microsc Microanal 19(3):617–628. https://doi.org/10.1017/S1431927613000159

28. De Graeve F, Debreuve E, Rahmoun S, Ecsedi S, Bahri A, Hubstenberger A, Descombes X, Besse F (2019) Detecting and quantifying stress granules in tissues of multicellular organisms with the Obj.MPP analysis tool. Traffic 20(9):697–711. https://doi.org/10.1111/tra.12678

29. List M, Schmidt S, Christiansen H, Rehmsmeier M, Tan Q, Mollenhauer J, Baumbach J (2016) Comprehensive analysis of high-throughput screens with HiTSeekR. Nucleic Acids Res 44(14):6639–6648. https://doi.org/10.1093/nar/gkw554

30. Dutta B, Azhir A, Merino LH, Guo Y, Revanur S, Madhamshettiwar PB, Germain RN, Smith JA, Simpson KJ, Martin SE, Buehler E, Fraser ID (2016) An interactive web-based application for comprehensive analysis of RNAi-screen data. Nat Commun 7:10578. https://doi.org/10.1038/ncomms10578

31. Kumar P, Goh G, Wongphayak S, Moreau D, Bard F (2013) ScreenSifter: analysis and visualization of RNAi screening data. BMC Bioinformatics 14:290. https://doi.org/10.1186/1471-2105-14-290

32. Echeverri CJ, Perrimon N (2006) High-throughput RNAi screening in cultured cells: a user's guide. Nat Rev Genet 7(5):373–384. https://doi.org/10.1038/nrg1836

Chapter 23

APEX Proximity Labeling of Stress Granule Proteins

Sara Elmsaouri, Sebastian Markmiller, and Gene W. Yeo

Abstract

Ascorbate peroxidase (APEX)-catalyzed proximity labeling has been recently established as a robust approach to uncover localized protein environments and transient protein-protein interactions occurring across mammalian cells. This molecular tool enables improved identification of individual proteins localized to and involved in specific cellular and subcellular pathways and functions. Engineering of an APEX2 fusion protein into the endogenous loci of proteins of interest enables directed biotinylation of neighboring polypeptides and mRNAs. This results in identification of subcellular and context-dependent proteomes or transcriptomes via quantitative mass spectrometry or RNA sequencing, respectively. Here, we describe the utility of APEX-mediated proximity labeling to recover components of stress granules (SGs) by endogenous tagging of well-established SG-associated proteins.

Key words APEX2, Proximity labeling, Stress granule, RNA binding protein, Biotinylation, Affinity purification, Tandem mass tag, Quantitative mass spectrometry

1 Introduction

RNA processing is essential for regulation of gene expression and is tightly orchestrated in space and time. RNA binding proteins (RBPs) and their RNA substrates are frequently organized into discrete areas of high local concentration, often referred to as RNA granules or membraneless organelles [1, 2]. RNA granules are highly dynamic and can respond rapidly to changes in cellular metabolism or external stimuli. A prominent example of dynamic RNA granules are stress granules (SGs), which rapidly assemble in the cytoplasm under conditions of cellular stress and disassemble once the stressor has been removed [3, 4]. In recent years, SGs have been the focus of intense research due to an emerging link between the biology of SGs and the insoluble protein aggregates that are the hallmarks of numerous late-onset neurodegenerative disorders such as Amyotrophic Lateral Sclerosis (ALS) and Frontotemporal Dementia (FTD) [5, 6]. Importantly, several SG-associated proteins have been identified as potential therapeutic targets in

Daniel Matějů and Jeffrey A. Chao (eds.), *The Integrated Stress Response: Methods and Protocols*,
Methods in Molecular Biology, vol. 2428, https://doi.org/10.1007/978-1-0716-1975-9_23,

neurodegenerative disorders [7–9]. SGs have been found to contain hundreds of different proteins [7, 10–12], but for most SG-associated proteins it is unclear if and how they contribute to SG formation and/or disassembly. Apart from a defined set of proteins that consistently associate with SGs under a broad range of conditions, SG composition can vary significantly between cell types and different stress conditions [7, 13]. In addition, SGs interact with numerous different cellular structures and machineries, including the endoplasmic reticulum [14], molecular chaperones [15], RNA helicases [12], proteasomes [16], autophagy [17], lysosomes [18], and the nucleocytoplasmic transport (NCT) machinery [19]. Furthermore, SG composition and behavior is regulated by posttranslational modifications, including ubiquitylation, SUMOylation, and arginine methylation [20, 21]. To elucidate how all these different factors shape the molecular interactions within SGs and to identify key regulators of SG biology, it is critically important to be able to profile dynamic changes in SG protein composition with high spatial and temporal resolution.

Proximity labeling of proteins with a covalent tag coupled with affinity purification and quantitative mass spectrometry has been developed in recent years to characterize protein landscapes of defined subcellular compartments. While there are several different methods for protein proximity labeling, the engineered peroxidase APEX2 has emerged as the most suitable enzyme for the characterization of dynamic RNA granules due to its rapid labeling kinetics [7, 10]. Upon activation with hydrogen peroxide (H_2O_2) for a period of only 30–60 s, APEX2 catalyzes the conversion of a biotin phenol substrate into short-lived and highly reactive phenoxyl radicals, which covalently biotinylate immediately proximal proteins within a very small diffusion radius [22–25]. By fusing APEX2 to strongly partitioned canonical SG proteins, it is thus possible to label and capture snapshots of the spatially restricted SG-associated proteome under a wide range of conditions. This approach is generalizable to other dynamic RNA granules, as long as APEX2 can be fused to a core component with a sufficiently high-partitioning coefficient.

We initially demonstrated the utility of proximity labeling combined with quantitative mass spectrometry to determine the composition of stress granules by using a G3BP1-APEX2-GFP fusion protein to simultaneously visualize SGs and biotinylate SG-proximal proteins in live cells. It is important to ensure that APEX2-fusion proteins are expressed at near-endogenous levels to avoid granulation artefacts that can arise from overexpression of RNA granule proteins such as G3BP1. We achieved this by tagging the endogenous G3BP1 locus with an APEX2-GFP fusion protein using CRISPR/Cas9 genome engineering in HEK293T cells and human iPSCs [7]. An alternative approach is to drive expression of a protein of interest (POI) fused to APEX2 from an inducible promoter that can be tuned to achieve near-endogenous expression levels [10].

Here, we provide a protocol for APEX proximity labeling of stress granule proteins based on our previous work using engineered HEK293T G3BP1-APEX2-GFP cells (Fig. 1). While our original work used a SILAC labeling approach for quantitative proteomics, the protocol presented here is compatible with a range of proteomics strategies that have been used in the context of APEX proximity labeling, including label-free proteomics and isobaric labeling methods such as TMT. It can also be easily adapted for proximity labeling of other types of RNA granules, provided that cell lines expressing suitable APEX2 fusion proteins are available.

2 Materials

2.1 APEX-Expressing Cell Line Culture

1. G3BP1-APEX2-GFP-expressing HEK293T cell line (*see* **Note 1**): single frozen vial usually contains 1×10^6 cells in 1 mL 10% DMSO freezing medium.
2. HEK293T Culture Medium: Dulbecco's Modified Eagle's Medium (DMEM) supplemented with 10% fetal bovine serum (FBS) and 1% Penicillin/Streptomycin (P/S) (optional). Sterile filter and store at 4 °C. Prewarm in a 37 °C water bath prior to use on cells (*see* **Note 2**).
3. *Optional (For SILAC labeling).*
   (a) *SILAC DMEM deficient in both arginine and lysine.*
   (b) *Dialyzed FBS.*
   (c) *L-arginine–HCl (MW 210.7 g/mol).*
   (d) *L-lysine–HCl (MW 182.7 g/mol).*
   (e) *L-arginine–HCl* $^{13}C_6$, $^{15}N_4$ *(MW 220.6 g/mol).*
   (f) *L-lysine–2HCl* $^{13}C_6$, $^{15}N_2$ *(MW 227.1 g/mol).*

   Prepare medium from SILAC DMEM deficient in both arginine and lysine. Supplement with 10% dialyzed FBS. For "light" medium, add L-arginine to 0.4 mM and L-lysine to 0.8 mM. For "heavy" medium, add L-arginine $^{13}C_6$, $^{15}N_4$ *at 0.5 mM and L-lysine* $^{13}C_6$, $^{15}N_2$ *at 0.8 mM (see* **Note 3***). Supplement both light and heavy medium with L-proline at 200 mg/mL (see* **Note 4***).*
4. 10 cm plastic tissue-culture dishes (*see* **Note 5**).
5. Dulbecco's phosphate-buffered saline (DPBS).
6. TrypLE Express Enzyme.

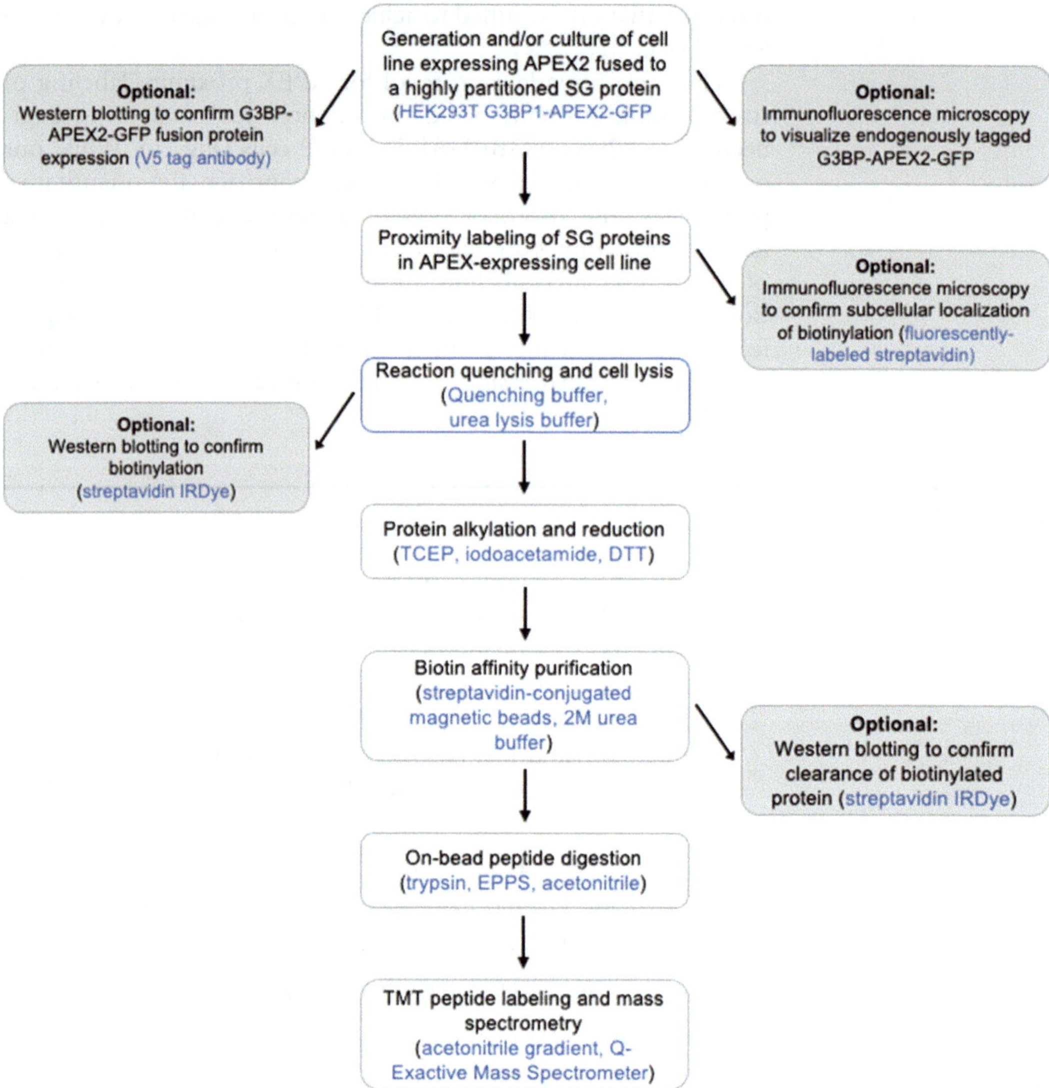

Fig. 1 Schematic outlining major steps for APEX proximity labeling with the end goal of mass spectrometry to identify stress granule associated proteins

2.2 Optional Validation of Cell Lines

1. Radioimmunoprecipitation assay (RIPA) lysis and extraction buffer.
2. BCA protein assay.
3. 4× lithium dodecyl sulfate (LDS) sample buffer.
4. DL-dithiothreitol (DTT) reducing agent.
5. 4–12% gradient bis-tris gel (NuPAGE Novex NP0321BOX).
6. 1× MOPS Running Buffer: dilute 50 mL of concentrated 20× MOPS (NuPAGE NP0001) with 950 mL deionized water to

make 1 L stock in an autoclaved glass bottle. Store at room temperature.

7. Broad range protein ladder (e.g., Spectra Broad Range, ThermoFisher).
8. PVDF membranes.
9. 1× Transfer buffer: dilute 50 mL of concentrated 20× transfer buffer (NuPAGE NP0006-1) with 950 mL deionized water to make 1 L stock in an autoclaved glass bottle. Store at 4 °C.
10. TBST: dilute 100 mL of 10× TBS and 1 mL Tween® 20 detergent in 900 mL of deionized water.
11. Membrane blocking buffer: 5% milk (w/v) in TBST.
12. V5 Tag antibody (37-7500-A488). (Alternatives: GFP antibody, G3BP1 antibody).
13. Chamber slides.
14. Paraformaldehyde (PFA): dilute 2 mL of 20% stock PFA in 8 mL DPBS for a 4% working solution.
15. IF wash buffer: dilute Triton X-100 to 0.01% (v/v) in DPBS.
16. IF blocking buffer: dilute Triton X-100 to 0.1% (v/v) and serum (of the same species as secondary antibody) to 5% (v/v) in DPBS.

2.3 APEX-Mediated Biotinylation

1. Biotin phenol ($C_{18}H_{25}N_3O_3S$), also sold as biotinyl-tyramide.
2. 0.05 mol/l sodium arsenite ($NaAsO_2$) solution (*see* **Note 6**).
3. 30% hydrogen peroxide solution (*see* **Note 7**).
4. Quenching solution.
   (a) Trolox (6-hydroxy-2,5,7,8- tetramethylchroman-2-carboxylic acid). Prepare a 500 mM stock solution by resuspending 1 g of Trolox in 8 mL of DMSO. Aliquot and store at −20 °C or −80 °C.
   (b) Sodium ascorbate ($C_6H_7NaO_6$). Prepare a 1 M stock solution by resuspending 1 g of sodium ascorbate in 5 mL of PBS. Prepare fresh immediately prior to each experiment.
   (c) Sodium azide. Prepare a 1 M stock solution by resuspending 650 mg in 10 mL of H_2O. Aliquot and store at room temperature protected from light (*see* **Note 8**).

   The labeling reaction is stopped by adding Trolox, sodium ascorbate and sodium azide to final concentrations of 5 mM, 10 mM and 10 mM, respectively. Prepare a 2× quenching solution containing 10 mM Trolox and 20 mM each of sodium ascorbate and sodium azide in ice-cold PBS (*see* **Note 9**).
5. Urea Lysis buffer: 8 M urea, 150 mM NaCl, 20 mM Tris pH 8.0 supplemented with protease inhibitors (e.g., Protease

Inhibitor Cocktail Set III, EDTA-Free, EMD Millipore) (*see* **Note 10**).

6. Dilution buffer: 150 mM NaCl, 20 mM Tris pH 8.0 supplemented with protease inhibitors.
7. Optional: Amine-free dilution buffer: 150 mM NaCl, 50 mM EPPS pH 8.5, supplemented with protease inhibitors (*see* **Note 10**).

2.4 Validation of Biotinylation by Western Blotting and/or Immunocytochemistry

1. 660 nm protein assay (Pierce).
2. IRDye 800CW Streptavidin antibody diluted 1:1000 in TBST.
3. Fluorescently labeled streptavidin, for example, streptavidin-Alexa488 (Thermo Fisher).

2.5 Streptavidin Affinity Purification and Preparation for Mass Spectrometry

1. Streptavidin-conjugated magnetic beads (Pierce).
2. Tris(2-carboxyethyl)phosphine (TCEP) reducing agent.
3. 2-Iodoacetamide alkylating agent.
4. Wash buffer: 2 M urea, 150 mM NaCl, 20 mM Tris pH 8.0 (alternatively use 50 mM EPPS pH 8.5, *see* **Note 10**).
5. Sequencing Grade Modified Trypsin (Promega).
6. 100% HPLC grade acetonitrile.
7. Quantitative Colorimetric Peptide Assay (Pierce).

3 Methods

3.1 Experimental Design Considerations

For successful identification of proteins associated with a particular type of RNA granule it is critical to (a) choose a primary bait that is highly partitioned into the RNA granule of interest, and (b) choose appropriate control conditions to calculate enrichment in granules vs background interaction with the bait.

To illustrate, for a comprehensive characterization of SG-associated proteins we performed 4 different pairwise enrichment analyses (Fig. 2) [1]. These included a comparison with a negative control (no biotin phenol substrate and no H_2O_2) to ensure that known SG proteins could be successfully enriched using the system. In addition, we analyzed the interactomes of the core SG protein G3BP1 under both stressed and unstressed conditions to determine dynamic changes in the G3BP1 interactome and to identify proteins that interact with G3BP1 only in the context of SGs. However, this comparison alone may miss substantial numbers of proteins that already interact with G3BP1 under unstressed conditions. To capture these G3BP1 interactors and to determine whether they maintain or lose their proximity to G3BP1 under stress conditions, we additionally compared enriched

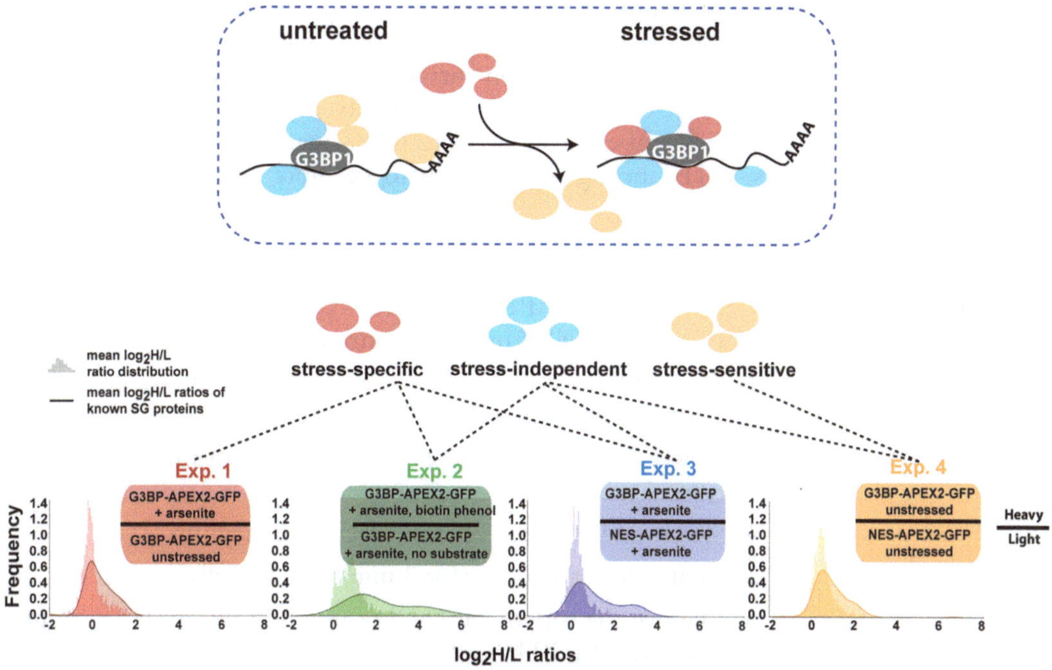

Fig. 2 Schematic of G3BP1 interactome changes upon stress and outlined experimental designs for detecting the G3BP1 interactome changes under different conditions, including log2 H/L ratio distributions of all proteins detected, overlaid with log2 H/L ratio distributions of known SG proteins. (Adapted with permission from Markmiller et al., Cell 2018)

proteins from G3BP1-APEX2-GFP fusions vs an NES-APEX2-GFP protein under both stressed and unstressed conditions. Ideally the control should consist of APEX2 being expressed at levels very similar to the bait protein and reflecting the subcellular distribution of the bait protein population that is not partitioned into the RNA granule of interest as closely as possible. The diffuse cytoplasmic localization of NES-APEX2-GFP closely mirrors the subcellular distribution of G3BP1-APEX2-GFP in the absence of stress and thus makes this a suitable control. Other types of RNA granules and bait proteins may require different controls, and it may be necessary to include more than one control condition in some cases.

In addition to careful selection of experimental and control conditions, it can be helpful to perform proximity labeling with additional bait proteins that partition strongly into the RNA granule of interest. The utility of this approach was recently demonstrated by Marmor-Kollet et al. (2020), who used three different SG-associated bait proteins to gain high-confidence sets of SG-associated proteins under various experimental conditions [2].

Lastly, the choice of mass spectrometry method is another important factor in experimental design. Early APEX proximity labeling studies, including our own work, used SILAC-based quantitative proteomics. This approach is robust and can be advantageous in cases where additional experimental sample manipulations after biotinylation (e.g., partial biochemical enrichment of subcellular compartments prior to affinity purification) might introduce sources of variability. However, for most protein proximity labeling applications other quantitative approaches such as TMT labeling have proven to be more versatile and easier to analyze.

3.2 Culture of APEX-Expressing Cell Lines

1. Remove 1 frozen vial each containing approximately 1×10^6 HEK293T G3BP1-APEX2-GFP cells and HEK293T NES-APEX2-GFP cells from liquid nitrogen storage.
2. Quickly thaw cells by placing vial into a 37 °C water bath for 1.5 to 2 min until about 90% thawed.
3. Slowly pipette 1 mL of cell suspension into a 15 mL conical tube.
4. Slowly add 9 mL of HEK293T culture medium.
5. Centrifuge at $200 \times g$ for 3 min to pellet cells.
6. Remove the supernatant and resuspend cells in 10 mL of prewarmed HEK293T culture medium.
7. Slowly pipette 10 mL of cell suspension into a 10 cm tissue culture dish and gently rock side to side in all directions to ensure even coverage and cell dispersion.
8. Place cells into a humidified incubator at 37 °C and 5% CO_2 and allow them to adhere overnight.
9. Replace spent medium with fresh, prewarmed every 2–3 days, dependent upon cell growth rate.
10. Once 80–90% confluency is reached, passage or expand cells at ratios between 1:10 to 1:20 to reach original confluency in approximately 3–5 days respectively (*see* **Note 11**).
11. Remove spent medium and rinse once with 5 mL of DPBS to remove debris before passaging cells.
12. Add 2 mL of TrypLE per 10 cm dish and place back into the incubator until cells begin to lift (*see* **Note 12**).
13. Triturate cells with pipette to further remove from the dish and transfer to a 15 mL conical tube.
14. Centrifuge at $200 \times g$ for 3 min to pellet cells.
15. Remove the supernatant, resuspend cells in 10 mL of HEK293T culture medium, and add 1 mL of cell suspension per 10 cm dish containing 9 mL of fresh culture medium.
16. Return the cells to a humidified incubator at 37 °C and 5% CO_2 and exchange spent medium every 2–3 days as needed.
17. Passage cells at minimum 2–3 times after thawing before utilizing in the final experiment.

18. *Optional: If a SILAC-based quantitative mass spectrometry approach will be pursued, split cells into two separate cultures immediately upon thawing and maintain in "light" and "heavy" SILAC medium (see* Subheading 2.1, **item 3** *and* **Notes 2–4***).* Perform SILAC label incorporation test after 4–5 passages to confirm >95% label incorporation (*see* **Note 13**).

3.3 Cell Line Validation

We have extensively characterized the HEK293T G3BP1-APEX2-GFGP cells described in Markmiller et al. (2018) and found them to be stable and to maintain robust expression of the G3BP1-APEX2-GFP fusion protein (Fig. 3). For less well-characterized or novel cell lines or for troubleshooting purposes we include several optional cell line validation steps below.

3.3.1 Fluorescence Microscopy to Visualize SG Localization of G3BP-APEX2-GFP

1. To visualize G3BP-GFP expression of cells directly via immunocytochemistry, seed 5-10 × 10^4 cells into each well of an 8-well chamber slide in ~250 μL of HEK293T culture medium.
2. After overnight incubation, add an equal volume of 1 mM arsenite to half of the chamber wells for a final concentration of 500 μM.
3. Incubate at 37 °C and 5% CO_2 for 1 h.
4. Remove medium (*see* **Note 6**), rinse once with DPBS and add 200 μL of 4% PFA to each chamber well.
5. Fix cells by incubating for 15–30 min at room temperature.
6. Remove PFA (*see* **Note 14**) and rinse cells twice with DPBS.
7. Simultaneously permeabilize and block cells by incubating in IF blocking buffer for 45 min at room temperature.
8. Rinse wells once with IF wash buffer.
9. Wash with DAPI for 15 min at room temperature.
10. Wash once with wash buffer.
11. Preserve in 50% glycerol/DPBS solution or mount and coverslip slides, drying overnight prior to imaging endogenous GFP signal.

3.3.2 Western Blotting to Confirm V5 Tag Expression of G3BP-GFP-APEX Fusion Protein

1. In 10 cm tissue culture dishes, seed 8 × 10^6 HEK293T G3BP1-APEX2-GFP cells in 8 mL of HEK293T culture medium.
2. After overnight incubation, aspirate spent medium and lyse the cells with 500 μL of cold RIPA buffer.
3. Harvest the lysed cells immediately using cell scrapers while keeping the plates on ice and transfer cell lysates to microcentrifuge tubes.

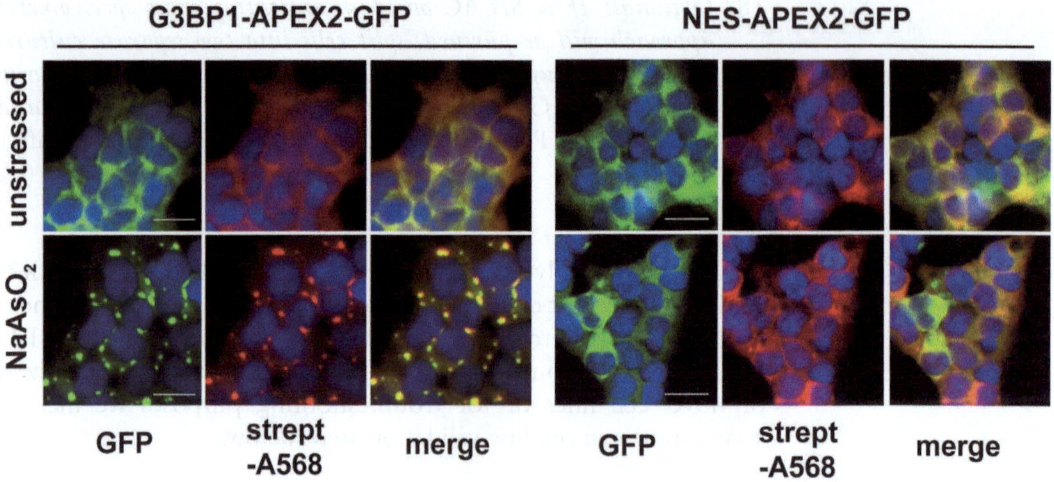

Fig. 3 Streptavidin staining of unstressed and $NaAsO_2$-treated HEK293T G3BP1-APEX2-GFP and hPGK-NES-APEX2-GFP cells. (Adapted with permission from Markmiller et al., Cell 2018)

4. Sonicate the samples for 5 cycles with 30 s on/off and at high intensity (*see* **Note 15**).
5. Centrifuge at (13,500 × *g*) for 10 min at 4 °C to clear the lysates and transfer the supernatant to a new microcentrifuge tube.
6. Determine the total protein concentration of each sample using the BCA protein assay.
7. Run a western blot using approximately 15 μL of each lysate, adjusted for equivalent protein concentration.
8. Add 4× LDS buffer at 4:1 v/v and DTT to a final concentration between 50 and 100 mM to each lysate sample. Pipette up and down to mix well. Briefly centrifuge tubes for 5–10 s.
9. Boil samples on a heat block at 95 °C for 5 min. Briefly centrifuge tubes and return samples to ice.
10. Load samples on a 4–12% gradient Bis-Tris gel with 1× MOPS running buffer. Buffer should be filled enough to submerge the sample wells completely as well as the electrode strips at the bottom of the gels.
11. Load 5 μL of protein ladder on both end wells.
12. Run the gel at 60 V for 20 min and then 150 V for 80 min.
13. Transfer gel onto methanol-activated PVDF membrane at 100 V for 2.5 h in 1× transfer buffer.
14. Block membrane in 5% milk in TBST for 1 h at room temperature.
15. Wash membrane for 10 min with TBST buffer. Repeat wash three times.

16. Probe with V5 tag fluorescent antibody diluted 1:500 in TBST for 1 h at room temperature or 4 °C overnight on a shaker. (Alternatively, probe with GFP fluorescent antibody diluted 1:2000 or G3BP1 antibody diluted 1:5000 to assess difference in size of the protein.)
17. Use a gel imager to image membrane for fluorescent signal at the appropriate channel (i.e., V5 Tag Antibody 37-7500-A488 is green).

3.4 APEX-Mediated Biotinylation of SG Proximal Proteins

1. In 10 cm tissue culture dishes, seed 8 × 10^6 HEK293T G3BP1-APEX2-GFP cells in 8 mL of HEK293T culture medium.
2. After overnight incubation, aspirate spent medium and add 4 mL of fresh medium containing biotin phenol (BP) at a final concentration of 500 μM (*see* **Note 2**).
3. For half of the plates, also add sodium arsenite ($NaAsO_2$) at a final concentration of 500 μM (*see* **Note 6**).
4. Incubate for 1 h at 37 °C.
5. Add 2 mL of DPBS containing 3 mM hydrogen peroxide (H_2O_2) for a final concentration of 1 mM and immediately mix by swirling the plate for 15–20 s.
6. After exactly 60 s, quench the reaction by adding one volume (6 mL) of ice cold 2× quenching solution (*see* **Notes 16** and **17**) and transferring the plate onto ice.
7. Aspirate quenched medium and rinse once with 4 mL of ice-cold 1× quenching solution.
8. Aspirate quenching solution and add 500 μL of ice-cold urea lysis buffer supplemented with protease inhibitors, 5 mM Trolox and 10 mM each of sodium ascorbate and sodium azide.
9. Harvest cells using a cell scraper and transfer to microcentrifuge tubes.
10. Complete cell lysis by sonicating the samples for 5 cycles of 30 s on/off at high intensity (*see* **Note 15**).
11. Centrifuge at (13,500 × *g*) for 10 min at 4 °C to clear the lysates and transfer the supernatant to a new microcentrifuge tube.
12. Determine the total protein concentration of each sample using the 660 nm protein assay (*see* **Note 18**).

3.5 Validation of APEX Labeling: Streptavidin Western Blotting and Immunocytochemistry

1. To confirm biotinylation, run a western blot using approximately 10 μL of each lysate, adjusted for equivalent protein concentration (Fig. 4).
2. Refer to **steps 8–15** under Subheading 3.3.2.
3. Probe with streptavidin IRDye diluted 1:1000 in TBST for 1 h at room temperature or 4 °C overnight.

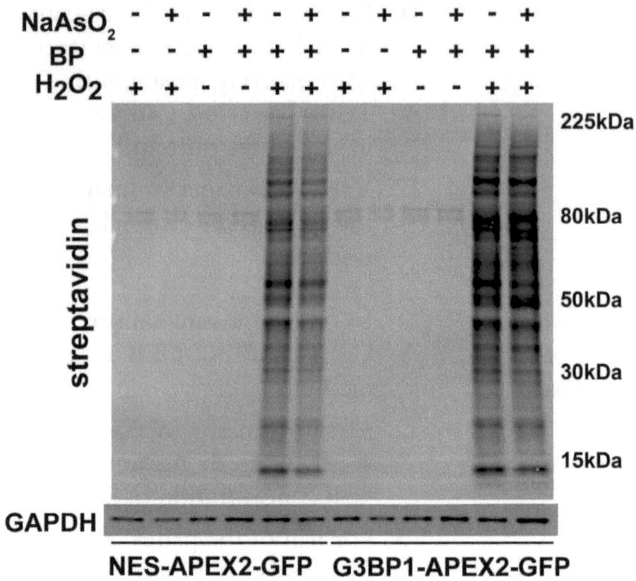

Fig. 4 Streptavidin-HRP western blot analysis of induced protein biotinylation in lysates from NES-APEX2-GFP and G3BP1-APEX2-GFP cells. (Adapted with permission from Markmiller et al., Cell 2018)

4. Use a gel imager to image membrane for fluorescent signal at the appropriate channel.
5. To visualize biotinylation of cells directly via immunocytochemistry (Fig. 3), seed 0.03×10^6 cells into each well of an 8-well chamber slide in 200 μL of HEK293T culture medium.
6. Repeat **steps 2–7** under Subheading 3.4. Do not lyse cells.
7. Wash with DPBS twice and fix in 150 μL of 4% PFA for 20 min at room temperature.
8. Refer to **steps 6–8** under Subheading 3.3.1.
9. Incubate in fluorescently labeled streptavidin diluted 1:1000 in IF wash buffer with 1% BSA (w/v) overnight at 4 °C or 1–2 h at room temperature.
10. Aspirate antibody and wash chamber wells with IF wash buffer twice.
11. Refer to **steps 9–11** under Subheading 3.3.1.

3.6 Affinity Purification of Biotinylated Samples and On-Bead Peptide Digestion

1. Add TCEP to a final concentration of 10 mM to each lysate and incubate at room temperature for 30 min (*see* **Note 19**).
2. Add freshly prepared iodoacetamide to a final concentration of 15 mM to each lysate and incubate in the dark at room temperature for 45 min (*see* **Note 19**).

3. Add DTT to a final concentration of 10 mM to each lysate and incubate in the dark at room temperature for 15 min (*see* **Note 19**).
4. Dilute 450 μL of each lysate with 1350 μL of EPPS buffer (150 mM NaCl, 50 mM EPPS pH 8.5) for a final urea concentration of 2 M.
5. Equilibrate 125 μL of streptavidin-conjugated magnetic beads in EPPS buffer containing 2 M urea (*see* **Note 20**).
6. Incubate samples with rotation at room temperature for 2 h at room temperature or overnight at 4 °C.
7. Briefly centrifuge to remove material from the inside of the lid and place tube on magnetic separation rack.
8. Wait until beads have completely separated and transfer the supernatant to a new microcentrifuge tube. Supernatant can be flash-frozen, stored at −20 °C or −80 °C, and analyzed to confirm complete clearance of biotinylated material during the affinity purification (*see* **Notes 21** and **22**).
9. Wash remaining streptavidin bead samples 6 times with 1 mL of cold EPPS buffer containing 1 M urea for 5 min per wash while rotating at 4 °C. Use a magnetic separation rack to collect streptavidin beads between each wash and remove the buffer with a pipette, being careful not to remove beads. After the final wash, remove as much supernatant as possible.
10. Add 1 μg of trypsin diluted in 50 mM EPPS pH 8.5 per sample. Use double volume or 250 μL trypsin per 125 μL of beads for each sample (*see* **Note 23**).
11. Incubate the trypsin digestion for 6 h at 37 °C with gentle shaking (*see* **Note 24**).
12. Add 100% HPLC grade acetonitrile at a final concentration of 10% (*see* **Note 25**).
13. Determine the peptide concentration of each sample using a quantitative colorimetric peptide assay. Subtract the trypsin concentration from the final peptide concentration to determine the accurate concentration of nontrypsin peptides.
14. Proceed to TMT labeling steps based on the amount of total peptides per sample.

4 Notes

1. In principle, APEX-mediated proximity labeling can be conducted with any cell type modified to express APEX2 fused to a protein of interest (POI) with a specific subcellular localization. For stress granules, APEX2 has been fused to several different

proteins, including G3BP1, FMR1 and FXR1 [7, 10]. We have had good results with CRISPR/Cas9-mediated insertion of APEX2 into the endogenous G3BP1 locus in HEK293T cells and induced pluripotent stem cells [7]. The homologous recombination donor plasmid for endogenous tagging is available from Addgene (Plasmid #105284) and the cell lines are available upon request. However, successful APEX labeling has also been performed in cells expressing POI-APEX2 fusion proteins from constitutive or tetracycline-inducible promoters upon either transient or stable transfection.

2. We have found APEX labeling to work well in a range of different cell culture media. However, we have repeatedly experienced poor labeling in mTeSR1 stem cell medium. We recommend substituting DMEM/F12 basal medium for the biotin phenol incubation period.
3. Amino acid concentrations given here are for DMEM but can vary depending on the cell culture medium used. We routinely prepare stock solutions of amino acids to supplement media as needed. For example, 0.4 mM of "heavy" L-arginine–HCl (MW 220.6 g/mol) = 88.24 mg/L. Thus, a 1000x stock solution of 88.24 mg/mL for DMEM can be stored at 4 °C or −20 °C and volumes can easily be adjusted to supplement other types of media (e.g., DMEM/F12 at 0.7 mM).
4. When grown in SILAC medium, many cell lines can convert substantial amounts of isotope-labeled arginine into proline. This isotope-labeled proline is then incorporated into newly synthesized proteins and impairs the accurate identification and quantification of heavy vs light peptides. This conversion can be prevented by supplementing the medium with excess proline at 200 mg/L [26]. We initially found this conversion to be a substantial issue in HEK293T cells and now routinely complement both heavy and light SILAC media with proline.
5. HEK293T cells are maintained on 10 cm plastic tissue-culture dishes throughout for the end goal of mass spectrometry. One 10 cm dish of ~80–90% confluent cells is used per final condition per replicate. This can be scaled up or scaled down based on alternative endpoints which require different amounts of starting material (e.g., different mass spectrometry techniques, RNA sequencing, immunocytochemistry, western blotting).
6. Sodium arsenite is toxic and potentially dangerous. Consult Material Safety Data Sheet and appropriate local laboratory safety guidelines for handling and waste disposal.
7. Once opened, hydrogen peroxide solution will lose activity within 3–6 months. Keep stock bottle closed and at 4 °C when not in use and keep track of the opening date.

8. Sodium azide is extremely toxic and must be handled accordingly. Consult Material Safety Data Sheet and appropriate local laboratory safety guidelines for handling and waste disposal.
9. Solubility of Trolox in aqueous solutions is low and Trolox may begin to precipitate in concentrations above the 10 mM used in 2× quenching solution.
10. For downstream applications that are not compatible with primary amine buffers such as Tris (e.g., TMT labeling), 50 mM EPPS buffer pH 8.5 (e.g., Fisher Scientific catalog no. AAJ61476AE) can be used instead of Tris.
11. For best results and reproducibility, it is critical that cells are cultured consistently under appropriate conditions. Do not allow cells to become overconfluent before passaging. We routinely use HEK293T cells for no more than 5–10 passages of thawing from a general frozen stock to maintain consistency between replicates and different experiments.
12. We routinely culture HEK293T cells without coating of tissue culture vessels. As a result, cells usually detach within 2–3 min of TrypLE treatment. Under some circumstances it may be desirable or necessary to grow HEK293T cells on dishes coated with poly-L-lysine. In this case, incubation for 5–10 min or longer may be necessary to detach cells for passaging.
13. *Optional (if performing SILAC labeling): In rapidly proliferating cells such as HEK293T, we generally observe near-complete label incorporation after 3–5 passages. By contrast, in other cell types with slower protein turnover* (e.g., *postmitotic neurons), extended SILAC labeling may be required for sufficient label incorporation.*
14. PFA is toxic and potentially carcinogenic. Consult Material Safety Data Sheet and appropriate local laboratory safety guidelines for handling and waste disposal.
15. We routinely use sonication to achieve complete sample lysis and nucleic acid fragmentation, but other physical disruption methods, including liquid homogenization or shearing through a small gauge needle are also acceptable. Complete homogenization and clearance of samples is important to avoid sample aggregation during streptavidin affinity purification.
16. Since MS/MS experiments are eventually done in triplicate samples and the APEX labeling requires several conditions for comparison (stressed and unstressed, with and without BP substrate), many plates must be handled at one time. The SG proteome is highly context dependent so slight variations in APEX labeling treatment times may lead to varying results across replicates. It is recommended to stagger the treatment times for BP and stressors in order to allow for a few minutes between each plate to perform the subsequent H_2O_2 treatment

followed by quenching and lysis. It is important to keep the BP and stressor treatment time (1 h) consistent across samples as well as the H_2O_2 treatment time precise (60 s) across samples to limit variability as much as possible.

17. Using a 2× quenching solution reduces hands-on time at the end of the 60 s incubation and allows for better control of the labeling time. It is also possible to aspirate the H_2O_2-containing medium and quench the reaction with ice-cold 1× quenching solution instead.
18. Urea lysis buffer supplemented with antioxidants (Trolox and sodium ascorbate) is not compatible with most other methods used to determine protein concentration (e.g., BCA assay).
19. Alkylation and reduction treatments with TCEP, iodoacetamide, and DTT can be done either predigestion at the protein level or postdigestion at the peptide level. Either TCEP or DTT can be used for the reduction step to expose free cysteine residues. Iodoacetamide then alkylates the free cysteines for easier detection via tandem mass spectrometry. The last DTT step inactivates any excess iodoacetamide.
20. We determined the appropriate amount of streptavidin beads per sample by titration and analyzing the amount of biotinylated proteins left behind in the supernatant after affinity purification. This amount may need to be adjusted depending on binding capacity of the specific beads used and the density, total protein content, and biotinylation efficiency of the cells that are being used.
21. If not proceeding directly to biotin affinity purification, lysates can be stored at −80 °C by first snap freezing the tubes in liquid nitrogen or ethanol over dry ice directly after clearing by centrifugation. Additionally, partial aliquots of each sample can be saved at this step for downstream protein quantification and streptavidin western blotting to confirm successful biotinylation of associated proteins.
22. Supernatants collected after the streptavidin bead incubation can be used to determine the efficacy of the biotin enrichment and streptavidin immunoprecipitation. Run 10 μL of each supernatant on a western blot and probe with streptavidin antibody. A successful enrichment of biotinylated proteins captured by the streptavidin beads would result in very little or no signal on the supernatant western blot. If a strong biotin signal is observed, the enrichment must be repeated and further optimized before proceeding.
23. Trypsin should be at a 1:100 enzyme to substrate ratio. However, the amount of protein captured by the beads will not be known at the time. Therefore, although 1 μg of trypsin per sample should be sufficient, further optimization may be

required and dependent on the abundance and interactions of the SG protein of interest.

24. The 6 h trypsin digestion at 37 °C can also be done overnight instead. The minimum digestion time suggested is 6 h and it is not recommended to go shorter.
25. Handle 100% HPLC grade acetonitrile carefully and only in a fume hood as it is highly flammable and toxic.

Acknowledgments

We acknowledge members of the Yeo lab for critical discussions and comments. We thank members of the Eric Bennett and Éric Lécuyer labs for contributions to our previous work utilizing the described methods and helpful advice. Previous work involving optimization of the described methods was partially supported by grants from the ALS Association and the NIH (NS103172 and HG004659) to G.W.Y. Continued work involving these methods is partially supported by an Allen Distinguished Investigator Award, a Paul G. Allen Frontiers Group advised grant of the Paul G. Allen Family Foundation to G.W.Y. and grants from the NIH (AG069098, HG004659) to G.W.Y.

Competing interests: G.W.Y. is on the Scientific Advisory Board of Jumpcode Genomics, and a cofounder, member of the Board of Directors, on the Scientific Advisory Board, equity holder and paid consultant for Locanabio and Eclipse BioInnovations. G.W.Y. is a visiting professor at the National University of Singapore. G.W.Y.'s interests have been reviewed and approved by the University of California San Diego, in accordance with its conflict-of-interest policies. The authors declare no other competing interests.

References

1. Anderson P, Kedersha N (2009) RNA granules: post-transcriptional and epigenetic modulators of gene expression. Nat Rev Mol Cell Biol 10(6):430–436. https://doi.org/10.1038/nrm2694
2. Tauber D, Tauber G, Parker R (2020) Mechanisms and regulation of RNA condensation in RNP granule formation. Trends Biochem Sci 45(9):764–778. https://doi.org/10.1016/j.tibs.2020.05.002
3. Hofmann S, Kedersha N, Anderson P, Ivanov P (1868) Molecular mechanisms of stress granule assembly and disassembly. Biochim Biophys Acta, Mol Cell Res 2021(1):118876. https://doi.org/10.1016/j.bbamcr.2020.118876
4. Wheeler JR, Matheny T, Jain S, Abrisch R, Parker R (2016) Distinct stages in stress granule assembly and disassembly. elife 5:e18413. https://doi.org/10.7554/eLife.18413
5. Wolozin B, Ivanov P (2019) Stress granules and neurodegeneration. Nat Rev Neurosci 20 (11):649–666. https://doi.org/10.1038/s41583-019-0222-5
6. Sidibe H, Vande Velde C (2019) RNA granules and their role in neurodegenerative diseases. Adv Exp Med Biol 1203:195–245. https://doi.org/10.1007/978-3-030-31434-7_8
7. Markmiller S, Soltanieh S, Server KL, Mak R, Jin W, Fang MY, Luo EC, Krach F, Yang D, Sen A, Fulzele A, Wozniak JM, Gonzalez DJ, Kankel MW, Gao FB, Bennett EJ, Lecuyer E,

Yeo GW (2018) Context-dependent and disease-specific diversity in protein interactions within stress granules. Cell 172(3):590–604. e13. https://doi.org/10.1016/j.cell.2017.12.032

8. Apicco DJ, Ash PEA, Maziuk B, LeBlang C, Medalla M, Al Abdullatif A, Ferragud A, Botelho E, Ballance HI, Dhawan U, Boudeau S, Cruz AL, Kashy D, Wong A, Goldberg LR, Yazdani N, Zhang C, Ung CY, Tripodis Y, Kanaan NM, Ikezu T, Cottone P, Leszyk J, Li H, Luebke J, Bryant CD, Wolozin B (2018) Reducing the RNA binding protein TIA1 protects against tau-mediated neurodegeneration in vivo. Nat Neurosci 21(1):72–80. https://doi.org/10.1038/s41593-017-0022-z
9. Becker LA, Huang B, Bieri G, Ma R, Knowles DA, Jafar-Nejad P, Messing J, Kim HJ, Soriano A, Auburger G, Pulst SM, Taylor JP, Rigo F, Gitler AD (2017) Therapeutic reduction of ataxin-2 extends lifespan and reduces pathology in TDP-43 mice. Nature 544 (7650):367–371. https://doi.org/10.1038/nature22038
10. Marmor-Kollet H, Siany A, Kedersha N, Knafo N, Rivkin N, Danino YM, Moens TG, Olender T, Sheban D, Cohen N, Dadosh T, Addadi Y, Ravid R, Eitan C, Toth Cohen B, Hofmann S, Riggs CL, Advani VM, Higginbottom A, Cooper-Knock J, Hanna JH, Merbl Y, Van Den Bosch L, Anderson P, Ivanov P, Geiger T, Hornstein E (2020) Spatiotemporal proteomic analysis of stress granule disassembly using APEX reveals regulation by SUMOylation and links to ALS pathogenesis. Mol Cell 80(5):876–891. e876. https://doi.org/10.1016/j.molcel.2020.10.032
11. Youn JY, Dunham WH, Hong SJ, Knight JDR, Bashkurov M, Chen GI, Bagci H, Rathod B, MacLeod G, Eng SWM, Angers S, Morris Q, Fabian M, Cote JF, Gingras AC (2018) High-density proximity mapping reveals the subcellular organization of mRNA-associated granules and bodies. Mol Cell 69(3):517–532. e11. https://doi.org/10.1016/j.molcel.2017.12.020
12. Jain S, Wheeler JR, Walters RW, Agrawal A, Barsic A, Parker R (2016) ATPase-modulated stress granules contain a diverse proteome and substructure. Cell 164(3):487–498. https://doi.org/10.1016/j.cell.2015.12.038
13. Advani VM, Ivanov P (2020) Stress granule subtypes: an emerging link to neurodegeneration. Cell Mol Life Sci 77(23):4827–4845. https://doi.org/10.1007/s00018-020-03565-0
14. Lee JE, Cathey PI, Wu H, Parker R, Voeltz GK (2020) Endoplasmic reticulum contact sites regulate the dynamics of membraneless organelles. Science 367(6477):eaay7108. https://doi.org/10.1126/science.aay7108
15. Mateju D, Franzmann TM, Patel A, Kopach A, Boczek EE, Maharana S, Lee HO, Carra S, Hyman AA, Alberti S (2017) An aberrant phase transition of stress granules triggered by misfolded protein and prevented by chaperone function. EMBO J 36(12):1669–1687. https://doi.org/10.15252/embj.201695957
16. Mazroui R, Di Marco S, Kaufman RJ, Gallouzi IE (2007) Inhibition of the ubiquitin-proteasome system induces stress granule formation. Mol Biol Cell 18(7):2603–2618. https://doi.org/10.1091/mbc.e06-12-1079
17. Buchan JR, Kolaitis RM, Taylor JP, Parker R (2013) Eukaryotic stress granules are cleared by autophagy and Cdc48/VCP function. Cell 153(7):1461–1474. https://doi.org/10.1016/j.cell.2013.05.037
18. Liao YC, Fernandopulle MS, Wang G, Choi H, Hao L, Drerup CM, Patel R, Qamar S, Nixon-Abell J, Shen Y, Meadows W, Vendruscolo M, TPJ K, Nelson M, Czekalska MA, Musteikyte G, Gachechiladze MA, Stephens CA, Pasolli HA, Forrest LR, St George-Hyslop P, Lippincott-Schwartz J, Ward ME (2019) RNA granules hitchhike on lysosomes for long-distance transport, using Annexin A11 as a molecular tether. Cell 179(1):147–164. e20. https://doi.org/10.1016/j.cell.2019.08.050
19. Zhang K, Daigle JG, Cunningham KM, Coyne AN, Ruan K, Grima JC, Bowen KE, Wadhwa H, Yang P, Rigo F, Taylor JP, Gitler AD, Rothstein JD, Lloyd TE (2018) Stress granule assembly disrupts nucleocytoplasmic transport. Cell 173(4):958–971.e17. https://doi.org/10.1016/j.cell.2018.03.025
20. Hofweber M, Dormann D (2019) Friend or foe-post-translational modifications as regulators of phase separation and RNP granule dynamics. J Biol Chem 294(18):7137–7150. https://doi.org/10.1074/jbc.TM118.001189
21. Protter DSW, Parker R (2016) Principles and properties of stress granules. Trends Cell Biol 26(9):668–679. https://doi.org/10.1016/j.tcb.2016.05.004
22. Hung V, Udeshi ND, Lam SS, Loh KH, Cox KJ, Pedram K, Carr SA, Ting AY (2016) Spatially resolved proteomic mapping in living cells with the engineered peroxidase APEX2. Nat Protoc 11(3):456–475. https://doi.org/10.1038/nprot.2016.018

23. Lam SS, Martell JD, Kamer KJ, Deerinck TJ, Ellisman MH, Mootha VK, Ting AY (2015) Directed evolution of APEX2 for electron microscopy and proximity labeling. Nat Methods 12(1):51–54. https://doi.org/10.1038/nmeth.3179

24. Hung V, Zou P, Rhee HW, Udeshi ND, Cracan V, Svinkina T, Carr SA, Mootha VK, Ting AY (2014) Proteomic mapping of the human mitochondrial intermembrane space in live cells via ratiometric APEX tagging. Mol Cell 55(2):332–341. https://doi.org/10.1016/j.molcel.2014.06.003

25. Rhee HW, Zou P, Udeshi ND, Martell JD, Mootha VK, Carr SA, Ting AY (2013) Proteomic mapping of mitochondria in living cells via spatially restricted enzymatic tagging. Science 339(6125):1328–1331. https://doi.org/10.1126/science.1230593

26. Bendall SC, Hughes C, Stewart MH, Doble B, Bhatia M, Lajoie GA (2008) Prevention of amino acid conversion in SILAC experiments with embryonic stem cells. Mol Cell Proteomics 7(9):1587–1597. https://doi.org/10.1074/mcp.M800113-MCP200

INDEX

Daniel Matějů and Jeffrey A. Chao (eds.), *The Integrated Stress Response: Methods and Protocols*, Methods in Molecular Biology, vol. 2428, https://doi.org/10.1007/978-1-0716-1975-9,

MIX
Papier aus verantwortungsvollen Quellen
Paper from responsible sources
FSC® C105338

If you have any concerns about our products,
you can contact us on
ProductSafety@springernature.com

In case Publisher is established outside the EU,
the EU authorized representative is:
Springer Nature Customer Service Center GmbH
Europaplatz 3, 69115 Heidelberg, Germany

Printed by Libri Plureos GmbH
in Hamburg, Germany